AN INTRODUCTION TO
THE MATHEMATICS OF FINANCE

An Introduction to the Mathematics of Finance

J. J. McCUTCHEON,

CBE, DSc, FFA, FRSE

and

W. F. SCOTT,

MA, Ph.D, FFA

Formerly of the
Department of Actuarial Mathematics and Statistics
Heriot-Watt University, Edinburgh

Published for the Institute of Actuaries
and the Faculty of Actuaries

BUTTERWORTH
HEINEMANN

OXFORD AMSTERDAM BOSTON LONDON NEW YORK PARIS
SAN DIEGO SAN FRANCISCO SINGAPORE SYDNEY TOKYO

Butterworth-Heinemann
An imprint of Elsevier Science
Linacre House, Jordan Hill, Oxford OX2 8DP
200 Wheeler Road, Burlington, MA 01803

First published 1986
Reprinted 1987, 1988, 1989, 1991, 1993, 1995, 1996, 1998, 1999, 2000, 2001, 2002, 2003
Transferred to digital printing 2004
Copyright © 1986, The Institute of Actuaries in Scotland.
All rights reserved

British Library Cataloguing in Publication Data
McCutcheon, J. J.
The mathematics of finance
1. Business mathematics
I. Title   II. Scott, W. F.
513'.93   HF5691

Library of Congress Cataloguing in Publication Data
A catalogue record for this book is available from the Library of Congress

ISBN 0 7506 0092 6

For information on all Butterworth-Heinemann publications
visit our website at www.bh.com

CONTENTS

CONTENTS

PREFACE

When writing this book we have primarily had in mind the requirements of students preparing for the professional examinations of the Institute of Actuaries or of the Faculty of Actuaries. At the same time, however, we have been well aware that many of the topics discussed are likely to be of interest to other groups of readers, such as investment analysts, economists, and accountants. Many of the subjects covered are of a highly practical nature. This important fact has influenced us considerably and we have tried throughout to illustrate our remarks with realistic worked examples. We hope that our presentation will enable all readers (and not simply mathematical specialists) to benefit from our discussions.

The classical theory of compound interest forms a major part of the book. The recent development of microcomputers has brought new life to this old subject. For example, it is now relatively simple to make a series of calculations to illustrate the consequences of changes in one or more of the factors relevant to any given problem. In this way valuable practical insights may often be gained with little effort. All our worked examples may be solved reasonably quickly with the aid of a pocket calculator and the tables given at the end of this book. For a complete understanding of the subject it is essential that the reader study these examples in detail.

We have tried to reflect relevant recent developments in both theory and practice. Thus, for example, we discuss index-linked stocks in some detail, and our final chapter provides an introduction to stochastic interest rate models, of which there is now a growing awareness.

We have obviously been influenced, consciously or otherwise, by several earlier writers. In particular, we must record our indebtedness to the well-known textbook by Mr D. W. A. Donald, which was for many years essential reading for all actuarial students.

We have benefited from numerous discussions with colleagues, from both business and academic worlds, who have offered constructive criticism on a wide range of topics. Dr L. W. G. Tutt and Mr A. V. Twigg, in an official capacity, kept friendly but critical eyes on the entire work and made many helpful suggestions. Mrs M. V. Butcher, Mr C. D. Daykin, and Professor J. B. H. Pegler commented in writing on specific sections of our draft manuscript.

Our greatest indebtedness is to Dr B. Johnston, Dr H. R. Waters, and

Professor A. D. Wilkie. Individually and collectively these gentlemen spent many hours with the authors discussing what must have appeared to be a never-ending sequence of draft manuscripts. Much of what is best in the book is to their credit, although any errors or inaccuracies which remain are entirely the responsibility of the authors. We must also thank Dr Johnston for assistance in preparing and checking the exercises and Professor Wilkie for advice relating to the list of references.

To all these people, and to many others, we are most grateful.

We thank the Councils of the Institute of Actuaries and of the Faculty of Actuaries for permission to use material from their examinations and tuition courses.

Finally, it is a pleasure to record our appreciation of the considerable skill of Mrs W. Hughes and Mrs J. Stewart, our secretaries, who so meticulously produced the manuscript. The patience and care with which they have handled the numerous corrections, amendments, and additions have made our task significantly lighter.

J. J. McCutcheon
W. F. Scott

Department of Actuarial Mathematics and Statistics,
Heriot-Watt University,
Edinburgh

1 July 1985

CHAPTER ONE

INTRODUCTION

1.1 The idea of interest

Interest may be regarded as a reward paid by one person or organization (the *borrower*) for the use of an asset, referred to as *capital*, belonging to another person or organization (the *lender*). The precise conditions of any transaction will be mutually agreed. For example, after a stated period the capital may be returned to the lender with the interest due. Alternatively, several interest payments may be made before the asset is finally returned by the borrower.

Capital and interest need not be measured in terms of the same commodity, but throughout this book, which relates primarily to problems of a financial nature, we shall assume that both are measured in the monetary units of a given currency. When expressed in monetary terms, capital is also referred to as *principal*.

If there is some risk of default (i.e. loss of capital or non-payment of interest) a lender would expect to be paid a higher rate of interest than would otherwise be the case. The additional interest in such a situation may be considered as a further reward for the lender's acceptance of the increased risk. (For example, a person who uses his money to finance the drilling for oil in a previously unexplored region would expect a relatively high return on his investment, if the drilling is successful – but might have to accept the loss of his capital if no oil were to be found.) Another factor which may influence the rate of interest on any transaction is an allowance for the possible depreciation or appreciation in the value of the currency in which the transaction is carried out. This factor is obviously very important in times of high inflation.

It is convenient to describe the operation of interest within the familiar context of a 'savings' (or 'investment') account, held in a bank, building society, or other similar organization. An investor who had opened such an account some time ago with an initial deposit of £100, and who had made no other payments to or from the account, would expect to withdraw more than £100 if he were now to close the account. Suppose, for example, that he receives £106 on closing his account. This sum may be regarded as consisting of £100 as the return of the initial deposit and £6 as interest. The interest is a payment by the bank to the investor for the use of his capital over the duration of the account.

1

The most elementary concept is that of *simple* interest. This leads naturally to the idea of *compound* interest, which is much more commonly found in practice – at least in relation to all but short-term investments. Both concepts are easily described within the framework of a savings account.

1.2 Simple interest

Suppose that an investor opens a savings account, which pays simple interest at the rate of 9% per annum, with a single deposit of £100. The account will be credited with £9 of interest for each complete year the money remains on deposit. If the account is closed after one year, the investor will receive £109; if the account is closed after two years, he will receive £118; and so on.

These brief illustrative remarks may be summarized more generally as follows. If an amount C is deposited in an account which pays simple interest at the rate of i per annum and the account is closed after n years – there being no intervening payments to or from the account – then the amount paid to the investor when the account is closed will be

$$C(1 + ni) \qquad (1.2.1)$$

This payment consists of a return of the initial deposit C, together with interest of amount

$$niC \qquad (1.2.2)$$

In our discussion so far we have implicitly assumed that, in each of these last two expressions, n is an integer. However, the normal commercial practice in relation to fractional periods of a year is to pay interest on a pro rata basis, so that expressions 1.2.1 and 1.2.2 may be considered as applying for *all* non-negative values of n.

Note that if the annual rate of interest is 12%, then $i = 0.12$; if the annual rate of interest is 9%, then $i = 0.09$; and so on.

Example 1.2.1

Suppose that £860 is deposited in a savings account which pays simple interest at the rate of $5\frac{3}{8}$% per annum. Assuming that there are no subsequent payments to or from the account, find the amount finally withdrawn if the account is closed after (a) six months, (b) ten months, (c) one year.

Solution

By letting $n = 1/2$, $10/12$, and 1 in expression 1.2.1 with $C = 860$ and $i = 0.05375$, we obtain the answers (a) £883.11, (b) £898.52, and (c) £906.22. (In each case we have given the answer to two decimal places of one pound, rounded down. This is quite a common commercial practice.)

Note In the above solution we have assumed that six months and ten months are periods of 1/2 and 10/12 of one year respectively. For accounts of duration less than one year it is usual to allow for the actual number of *days* an account is held, so, for example, two six-month periods are not necessarily regarded as being of equal length. In this case expression 1.2.1 becomes

$$C\left(1 + \frac{di}{365}\right) \tag{1.2.3}$$

where d is the duration of the account, measured in days, and i is the annual rate of interest.

The essential feature of simple interest, as expressed algebraically by expression 1.2.1, is that interest, once credited to an account, does not itself earn further interest. This leads to inconsistencies which are avoided by the application of compound interest theory.

1.3 Compound interest

Suppose now that a certain type of savings account pays simple interest at the rate of i per annum. Suppose further that this rate is guaranteed to apply throughout the next two years and that accounts may be opened and closed at any time.

Consider an investor who opens an account at the present time with an initial deposit of C. The investor may close this account after one year, at which time he will withdraw $C(1 + i)$ (see expression 1.2.1). He may then place this sum on deposit in a new account and close this second account after one further year. When this latter account is closed the sum withdrawn (again see expression 1.2.1) will be $[C(1 + i)](1 + i)$, i.e. $C(1 + i)^2$ or $C(1 + 2i + i^2)$.

If, however, the investor chooses not to switch accounts after one year and leaves his money in the original account, on closing this account after two years he will receive $C(1 + 2i)$. Thus, simply by switching accounts in the middle of the two-year period, the investor will receive an additional amount i^2C at the end of the period. This extra payment is, of course, equal to $i(iC)$ and arises as interest paid (at the end of the second year) on the interest credited to the original account at the end of the first year.

From a practical viewpoint it would be difficult to prevent an investor switching accounts in the manner described above (or with even greater frequency). Furthermore, the investor, having closed his second account after one year, could then deposit the entire amount withdrawn in yet another account.

Any bank would find it administratively most inconvenient to have to keep opening and closing accounts in the manner described above! Moreover, on closing one account, the investor might choose to deposit his money elsewhere. Thus, partly to encourage long-term investment and partly for other practical reasons, it is common commercial practice – at least in relation to investments of duration greater than one year – to pay compound interest on savings accounts. Moreover, the concepts of compound interest are always used in the assessment and evaluation of investments.

The essential feature of compound interest is that interest itself earns interest. The operation of compound interest may be described as follows. Consider a savings account, which pays compound interest at rate i per annum, into which is placed an initial deposit C. (We assume that there are no further payments to or from the account.) If the account is closed after one year, the investor will receive $C(1 + i)$. More generally, let A_n be the amount which will be received by the investor if he closes the account after n years. Thus $A_1 = C(1 + i)$. By definition, the amount received by the investor on closing the account at the end of any year is equal to the amount he would have received, if he had closed the account one year previously, plus further interest of i times this amount. Thus the interest credited to the account up to the start of the final year itself earns interest (at rate i per annum) over the final year.

Expressed algebraically, the above definition becomes

$$A_{n+1} = A_n + iA_n$$

or

$$A_{n+1} = (1 + i)A_n \qquad n \geq 1 \qquad (1.3.1)$$

Since (by definition)

$$A_1 = C(1 + i)$$

equation 1.3.1 implies that, for $n = 1, 2, \ldots,$

$$A_n = C(1 + i)^n \qquad (1.3.2)$$

Thus, if the investor closes the account after n years, he will receive

$$C(1 + i)^n \qquad (1.3.3)$$

This payment consists of a return of the initial deposit C, together with *accumulated* interest (i.e. interest which, if $n > 1$, has itself earned further interest) of amount

$$C[(1 + i)^n - 1] \tag{1.3.4}$$

In our discussion so far we have assumed that in both these last expressions n is an integer. In chapter 2 we widen the discussion and show that, under very general conditions, expressions 1.3.3 and 1.3.4 remain valid for *all* non-negative values of n.

Since

$$[C(1 + i)^{t_1}](1 + i)^{t_2} = C(1 + i)^{t_1 + t_2}$$

an investor who is able to switch his money between two accounts, both of which pay compound interest at the same rate, is not able to profit by such action. This is in contrast with the somewhat anomalous situation, described at the beginning of this section, which may prevail if simple interest is paid.

Expressions 1.3.3 and 1.3.4 should be compared with the corresponding expressions under the operation of simple interest (i.e. expressions 1.2.1 and 1.2.2). If interest *compounds* (i.e. earns further interest), the effect on the accumulation of an account can be very significant, especially if the duration of the account or the rate of interest is great. This is illustrated by the following example.

Example 1.3.1

Suppose that £100 is deposited in a savings account. Construct a table to show the accumulated amount of the account after 5, 10, 20, and 40 years on the assumption that compound interest is paid at the rate of (a) 4% per annum, and (b) 8% per annum. Give also the corresponding figures on the assumption that only simple interest is paid.

Solution

From expressions 1.2.1 and 1.3.3 we obtain the values in table 1.3.1. The reader should verify the figures using the compound interest tables at the end of this book.

Table 1.3.1 *Accumulated amount of £100*

Term (years)	Annual rate of interest 4%		Annual rate of interest 8%	
	Simple	Compound	Simple	Compound
5	£120	£121·66	£140	£146·93
10	£140	£148·02	£180	£215·89
20	£180	£219·11	£260	£466·09
40	£260	£480·10	£420	£2172·45

Note, for example, that over 40 years at 8% per annum interest the account with compound interest accumulates to more than five times the amount of the corresponding account with simple interest.

The exponential growth of money under compound interest and its linear growth under simple interest are illustrated in Figure 1.3.1 for the case when $i = 0.08$.

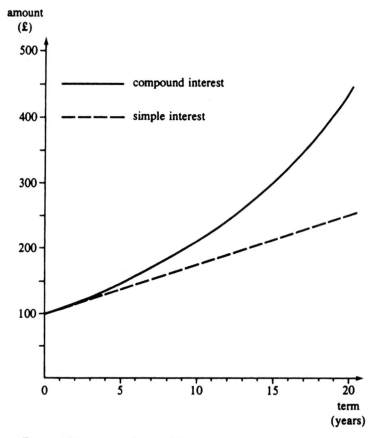

FIGURE 1.3.1   *Accumulation of £100 with interest at 8% per annum*

As we have already indicated, the concepts of compound interest are always used in the assessment and evaluation of investments. In the final section of this chapter we describe briefly several kinds of situation which typically arise in practice. The analyses of these types of problem are among those discussed later in this book.

1.4 Some practical illustrations

(a) As a simple illustration, consider an investor who is offered a contract with a financial institution which provides £22 500 at the end of ten years in return for a single payment of £10 000 now. If the investor is willing to tie up this amount of capital for ten years, the decision as to whether or not he enters into the contract will depend upon the alternative investments available. For example, if the investor can obtain elsewhere a guaranteed compound rate of interest for the next ten years of 10% per annum, then he should not effect the contract. (Why? See expression 1.3.3.) On the other hand, if he can obtain this rate of interest with certainty only for the next six years, in deciding whether or not to effect the contract he will have to make a judgement about the rates of interest he is likely to be able to obtain over the four-year period commencing six years from now. (We ignore for the present further complications, such as the effect of taxation or the reliability of the company offering the contract.)

(b) Similar considerations would apply in relation to a contract which offered to provide a specified lump sum at the end of a given period in return for the payment of a series of premiums of stated (and often constant) amount at regular intervals throughout the period. Would the reader consider favourably a contract which provides £3500 tax free at the end of ten years in return for ten annual premiums, each of £200, payable at the start of each year?

(c) As a further example, consider a business venture, requiring an initial outlay of £500 000, which will provide a return of £550 000 after five years and £480 000 after a further three years, both these sums being paid free of tax. A person or organization with £500 000 of spare cash might compare this opportunity with other available investments of a similar term. An investor who had no spare cash might consider financing the venture by borrowing the initial outlay from a bank. Whether or not he should do so depends upon the rate of interest charged for the loan. If the rate charged is more than a certain 'critical' value, it will not be profitable to finance the investment in this way.

(d) Another practical illustration of compound interest is provided by *mortgage loans*, i.e. loans which are made for the specific purpose of house purchase. (Such loans are very common in the UK and many other countries. The property for the purchase of which the loan is made usually acts as security for the loan.) Suppose, for example, that a person wishes to borrow £35 000 for the purchase of a house with the intention of repaying the loan by regular periodic payments of a fixed amount over 25 years. What should be the amount of each regular repayment? Obviously this will depend on both the rate of interest charged by the lender and the precise frequency of the repayments (monthly, half-yearly, annual, etc.). It should also be noted that, under modern conditions in the UK, most lenders would be unwilling to quote a fixed rate of interest for such a long period. During the course of such a loan the rate of interest might well be revised several times (according to market conditions) and on each revision there would be a corresponding change in the amount of the borrower's regular repayment or in the outstanding term of the loan. Compound interest techniques enable the revised amount of the repayment or the new outstanding term to be found in such cases.

(e) One of the most important applications of compound interest lies in the analysis and evaluation of stock exchange investments, particularly *fixed-interest* securities. For example, assume that any *one* of the following series of payments may be purchased for £1000 by an investor who is not liable to tax:

   (i) Income of £120 per annum payable in arrear at yearly intervals for eight years together with a payment of £1000 at the end of eight years;

   (ii) Income of £90 per annum payable in arrear at yearly intervals for eight years together with a payment of £1300 at the end of eight years;

   (iii) A series of eight payments, each of amount £180, payable annually in arrear.

   (The first two of the above may be considered as typical fixed-interest securities. The third is generally known as a level *annuity* (or, more precisely, *annuity-certain*), in this case payable for eight years.)

   In an obvious sense the *yield* on the first investment is 12% per annum. Each year the investor receives an income of 12% of his outlay until such time as this outlay is repaid. However, it is less clear what is meant by the yield on the second or third investments. For the second investment the annual income is 9% of the purchase price, but the final payment after eight years exceeds the purchase price. Intuitively, therefore, one would consider the second investment as providing a yield greater than 9% per annum. How much greater? Does the yield on the second investment exceed that on the first? What is the yield on the third investment? Is the investment with the highest yield likely to be the most profitable?

The answers to these and similar questions, the analyses of the practical situations described above, and the solution of many similar problems are provided by the theory of compound interest.

CHAPTER TWO

THEORY OF INTEREST RATES

2.1 The rate of interest

We begin by considering investments in which capital and interest are paid at the end of a fixed term, there being no intermediate interest or capital payments. An example of this kind of investment is a short-term deposit in which the lender invests £1000 and receives a return of £1035 six months later; £1000 may be considered to be a repayment of capital and £35 a payment of interest, i.e. a reward for the use of the capital for six months.

It is essential in any compound interest problem to define the *unit of time*. This may be, for example, a month or a year, the latter period being frequently used in practice. In certain situations, however, it is more appropriate to choose a different period (e.g. six months) as the basic time unit.

Consider an investment of 1 for a period of 1 time unit, commencing at time t, and suppose that $1 + i(t)$ is returned at time $t + 1$. We call $i(t)$ the *rate of interest* for the period t to $t + 1$. One sometimes refers to $i(t)$ as the *effective rate of interest* for the period, to distinguish it from *nominal* and *flat* rates of interest which will be discussed below. If it is assumed that the rate of interest does not depend on the amount invested, the cash returned at time $t + 1$ from an investment of C at time t is $C[1 + i(t)]$. (In practice a rather higher rate of interest may be obtained from a large investment than from a small one, but we ignore this point here.) It may easily be seen that the accumulation of C from time 0 to time n (where n is some positive integer) is

$$C[1 + i(0)][1 + i(1)]\cdots[1 + i(n - 1)] \tag{2.1.1}$$

since the proceeds $C[1 + i(0)]$ at time 1 may be invested at this time to produce $C[1 + i(0)][1 + i(1)]$ at time 2, and so on.

Rates of interest are often quoted *per cent*. Thus, for example, we may speak of an effective rate of interest (for a given period) of $12\frac{3}{4}\%$. This means that the effective rate of interest for the period is 0·1275.

If the rate of interest per period does not depend on the time t at which the investment is made, we write $i(t) = i$ for all t. In this case the accumulation of an investment of C for *any* period of length n time units is, by formula 2.1.1,

$$C(1 + i)^n \tag{2.1.2}$$

This formula, which will be shown later to hold (under certain assumptions) even when n is not an integer, is referred to as the *accumulation* of C for n time units under *compound interest* at rate i per time unit. The corresponding accumulation under *simple interest* at rate i per time unit is defined, as in chapter 1, as

$$C(1 + in) \qquad (2.1.3)$$

This last formula may also be considered to hold for any positive n, not necessarily an integer. A comparison of the accumulations under simple and compound interest is given in example 1.3.1 and exercise 2.1.

Example 2.1.1

The rate of interest on a certain bank deposit account is $4\frac{1}{2}\%$ per annum effective. Find the accumulation of £5000 after seven years in this account.

Solution

By formula 2.1.2, the accumulation is

$$5000(1{\cdot}045)^7 = 5000 \times 1{\cdot}360\,86 = £6804{\cdot}30$$

Example 2.1.2

The effective rate of interest per annum on a certain building society account is at present 7%, but in two years' time it will be reduced to 6%. Find the accumulation in five years' time of an investment of £4000 in this account.

Solution

By formula 2.1.1, the accumulation is

$$4000(1{\cdot}07)^2(1{\cdot}06)^3 = £5454{\cdot}38$$

2.2 Nominal rates of interest

Now consider transactions for a term of length h time units, where $h > 0$ and h need not be an integer. We define $i_h(t)$, the *nominal rate of interest* per unit time on transactions of term h beginning at time t, to be such that the effective rate of interest for the period of length h beginning at time t is $hi_h(t)$. Thus, if the sum of C is invested at time t for a term h, the sum to be received at time $t + h$ is, *by definition*,

$$C[1 + hi_h(t)] \qquad (2.2.1)$$

If $h = 1$, the nominal rate of interest coincides with the effective rate of interest for the period t to $t + 1$, so

$$i_1(t) = i(t) \qquad (2.2.2)$$

In many practical applications $i_h(t)$ does not depend on t, in which case we may write

$$i_h(t) = i_h \qquad \text{for all } t \qquad (2.2.3)$$

If, in this case, we also have $h = 1/p$, where p is a positive integer (i.e. h is a simple fraction of a time unit), it is more usual to write $i^{(p)}$ rather than $i_{1/p}$. Thus, as a definition, we have

$$i^{(p)} = i_{1/p} \qquad (2.2.4)$$

It follows that, in this case, an investment of 1 for any period of length $1/p$ will produce a return of

$$1 + \frac{i^{(p)}}{p} \qquad (2.2.5)$$

Note that $i^{(p)}$ is often referred to as a nominal rate of interest per unit time *payable* pthly, or *convertible* pthly, or *with* pthly *rests*. (See chapter 4 for a fuller discussion of this topic.) Note that $i^{(1)}$ coincides with the effective rate of interest per unit time, i.

Nominal rates of interest are often quoted in practice, as in the following example.

Example 2.2.1

The nominal rates of interest per annum quoted in the financial press for local authority deposits on a particular day are as follows:

Term	Nominal rate of interest (%)
1 day	$11\frac{3}{4}$
2 days	$11\frac{5}{8}$
7 days	$11\frac{1}{2}$
1 month	$11\frac{3}{8}$
3 months	$11\frac{1}{4}$

(Investments of term one day are often referred to as *overnight money*.) Find the accumulation of an investment at this time of £1000 for (a) one week and (b) one month.

Solution

To express the above information in terms of our notation, we draw up the following table (in which the unit of time is one year and the particular time is taken as t_0):

Term h	1/365	2/365	7/365	1/12	1/4
$i_h(t_0)$	0·1175	0·11625	0·115	0·11375	0·1125

By formula 2.2.1, the accumulations are $1000[1 + hi_h(t_0)]$ where (a) $h = 7/365$ and

(b) $h = 1/12$. This gives the answers

$$\text{(a)} \quad 1000(1 + \tfrac{7}{365}0 \cdot 115) = £1002 \cdot 21$$
$$\text{(b)} \quad 1000(1 + \tfrac{1}{12}0 \cdot 11375) = £1009 \cdot 48$$

Note The nominal rates of interest for varying terms (as illustrated by example 2.2.1) are liable to vary from day to day: they should not be assumed to be fixed. If they were constant with time and equal to the above values, an investment of £1 000 000 for two successive one-day periods would accumulate to £1 000 000 × $[1 + (0 \cdot 1175/365)]^2 = £1 000 644$, whereas an investment for a single two-day term would give £1 000 000 × $[1 + (0 \cdot 116\,25 \times 2/365)] = £1 000 637$. This apparent inconsistency may be explained (partly) by the fact that the market expects interest rates to change in the future.

2.3 Accumulation factors

Let time be measured in suitable units (e.g. years). For $t_1 \leqslant t_2$ we define $A(t_1, t_2)$ to be the accumulation at time t_2 of an investment of 1 at time t_1 for a term of $(t_2 - t_1)$. Thus $A(t_1, t_2)$ is the amount which will be repaid at time t_2 in return for an investment of 1 at time t_1. It follows by the definition of $i_h(t)$ that, for all t and for all $h > 0$,

$$A(t, t + h) = 1 + h i_h(t) \qquad (2.3.1)$$

and hence that

$$i_h(t) = \frac{A(t, t + h) - 1}{h} \qquad h > 0 \qquad (2.3.2)$$

We also define $A(t, t) = 1$ for all t. The number $A(t_1, t_2)$ is often called an accumulation factor, since the accumulation at time t_2 of an investment of the sum C at time t_1 is, by proportion,

$$C A(t_1, t_2) \qquad (2.3.3)$$

In relation to the past, i.e. when the present moment is taken as time 0 and t and $t + h$ are both less than or equal to 0, the factors $A(t, t + h)$ and the nominal rates of interest $i_h(t)$ are a matter of recorded fact in respect of any given transaction. As for their values in the future, estimates must be made (unless one invests in fixed-interest securities with guaranteed rates of interest applying both now and in the future).

Now let $t_0 \leqslant t_1 \leqslant t_2$ and consider an investment of 1 at time t_0. The proceeds at time t_2 will be $A(t_0, t_2)$ if one invests at time t_0 for term $t_2 - t_0$, or $A(t_0, t_1) A(t_1, t_2)$ if one invests at time t_0 for term $t_1 - t_0$ and then, at time t_1, reinvests the proceeds for term $t_2 - t_1$. In a consistent market these proceeds should not depend on the course of action taken by the investor. Accordingly, we say that under the *principle of consistency*

$$A(t_0, t_2) = A(t_0, t_1)A(t_1, t_2) \qquad (2.3.4)$$

for all $t_0 \leqslant t_1 \leqslant t_2$. It follows easily by induction that, if the consistency principle holds,

$$A(t_0, t_n) = A(t_0, t_1)A(t_1, t_2) \cdots A(t_{n-1}, t_n) \qquad (2.3.5)$$

for any n and any increasing set of numbers $t_0, t_1, \ldots t_n$.

Unless it is stated otherwise, we shall assume that the principle of consistency holds. In practice, however, it is unlikely to be realized exactly because of dealing expenses, taxation and other factors. Moreover, it is sometimes true that the accumulation factors implied by certain mathematical models (e.g. the reinvestment model of section 6.7) do not in general satisfy the principle of consistency. It will be shown in section 2.4 that, under very general conditions, accumulation factors satisfying the consistency principle must have a particular form (see formula 2.4.3).

Example 2.3.1

Let time be measured in years, and suppose that, for all $t_1 \leqslant t_2$,

$$A(t_1, t_2) = \exp[0.05(t_2 - t_1)]$$

(a) Verify that the consistency principle holds.
(b) Find the accumulation 15 years later of an investment of £600 at any time.

Solution

(a) This is left as a straightforward exercise.
(b) By formula 2.3.3, the accumulation is

$$600 \exp(0.05 \times 15) = £1270.20$$

2.4 The force of interest

Equation 2.3.2 indicates how $i_h(t)$ is defined in terms of the accumulation factor $A(t, t + h)$. In example 2.2.1 we gave (in relation to a particular time t_0) the values of $i_h(t_0)$ for a series of values of h, varying from 1/4 (i.e. three months) to 1/365 (i.e. one day). The trend of these values should be noted. In practical situations it is not unreasonable to assume that, as h becomes smaller and smaller, $i_h(t)$ tends to a limiting value. In general, of course, this limiting value will depend on t.

We therefore assume that for each value of t there is number $\delta(t)$ such that

$$\lim_{h \to 0^+} i_h(t) = \delta(t) \qquad (2.4.1)$$

(The notation $h \to 0^+$ indicates that the limit is considered as h tends to zero 'from above' or through positive values.)

It is usual to call $\delta(t)$ the *force of interest per unit time at time t*. In view of formula 2.4.1, $\delta(t)$ is sometimes called the nominal rate of interest per unit time at time t *convertible momently*.

Although it is a mathematical idealization of reality, the force of interest plays a crucial role in compound interest theory. Note that by combining equations 2.3.2 and 2.4.1 we may define $\delta(t)$ directly in terms of the accumulation factor as

$$\delta(t) = \lim_{h \to 0^+} \left[\frac{A(t, t+h) - 1}{h} \right] \qquad (2.4.2)$$

The force of interest function $\delta(t)$ is defined in terms of the accumulation function $A(t_1, t_2)$, but when the principle of consistency holds it is possible, under very general conditions, to express the accumulation factor in terms of the force of interest. This result is contained in the following theorem.

Theorem 2.4.1

If $\delta(t)$ and $A(t_0, t)$ are continuous functions of t for $t \geqslant t_0$, and the principle of consistency holds, then, for $t_0 \leqslant t_1 \leqslant t_2$,

$$A(t_1, t_2) = \exp \left[\int_{t_1}^{t_2} \delta(t) dt \right] \qquad (2.4.3)$$

The proof of this theorem is given in appendix 1.

Equation 2.4.3 indicates the vital importance of the force of interest. As soon as $\delta(t)$, the force of interest per unit time, is specified, the accumulation factors $A(t_1, t_2)$ can be determined by formula 2.4.3. We may also find $i_h(t)$ by formulae 2.4.3 and 2.3.2. Thus

$$i_h(t) = \frac{\exp \left[\int_t^{t+h} \delta(s) ds \right] - 1}{h} \qquad (2.4.4)$$

The following examples illustrate the above discussion.

Example 2.4.1

Assume that $\delta(t)$, the force of interest per unit time at time t, is given (a) by $\delta(t) = \delta$ (where δ is some constant), and (b) by $\delta(t) = a + bt$. Find formulae for the accumulation of 1 from time t_1 to time t_2 in each case.

Solution

In case (a), formula 2.4.3 gives

$$A(t_1, t_2) = \exp \left[\delta(t_2 - t_1) \right]$$

and in case (b) this formula gives

$$A(t_1, t_2) = \exp\left[\int_{t_1}^{t_2} (a + bt)dt\right]$$

$$= \exp\left[(at_2 + \tfrac{1}{2}bt_2^2) - (at_1 + \tfrac{1}{2}bt_1^2)\right]$$

Note The case when $\delta(t) = \delta$ for all t is of very great practical importance. It is clear that in this case

$$A(t_0, t_0 + n) = e^{\delta n} \tag{2.4.5}$$

for *all* t_0 and $n \geqslant 0$. By formula 2.4.4, the effective rate of interest per time unit is

$$i = e^{\delta} - 1 \tag{2.4.6}$$

and hence

$$e^{\delta} = 1 + i \tag{2.4.7}$$

The accumulation factor $A(t_0, t_0 + n)$ may thus be expressed in the alternative form

$$A(t_0, t_0 + n) = (1 + i)^n \tag{2.4.8}$$

We thus have a generalization of formula 2.1.2 to *all* $n \geqslant 0$, not merely the positive integers. Notation and theory may be simplified when $\delta(t) = \delta$ for all t. This case will be considered in detail in chapter 3.

Example 2.4.2

The force of interest per unit time, $\delta(t)$, where time is measured in years, equals 0·12 for all t. Find the nominal rate of interest per annum on deposits of term (a) seven days, (b) one month, and (c) six months.

Solution

Using formula 2.4.4 with $\delta(t) = 0.12$ for all t, we obtain, for all t,

$$i_h = i_h(t) = \frac{\exp(0.12h) - 1}{h}$$

Substituting (a) $h = 7/365$, (b) $h = 1/12$, and (c) $h = 1/2$, we obtain the nominal rates of interest (a) 12·01%, (b) 12·06%, and (c) 12·37%.

Let us now define

$$F(t) = A(t_0, t) \tag{2.4.9}$$

where t_0 is fixed and $t_0 \leqslant t$. Thus $F(t)$ is the accumulation at time t of an investment of 1 at time t_0. By formula 2.4.3,

$$\log F(t) = \int_{t_0}^{t} \delta(s)ds \qquad (2.4.10)$$

and hence, for $t > t_0$,

$$\delta(t) = \frac{d}{dt}\log F(t) = \frac{F'(t)}{F(t)} \qquad (2.4.11)$$

Example 2.4.3

A bank credits interest on deposits using accumulation factors based on a variable force of interest. On 1 July 1983 a customer deposited £50 000 with the bank. On 1 July 1985 his deposit had grown to £59 102. Assuming that the force of interest per annum was a linear function of time during the period from 1 July 1983 to 1 July 1985, find the force of interest per annum on 1 July 1984.

Solution

Using the above notation and measuring time in years from 1 July 1983, we have $F(0) = 1$ and $F(2) = 59\ 102/50\ 000 = 1·182\ 040$. Since (by assumption) $\delta(t)$ is a linear function of t, it follows from equation 2.4.10 that $\log F(t)$ is a quadratic function of t for $0 \leqslant t \leqslant 2$.

The reader should verify that, for any quadratic function $g(t)$, $a - h \leqslant t \leqslant a + h$,

$$g'(a) = \frac{g(a + h) - g(a - h)}{2h}$$

This property is useful in a variety of practical applications.

It now follows from equation 2.4.11 that

$$\delta(1) = \tfrac{1}{2}[\log F(2) - \log F(0)]$$
$$= \tfrac{1}{2}[0·167\ 242 - 0]$$
$$= 0·083\ 621$$

Thus the force of interest per annum on 1 July 1984 was 0·083 621.

Note In practice, the force of interest (which is approximately equal to the nominal rate of interest on 'overnight money') may behave much more erratically than is assumed in example 2.4.3.

Although we have assumed so far that $\delta(t)$ is a continuous function of time t, in certain practical problems we may wish to consider rather more general functions $\delta(t)$. In particular, we sometimes consider $\delta(t)$ to be piecewise constant; e.g. measuring time in years, we might assume that

$$\delta(t) = \begin{cases} 0·06 & \text{for } t < 5 \\ 0·05 & \text{for } 5 \leqslant t < 10 \\ 0·03 & \text{for } t \geqslant 10 \end{cases}$$

In such cases theorem 2.4.1 and other results are still valid. They may be established by considering $\delta(t)$ to be the limit, in a certain sense, of a sequence of continuous functions.

2.5 Present values

Let $t_1 \leqslant t_2$. It follows by formula 2.3.3 that an investment of $C/A(t_1, t_2)$, i.e. $C\exp[-\int_{t_1}^{t_2}\delta(t)dt]$, at time t_1 will produce a return of C at time t_2. We therefore say that the *discounted value* at time t_1 of C due at time t_2 is

$$C\exp\left[-\int_{t_1}^{t_2}\delta(t)dt\right] \qquad (2.5.1)$$

This is the sum of money which, if invested at time t_1, will give C at time t_2. In particular, the discounted value at time 0 (the present time) of C due at time $t \geqslant 0$ is called its *discounted present value* (or, more briefly, its *present value*); it is equal to

$$C\exp\left[-\int_0^t\delta(s)ds\right] \qquad (2.5.2)$$

We now define the function

$$v(t) = \exp\left[-\int_0^t\delta(s)ds\right] \qquad (2.5.3)$$

When $t \geqslant 0$, $v(t)$ is the *(discounted) present value* of 1 due at time t. When $t < 0$, the convention $\int_0^t\delta(s)ds = -\int_t^0\delta(s)ds$ shows that $v(t)$ is the *accumulation* of 1 from time t to time 0. It follows by formulae 2.5.2 and 2.5.3 that the discounted present value of C due at a non-negative time t is

$$Cv(t) \qquad (2.5.4)$$

In the important practical case in which $\delta(t) = \delta$ for all t, we may write

$$v(t) = v^t \qquad \text{for all } t \qquad (2.5.5)$$

where $v = v(1) = e^{-\delta}$ (see chapter 3). The values of v^t $(t = 1, 2, 3, ...)$ at various interest rates are included in the compound interest tables at the back of this book.

Example 2.5.1

Measure time in years from the present, and suppose that $\delta(t) = 0.06(0.9)^t$ for all t.

Find a simple expression for $v(t)$, and hence find the discounted present value of £100 due in 3·5 years' time.

Solution

By formula 2.5.3,

$$v(t) = \exp\left[-\int_0^t 0{\cdot}06(0{\cdot}9)^s \, ds \right]$$
$$= \exp[-0{\cdot}06(0{\cdot}9^t - 1)/\log(0{\cdot}9)]$$

Hence the present value of £100 due in 3·5 years' time is, by formula 2.5.4,

$$100 \exp[-0{\cdot}06(0{\cdot}9^{3{\cdot}5} - 1)/\log(0{\cdot}9)] = \text{£}83{\cdot}89$$

Example 2.5.2

Suppose that

$$\delta(t) = \begin{cases} 0{\cdot}09 & \text{for } 0 \leqslant t < 5 \\ 0{\cdot}08 & \text{for } 5 \leqslant t < 10 \\ 0{\cdot}07 & \text{for } \quad t \geqslant 10 \end{cases}$$

Find simple expressions for $v(t)$ when $t \geqslant 0$.

Solution

Note that for $5 \leqslant t < 10$ we may evaluate $\int_0^t \delta(s)ds$ as $[\int_0^5 \delta(s)ds + \int_5^t \delta(s)ds]$ and that for $t \geqslant 10$ we may use the form $[\int_0^{10} \delta(s)ds + \int_{10}^t \delta(s)ds]$, with the appropriate formula for $\delta(s)$ in each case. This gives immediately

$$v(t) = \begin{cases} \exp(-0{\cdot}09t) & \text{for } 0 \leqslant t < 5 \\ \exp(-0{\cdot}05 - 0{\cdot}08t) & \text{for } 5 \leqslant t < 10 \\ \exp(-0{\cdot}15 - 0{\cdot}07t) & \text{for } \quad t \geqslant 10 \end{cases}$$

Note When $\delta(t)$ is piecewise constant, as in this case, we may apply formulae applicable when $\delta(t)$ is constant to value payments in each of the relevant time intervals; examples of this will be given in chapter 6.

2.6 Stoodley's formula for the force of interest

An important example of a mathematical formula for $\delta(t)$ is *Stoodley's formula* (see references [31], [47]), which may be written

$$\delta(t) = p + \frac{s}{1 + re^{st}} \tag{2.6.1}$$

The parameters p, r, and s may be chosen, by methods described in section 6.6, so as to model a smoothly decreasing (or, in some circumstances, smoothly increasing) force of interest. If Stoodley's formula holds, we have

$$v(t) = \exp\left[-\int_0^t \delta(y)dy \right]$$

$$= \exp\left[-\int_0^t \left(p + \frac{s}{1 + re^{sy}} \right) dy \right]$$

$$= \exp\left[-\int_0^t \left(p + s - \frac{rse^{sy}}{1 + re^{sy}} \right) dy \right]$$

$$= \exp\left\{ -(p + s)t + [\log(1 + re^{sy})]_0^t \right\}$$

$$= \exp\left[-(p + s)t \right] \frac{1 + re^{st}}{1 + r}$$

$$= \frac{1}{1 + r} e^{-(p+s)t} + \frac{r}{1 + r} e^{-pt} \qquad (2.6.2)$$

If we define $v_1 = e^{-(p+s)}$ and $v_2 = e^{-p}$, we may write

$$v(t) = \frac{1}{1 + r} v_1^t + \frac{r}{1 + r} v_2^t \qquad (2.6.3)$$

If Stoodley's formula holds, the present value of any cash flow (see section 2.7) may be expressed as a weighted average of the corresponding present values at two fixed rates of interest. This feature extends to the valuation of annuities and assurances in life contingencies, and is of considerable practical importance.

Example 2.6.1

The force of interest per annum $\delta(t)$ follows Stoodley's formula with $p = 0.076\,961$, $r = 0.5$, and $s = 0.121\,890$, i.e

$$\delta(t) = 0.076\,961 + \frac{0.121\,890}{1 + 0.5\exp(0.121\,890t)}$$

Find a formula for $v(t)$, and use it to find the present value of 1 due in ten years' time.

Solution

By equation 2.6.2,

$$v(t) = \tfrac{2}{3}\exp\left[-(0.076\,961 + 0.121\,890)t \right] + \tfrac{1}{3}\exp(-0.076\,961t)$$
$$= \tfrac{2}{3}(1.22)^{-t} + \tfrac{1}{3}(1.08)^{-t}$$

Hence the present value of 1 due in ten years' time is

$$v(10) = \tfrac{2}{3}(1.22)^{-10} + \tfrac{1}{3}(1.08)^{-10} = 0.245\,66$$

2.7 Present values of cash flows

In many compound interest problems one must find the discounted present value of cash payments (or, as they are often called, cash flows) due in the future. It is important to distinguish between (a) *discrete* and (b) *continuous* payments.

Discrete cash flows

The present value of the sums $c_{t_1}, c_{t_2}, \ldots, c_{t_n}$ due at times $t_1, t_2, \ldots, t_n$ (where $0 \leqslant t_1 < t_2 < \cdots < t_n$) is, by formula 2.5.4,

$$c_{t_1}v(t_1) + c_{t_2}v(t_2) + \cdots + c_{t_n}v(t_n) = \sum_{j=1}^{n} c_{t_j}v(t_j) \qquad (2.7.1)$$

If the number of payments is infinite, the present value is defined to be

$$\sum_{j=1}^{\infty} c_{t_j}v(t_j) \qquad (2.7.2)$$

(provided that this series converges; it usually will in practical problems).

The process of finding discounted present values may be illustrated as in Figure 2.7.1. The discounting factors $v(t_1)$, $v(t_2)$, $v(t_3)$ are applied to bring the cash payments 'back to the present time'.

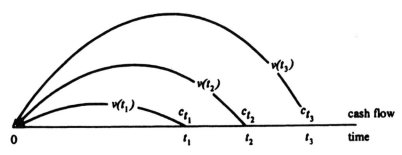

FIGURE 2.7.1 *Discounted cash flow*

Continuously payable cash flows (payment streams)

The concept of a continuously payable cash flow, although essentially theoretical, is important. (For example, for many practical purposes a pension which is payable weekly may be considered as payable continuously.) Suppose that $T > 0$ and that between times 0 and T an investor will be paid money continuously, the rate of payment at time t being £$\rho(t)$ per unit time. What is the present value of this cash flow?

In order to answer this question it is essential to understand what is meant by the *rate of payment* of the cash flow at time t. If $M(t)$ denotes the *total* payment made between time 0 and time t, then, *by definition*,

$$\rho(t) = M'(t) \qquad \text{for all } t \tag{2.7.3}$$

Then, if $0 \leqslant \alpha < \beta \leqslant T$, the total payment received between time α and time β is

$$M(\beta) - M(\alpha) = \int_\alpha^\beta M'(t)\mathrm{d}t$$

$$= \int_\alpha^\beta \rho(t)\mathrm{d}t \tag{2.7.4}$$

Thus the rate of payment at any time is simply the derivative of the *total* amount paid up to that time, and the total amount paid between any two times is the integral of the rate of payment over the appropriate time interval.

Between times t and $t + \mathrm{d}t$ the total payment received is $M(t + \mathrm{d}t) - M(t)$. If $\mathrm{d}t$ is very small this is approximately $M'(t)\mathrm{d}t$ or $\rho(t)\mathrm{d}t$. Theoretically, therefore, we may consider the present value of the money received between times t and $t + \mathrm{d}t$ as $v(t)\rho(t)\mathrm{d}t$. The present value of the entire cash flow is obtained by integration as

$$\int_0^T v(t)\rho(t)\mathrm{d}t \tag{2.7.5}$$

A rigorous proof of this result is given in textbooks on elementary analysis; $\rho(t)$ will be assumed to satisfy an appropriate condition (e.g. that it is piecewise continuous).

If T is infinite we obtain, by a similar argument, the present value

$$\int_0^\infty v(t)\rho(t)\mathrm{d}t \tag{2.7.6}$$

We may regard formula 2.7.5 as a special case of formula 2.7.6 with $\rho(t) = 0$ for $t > T$. By combining the results for discrete and continuous cash flows, we obtain the formula

$$\sum c_t v(t) + \int_0^\infty v(t)\rho(t)\mathrm{d}t \tag{2.7.7}$$

for the present value of a general cash flow (the summation being over those values of t for which c_t, the discrete cash flow at time t, is non-zero).

So far we have assumed that all payments, whether discrete or continuous, are positive. If one has a series of incoming payments (which may be regarded as positive) and a series of outgoings (which may be regarded as negative), their *net present value* is defined as the difference between the value of the positive cash flow and the value of the negative cash flow. (See chapter 5, in which $\delta(t)$ is assumed to be constant.)

Example 2.7.1

Assume that time is measured in years, and that

$$\delta(t) = \begin{cases} 0{\cdot}04 & \text{for } t < 10 \\ 0{\cdot}03 & \text{for } t \geqslant 10 \end{cases}$$

Find $v(t)$ for all t, and hence find the present value of a continuous payment stream at the rate of 1 p.a. for 15 years, beginning at time 0.

Solution

By formula 2.5.3,

$$v(t) = \begin{cases} \exp\left(-\int_0^t 0{\cdot}04ds\right) & \text{for } t < 10 \\ \exp\left(-\int_0^{10} 0{\cdot}04ds - \int_{10}^t 0{\cdot}03ds\right) & \text{for } t \geqslant 10 \end{cases}$$

i.e.

$$v(t) = \begin{cases} \exp(-0{\cdot}04t) & \text{for } t < 10 \\ \exp(-0{\cdot}1 - 0{\cdot}03t) & \text{for } t \geqslant 10 \end{cases}$$

The present value of the payment stream is, by formula 2.7.5,

$$\int_0^{15} 1 \times v(t)dt = \int_0^{10} \exp(-0{\cdot}04t)dt + \int_{10}^{15} \exp(-0{\cdot}1 - 0{\cdot}03t)dt$$

$$= \frac{1 - \exp(-0{\cdot}4)}{0{\cdot}04} + \exp(-0{\cdot}1)\frac{\exp(-0{\cdot}3) - \exp(-0{\cdot}45)}{0{\cdot}03}$$

$$= 11{\cdot}3543$$

2.8 Valuing cash flows

Consider times t_1 and t_2, where t_2 is not necessarily greater than t_1. The *value at time t_1 of the sum C due at time t_2* is defined as:

(a) If $t_1 \geqslant t_2$, the accumulation of C from time t_2 until time t_1; or
(b) If $t_1 < t_2$, the discounted value at time t_1 of C due at time t_2.

It follows by formulae 2.4.3 and 2.5.1 that in both cases the value at time t_1 of C due at time t_2 is

$$C \exp\left[-\int_{t_1}^{t_2} \delta(t)dt\right] \qquad (2.8.1)$$

(Note the convention that, if $t_1 > t_2$, $\int_{t_1}^{t_2}\delta(t)dt = -\int_{t_2}^{t_1}\delta(t)dt$.)

Since

$$\int_{t_1}^{t_2} \delta(t)dt = \int_0^{t_2} \delta(t)dt - \int_0^{t_1} \delta(t)dt$$

it follows immediately from equations 2.5.3 and 2.8.1 that the value at time t_1 of C due at time t_2 is

$$C\frac{v(t_2)}{v(t_1)} \tag{2.8.2}$$

The value at a general time t_1 of a discrete cash flow of c_t at time t (for various values of t) and a continuous payment stream at rate $\rho(t)$ per time unit may now be found, by the methods given in section 2.7, as

$$\sum c_t \frac{v(t)}{v(t_1)} + \int_{-\infty}^{\infty} \rho(t)\frac{v(t)}{v(t_1)}dt \tag{2.8.3}$$

where the summation is over those values of t for which $c_t \neq 0$. We note that in the special case when $t_1 = 0$ (the present time), the value of the cash flow is

$$\sum c_t v(t) + \int_{-\infty}^{\infty} \rho(t)v(t)dt \tag{2.8.4}$$

where the summation is over those values of t for which $c_t \neq 0$. This is a generalization of formula 2.7.7 to cover past as well as present or future payments. If there are incoming and outgoing payments, the corresponding *net value* may be defined, as in section 2.7, as the difference between the value of the *positive* and the *negative* cash flows. If all the payments are due at or after time t_1, their value at time t_1 may also be called their *discounted value*, and if they are due at or before time t_1, their value may be referred to as their *accumulation*. It follows that any value may be expressed as the sum of a discounted value and an accumulation; this fact is helpful in certain problems. Also, if $t_1 = 0$ and all the payments are due at or after the present time, their value may also be described as their *(discounted) present value*, as defined by formula 2.7.7.

It follows from formula 2.8.3 that the value at any time t_1 of a cash flow may be obtained from its value at another time t_2 by applying the factor $v(t_2)/v(t_1)$, i.e.

$$\begin{bmatrix} \text{value at time } t_1 \\ \text{of cash flow} \end{bmatrix} = \begin{bmatrix} \text{value at time } t_2 \\ \text{of cash flow} \end{bmatrix} \begin{bmatrix} \dfrac{v(t_2)}{v(t_1)} \end{bmatrix} \tag{2.8.5}$$

or

$$\begin{bmatrix} \text{value at time } t_1 \\ \text{of cash flow} \end{bmatrix}[v(t_1)] = \begin{bmatrix} \text{value at time } t_2 \\ \text{of cash flow} \end{bmatrix}[v(t_2)] \tag{2.8.6}$$

Each side of equation 2.8.6 is the value of the cash flow at the present time (time 0).

In particular, by choosing time t_2 as the present time and letting $t_1 = t$, we obtain the result:

$$\begin{bmatrix} \text{value at time } t \\ \text{of cash flow} \end{bmatrix} = \begin{bmatrix} \text{value at the present} \\ \text{time of cash flow} \end{bmatrix} \begin{bmatrix} \dfrac{1}{v(t)} \end{bmatrix} \qquad (2.8.7)$$

These results are useful in many practical examples. The time 0 and the unit of time may be chosen so as to simplify the calculations.

Example 2.8.1

A businessman is owed the following amounts: £1000 on 1 January 1986; £2500 on 1 January 1987; and £3000 on 1 July 1987. Assuming a constant force of interest of 0.06 per annum, find the value of these payments on (a) 1 January 1984 and (b) 1 March 1985.

Solution

Let time be measured in years from 1 January 1984. The value of the debts at that date is, by formula 2.7.1,

$$1000v(2) + 2500v(3) + 3000v(3 \cdot 5)$$
$$= 1000 \exp(-0 \cdot 12) + 2500 \exp(-0 \cdot 18) + 3000 \exp(-0 \cdot 21)$$
$$= £5406 \cdot 85$$

The value at 1 March 1985 of the same debts is, by formula 2.8.7,

$$5406 \cdot 85 \exp(0 \cdot 06 \times \tfrac{14}{12}) = £5798 \cdot 89$$

Example 2.8.2

Suppose that time is measured in years and that $\delta(t)$, the force of interest per unit time, is given by Stoodley's formula with the parameters p, r, and s as specified in example 2.6.1.

(a) Find the single payment which, if invested at time 10, will accumulate to £30 000 at time 20.
(b) Find the accumulated amount after ten years of ten annual payments each of £1000, the first payment being made at time 0.

Solution

(a) This is simply the value at time 10 of £30 000 due at time 20. From equation 2.8.2 the required payment is

$$30\,000\, v(20)/v(10)$$

where (see example 2.6.1)

$$v(t) = \tfrac{2}{3}(1 \cdot 22)^{-t} + \tfrac{1}{3}(1 \cdot 08)^{-t}$$

This gives the value as £10 259.

(b) We have a cash flow with ten equal payments of £1000. The value of the cash flow at time 0 is

$$1000 \sum_{t=0}^{9} v(t)$$

and hence the value at time 10 is (by equation 2.8.7)

$$1000 \sum_{t=0}^{9} v(t)/v(10) = 22\,822$$

The accumulated amount is thus £22 822.

2.9 Interest income

Consider now an investor who wishes not to accumulate money but to receive an income while keeping his capital fixed at C. If the rate of interest is fixed at i per time unit, and if the investor wishes to receive his income at the end of each time unit, it is clear that his income will be iC per time unit, payable in arrear, until such time as he withdraws his capital. This case is discussed further in chapter 3.

More generally, suppose that $t > t_0$ and that the investor wishes to deposit C at time t_0 for withdrawal at time t. Suppose further that $n > 1$ and that the investor wishes to receive interest on his deposit at the n equally spaced times $t_0 + h$, $t_0 + 2h, \ldots, t_0 + nh$, where $h = (t - t_0)/n$. The interest payable at time $t_0 + (j + 1)h$, for the period $t_0 + jh$ to $t_0 + (j + 1)h$, will be

$$Chi_h(t_0 + jh)$$

so that the total interest income payable between times t_0 and t will be

$$C \sum_{j=0}^{n-1} hi_h(t_0 + jh) \tag{2.9.1}$$

Since, by assumption, $i_h(t)$ tends to $\delta(t)$ as h tends to 0, it is fairly easily shown (provided that $\delta(t)$ is continuous) that as n increases (so that h tends to 0) the total interest received between times t_0 and t converges to

$$I(t) = C \int_{t_0}^{t} \delta(s)ds \tag{2.9.2}$$

Hence, in the limit, the *rate of payment of interest income* per unit time at time t, $I'(t)$, equals

$$C\delta(t) \tag{2.9.3}$$

The position may be illustrated by Figure 2.9.1. The cash C in the 'tank' remains constant at C, while interest income is decanted continuously at the

FIGURE 2.9.1 *Interest income flow*

instantaneous rate $C\delta(t)$ per unit time at time t. Thus, if interest is paid very frequently from a variable-interest deposit account, the position may be idealized to that shown in the figure, which depicts a continuous flow of interest income. Of course, if $\delta(t) = \delta$ for all t, interest is received at the constant rate $C\delta$ per time unit.

If the investor withdraws his capital at time T, the present values of his income and capital are, by formulae 2.5.4 and 2.7.5,

$$C \int_0^T \delta(t)v(t)\mathrm{d}t \tag{2.9.4}$$

and

$$Cv(T) \tag{2.9.5}$$

respectively. Since

$$\int_0^T \delta(t)v(t)\mathrm{d}t = \int_0^T \delta(t)\exp\left[-\int_0^t \delta(s)\mathrm{d}s\right]\mathrm{d}t$$

$$= \left[-\exp\left(-\int_0^t \delta(s)\mathrm{d}s\right)\right]_0^T$$

$$= 1 - v(T)$$

we obtain

$$C = C \int_0^T \delta(t)v(t)\mathrm{d}t + Cv(T) \tag{2.9.6}$$

as one would expect by general reasoning. In the case when $T = \infty$ (in which the investor never withdraws his capital) a similar argument gives the result that

$$C = C \int_0^\infty \delta(t)v(t)\mathrm{d}t \tag{2.9.7}$$

where the expression on the right-hand side is the present value of the interest income. The case when $\delta(t) = \delta$ for all t is discussed further in chapter 3.

2.10 Capital gains and losses, and taxation

So far we have described the difference between money returned at the end of the term and the cash originally invested as 'interest'. In practice, however, this quantity may be divided into *interest income* and *capital gains*, the term *capital loss* being used for a negative capital gain.

Some investments, known as zero-coupon bonds, bear no interest income. A number of these securities have been issued; for example, in the United States a zero-coupon stock providing a return of $100 in 1988 was issued seven years earlier at a price of $39·164. Many other securities provide both interest income and capital gains; these will be considered later in this book.

Since the basis of taxation of capital gains is usually different from that of interest income the distinction between interest income and capital gains is of importance for tax-paying investors. Income tax may usually be assumed to be deducted from interest income at the rate applicable when the income is received, although in certain cases it is charged some time later (for example, at the end of the financial year). Capital gains tax is a little more complicated and will be dealt with in chapter 8.

The theory developed in the preceding sections is unaltered if we replace the term 'interest' by 'interest and capital gains less any income and capital gains taxes'. The force of interest (which, to avoid any confusion of terminology, should perhaps be called the *force of growth*) will include an allowance for capital appreciation or depreciation as well as interest income, and will also allow for the incidence of income and capital gains taxes on the investor.

2.11 Further references

There exists a considerable number of books and papers relating to interest rates and their *term structure* (i.e. the variation of interest rates according to the term of investment). The interested reader may refer, for example, to references [7], [14], [17], [18], [19], [20], [21], [22], [29], [33], [34], [40], and [50].

Exercises

2.1 With *compound* interest at an effective rate i per annum the accumulation of 1 after t years is $(1 + i)^t$ (see equation 2.4.8), while with *simple* interest at the same rate the corresponding accumulation is $(1 + ti)$ (see equation 2.1.3).

Show, for any positive value of i, that (a) the accumulation with simple interest exceeds that with compound interest if $0 < t < 1$, and (b) the opposite is true if $t > 1$.

Hint   Let $f(i) = (1 + i)^t - (1 + ti)$. Note that $f(0) = 0$ and consider the sign of $f'(i)$ for $i > 0$.

2.2   Over a given year the force of interest per annum is a linear function of time, falling from 0·15 at the start of the year to 0·12 at the end of the year. Find the value at the start of the year of the nominal rate of interest per annum on transactions of term (a) three months, (b) one month, and (c) one day. Find also the corresponding values midway through the year. (Note how these values tend to the force of interest at the appropriate time.)

2.3   A bank credits interest on deposits using a variable force of interest. At the start of a given year an investor deposited £20 000 with the bank. The accumulated amount of the investor's account was £20 596·21 midway through the year and £21 183·70 at the end of the year.
      Measuring time in years from the start of the given year and assuming that over the year the force of interest per annum was a linear function of time, derive an expression for the force of interest per annum at time t ($0 \leqslant t \leqslant 1$) and find the accumulated amount of the account three-quarters of the way through the year.

2.4   A borrower is under an obligation to repay a bank £6280 in four years' time, £8460 in seven years' time and £7350 in thirteen years' time.
      As part of a review of his future commitments the borrower now offers either

(a) To discharge his liability for these three debts by making an appropriate single payment five years from now; or
(b) To repay the total amount owed (i.e. £22 090) in a single payment at an appropriate future time.

On the basis of a constant force of interest per annum of 0·076 961 (i.e. log 1·08), find the appropriate single payment if offer (a) is accepted by the bank, and the appropriate time to repay the entire indebtedness if offer (b) is accepted. (Your answers should be derived on the basis that the present value of the single payment under the revised arrangement should equal the present value of the three payments due under the current obligation.)

2.5   For certain bank deposits over a given year, the force of interest per annum was 0·15 at the start of the year, 0·10 midway through the year, and 0·08 at the end of the year.
      Find the accumulated amount at the end of the year of a deposit of £5000 at the start of the year, assuming that the force of interest per annum was

(a) A quadratic function of time over the year;
(b) A linear function of time on the first half of the year and also a linear function of time on the second half of the year.

2.6   (Application of Stoodley's formula)
      Assume that $\delta(t)$, the force of interest per annum at time t (measured in years from the present), is given by Stoodley's formula

$$\delta(t) = p + \frac{s}{1 + re^{st}}$$

with $p = 0.058\,269$, $s = 0.037\,041$, and $r = 1/3$.

(a) Show that

$$v(t) = \tfrac{1}{4}(1.06)^{-t} + \tfrac{3}{4}(1.1)^{-t}$$

(b) An investor agrees to make 12 annual payments, each of £600, the first payment being made now. In return the investor will receive either
   (i) The accumulated amount of his payments 12 years from now; or
   (ii) A series of 12 annual payments, each of the same amount, the first payment being made 12 years from now.
   Find the amount of the lump sum and of the alternative level annuity offered to the investor.

Note In the compound interest tables at the back of this book the value of $\sum_{t=1}^{n}(1+i)^{-t}$ is denoted by $a_{\overline{n}|}$. Use the tables with interest rates of 6% and 10%.

2.7 The force of interest $\delta(t)$ per annum at time t (measured in years from now) will be a linear function of t for m years and thereafter will be constant at the level attained at time m.

(a) Considering separately the cases when $n \leqslant m$ and $n > m$, derive an expression in terms of $n, m, \delta(0),$ and $\delta(m)$ for the accumulation of 1 from time 0 to time n.
(b) Given that $m = 16$, $\delta(0) = 0.08$, and $\delta(16) = 0.048$, evaluate your expression when (i) $n = 15$ and (ii) $n = 40$.
(c) Find the constant force of interest which would produce the same accumulation (i) over 15 years and (ii) over 40 years.

2.8 (Piecewise constant force of interest)
Assume that $\delta(t)$, the force of interest per annum at time t (years), is given by the formula

$$\delta(t) = \begin{cases} 0.08 & \text{for } 0 \leqslant t < 5 \\ 0.06 & \text{for } 5 \leqslant t < 10 \\ 0.04 & \text{for } \quad t \geqslant 10 \end{cases}$$

(a) Derive expressions for $v(t)$, the present value of 1 due at time t.
(b) An investor effects a contract under which he will pay 15 premiums annually in advance into an account which will accumulate according to the above force of interest. Each premium will be of amount £600 and the first premium will be paid at time 0. In return the investor will receive either
   (i) The accumulated amount of the account one year after the final premium is paid; or
   (ii) A level annuity payable annually for eight years, the first payment being made one year after the final premium is paid.
   Find the lump sum payment under option (i) and the amount of the annual annuity under option (ii).

2.9 (a) In valuing future payments an investor uses the formula

$$v(t) = \frac{\alpha(\alpha + 1)}{(\alpha + t)(\alpha + t + 1)} \qquad t \geqslant 0$$

where α is a given positive constant, for the value at time 0 of 1 due at time t (measured in years).

Show that the above formula implies that

(i) The force of interest per annum at time t will be

$$\delta(t) = \frac{2t + 2\alpha + 1}{(\alpha + t)(\alpha + t + 1)} \qquad (1)$$

(ii) The effective rate of interest for the period n to $n + 1$ will be

$$i(n) = \frac{2}{n + \alpha}$$

and

(iii) The present value of a series of n payments, each of amount 1 (the rth payment being due at time r) is

$$a(n) = \frac{n\alpha}{(n + \alpha + 1)}$$

(b) Suppose now that $\delta(t)$ is given by equation 1 with $\alpha = 15$.

Find the level annual premium, payable in advance for twelve years, which will provide an annuity of £1800 per annum, payable annually for ten years, the first annuity payment being made one year after payment of the final premium.

What is the value at time 12 of the series of annuity payments?

2.10 (A continuous payment stream)

An investor purchases an annuity which is payable continuously for n years, where n is an integer. The rate of payment of the annuity is a linear function of time.

(a) Measuring time in years from the date of purchase of the annuity and letting $I_r (r = 1, 2, \ldots, n)$ denote the amount of annuity paid in the rth year, derive an expression in terms of I_1 and I_2 for the rate of payment per annum at time t. Find also the total amount of annuity paid up to time $t (0 < t \leqslant n)$.

(b) (i) On the basis of a constant force of interest per annum δ, derive an expression for the present value of the annuity in terms of n, δ, I_1, and I_2.

(ii) Given that $n = 20$, $I_2 = 1 \cdot 07 I_1$, and that when $\delta = \log 1 \cdot 06$ the present value of the annuity is £9047, find the value of I_1 and the amount of annuity paid in the final year.

2.11 Suppose that the force of interest per annum at time t years is

$$\delta(t) = ae^{-bt} \qquad (1)$$

(a) Show that the present value of 1 due at time t is

$$v(t) = \exp\left[\frac{a}{b}(e^{-bt} - 1)\right]$$

(b) (i) Assuming that the force of interest per annum is given by equation 1 and that it will fall by 50% over ten years from the value 0·10 at time 0, find the present value of a series of four annual payments, each of amount £1000, the first payment being made at time 1.

(ii) At what *constant* force of interest per annum does this series of payments have the same present value as that found in (i)?

2.12 (Present value of a continuously payable cash flow)
Suppose that the force of interest per annum at time t years is

$$\delta(t) = r + se^{-rt}$$

(a) Show that the present value of 1 due at time t is

$$v(t) = \exp\left(\frac{-s}{r}\right)\exp(-rt)\exp\left(\frac{s}{r}e^{-rt}\right)$$

(b) (i) Hence show that the present value of an annuity payable continuously for n years at the constant rate of £1000 per annum is

$$\frac{1000}{s}\left\{1 - \exp\left[\frac{s}{r}(e^{-rn} - 1)\right]\right\}$$

(ii) Evaluate the last expression when $n = 50$, $r = \log 1{\cdot}01$ and $s = 0{\cdot}03$.

CHAPTER THREE

THE BASIC COMPOUND INTEREST FUNCTIONS

3.1 Introduction

The particular case in which $\delta(t)$, the force of interest per unit time at time t, does *not* depend on t is of special importance. In this situation we assume that, for all values of t,

$$\delta(t) = \delta \qquad (3.1.1)$$

where δ is some constant. Throughout this chapter we shall assume that equation 3.1.1 is valid, unless otherwise stated.

The value at time s of 1 due at time $s + t$ is (see equation 2.8.1)

$$\exp\left[-\int_s^{s+t} \delta(r)dr\right] = \exp\left(-\int_s^{s+t} \delta dr\right)$$

$$= \exp(-\delta t)$$

which does *not* depend on s. Thus the value at *any* given time of 1 due after a further period t is

$$v(t) = e^{-\delta t} \qquad (3.1.2)$$

$$= v^t \qquad (3.1.3)$$

$$= (1 - d)^t \qquad (3.1.4)$$

where v and d are defined in terms of δ by the equations

$$v = e^{-\delta} \qquad (3.1.5)$$

and

$$1 - d = e^{-\delta} \qquad (3.1.6)$$

Thus, in return for a repayment of 1 at time 1, an investor will lend an amount $(1 - d)$ at time 0. The sum of $(1 - d)$ may be considered as a loan of 1 (to be repaid after 1 unit of time) on which interest of amount d is payable *in advance*. For this reason d is called the 'rate of discount' per unit time. Sometimes, in order to avoid confusion with nominal rates of discount (see chapter 4), d is called the *effective rate of discount* per unit time.

Similarly, it follows immediately from equation 2.4.9 that the accumulated amount at time $s + t$ of 1 invested at time s does *not* depend on s and is given by

$$F(t) = e^{\delta t} \tag{3.1.7}$$

$$= (1 + i)^t \tag{3.1.8}$$

where i is defined by the equation

$$1 + i = e^{\delta} \tag{3.1.9}$$

Thus an investor will lend an amount 1 at time 0 in return for a repayment of $(1 + i)$ at time 1. Accordingly i is called the *rate of interest* (or the *effective rate of interest*) per unit time.

Although we have chosen to define i, v, and d in terms of the force of interest δ, any three of i, v, d, and δ are uniquely determined by the fourth. Thus, for example, if we choose to regard i as the basic parameter, then it follows from equation 3.1.9 that

$$\delta = \log(1 + i)$$

In addition, equations 3.1.5 and 3.1.9 imply that

$$v = (1 + i)^{-1}$$

while equations 3.1.6 and 3.1.9 imply that

$$d = 1 - (1 + i)^{-1}$$

$$= \frac{i}{1 + i}$$

These last three equations define δ, v, and d in terms of i.

The last equation may be written as

$$d = iv$$

which confirms that a payment of i at time 1 has the same value as a payment of d at time 0. What sum paid *continuously* (at a constant rate) over the time interval $[0, 1]$ has the same value as either of these payments? Let the required sum be σ. Then, taking values at time 0, we have

$$d = \int_0^1 \sigma e^{-\delta t} dt$$

$$= \sigma \left(\frac{1 - e^{-\delta}}{\delta} \right) \qquad \text{(if } \delta \neq 0\text{)}$$

$$= \sigma \left(\frac{d}{\delta} \right) \qquad \text{(by equation 3.1.6)}$$

Hence $\sigma = \delta$. This result is also true, of course, when $\delta = 0$. This establishes the important fact that a payment of δ made continuously over the period $[0, 1]$ has the same value as a payment of d at time 0 or a payment of i at time 1. Each of the three payments may be regarded as alternative methods of paying interest on a loan of 1 over the period.

In certain situations it may be natural to regard the force of interest as the basic parameter, with its implied values for i, v, and d. In other cases it may be preferable to assume a certain value for i (or d or v) and to calculate, if necessary, the values implied for the other three parameters. It is left as a simple but important exercise for the reader to verify the relationships illustrated by table 3.1.1. The reader should verify that, if i is small, then i, δ, and d are all of the same order of magnitude and that $v \simeq 1 - i$.

Table 3.1.1 *Relationships between δ, i, v, and d*

In terms of \ Value of	δ	i	v	d
δ		$e^{\delta} - 1$	$e^{-\delta}$	$1 - e^{-\delta}$
i	$\log(1 + i)$		$(1 + i)^{-1}$	$i(1 + i)^{-1}$
v	$-\log v$	$v^{-1} - 1$		$1 - v$
d	$-\log(1 - d)$	$(1 - d)^{-1} - 1$	$1 - d$	

When i is small, approximate formulae for d and δ in terms of i may be obtained from well-known series by neglecting the remainder after a small number of terms. For example, since

$$\delta = \log(1 + i)$$
$$= i - \tfrac{1}{2}i^2 + \tfrac{1}{3}i^3 - \tfrac{1}{4}i^4 + \cdots \qquad (\text{if } |i| < 1)$$

it follows that, for small values of i,

$$\delta \simeq i - \tfrac{1}{2}i^2$$

Similarly

$$d = i(1 + i)^{-1}$$
$$= i(1 - i + i^2 - i^3 + \cdots) \qquad (\text{if } |i| < 1)$$
$$= i - i^2 + i^3 - i^4 + \cdots$$

so, if i is small,

$$d \simeq i - i^2$$

Using the relationships contained in table 3.1.1, the reader should verify that, if δ is small,

$$i \simeq \delta + \tfrac{1}{2}\delta^2$$

and

$$d \simeq \delta - \tfrac{1}{2}\delta^2$$

An analysis of the errors contained in the above approximations is provided by a more careful study of the remainder term in the appropriate series. This is illustrated by the following example.

Example 3.1.1

Show that, if $0 < \delta < 0.1$, then

$$\delta + \tfrac{1}{2}\delta^2 < i < \delta + \tfrac{1}{2}\delta^2 + 0.185\delta^3$$

Solution

Let $f(x) = e^x$. Since (by Taylor's theorem)

$$f(x) = f(0) + xf'(0) + \frac{x^2}{2!}f''(0) + \frac{x^3}{3!}f'''(\varepsilon)$$

where ε lies between 0 and x, it follows that, for all positive values of δ,

$$e^\delta = 1 + \delta + \frac{\delta^2}{2} + \frac{\delta^3}{6}e^\varepsilon \tag{1}$$

where $0 < \varepsilon < \delta$. Now

$$i = e^\delta - 1$$

$$= \delta + \frac{\delta^2}{2} + \delta^3\frac{e^\varepsilon}{6} \qquad \text{(by equation 1)}$$

If $0 < \delta < 0.1$ and $0 < \varepsilon < \delta$, then $e^\varepsilon/6 < e^{0.1}/6 = 0.1842 < 0.185$. The last inequality establishes the required result.

3.2 The equation of value and the yield on a transaction

Suppose that in return for an outlay of X at time 0 an investor will receive n payments, each of amount jX, at times $1, 2, \ldots, n$ in addition to the repayment at time n of his original investment. Thus, until it is repaid, the investment of X generates income of jX at the end of each period. Intuitively we may speak of j as the 'yield' per unit time for the investment.

We wish to define the concept of the yield for a wider class of investments. In order to do so, it is first necessary to define the *equation of value* associated with any transaction.

Consider a transaction which provides that, in return for outlays of amount $a_{t_1}, a_{t_2}, \ldots, a_{t_n}$ at times $t_1, t_2, \ldots, t_n$, an investor will receive payments of $b_{t_1}, b_{t_2}, \ldots, b_{t_n}$ at these times respectively. (In most situations only *one* of a_{t_r} and b_{t_r} will be non-zero.) At what force or rate of interest does the series of outlays have the same value as the series of receipts? At force of interest δ the two series are of equal value if and only if

$$\sum_{r=1}^{n} a_{t_r} e^{-\delta t_r} = \sum_{r=1}^{n} b_{t_r} e^{-\delta t_r} \tag{3.2.1}$$

This equation may be written as

$$\sum_{r=1}^{n} c_{t_r} e^{-\delta t_r} = 0 \tag{3.2.2}$$

where

$$c_{t_r} = b_{t_r} - a_{t_r}$$

is the amount of the *net cash flow* at time t_r. (We adopt the convention that a negative cash flow corresponds to a payment *by* the investor and a positive cash flow represents a payment *to* him.)

Equation 3.2.2, which expresses algebraically the condition that, at force of interest δ, the total value of the net cash flows is 0, is called the *equation of value* for the force of interest implied by the transaction. If (following chapter 2) we let $e^\delta = 1 + i$, the equation may be written as

$$\sum_{r=1}^{n} c_{t_r} (1 + i)^{-t_r} = 0 \tag{3.2.3}$$

The latter form is known as the equation of value for the rate of interest or the *yield equation*. Alternatively, the equation may be written as

$$\sum_{r=1}^{n} c_{t_r} v^{t_r} = 0$$

(Note that in the above equations n may be infinite.)

In relation to continuous payment streams, if we let $\rho_1(t)$ and $\rho_2(t)$ be the rates of paying and receiving money at time t respectively, we call $\rho(t) = \rho_2(t) - \rho_1(t)$ the *net rate of cash flow* at time t. The equation of value (corresponding to equation 3.2.2) for the force of interest is

$$\int_0^\infty \rho(t) e^{-\delta t} dt = 0 \tag{3.2.4}$$

When both discrete and continuous cash flows are present, the equation of value is

$$\sum_{r=1}^{n} c_{t_r} e^{-\delta t_r} + \int_0^\infty \rho(t) e^{-\delta t} dt = 0 \tag{3.2.5}$$

and the equivalent yield equation is

$$\sum_{r=1}^{n} c_{t_r}(1 + i)^{-t_r} + \int_0^\infty \rho(t)(1 + i)^{-t}dt = 0 \qquad (3.2.6)$$

For any given transaction, equation 3.2.5 may have no roots, a unique root, or several roots. (We consider only *real* roots.) If there is a unique root, δ_0 say, it is known as the force of interest implied by the transaction, and the corresponding rate of interest $i_0 = e^{\delta_0} - 1$ is called the *yield* per unit time. (Alternative terms for the yield are the *internal rate of return* and the *money-weighted rate of return* for the transaction.) Thus the yield is defined if and only if equation 3.2.6 has precisely one root greater than -1 and, when such a root exists, it is the yield.

Several authors have considered the problem of determining whether or not equation 3.2.6 has a unique root for particular classes of transaction. The interested reader may turn to references [4], [42], and [48] for a more detailed discussion of this topic.

Although for certain investments the yield does not exist (since the equation of value 3.2.2 has no roots or more than one root: see exercise 3.6 and chapter 5), there is one important class of transaction for which the yield always exists. This is described in the following theorem.

Theorem 3.2.1

For any transaction in which *all* the negative net cash flows precede *all* the positive net cash flows (or vice versa) the yield is well-defined.

Proof

For such a transaction we may assume without loss of generality that all the negative net cash flows precede all the positive net cash flows. (If the opposite holds, we simply multiply the equation of value by -1 to revert to this situation.) There is thus an index l such that the equation of value 3.2.2 can be written as

$$-\{\alpha_1 e^{-\delta t_1} + \alpha_2 e^{-\delta t_2} + \cdots + \alpha_l e^{-\delta t_l}\}$$
$$+ \{\alpha_{l+1} e^{-\delta t_{l+1}} + \alpha_{l+2} e^{-\delta t_{l+2}} + \cdots + \alpha_n e^{-\delta t_n}\} = 0 \qquad (3.2.7)$$

where

$$t_1 < t_2 < \cdots < t_n \qquad (3.2.8)$$

and each $\alpha_i > 0$ $(1 \leqslant i \leqslant n)$.

After multiplication by $-e^{\delta t_l}$ equation 3.2.7 becomes

$$g(\delta) \equiv g_1(\delta) - g_2(\delta) = 0$$

where

$$g_1(\delta) = \sum_{r=1}^{l} \alpha_r e^{\delta(t_l - t_r)}$$

$$g_2(\delta) = \sum_{r=l+1}^{n} \alpha_r e^{-\delta(t_r - t_l)}$$

Since each α_i is positive, the order condition 3.2.8 implies that g_1 is an increasing function of δ and that g_2 is a strictly decreasing function. Hence g is a strictly increasing function. Since $\lim_{\delta \to \infty} g(\delta) = \infty$ and $\lim_{\delta \to -\infty} g(\delta) = -\infty$, it follows that the equation of value has a unique root (which may be positive, zero, or negative). This completes the proof of the theorem in the discrete case. The proof when continuous cash flows are present follows similarly.

Although the yield is defined only when equation 3.2.3 has a unique root greater than -1, it is sometimes of interest to consider transactions in which the yield equation has a unique *positive* root (even though there may also be negative roots). There is one easily described class of transaction for which the yield equation always has precisely one positive root. This is described in the following theorem.

Theorem 3.2.2

Suppose that $t_0 < t_1 < \cdots < t_n$ and consider a transaction for which the investor's net cash flow at time t_i is of amount c_{t_i}. (Some of the $\{c_{t_i}\}$ will be positive and some negative, according to the convention described above.) For $i = 0, 1, \ldots, n$ let $A_i = \sum_{r=0}^{i} c_{t_r}$, so that A_i denotes the *cumulative* total amount received by the investor after the cash flow at time t_i has occurred. Suppose that A_0 and A_n are both non-zero and that, when any zero values are excluded, the sequence $\{A_0, A_1, \ldots, A_n\}$ contains precisely one change of sign. Then the yield equation has exactly one positive root.

It should be noted that the theorem gives no information on the existence of negative roots.

We omit the proof of this theorem, which we believe is originally due to Steffensen.

One particular example of this situation is provided by a transaction in which all the investor's outlays precede all his receipts and the total amount received exceeds the total outlays. The existence of the yield for this type of transaction has been established by the previous theorem, the proof of which shows that the yield is positive. A more general example is given by a transaction which provides that, in return for payments of amount 5 and 3 at times 0 and 2 respectively, an investor will receive returns of 1, 8, and 4 at times 1, 3, and 4 respectively. For this investment the net cash flows are given chronologically by the sequence

$$\{c_{t_i}\} = \{-5, 1, -3, 8, 4\}$$

and the *cumulative* total cash flows (in chronological order) by the sequence

$$\{A_i\} = \{-5, -4, -7, 1, 5\}$$

Since the latter sequence contains only one sign change (in this case from negative to positive), the yield equation has only one positive root. (The root is in fact 0·221 09 or 22·109%.)

Consider in particular the simple transaction described in the first paragraph of this section. Here the investor makes a single payment of X at time 0 in return for subsequent receipts. The above remarks imply that the yield is well defined and is the unique root of the equation

$$f(i) = 0$$

where

$$f(i) = -X + jX(1 + i)^{-1} + jX(1 + i)^{-2} + \cdots + jX(1 + i)^{-n} + X(1 + i)^{-n}$$

The reader should verify that $f(j) = 0$. The yield is thus j, so that for this investment the wider mathematical definition of yield is consistent with our intuition.

The analysis of the equation of value for a given transaction may be somewhat complex. (See appendix 2 for possible methods of solution.) However, when the equation $f(i) = 0$ is such that f is a monotonic function, its analysis is particularly simple. The equation has a root if and only if we can find i_1 and i_2 with $f(i_1)$ and $f(i_2)$ of opposite sign. In this case, the root is unique and lies between i_1 and i_2. By choosing i_1 and i_2 to be 'tabulated' rates sufficiently close to each other, we may determine the yield to any desired degree of accuracy. (See Figure 3.2.1.)

It should be noted that, after multiplication by $(1 + i)^{t_0}$, equation 3.2.3 takes the equivalent form

$$\sum_{r=1}^{n} c_{t_r}(1 + i)^{t_0 - t_r} = 0 \tag{3.2.9}$$

This slightly more general form may be called *the equation of value at time t_0*. It is of course directly equivalent to the original equation (which is now seen

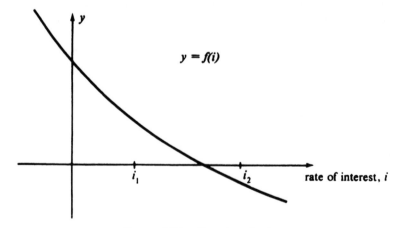

$$y = f(i)$$

FIGURE 3.2.1   *Equation of value*

to be the equation of value at time 0). In certain problems a particular choice of t_0 may simplify the solution. The student should always consider whether or not a value of t_0 other than 0 may be appropriate in any given problem. The unit of time should be chosen so as to simplify the calculations.

Example 3.2.1

In return for an immediate payment of £500 and a further payment of £200 two years from now an investor will receive £1000 after five years. Find the yield for the transaction.

Solution

Choose one year as the unit of time. The equation of value (at time 0) is

$$f(i) = -500 - 200(1 + i)^{-2} + 1000(1 + i)^{-5} = 0$$

Our earlier discussion indicates that there is a unique root. Since $f(0.08) = 9.115$ and $f(0.09) = -18.405$, the yield is between 8% and 9% per annum. A first approximation for the yield, obtained by linear interpolation, is

$$0.08 + (0.09 - 0.08)\frac{9.115 - 0}{9.115 - (-18.405)} = 0.0833$$

or 8.33% per annum. If it is desired to find the yield to a greater degree of accuracy, one might evaluate $f(0.085)$ using functions tabulated at $8\frac{1}{2}\%$ or use more accurate methods (see appendix 2). (The yield per cent per annum to four decimal places is in fact 8.3248.)

The interpretation of the yield in this case is simple. If the outlays of £500 and £200 were to be deposited in an account on which interest is payable at the rate of 8.3248% per annum, at the end of five years the accumulated amount of the account would be £1000. If the investor expects to be able to make deposits over the next five years at a greater rate of interest than 8.3248% p.a., he will not choose the investment.

Note Equivalently, we may work with the equation of value at time 5. This is

$$-500(1 + i)^5 - 200(1 + i)^3 + 1000 = 0$$

Example 3.2.2

In return for a loan of £100 a borrower agrees to repay £110 after seven months. Find

(a) The rate of interest per annum;
(b) The rate of discount per annum;
(c) The force of interest per annum for the transaction.

Shortly after receiving the loan the borrower requests that he be allowed to repay the loan by a payment of £50 on the original settlement date and a second payment six months after this date. Assuming that the lender agrees to the request and that the calculation is made on the original interest basis, find the amount of the second payment under the revised transaction.

Solution

For illustrative purposes we find each of i, d, and δ from first principles.

(a) The rate of interest per annum is given by the equation

$$100(1 + i)^{7/12} = 110$$

from which it follows that $i = 0\cdot177\,49$ or $17\cdot749\%$
(b) The rate of discount per annum d is given by the equation

$$100 = 110(1 - d)^{7/12}$$

from which we obtain $d = 0\cdot150\,74$ or $15\cdot074\%$.
(c) The force of interest per annum δ is given by the equation

$$100e^{(7/12)\delta} = 110$$

so that $\delta = 0\cdot163\,39$ or $16\cdot339\%$.
    The reader may check that these values are consistent with the relationships in table 3.1.1.
    Let the amount of the second payment required under the revised transaction be £X. Then, applying the equation of value at the final repayment date (which is 13 months after the loan was made), we obtain

$$100e^{(13/12)\delta} - 50e^{(1/2)\delta} - X = 0$$

so that

$$X = 100e^{(13/12)\delta} - 50e^{(1/2)\delta}$$

Since it is assumed that the original basis applies, $\delta = 0\cdot163\,39$ (as found above) and the last equation then implies that $X = 65\cdot11$.
    An alternative approach to finding the value of X is provided by noting that the borrower wishes to replace the originally agreed payment of £110 by a payment of £50 – i.e. he wishes to defer a payment of £60 due on the original repayment date by a period of six months. The deferred payment (of amount £X) must have the same value as the £60 it replaces. Equating these values at the final repayment date, we obtain

$$60e^{(1/2)\delta} = X$$

from which $X = 65.11$, as before.

Example 3.2.3 (The equated time of a series of payments)

A borrower who is due to repay amounts $x_1, x_2, \ldots, x_k$ at distinct times $t_1, t_2, \ldots, t_k$ respectively, wishes instead to make a single repayment of amount $(x_1 + x_2 + \cdots + x_k)$. He suggests that this single repayment be made at time t^*, where t^* is a weighted average of the original times defined by

$$t^* = \frac{x_1 t_1 + x_2 t_2 + \cdots + x_k t_k}{x_1 + x_2 + \cdots + x_k}$$

The lender proposes alternatively that the single repayment be made at time T, where T is determined by an equation of value based on a given positive force of interest δ.

Thus T is defined by the equation

$$(x_1 + x_2 + \cdots + x_k)e^{-\delta T} = x_1 e^{-\delta t_1} + x_2 e^{-\delta t_2} + \cdots + x_k e^{-\delta t_k}$$

Prove that $t^* > T$.

T is called the *equated time* for the series of payments (on the basis of the given force of interest δ). The quantity t^* (which does not depend on δ) is a commonly – used approximation for T. The above result shows that, when the force of interest is positive, the use of the approximation favours the borrower, in that the approximate value t^* is greater than the true equated time T.

Solution

By working in the smallest unit of currency we may assume that $x_1, x_2, \ldots, x_k$ are all integers. (This assumption avoids a slightly tricky analytical argument, although the result is true for all positive values $\{x_i\}$.) Consider now a set consisting of x_1 real numbers each equal to $e^{-\delta t_1}$, x_2 real numbers each equal to $e^{-\delta t_2}, \ldots, x_k$ real numbers each equal to $e^{-\delta t_k}$. The numbers of the set are not all equal (since $t_1, \ldots, t_k$ are distinct) and hence (by a well-known algebraic result) their arithmetic mean exceeds their geometric mean. Thus

$$
\begin{aligned}
e^{-\delta T} &= \frac{x_1 e^{-\delta t_1} + x_2 e^{-\delta t_2} + \cdots + x_k e^{-\delta t_k}}{x_1 + x_2 + \cdots + x_k} \\
&> \left[(e^{-\delta t_1})^{x_1}(e^{-\delta t_2})^{x_2} \cdots (e^{-\delta t_k})^{x_k} \right]^{1/(x_1 + x_2 + \cdots + x_k)} \\
&= \exp\left[-\delta\left(\frac{x_1 t_1 + x_2 t_2 + \cdots + x_k t_k}{x_1 + x_2 + \cdots + x_k} \right) \right] \\
&= e^{-\delta t^*}
\end{aligned}
$$

Hence

$$e^{-\delta T} > e^{-\delta t^*}$$

which (since $\delta > 0$ by hypothesis) implies that $t^* > T$, as required

Example 3.2.4 (Commercial discount)

A lender bases his short-term transactions on a *rate of commercial discount D*, where $0 < D < 1$. This means that, if $0 < t \leqslant 1$, in return for a repayment of X after a period t he will lend $X(1 - Dt)$ at the start of the period.

For such a transaction over an interval of length t $(0 < t \leqslant 1)$ derive an expression (in terms of D and t) for d, the effective rate of discount per unit time. Hence show that, regarded as a function of t, d is increasing on the interval $0 < t \leqslant 1$.

Solution

The equation of value for the transaction at the time the loan is made may be written in the form

$$(1 - d)^t = 1 - Dt$$

so

$$d = 1 - (1 - Dt)^{1/t}$$

which is an expression for d in terms of D and t.

From the first equation it follows that

$$t \log(1 - d) = \log(1 - Dt)$$

$$= -Dt - \frac{(Dt)^2}{2} - \frac{(Dt)^3}{3} - \cdots$$

(by the well-known series for $\log(1 - x)$ with $|x| < 1$). Hence

$$\log(1 - d) = -D - D^2\frac{t}{2} - D^3\frac{t^2}{3} - \cdots$$

Now, as t increases, the right-hand side of the last equation decreases. (The absolute value of each term on the right-hand side of the equation increases, but the terms are all negative.) Thus $\log(1 - d)$ decreases. This implies that $(1 - d)$ decreases, which in turn implies that d increases.

As an illustration of the result of example 3.2.4, consider the situation in which the commercial rate of discount per annum is 18%, i.e. in the notation above, $D = 0.18$, and the unit of time is one year. Suppose that we borrow two amounts, where each loan is to be settled by a payment of £1000, one after three months and the other after nine months.

For the shorter of the loans the amount borrowed is $1000[1 - \frac{1}{4}(0.18)]$, i.e. £955, and the annual rate of discount d is given by the equation

$$955 = 1000(1 - d)^{1/4}$$

from which it follows that $d = 0.168\,21$ or 16.821%.

For the longer of the loans the amount borrowed is $1000[1 - \frac{3}{4}(0.18)]$, i.e. £865, and the annual rate of discount d is given by the equation

$$865 = 1000(1 - d)^{3/4}$$

from which $d = 0.175\,82$ or 17.582%.

It should be noted that the longer loan has the greater effective annual rate of discount.

3.3 Annuities-certain: present values and accumulations

Consider a series of n payments, each of amount 1, to be made at time intervals of one unit, the first payment being made at time $t + 1$.

	1	1	1	$\cdots$	1	1	payment
t	$t+1$	$t+2$	$t+3$	$\cdots$	$t+n-1$	$t+n$	time

Such a sequence of payments is illustrated in the diagram above, in which the rth payment is made at time $t + r$.

The value of this series of payments *one unit of time before the first payment is made* is denoted by $a_{\overline{n}|}$. (For the series of payments illustrated in the diagram above, the value relates to time t.) Clearly, if $i = 0$, then $a_{\overline{n}|} = n$; otherwise

$$a_{\overline{n}|} = v + v^2 + v^3 + \cdots + v^n$$

$$= \frac{v(1 - v^n)}{1 - v}$$

$$= \frac{1 - v^n}{v^{-1} - 1}$$

$$= \frac{1 - v^n}{i} \tag{3.3.1}$$

If $n = 0$, $a_{\overline{n}|}$ is defined to be zero.

Thus $a_{\overline{n}|}$ is the value at the start of any period of length n of a series of n payments, each of amount 1, to be made *in arrear* at unit time intervals over the period. It is common to refer to such a series of payments, made in arrear, as an *immediate annuity-certain* and to call $a_{\overline{n}|}$ the present value of the immediate annuity-certain. When there is no possibility of confusion with a life annuity (i.e. a series of payments dependent on the survival of one or more human lives), the term *annuity* may be used as an alternative to annuity-certain.

The value of this series of payments *at the time the first payment is made* is denoted by $\ddot{a}_{\overline{n}|}$. If $i = 0$, then $\ddot{a}_{\overline{n}|} = n$; otherwise

$$\ddot{a}_{\overline{n}|} = 1 + v + v^2 + \cdots + v^{n-1}$$

$$= \frac{1 - v^n}{1 - v}$$

$$= \frac{1 - v^n}{d} \tag{3.3.2}$$

Thus $\ddot{a}_{\overline{n}|}$ is the value at the start of any given period of length n of a series of n payments, each of amount 1, to be made *in advance* at unit time intervals over the period. It is common to refer to such a series of payments, made in advance, as an *annuity-due* and to call $\ddot{a}_{\overline{n}|}$ the present value of the annuity-due.

It follows directly from the above definitions that

$$\left. \begin{array}{c} \ddot{a}_{\overline{n}|} = (1 + i)a_{\overline{n}|} \\[2mm] \ddot{a}_{\overline{n}|} = 1 + a_{\overline{n-1}|} \end{array} \right\} \tag{3.3.3}$$

and that, for $n \geqslant 2$,

The reader should verify these relationships algebraically and by general reasoning.

The value of the series of payments *at the time the last payment is made* is denoted by $s_{\overline{n}|}$. The value *one unit of time after the last payment is made* is denoted by $\ddot{s}_{\overline{n}|}$. If $i = 0$ then $s_{\overline{n}|} = \ddot{s}_{\overline{n}|} = n$; otherwise

$$s_{\overline{n}|} = (1 + i)^{n-1} + (1 + i)^{n-2} + (1 + i)^{n-3} + \cdots + 1$$
$$= (1 + i)^n a_{\overline{n}|}$$
$$= \frac{(1 + i)^n - 1}{i} \tag{3.3.4}$$

and

$$\ddot{s}_{\overline{n}|} = (1 + i)^n + (1 + i)^{n-1} + (1 + i)^{n-2} + \cdots + (1 + i)$$
$$= (1 + i)^n \ddot{a}_{\overline{n}|}$$
$$= \frac{(1 + i)^n - 1}{d} \tag{3.3.5}$$

Thus $s_{\overline{n}|}$ and $\ddot{s}_{\overline{n}|}$ are the values at the end of any period of length n of a series of n payments, each of amount 1, made at unit time intervals over the period, where the payments are made in arrear and in advance respectively. Sometimes $s_{\overline{n}|}$ and $\ddot{s}_{\overline{n}|}$ are called the *accumulation* (or the *accumulated amount*) of an immediate annuity and an annuity-due respectively. When $n = 0$, $s_{\overline{n}|}$ and $\ddot{s}_{\overline{n}|}$ are defined to be zero.

It is an immediate consequence of the above definition that

$$\left. \begin{array}{c} \ddot{s}_{\overline{n}|} = (1 + i)s_{\overline{n}|} \\[2ex] s_{\overline{n+1}|} = 1 + \ddot{s}_{\overline{n}|} \\[2ex] \ddot{s}_{\overline{n}|} = s_{\overline{n+1}|} - 1 \end{array} \right\} \tag{3.3.6}$$

and that

or

The reader should verify these relationships algebraically and by general reasoning.

Equations 3.3.1, 3.3.2, 3.3.4, and 3.3.5 may be expressed in the form

$$\left. \begin{array}{c} 1 = ia_{\overline{n}|} + v^n \\[1ex] 1 = d\ddot{a}_{\overline{n}|} + v^n \\[1ex] (1 + i)^n = is_{\overline{n}|} + 1 \\[1ex] (1 + i)^n = d\ddot{s}_{\overline{n}|} + 1 \end{array} \right\} \tag{3.3.7}$$

respectively. The reader should be able to write down these last four equations 3.3.7 immediately. The first equation is simply the equation of value at time 0 for a loan of amount 1 over the period from time 0 to time n, when interest is payable in arrear. The other three equations may be similarly interpreted, the last two being equations of value at time n.

As the rate of interest i increases, v decreases, so $\sum_{r=1}^{n} v^r$ decreases. Thus, for a fixed value of n, $a_{\overline{n}|}$ is a decreasing function of i. Similarly, $\ddot{a}_{\overline{n}|}$ is a decreasing function of i, while $s_{\overline{n}|}$ and $\ddot{s}_{\overline{n}|}$ are increasing functions.

For a fixed rate of interest, $a_{\overline{n}|}$, $\ddot{a}_{\overline{n}|}$, $s_{\overline{n}|}$ and $\ddot{s}_{\overline{n}|}$ are all increasing functions of n. When n becomes infinite, the corresponding annuity (or annuity-due) is known as a *perpetuity* (or *perpetuity-due*). The notation $a_{\overline{\infty}|}$ and $\ddot{a}_{\overline{\infty}|}$ is used to denote the corresponding present values. Thus, if $i > 0$,

$$a_{\overline{\infty}|} = \lim_{n \to \infty} a_{\overline{n}|} = \frac{1}{i} \tag{3.3.8}$$

and

$$\ddot{a}_{\overline{\infty}|} = \lim_{n \to \infty} \ddot{a}_{\overline{n}|} = \frac{1}{d} \tag{3.3.9}$$

(If $i \leqslant 0$, both $a_{\overline{\infty}|}$ and $\ddot{a}_{\overline{\infty}|}$ are infinite.)

It is convenient to have tables of annuity and accumulation values at various rates of interest. In view of the relationship 3.3.3 it is not necessary to give the values of both $a_{\overline{n}|}$ and $\ddot{a}_{\overline{n}|}$; similarly, the relationship 3.3.6 obviates the need to tabulate both $s_{\overline{n}|}$ and $\ddot{s}_{\overline{n}|}$. The compound interest tables at the back of this book give values of $a_{\overline{n}|}$ and $s_{\overline{n}|}$ for a range of values of n at several rates of interest.

Example 3.3.1

A loan of £2400 is to be repaid by 20 equal annual instalments. The rate of interest for the transaction is 10% per annum. Find the amount of each annual repayment, assuming that payments are made (a) in arrear and (b) in advance.

Solution

(a) Let the annual repayment be £X. Then

$$2400 = X(v + v^2 + \cdots + v^{20})$$
$$= X a_{\overline{20}|} \qquad \text{at } 10\%$$

so

$$X = 2400/a_{\overline{20}|} = 2400/8 \cdot 5136 = 281 \cdot 90$$

(b) Let the annual repayment be £Y. Then

$$2400 = Y(1 + v + \cdots + v^{19})$$
$$= Y \ddot{a}_{\overline{20}|}$$

so

$$Y = 2400/\ddot{a}_{\overline{20}|} \qquad \text{at } 10\%$$
$$= 2400/(1 + a_{\overline{19}|}) = 2400/9 \cdot 3649 = 256 \cdot 28$$

Example 3.3.2

On 15 November in each of the years 1964 to 1979 inclusive an investor deposited £500 in a special bank savings account. On 15 November 1983 the investor withdrew his savings. Given that over the entire period the bank used an annual interest rate of 7% for its special savings accounts, find the sum withdrawn by the investor.

Solution

Two alternative solutions are considered.

(a) The investor made 16 deposits in his account. On 15 November 1979 (i.e. the date of the final deposit) the amount of the account was therefore $500\,s_{\overline{16}|}$ at 7%, i.e. £500 × 27·888 05 = £13 944·03. Four years later, i.e. on 15 November 1983, the amount of the account was £13 944·03 × $(1·07)^4$ = £18 277·78, which was the sum withdrawn.

(b) Although the investor made no deposits in the years 1980 to 1983, we shall value his account on the basis that the payments of £500 continued in these years. We then deduct from the accumulated account, calculated on this basis, the accumulated value of the payments which he did *not* make in the years 1980 to 1983. This leads to a value for the sum withdrawn of

$$500(s_{\overline{20}|} - s_{\overline{4}|}) \qquad \text{at } 7\%$$
$$= 500(40·995\,49 - 4·439\,94) = £18\,277·78$$

Example 3.3.3

A borrower agrees to repay a loan of £3000 by 15 annual repayments of £500, the first repayment being due after five years. Find the annual yield for this transaction.

Solution

The equation of value may be written as

$$3000 = 500(v^5 + v^6 + \cdots + v^{19})$$
$$= 500[(v + v^2 + \cdots + v^{19}) - (v + v^2 + v^3 + v^4)]$$
$$= 500(a_{\overline{19}|} - a_{\overline{4}|})$$

We are required to solve this equation for the rate of interest *i*. (Our remarks in section 3.2 indicate that there is a unique root.) Since the right-hand side of the equation is a monotonic function of *i*, the solution may be found quite simply to any desired degree of accuracy.

At 8% the right-hand side has value

$$500(9·6036 - 3·3121) = 3145·75$$

and at 9% its value is

$$500(8·9501 - 3·2397) = 2855·20$$

Since

$$3145·75 > 3000 > 2855·20$$

the value of i is between 8% and 9%. For our present purposes we shall consider it sufficient to estimate i by linear interpolation as

$$0.08 + (0.09 - 0.08)\frac{3145.75 - 3000}{3145.75 - 2855.20} = 0.085\,02$$

or 8·5%.

Note If a more accurate solution is required, we may either use rates tabulated at smaller intervals (e.g. 8% and 8·5%) or, alternatively, adopt a more accurate method of solution (see appendix 2). The solution, to five decimal places, is in fact $i = 0.08\,486$: say, 8·49%.

Example 3.3.4

Prove that, at rate of interest i,

$$\frac{1}{a_{\overline{n}|}} = \frac{1}{s_{\overline{n}|}} + i \tag{3.3.10}$$

and interpret this result by general reasoning.

Solution

Since $s_{\overline{n}|} = [(1 + i)^n - 1]/i$, we have

$$\frac{1}{s_{\overline{n}|}} + i = \frac{i}{(1 + i)^n - 1} + i$$

$$= \frac{i(1 + i)^n}{(1 + i)^n - 1} = \frac{i}{[1 - (1 + i)^{-n}]} = \frac{1}{a_{\overline{n}|}}$$

as required.

Equation 3.3.10 has the following interpretation. Consider a loan of 1 which is to be repaid by n equal instalments annually in arrear. If the annual rate of interest is i, then the amount of each repayment is X, where $Xa_{\overline{n}|} = 1$. Thus $X = 1/a_{\overline{n}|}$. When the lender receives each repayment he may consider an amount i of the payment as interest (for his immediate use) and deposit the balance of the payment, i.e. $(1/a_{\overline{n}|}) - i$, in a savings account which earns interest at rate i per annum.

After n years the accumulated savings account must be sufficient to repay the original loan (i.e. 1). Thus, at rate of interest i,

$$\left(\frac{1}{a_{\overline{n}|}} - i\right)s_{\overline{n}|} = 1$$

or

$$\frac{1}{a_{\overline{n}|}} = \frac{1}{s_{\overline{n}|}} + i$$

as required.

3.4 Deferred annuities

Suppose that m and n are non-negative integers. The value at time 0 of a series of n payments, each of amount 1, due at times $(m + 1)$, $(m + 2)$,..., $(m + n)$ is denoted by $_m|a_{\overline{n}|}$ (see the figure below).

				1	1	$\cdots$	1	payment
0	1	$\cdots$	m	$m+1$	$m+2$	$\cdots$	$m+n$	time

Such a series of payments may be considered as an immediate annuity, deferred for m time units. When $n > 0$,

$$_m|a_{\overline{n}|} = v^{m+1} + v^{m+2} + v^{m+3} + \cdots + v^{m+n} \tag{3.4.1}$$
$$= (v + v^2 + v^3 + \cdots + v^{m+n}) - (v + v^2 + v^3 + \cdots + v^m)$$
$$= v^m(v + v^2 + v^3 + \cdots + v^n)$$

The last two equations show that

$$_m|a_{\overline{n}|} = a_{\overline{m+n}|} - a_{\overline{m}|} \tag{3.4.2}$$
$$= v^m a_{\overline{n}|} \tag{3.4.3}$$

Either of these two equations may be used to determine the value of a deferred immediate annuity. Together they imply that

$$a_{\overline{m+n}|} = a_{\overline{m}|} + v^m a_{\overline{n}|} \tag{3.4.4}$$

This equation is sometimes useful.

At this stage it is perhaps worth pointing out that the definition 3.4.1 may be used for $_m|a_{\overline{n}|}$ when m is any non-negative number, not only an integer. In this case equation 3.4.3 is valid, but equations 3.4.2 and 3.4.4 have as yet no meaning, since $a_{\overline{k}|}$ has been defined only when k is an integer. (In chapter 4 for completeness we shall extend the definition of $a_{\overline{k}|}$ to non-integral values of k, and it will be seen that equations 3.4.2 and 3.4.4 are always valid.)

We may define the corresponding deferred annuity-due as

$$_m|\ddot{a}_{\overline{n}|} = v^m \ddot{a}_{\overline{n}|} \tag{3.4.5}$$

3.5 Continuously payable annuities

Let n be a non-negative number. The value at time 0 of an annuity payable continuously between time 0 and time n, where the rate of payment per unit time is constant and equal to 1, is denoted by $\bar{a}_{\overline{n}|}$. Clearly

$$\bar{a}_{\overline{n}|} = \int_0^n e^{-\delta t} dt$$

$$= \frac{1 - e^{-\delta n}}{\delta}$$

$$= \frac{1 - v^n}{\delta} \qquad \text{(if } \delta \neq 0\text{)} \qquad (3.5.1)$$

Note that $\bar{a}_{\overline{n}|}$ is defined even for non-integral values of n. If $\delta = 0$ (or, equivalently, $i = 0$), $\bar{a}_{\overline{n}|}$ is of course equal to n.

If m is a non-negative number, we use the symbol $_m|\bar{a}_{\overline{n}|}$ to denote the present value of a continuously payable annuity of 1 per unit time for n time units, deferred for m time units. Thus

$$_m|\bar{a}_{\overline{n}|} = \int_m^{m+n} e^{-\delta t} dt$$

$$= e^{-\delta m} \int_0^n e^{-\delta s} ds$$

$$= \int_0^{m+n} e^{-\delta t} dt - \int_0^m e^{-\delta t} dt$$

Hence

$$_m|\bar{a}_{\overline{n}|} = \bar{a}_{\overline{m+n}|} - \bar{a}_{\overline{m}|} \qquad (3.5.2)$$

$$= v^m \bar{a}_{\overline{n}|} \qquad (3.5.3)$$

Since equation 3.5.1 may be written as

$$\bar{a}_{\overline{n}|} = \frac{i}{\delta} \left(\frac{1 - v^n}{i} \right)$$

it follows immediately that, if n is an integer,

$$\bar{a}_{\overline{n}|} = \frac{i}{\delta} a_{\overline{n}|} \qquad \text{(if } \delta \neq 0\text{)} \qquad (3.5.4)$$

The reader will note that equations 3.5.2 and 3.5.3 are analogous to equations 3.4.2 and 3.4.3. (In chapter 4 we shall show that, with the appropriate definition of $a_{\overline{n}|}$ for non-integral n, equation 3.5.4 is valid for all non-negative n.)

3.6 Varying annuities

In sections 3.3 and 3.4 we have considered annuities for which the amount of each payment is constant. For an annuity in which the payments are not all of an equal amount it is a simple matter to find the present (or accumulated) value from first principles. Thus, for example, the present value of such an annuity may always be evaluated as

$$\sum_{i=1}^{n} X_i v^{t_i}$$

where the ith payment, of amount X_i, is made at time t_i.

In the particular case when $X_i = t_i = i$ the annuity is known as an *increasing annuity* and its present value is denoted by $(Ia)_{\overline{n}|}$. Thus

$$(Ia)_{\overline{n}|} = v + 2v^2 + 3v^3 + \cdots + nv^n \tag{3.6.1}$$

Hence

$$(1 + i)(Ia)_{\overline{n}|} = 1 + 2v + 3v^2 + \cdots + nv^{n-1}$$

By subtraction, we obtain

$$i(Ia)_{\overline{n}|} = 1 + v + v^2 + \cdots + v^{n-1} - nv^n$$
$$= \ddot{a}_{\overline{n}|} - nv^n$$

so

$$(Ia)_{\overline{n}|} = \frac{\ddot{a}_{\overline{n}|} - nv^n}{i} \tag{3.6.2}$$

The last equation need *not* be memorized, as it may be rapidly derived from first principles. A simple way of recalling equation 3.6.2 is to express it in the form

$$\ddot{a}_{\overline{n}|} = i(Ia)_{\overline{n}|} + nv^n \tag{3.6.3}$$

This equation is simply the equation of value for a transaction in which an investor lends 1 at the start of each year for n years in return for interest at the end of each year of amount i times the outstanding loan and a repayment of the total amount lent (i.e. n) after n years. The two sides of the equation represent the value (at the start of the transaction) of the payments made by the lender and the borrower respectively. The function $(Ia)_{\overline{n}|}$ is included in the compound interest tables at the end of this book.

The present value of any annuity payable in arrear for n time units for which the amounts of successive payments form an arithmetic progression can be expressed in terms of $a_{\overline{n}|}$ and $(Ia)_{\overline{n}|}$. If the first payment of such an annuity is P and the second payment is $(P + Q)$, the tth payment is $(P - Q) + Qt$, and the present value of the annuity is therefore

$$(P - Q)a_{\overline{n}|} + Q(Ia)_{\overline{n}|}$$

Alternatively, the present value of the annuity can be derived from first principles.

The notation $(I\ddot{a})_{\overline{n}|}$ is used to denote the present value of an increasing annuity-due payable for n time units, the tth payment (of amount t) being made at time $t - 1$. Thus

$$(I\bar{a})_{\overline{n}|} = 1 + 2v + 3v^2 + \cdots + nv^{n-1}$$

$$= (1 + i)(Ia)_{\overline{n}|} \tag{3.6.4}$$

$$= 1 + a_{\overline{n-1}|} + (Ia)_{\overline{n-1}|} \tag{3.6.5}$$

For increasing annuities which are payable continuously it is important to distinguish between an annuity which has a constant rate of payment r (per unit time) throughout the rth period and an annuity which has a rate of payment t at time t. For the former the rate of payment is a step function, taking the discrete values $1, 2, \ldots$. For the latter the rate of payment itself increases continuously. If the annuities are payable for n time units, their present values are denoted by $(I\bar{a})_{\overline{n}|}$ and $(\bar{I}\bar{a})_{\overline{n}|}$ respectively. Clearly

$$(I\bar{a})_{\overline{n}|} = \sum_{r=1}^{n} \left(\int_{r-1}^{r} rv^t \, dt \right)$$

and

$$(\bar{I}\bar{a})_{\overline{n}|} = \int_{0}^{n} tv^t \, dt$$

Using integration by parts in the second case, the reader should verify that

$$(I\bar{a})_{\overline{n}|} = \frac{\ddot{a}_{\overline{n}|} - nv^n}{\delta} \tag{3.6.6}$$

and

$$(\bar{I}\bar{a})_{\overline{n}|} = \frac{\bar{a}_{\overline{n}|} - nv^n}{\delta} \tag{3.6.7}$$

Each of the last two equations, expressed in a form analogous to equation 3.6.3, may be easily written down as the equation of value for an appropriate transaction.

Corresponding to the present values $(Ia)_{\overline{n}|}, (I\bar{a})_{\overline{n}|}, (I\bar{a})_{\overline{n}|}$, and $(\bar{I}\bar{a})_{\overline{n}|}$ are the accumulations at time n of the relevant series of payments. These accumulations are denoted by $(Is)_{\overline{n}|}, (I\bar{s})_{\overline{n}|}, (I\bar{s})_{\overline{n}|}$, and $(\bar{I}\bar{s})_{\overline{n}|}$ respectively. Since these symbols denote the values n time units later of the same series of payments as before, it follows that

$$\left.\begin{array}{l} (Is)_{\overline{n}|} = (1 + i)^n (Ia)_{\overline{n}|} \\ (I\bar{s})_{\overline{n}|} = (1 + i)^n (I\bar{a})_{\overline{n}|} \\ (I\bar{s})_{\overline{n}|} = (1 + i)^n (I\bar{a})_{\overline{n}|} \\ (\bar{I}\bar{s})_{\overline{n}|} = (1 + i)^n (\bar{I}\bar{a})_{\overline{n}|} \end{array}\right\} \tag{3.6.8}$$

The present values of deferred increasing annuities are defined in the obvious manner: for example,

$$_m|(Ia)_{\overline{n}|} = v^m (Ia)_{\overline{n}|}.$$

Example 3.6.1

An annuity is payable annually in arrear for 20 years. The first payment is of amount £8000 and the amount of each subsequent payment decreases by £300 each year. Find the present value of the annuity on the basis of an interest rate of 5% per annum.

Solution

(a) *From first principles*   Let the present value be £X. Then

$$X = 8000v + 7700v^2 + 7400v^3 + \cdots + 2600v^{19} + 2300v^{20}$$

so $(1 + i)X = 8000 + 7700v + 7400v^2 + \cdots + 2600v^{18} + 2300v^{19}$

By subtraction we obtain

$$iX = 8000 - 300(v + v^2 + \cdots + v^{19}) - 2300v^{20}$$

so $X = \dfrac{8000 - 300a_{\overline{19}|} - 2300v^{20}}{i}$ at 5% $= 70\,151$

Note   The first annuity payment is £8000 and the last payment is £2300. The average payment is thus £5150. Accordingly a rough approximation for the value of the annuity is $5150a_{\overline{20}|} = 64\,180$. Because of the increasing effect of discount with time, this approximation understates the true value, but confirms the order of magnitude of our answer. It is very important in practical work to incorporate checks to ensure, for example, that one does not misplace a decimal point or make some other simple error.

(b) *Using increasing annuity functions*   We consider the annuity to be a level annuity of £8300 per annum less an increasing annuity for which the rth payment is of amount £300r.

Hence $X = 8300(v + v^2 + v^3 + \cdots + v^{20}) - 300(v + 2v^2 + 3v^3 + \cdots + 20v^{20})$

$$= 8300a_{\overline{20}|} - 300(Ia)_{\overline{20}|}$$

$$= 8300a_{\overline{20}|} - 300\frac{\ddot{a}_{\overline{20}|} - 20v^{20}}{i}   \text{ at 5\%}   = 70\,151$$

It is important to realize that in general there is no one correct or 'best' method of solution for many compound interest problems. Provided that the reader has a good grasp of the underlying principles, he will be able to use that method which is most suited to his own approach.

Example 3.6.2

An annuity is payable half-yearly for six years, the first half-yearly payment of amount £1800 being due after two years. The amount of subsequent payments decreases by £30 each half-year. On the basis of an interest rate of 5% per half-year, find the present value of the annuity.

Solution

As before, we give two solutions which illustrate different approaches. It is quite possible that the reader may prefer yet another method.

   Choose the half-year as our basic time unit. There are 12 annuity payments, the last of amount £[1800 − (11 × 30)], i.e. £1470.

(a) *From first principles* The present value is

$$X = 1800v^4 + 1770v^5 + 1740v^6 + \cdots + 1500v^{14} + 1470v^{15} \qquad \text{at } 5\%$$

Therefore

$$(1 + i)X = 1800v^3 + 1770v^4 + 1740v^5 + \cdots + 1500v^{13} + 1470v^{14}$$

Hence

$$iX = 1800v^3 - 30(v^4 + v^5 + \cdots + v^{14}) - 1470v^{15}$$

and

$$X = \frac{1800v^3 - 30(a_{\overline{14}|} - a_{\overline{1}|}) - 1470v^{15}}{i} = 12\,651$$

(b) We may write

$$X = v^3[1830(v + v^2 + v^3 + \cdots + v^{12}) - 30(v + 2v^2 + 3v^3 + \cdots + 12v^{12})]$$
$$= v^3[1830a_{\overline{12}|} - 30(Ia)_{\overline{12}|}]$$
$$= v^3\left(1830a_{\overline{12}|} - 30\frac{\ddot{a}_{\overline{12}|} - 12v^{12}}{i}\right) = 12\,651$$

A remark on notation

When there is no ambiguity as to the value of i, the rate of interest, functions such as $a_{\overline{n}|}$, $s_{\overline{n}|}$, and d are clearly defined. In this case an equation such as

$$Xa_{\overline{n}|} = 1000$$

may be solved immediately for X. If it is desired to emphasize the rate of interest, the equation may be written in the form

$$Xa_{\overline{n}|} = 1000 \qquad \text{at rate } i$$

The solution of certain problems may involve more than one rate of interest and in such cases there may be doubt as to the interest rate implicit in a particular function. In order to avoid ambiguity, we may attach the rate of interest as a suffix to the standard functions. For example, we may write $a_{\overline{n}|i}$, v_i^n, $s_{\overline{n}|i}$, and d_i for $a_{\overline{n}|}$, v^n, $s_{\overline{n}|}$ and d at rate i. With this notation, an equation such as

$$Xs_{\overline{10}|0.04} = 100a_{\overline{15}|0.03}$$

is quite precise.

This notation is readily extended to the functions defined in chapter 4.

3.7 The general loan schedule

Suppose that at time 0 an investor lends L in return for a series of n payments, the rth payment, of amount x_r, being due at time r ($1 \leqslant r \leqslant n$). Suppose further that the amount lent is calculated on the basis of an effective annual interest rate i, for the rth year $(1 \leqslant r \leqslant n)$. (In many situations i, may not depend on r, but at this stage it is convenient to consider the more general case.)

The amount lent is simply the present value, on the stated interest basis, of the repayments. Thus

$$
\begin{aligned}
L = & x_1(1 + i_1)^{-1} + x_2(1 + i_1)^{-1}(1 + i_2)^{-1} \\
& + x_3(1 + i_1)^{-1}(1 + i_2)^{-1}(1 + i_3)^{-1} + \cdots \\
& \cdots + x_n(1 + i_1)^{-1}(1 + i_2)^{-1} \cdots (1 + i_n)^{-1}
\end{aligned}
\tag{3.7.1}
$$

The investor may consider part of each payment as interest (for the latest period) on the outstanding loan and regard the balance of each payment as a capital repayment, which is used to reduce the amount of the loan outstanding. If any payment is insufficient to cover the interest on the outstanding loan, the shortfall in interest is added to the amount of the outstanding loan. In this situation the investor may draw up a schedule which shows the amount of interest contained in each payment and also the amount of the loan outstanding immediately after each payment has been received. It is desirable to consider this schedule in greater detail. (The division of each payment into interest and capital is frequently necessary for taxation purposes. Further, in the event of default by the borrower, it may be necessary for the lender to know the amount of loan outstanding at the time of default.)

Let $F_0 = L$ and, for $t = 1, 2, \ldots, n$, let F_t be the loan outstanding immediately *after* the payment due at time t has been made. The loan repaid at time t is simply the amount by which the payment then made, x_t, exceeds the interest then due, $i_t F_{t-1}$. Also, the loan outstanding immediately after the tth payment equals the loan outstanding immediately after the previous payment *minus* the amount of loan repaid at time t. Hence

$$
F_t = F_{t-1} - (x_t - i_t F_{t-1}) \qquad 1 \leqslant t \leqslant n
\tag{3.7.2}
$$

(This equation holds for $t = 1$, since we have defined $F_0 = L$.) Thus

$$
F_t = (1 + i_t)F_{t-1} - x_t \qquad t \geqslant 1
\tag{3.7.3}
$$

Hence

$$
\begin{aligned}
F_1 &= (1 + i_1)F_0 - x_1 \\
&= (1 + i_1)L - x_1
\end{aligned}
$$

Then

$$
\begin{aligned}
F_2 &= (1 + i_2)F_1 - x_2 \\
&= (1 + i_2)[(1 + i_1)L - x_1] - x_2 \\
&= (1 + i_1)(1 + i_2)L - (1 + i_2)x_1 - x_2
\end{aligned}
$$

and so on.

More generally, it is easily seen that

$$F_t = (1 + i_1)(1 + i_2)\cdots(1 + i_t)L - (1 + i_2)(1 + i_3)\cdots(1 + i_t)x_1$$
$$- (1 + i_3)(1 + i_4)\cdots(1 + i_t)x_2 - \cdots - (1 + i_t)x_{t-1} - x_t \qquad (3.7.4)$$

Thus F_t is simply the amount lent *accumulated to time t* less the repayments received by time t *accumulated to that time*, the accumulations being made on the appropriate varying interest basis.

An alternative expression for F_t is obtained by multiplying equation 3.7.1 by $(1 + i_1)(1 + i_2)\cdots(1 + i_t)$. This gives

$$L(1 + i_1)(1 + i_2)\cdots(1 + i_t) = (1 + i_2)(1 + i_3)\cdots(1 + i_t)x_1 + \cdots + x_t$$
$$+ (1 + i_{t+1})^{-1}x_{t+1} + \cdots$$
$$\cdots + (1 + i_{t+1})^{-1}(1 + i_{t+2})^{-1}\cdots(1 + i_n)^{-1}x_n$$

By combining this equation with equation 3.7.4 we immediately obtain

$$F_t = (1 + i_{t+1})^{-1}x_{t+1} + (1 + i_{t+1})^{-1}(1 + i_{t+2})^{-1}x_{t+2} + \cdots$$
$$+ (1 + i_{t+1})^{-1}(1 + i_{t+2})^{-1}\cdots(1 + i_n)^{-1}x_n \qquad (3.7.5)$$

This shows that F_t is simply *the value at time t of the outstanding repayments*. Equation 3.7.3 is

$$F_t = (1 + i_t)F_{t-1} - x_t$$

Similarly

$$F_{t+1} = (1 + i_{t+1})F_t - x_{t+1}$$

If $i_t = i_{t+1}$, it follows by subtraction that

$$(F_t - F_{t+1}) = (1 + i)(F_{t-1} - F_t) + x_{t+1} - x_t \qquad (3.7.6)$$

where i denotes the common value of i_t and i_{t+1}. Letting f_t denote the amount of loan repaid at time t, we may write equation 3.7.6 as

$$f_{t+1} = (1 + i)f_t + x_{t+1} - x_t \qquad (3.7.7)$$

It is important to realize that equation 3.7.7 holds *only* when the same rate of interest i is applicable to *both* the tth year and the $(t + 1)$th year. In particular, when the rate of interest is constant throughout the transaction and all the repayments are of equal size, the amounts of successive loan repayments form a geometric progression with common ratio $(1 + i)$. We consider this case briefly in section 3.8.

3.8 The loan schedule for a level annuity

Consider the particular case where, on the basis of an interest rate of i per unit time, a loan of amount $a_{\overline{n}|}$ is made at time 0 in return for n repayments, each of amount 1, to be made at times $1, 2, \ldots, n$. The lender may construct a schedule showing the division of each payment into capital and interest.

Immediately after the tth repayment has been made there remain $(n - t)$ outstanding payments, and equation 3.7.5 shows that the outstanding loan is simply $a_{\overline{n-t}|}$. Thus, in the notation of section 3.7,

$$F_t = a_{\overline{n-t}|} \tag{3.8.1}$$

Then the amount of loan repaid at time t is

$$f_t = F_{t-1} - F_t = a_{\overline{n-t+1}|} - a_{\overline{n-t}|}$$
$$= v^{n-t+1} \tag{3.8.2}$$

The lender's schedule may be presented in the form of table 3.8.1. More generally, if an amount L is lent in return for n repayments, each of amount $L/a_{\overline{n}|}$, the monetary amounts in the lender's schedule are simply those in the schedule of table 3.8.1 multiplied by the constant factor $L/a_{\overline{n}|}$.

Table 3.8.1 *Schedule for a level annuity (amount of loan $a_{\overline{n}|}$)*

Payment	Interest content of payment	Capital repaid	Loan outstanding after payment
1	$ia_{\overline{n}|} = 1 - v^n$	v^n	$a_{\overline{n}|} - v^n = a_{\overline{n-1}|}$
2	$ia_{\overline{n-1}|} = 1 - v^{n-1}$	v^{n-1}	$a_{\overline{n-1}|} - v^{n-1} = a_{\overline{n-2}|}$
$\vdots$			
t	$ia_{\overline{n-t+1}|} = 1 - v^{n-t+1}$	v^{n-t+1}	$a_{\overline{n-t+1}|} - v^{n-t+1} = a_{\overline{n-t}|}$
$\vdots$			
$n-1$	$ia_{\overline{2}|} = 1 - v^2$	v^2	$a_{\overline{2}|} - v^2 = a_{\overline{1}|}$
n	$ia_{\overline{1}|} = 1 - v$	v	$a_{\overline{1}|} - v = 0$

Example 3.8.1

A loan of £10 000 is to be repaid over ten years by a level annuity payable monthly in arrear. The amount of the monthly payment is calculated on the basis of an interest rate of 1% per month effective. Find

(a) The monthly repayment;
(b) The total capital repaid and interest paid in (i) the first year and (ii) the final year;
(c) After which monthly repayment the outstanding loan is first less than £5000; and
(d) For which monthly repayment the capital repaid first exceeds the interest content.

Solution

We choose one month as our time unit and let $i = 0.01$.

(a) Since the period of the loan is 120 months we need the value of $a_{\overline{120}|}$ at 1%. This may be obtained from one of the forms

$$a_{\overline{120}|} = a_{\overline{100}|} + v^{100}a_{\overline{20}|}$$

$$= \frac{1 - v^{120}}{i}$$

either of which give $a_{\overline{120}|} = 69.700\,522$. The monthly repayment is £x, where

$$xa_{\overline{120}|} = 10\,000$$

from which it follows that $x = 143.47$.

Note that the total payment in any year is $12x = 1721.64$.

(b) (i) To find the capital repaid in the first year we may use either of two approaches.

The loan outstanding at the end of the first year (i.e. immediately after the 12th monthly payment) is simply the value then of the remaining repayments, i.e.

$$xa_{\overline{108}|} = 9448.62 \qquad \text{at } 1^\circ{}_0$$

This means that the capital repaid in the first year is $(10\,000 - 9448.62)$. i.e. £551.38. Hence the interest paid in the first year is $(1721.64 - 551.38)$, i.e. £1170.26.

Alternatively, note that the capital repaid in the first monthly repayment is $143.47 - (0.01 \times 10\,000)$, i.e. 43.47. Since we are dealing with a level annuity, successive capital payments form a geometric progression with common ratio $(1 + i)$, i.e., 1.01. The total of the first 12 capital payments is therefore $43.47s_{\overline{12}|} = £551.31$. (Because of rounding errors, this differs slightly from the value of £551.38 found by the first method. In practice, to avoid such errors, the capital repayments may be adjusted slightly in order that their total exactly equals the original amount lent.)

(ii) The capital repaid in the final year is simply the loan outstanding at the start of the final year. This is $143.47a_{\overline{12}|} = £1614.77$. The interest paid in the final year is therefore $(1721.64 - 1614.77)$, i.e. £106.87.

(c) After the tth monthly repayment the outstanding loan is $143.47\,a_{\overline{120-t}|}$. Consider the equation

$$143.47a_{\overline{120-t}|} = 5000$$

i.e.

$$a_{\overline{120-t}|} = 34.850 \qquad \text{at } 1\%$$

Since $a_{\overline{43}|} = 34.8100$ and $a_{\overline{44}|} = 35.4555$, the outstanding loan is first less than £5000 when $120 - t = 43$ (so $t = 77$), i.e. after the 77th monthly repayment has been made.

(d) The tth capital payment is of amount $43.47(1.01)^{t-1}$. We need to know when this is first greater than one-half of the total monthly payment.

Thus we seek the least integer t for which

$$43.47(1.01)^{t-1} > \frac{143.47}{2}$$

i.e.

$$1 \cdot 01^{t-1} > 1 \cdot 6502$$

i.e.

$$t - 1 > \frac{\log 1 \cdot 6502}{\log 1 \cdot 01} = 50 \cdot 34$$

Hence the required value of t is 52.

Example 3.8.2

An investor purchases an annuity payable annually in arrear for 20 years. The first annuity payment is £2000 and subsequent payments increase by £100 each year. The investor, who calculates his purchase price on the basis of an interest rate of 4% per annum, draws up a schedule showing the division of each repayment into capital and interest.

Find the purchase price. Derive expressions for the capital and interest content of the tth payment and for the loan outstanding after the tth payment has been made.

Solution

The purchase price is

$$1900 a_{\overline{20}|} + 100(Ia)_{\overline{20}|} \qquad \text{at } 4\% \qquad = 38\,337 \cdot 12$$

Let f_t denote the capital repaid in the tth annuity payment. Then

$$f_1 = 2000 - (0 \cdot 04 \times 38\,337 \cdot 12) = 466 \cdot 515$$

(At this stage, to avoid rounding errors, we work with three decimal places.) Since each payment is £100 greater than the previous payment, equation 3.7.7 implies that

$$f_{t+1} = 1 \cdot 04 f_t + 100$$

from which it follows easily by induction that, for $t > 1$,

$$f_t = 1 \cdot 04^{t-1} f_1 + 100(1 \cdot 04^{t-2} + 1 \cdot 04^{t-3} + \cdots + 1)$$

$$= (1 \cdot 04^{t-1} \times 466 \cdot 515) + 100 \frac{1 \cdot 04^{t-1} - 1}{0 \cdot 04}$$

$$= 2966 \cdot 515 \times 1 \cdot 04^{t-1} - 2500$$

Since the tth payment is $(1900 + 100t)$, the interest content is

$$1900 + 100t - [(2966 \cdot 515 \times 1 \cdot 04^{t-1}) - 2500]$$

$$= 4400 + 100t - 2966 \cdot 515 \times 1 \cdot 04^{t-1}$$

The loan repaid in the first t payments is

$$\sum_{r=1}^{t} f_r = \sum_{r=1}^{t} [(2966 \cdot 515 \times 1 \cdot 04^{r-1}) - 2500]$$

$$= 2966 \cdot 515 \frac{1 \cdot 04^t - 1}{0 \cdot 04} - 2500t$$

$$= 74\,162 \cdot 875(1 \cdot 04^t - 1) - 2500t$$

The loan outstanding after t payments have been received is thus

$$38\,337{\cdot}12 - [74\,162{\cdot}875(1{\cdot}04^t - 1) - 2500t]$$

$$= 112\,500 + 2500t - 74\,162{\cdot}875 \times 1{\cdot}04^t$$

Note   A check on this final expression is obtained by observing that the loan outstanding immediately after the repayment at time t is the value then of the remaining payments. Since the payment at time $t+r$ is £$(1900 + 100t + 100r)$, the value at time t of the outstanding payments is

$$(1900 + 100t)\,a_{\overline{20-t}|} + 100(Ia)_{\overline{20-t}|}$$

at 4%. The reader should verify that this equals the expression above for the outstanding loan at time t. (Both expressions are equal to $112\,500 + 2500t - 162\,500 \times (1{\cdot}04)^{t-20}$.)

Exercises

3.1 (a) Using the power series for $\log(1 + x)$, show that with compound interest at $K\%$ per annum the time taken for a sum of money to double itself is approximately $70/K$ years, and find a corresponding rule-of-thumb for the time taken for money to treble itself.
   (b) Using compound interest tables, compare the approximate values with the true values when (i) $K = 5$, (ii) $K = 10$, and (iii) $K = 20$.
   (c) What can you infer about the accuracy of the approximations?

3.2 (a) Establish from first principles all the relationships between δ, i, v and d in table 3.1.1.
   (b) (i) Given that $\delta = 0{\cdot}08$, find the values of i, d, and v.
      (ii) Given that $d = 0{\cdot}08$, find the values of v, i, and δ.
      (iii) Given that $i = 0{\cdot}08$, find the values of v, d, and δ.
      (iv) Given that $v = 0{\cdot}95$, find the values of d, i, and δ.

3.3 In return for a single payment of £1000 a building society offers the following alternative benefits:

   (i) A lump sum of £1330 after three years;
   (ii) A lump sum of £1550 after five years; or
   (iii) Four annual payments, each of amount £425, the first payment being made after five years.

   Any investor must specify which benefit he is choosing when he makes the single payment.

   (a) Write down an equation of value for each savings plan and hence find the yield for each.
   (b) Assume that an investor opts for plan (i) and that after three years he invests the proceeds of the plan for a further two years at a fixed rate of interest. How large must this rate of interest be in order for him to receive £1550 from this further investment?
   (c) Assume that an investor opts for plan (ii) and that after five years he uses the proceeds of the plan to buy a level annuity-due payable for four years, the amount of the annuity payment being calculated on the basis of a fixed

interest rate. How large must this rate of interest be in order for the annuity payment to be £425?

3.4   An investor has the choice of either of the following savings plans:

   (i) Ten annual premiums, each of £100 and payable in advance, will give £1700 after ten years; or
   (ii) Fifteen annual premiums, each of £100 and payable in advance, will give £3200 after fifteen years.

The investor must declare which plan he is choosing when he pays the first premium.

   (a) Find the yield on each plan.
   (b) Assume that an investor has chosen plan (i). Assume further that after ten years he deposits the proceeds of the plan in an account which will earn interest at a fixed rate and that he also makes five annual payments of £100 to this account, the first payment being made at the time the original savings plan matures. How large must the fixed rate of interest be in order that finally, after fifteen years, the investor may receive £3200?

3.5   A savings plan provides that in return for n annual premiums of £X (payable annually in advance), an investor will receive m annual payments of £Y, the first such payment being made one year after payment of the last premium.

   (a) Show that the equation of value for the transaction can be expressed in either of the following forms:
      (i) $Ya_{\overline{n+m}|} - (X + Y)a_{\overline{n}|} = 0$; or
      (ii) $(X + Y)s_{\overline{m}|} - Xs_{\overline{n+m}|} = 0$.
   (b) Suppose that $X = 1000$, $Y = 2000$, and $n = 10$.
      (i) Find the yield per annum on the transaction, if $m = 10$.
      (ii) For what values of m is the annual yield on the transaction between 8% and 10%?
   (c) Suppose that $X = 1000$, $Y = 2000$, and $m = 20$. For what values of n is the annual yield on the transaction between 8% and 10%?

3.6 (Multiple roots of the equation of value)
      In certain mining projects an investor must make an initial outlay at the present time and a further outlay after two years to finance the re-landscaping of the used mine area. In return the investor will receive a payment after one year.
      For both of the following transactions write down the equation of value in terms of i, the rate of interest per annum. What information (if any) about the roots is provided by theorems 3.2.1 and 3.2.2? In each case find all the roots of the equation of value.

Alternative mining projects	Initial outlay (£000s)	Outlay after two years (£000s)	Income after one year (£000s)
Project (a)	10 000	11 550	21 500
Project (b)	10 000	10 395	20 400

3.7 (a) The manufacturer of a certain toy sells to retailers on either of the following terms:

(i) Cash payment: 30% below recommended retail price;

(ii) Six months credit: 25% below recommended retail price.

Find the effective annual rate of discount offered by the manufacturer to retailers who pay cash. Express this as an effective annual interest rate charged to those retailers who accept the credit terms.

(b) The manufacturer is considering changing his credit terms. Credit for six months will no longer be available, but for three months' credit a discount of 27·5% below the recommended retail price will be allowed. The terms for cash payment will be unaltered.

Does this new arrangement offer a greater or lower effective annual rate of discount to cash purchasers?

3.8 (a) Find the value at an interest rate of 4% of the following:

$$_5|a_{\overline{32}|}, \quad \ddot{a}_{\overline{62}|}, \quad \bar{a}_{\overline{62}|}, \quad _{12}|\ddot{a}_{\overline{50}|}, \quad s_{\overline{62}|}, \quad \ddot{s}_{\overline{61}|},$$

$$(I\ddot{a})_{\overline{62}|}, \quad _5|(Ia)_{\overline{20}|}, \quad (I\bar{a})_{\overline{23}|}, \quad (\bar{I}\bar{a})_{\overline{23}|}$$

(b) Given that $\ddot{a}_{\overline{n}|} = 7{\cdot}029\,584$ and $\ddot{a}_{\overline{3n}|} = 10{\cdot}934\,563$, find the rate of interest and n.

3.9 An annuity-certain is payable annually for 20 years. The annual payment is £5 for six years, then £7 for nine years, and finally £10 for five years.

(a) Show that the value of the annuity at the time of the first payment may be expressed as:

(i) $5\ddot{a}_{\overline{6}|} + 7_6|\ddot{a}_{\overline{9}|} + 10_{15}|\ddot{a}_{\overline{5}|}$;

(ii) $10\ddot{a}_{\overline{20}|} - 3\ddot{a}_{\overline{15}|} - 2\ddot{a}_{\overline{6}|}$; or

(iii) $5 + 10a_{\overline{19}|} - 3a_{\overline{14}|} - 2a_{\overline{5}|}$.

(b) Show that the value of the annuity at the time of the last payment may be expressed as:

(i) $5(1 + i)^{14}s_{\overline{6}|} + 7(1 + i)^5 s_{\overline{9}|} + 10s_{\overline{5}|}$; or

(ii) $5s_{\overline{20}|} + 2s_{\overline{14}|} + 3s_{\overline{5}|}$.

3.10 (a) An annuity-certain is payable annually in advance for n years. The first payment of the annuity is 1. Thereafter the amount of each payment is $(1 + r)$ times that of the preceding payment.

Show that, on the basis of an interest rate of i per annum, the present value of the annuity is $\ddot{a}_{\overline{n}|}$ at rate j, where

$$j = \frac{i - r}{1 + r}$$

(b) Suppose instead that the annuity is payable annually in arrear. Is its present value (at rate i) now equal to $a_{\overline{n}|}$ at rate j?

(c) In return for a single premium of £10 000 a man will receive an annuity payable annually in arrear for 20 years. The annuity payments increase from year to year at the (compound) rate of 5% per annum.

Given that the initial amount of the annuity is determined on the basis of an interest rate of 9% per annum, find the amount of the first payment.

3.11 An investor agrees to pay 20 premiums, annually in advance. At the end of 20 years the investor will receive the accumulated amount of his payments. This amount is calculated on the basis of an effective annual interest rate of 8% for the first five years, 6% for the next seven years, and 5% for the final eight years.

    Find the amount which the investor will receive in return for an annual premium of £100. Find also his yield per annum on the complete transaction.

3.12 (Loan repayable by an annuity calculated on the basis of two rates of interest) In return for a loan of L a lender will receive from a borrower a series of payments, each of amount X, annually in arrear for n years. Each year the lender will use an amount jL of the annual payment as 'income' and will deposit the balance of the payment in an account which earns interest at an effective annual rate i (where $i < j$). At the end of n years the lender will withdraw the proceeds of the accumulated account.

(a) Given that after n years the accumulated account will be of amount L, show that

$$X = L\left(j - i + \frac{1}{a_{\overline{n}|i}}\right)$$

Deduce that the effective annual rate of interest paid by the *borrower* is greater than j. What is the annual yield to the lender on the completed transaction?

(b) In the particular case when $n = 10$ and $i = 0.04$, find the effective rate of interest paid by the borrower when (i) $j = 0.06$ and (ii) $j = 0.07$.

3.13 A loan of £3000 is to be repaid by a level annuity-certain, payable annually in arrear for 25 years and calculated on the basis of an interest rate of 12% per annum.

(a) Find
    (i) The annual repayment;
   (ii) The capital repayment and interest paid at the end of (1) the tenth year and (2) the final year;
  (iii) After which repayment the outstanding loan will first be less than £1800; and
  (iv) For which repayment the capital content will first exceed the interest content.
(b) Immediately after making the fifteenth repayment the borrower requests that the term of the loan be extended by six years, the annual repayment being reduced appropriately. Assuming that the lender agrees to the request and carries out his calculations on the original interest basis, find the amount of the revised annual repayment.

3.14 A loan of £16 000 was issued to be repaid by a level annuity-certain payable annually in arrear over ten years and calculated on the basis of an interest rate of 8% per annum. The terms of the loan provided that at any time the lender could alter the rate of interest, in which case the amount of the annual repayment would be revised appropriately.

(a) Find the initial amount of the annual repayment.

(b) Immediately after the fourth repayment was made the annual rate of interest was increased to 10%. Find the revised amount of the level annual repayment.

(c) Immediately after the seventh repayment was made the annual rate of interest was reduced to 9%. There was no further change to the rate of interest. Find the final amount of the level annual repayment and the effective rate of interest paid by the borrower on the completed transaction.

3.15 A loan of £2000 is repayable by a level annuity-certain, payable annually in arrear for eighteen years. The amount of the annual repayment is calculated on the basis of an annual interest rate of 10% for the first six years and 9% thereafter.

(a) Find (i) the amount of the annual repayment and (ii) the amount of capital contained in (1) the fourth repayment and (2) the twelfth repayment.

(b) Immediately after making the twelfth repayment the borrower makes an additional capital repayment of £100, the amount of the annual repayment being appropriately reduced. Assuming that the interest basis is unaltered, find the amount of the revised repayment.

3.16 A loan of £1000 is repayable by an annuity-certain, payable in arrear for ten years. The annual amount of the annuity doubles after the first five years.

Given that the transaction is based on an interest rate of 10% per annum effective, find the initial amount of the annual repayment. Find also an expression for the amount of loan repaid at time t $(t = 1, 2, \ldots, 10)$.

3.17 An annuity is payable continuously between time 0 and time n (where n is not necessarily an integer). The rate of payment of the annuity at time t $(0 \leqslant t \leqslant n)$ is t per unit time.

On the basis of an effective interest rate of 5% per annum, the value of the annuity at time 0 is equal to one-half of the total amount which will be paid. Find n.

CHAPTER FOUR

NOMINAL RATES OF INTEREST: ANNUITIES PAYABLE pTHLY

4.1 Interest payable pthly

Suppose that, as in the preceding chapter, the force of interest per unit time is constant and equal to δ. Let i and d be the corresponding rates of interest and discount respectively.

In chapter 3 we showed that d payable at time 0, i payable at time 1, and δ payable continuously at a constant rate over the time interval $[0, 1]$ all have the same value (on the basis of the force of interest δ). Each of these payments may be regarded as the interest for the period $[0, 1]$ payable on a loan of 1 made at time 0.

Suppose, however, that a borrower, who is lent 1 at time 0 for repayment at time 1, wishes to pay the interest on his loan in p equal instalments over the interval. How much interest should he pay? We define $i^{(p)}$ to be that *total* amount of interest, payable in equal instalments at the *end* of each pth subinterval (i.e. at times $1/p, 2/p, 3/p, \ldots, 1$), which has the same value as each of the interest payments described in the previous paragraph. Likewise we define $d^{(p)}$ to be that *total* amount of interest, payable in equal instalments at the *start* of each pth subinterval (i.e. at times $0, 1/p, 2/p, \ldots$ $\ldots (p-1)/p$), which has the same value as each of these other payments.

We may easily express $i^{(p)}$ in terms of i. Since $i^{(p)}$ is the total interest paid, each interest payment is of amount $i^{(p)}/p$ and our definition implies that

$$\sum_{t=1}^{p} \frac{i^{(p)}}{p}(1 + i)^{(p - t)/p} = i \qquad (4.1.1)$$

or, if $i \neq 0$,

$$\frac{i^{(p)}}{p}\left[\frac{(1 + i) - 1}{(1 + i)^{1/p} - 1}\right] = i$$

Hence

$$i^{(p)} = p[(1 + i)^{1/p} - 1] \qquad (4.1.2)$$

and

$$\left[1 + \frac{i^{(p)}}{p}\right]^{p} = 1 + i \qquad (4.1.3)$$

(The last two equations are valid even when $i = 0$.)

Equations 4.1.2 and 4.1.3 are most important. Indeed, either equation may be regarded as providing a definition of $i^{(p)}$. If such a definition is used, it is a trivial matter to establish equation 4.1.1, which shows that $i^{(p)}$ may be interpreted as the total interest payable p-thly in arrear in equal instalments for a loan of 1 over one time unit.

Likewise it is a consequence of our definition of $d^{(p)}$ that

$$\sum_{t=1}^{p} \frac{d^{(p)}}{p}(1-d)^{(t-1)/p} = d \qquad (4.1.4)$$

or, if $d \neq 0$,

$$\frac{d^{(p)}}{p}\left[\frac{1-(1-d)}{1-(1-d)^{1/p}}\right] = d$$

Hence

$$d^{(p)} = p[1 - (1-d)^{1/p}] \qquad (4.1.5)$$

and

$$\left[1 - \frac{d^{(p)}}{p}\right]^{p} = 1 - d \qquad (4.1.6)$$

Again the last two equations are important. They are valid even when $d = 0$. Either may be used to define $d^{(p)}$, in which case equation 4.1.4 is readily verified and our original definition is confirmed. Note that $i^{(1)} = i$ and $d^{(1)} = d$. It is usual to include values of $i^{(p)}$ and $d^{(p)}$, at least for $p = 2, 4$, and 12, in compound interest tables. These quantities and various others are given on the left-hand side of the compound interest tables at the back of this book.

It is essential to appreciate that, at force of interest δ per unit time, the five series of payments illustrated in Figure 4.1.1 all have the same value.

If we choose to regard $i^{(p)}$ or $d^{(p)}$ as the basic quantity, equation 4.1.3 or 4.1.6 may be used to define i in terms of $i^{(p)}$ or d in terms of $d^{(p)}$. It is customary to refer to $i^{(p)}$ and $d^{(p)}$ as *nominal* rates of interest and discount convertible *p*thly. For example, if we speak of a rate of interest of 12% per annum convertible quarterly, we have $i^{(4)} = 0.12$ (with one year as the unit of time). Since $(1 + i) = \{1 + [i^{(4)}/4]\}^{4}$, this means that $i = 0.125\,509$. Thus the equivalent annual rate of interest is 12·5509 %. As has been mentioned in previous chapters, when interest rates are expressed in nominal terms it is customary to refer to the equivalent rate per unit time as an *effective* rate. Thus, if the nominal rate of interest convertible quarterly is 12%, the effective rate per annum is 12·5509%.

	0	$\frac{1}{p}$	$\frac{2}{p}$	$\frac{3}{p}$	$\cdots$	$\frac{p-1}{p}$	1	time
(1)	d							equivalent payments
(2)	$\frac{d^{(p)}}{p}$	$\frac{d^{(p)}}{p}$	$\frac{d^{(p)}}{p}$	$\frac{d^{(p)}}{p}$	$\cdots$	$\frac{d^{(p)}}{p}$		
(3)		$\frac{i^{(p)}}{p}$	$\frac{i^{(p)}}{p}$	$\frac{i^{(p)}}{p}$	$\cdots$	$\frac{i^{(p)}}{p}$	$\frac{i^{(p)}}{p}$	
(4)							i	
(5)	←				δ		→	

FIGURE 4.1.1 *Equivalent payments*

The treatment of problems involving nominal rates of interest (or discount) is almost always considerably simplified by an appropriate choice of the time unit. For example, on the basis of a nominal rate of interest of 12% per annum convertible quarterly, the present value of 1 due after t years is

$$\begin{aligned}(1+i)^{-t} &= \left[1+\frac{i^{(4)}}{4}\right]^{-4t} && \text{(by equation 4.1.3)}\\ &= \left(1+\frac{0\cdot 12}{4}\right)^{-4t} && \text{(since } i^{(4)}=0\cdot 12)\\ &= 1\cdot 03^{-4t}\end{aligned}$$

Thus if we adopt a quarter-year as our basic time unit and use 3% as the effective rate of interest, we correctly value future payments.

The general rule to be used in conjunction with nominal rates is very simple. Choose as the basic time unit the period corresponding to the frequency with which the nominal rate of interest is convertible and use $i^{(p)}/p$ as the effective rate of interest per unit time. For example, if we have a nominal rate of interest of 18 % per annum convertible monthly, we should take one month as the unit of time and $1\frac{1}{2}$% as the rate of interest per unit time.

Example 4.1.1

Given that $\delta = 0\cdot1$, find the values of
(i) $i, i^{(4)}, i^{(12)}, i^{(52)}, i^{(365)}$
(ii) $d, d^{(4)}, d^{(12)}, d^{(52)}, d^{(365)}$

Solution

$$i^{(p)} = p[(1 + i)^{1/p} - 1]$$
$$= p(e^{\delta/p} - 1)$$
$$= p(e^{0\cdot1/p} - 1) \qquad \text{(since } \delta = 0\cdot1\text{)}$$

Also,

$$d^{(p)} = p[1 - (1 - d)^{1/p}]$$
$$= p(1 - e^{-\delta/p})$$
$$= p(1 - e^{-0\cdot1/p}) \qquad \text{(since } \delta = 0\cdot1\text{)}$$

Hence we have the following table for the required nominal rates of interest and discount when $\delta = 0\cdot1$:

p	1	4	12	52	365
$i^{(p)}$	0·105 171	0·101 260	0·100 418	0·100 096	0·100 014
$d^{(p)}$	0·095 163	0·098 760	0·099 584	0·099 904	0·099 986

Example 4.1.2

Given that $i = 0\cdot08$, find the values of $i^{(12)}$, $d^{(4)}$ and δ.

Solution

$$i^{(12)} = 12[(1 + i)^{1/12} - 1] = 0\cdot077\,208$$
$$d^{(4)} = 4[1 - (1 - d)^{1/4}] = 4[1 - (1 + i)^{-1/4}] = 0\cdot076\,225$$
$$\delta = \log(1 + i) = 0\cdot076\,961$$

Example 4.1.3

Suppose that l and m are positive integers. Express $i^{(m)}$ in terms of l, m, and $d^{(l)}$. Hence find $i^{(12)}$ when $d^{(4)} = 0\cdot057\,847$.

Solution

$$\left[1 - \frac{d^{(l)}}{l}\right]^l = e^{-\delta} = \left[1 + \frac{i^{(m)}}{m}\right]^{-m}$$

Hence

$$d^{(l)} = l\left\{1 - \left[1 + \frac{i^{(m)}}{m}\right]^{-m/l}\right\}$$

and

$$i^{(m)} = m\left\{\left[1 - \frac{d^{(l)}}{l}\right]^{-l/m} - 1\right\}$$

In particular,

$$i^{(12)} = 12\left\{\left[1 - \frac{d^{(4)}}{4}\right]^{-1/3} - 1\right\}$$

$$= 0.058\,411 \qquad \text{(when } d^{(4)} = 0.057\,847)$$

Alternatively, we may note that $(1 + i) = \{1 - [d^{(4)}/4]\}^{-4} = 1.06$, from which the value of $i^{(12)}$ follows immediately.

Note that $i^{(p)}$ and $d^{(p)}$ are given directly in terms of the force of interest δ by the equations

and
$$\left.\begin{aligned} i^{(p)} &= p(e^{\delta/p} - 1) \\ d^{(p)} &= p(1 - e^{-\delta/p}) \end{aligned}\right\} \tag{4.1.7}$$

Since

$$\lim_{x \to \infty} x(e^{\delta/x} - 1) = \lim_{x \to \infty} x(1 - e^{-\delta/x}) = \delta$$

it follows immediately from the equations 4.1.7 that

$$\lim_{p \to \infty} i^{(p)} = \lim_{p \to \infty} d^{(p)} = \delta \tag{4.1.8}$$

This is intuitively obvious from our original definitions, since a continuous payment stream may be regarded as the limit as p tends to infinity of a corresponding series of payments at intervals of time $1/p$ (see also section 2.4).

It is easy to establish (see exercise 4.4) that

$$i > i^{(2)} > i^{(3)} > \cdots$$

and

$$d < d^{(2)} < d^{(3)} < \cdots$$

so that the sequences $\{i^{(p)}\}$ and $\{d^{(p)}\}$ tend monotonically to the common limit δ from above and below respectively.

Various approximations for $i^{(p)}$ and $d^{(p)}$ in terms of p and i, d, or δ may be obtained, as in the following example.

Example 4.1.4

Show that, if δ is small,

$$d^{(p)} \simeq \delta - \frac{\delta^2}{2p}$$

and derive a corresponding approximation for $i^{(p)}$.

Solution

$$d^{(p)} = p(1 - e^{-\delta/p}) \qquad \text{(by equation 4.1.7)}$$

$$= p\left[1 - \left(1 - \frac{\delta}{p} + \frac{\delta^2}{2p^2} - \frac{\delta^3}{6p^3} + \cdots\right)\right]$$

$$= \delta - \frac{\delta^2}{2p} + \frac{\delta^3}{6p^2} - \cdots$$

$$\simeq \delta - \frac{\delta^2}{2p} \qquad \text{(if } \delta \text{ is small)}$$

Similarly,

$$i^{(p)} = p(e^{\delta/p} - 1)$$

$$= p\left[\left(1 + \frac{\delta}{p} + \frac{\delta^2}{2p^2} + \frac{\delta^3}{6p^3} + \cdots\right) - 1\right]$$

$$= \delta + \frac{\delta^2}{2p} + \frac{\delta^3}{6p^2} + \cdots$$

$$\simeq \delta + \frac{\delta^2}{2p} \qquad \text{(if } \delta \text{ is small)}$$

4.2 Annuities payable *p*thly: present values and accumulations

The nominal rates of interest and discount introduced in the previous section are of particular importance in relation to annuities which are payable more frequently than once per unit time. We shall refer to an annuity which is payable p times per unit time as *payable* $p$thly.

If p and n are positive integers, the notation $a_{\overline{n}|}^{(p)}$ is used to denote the value at time 0 of a level annuity payable $p$thly in arrear at the rate of 1 per unit time over the time interval $[0, n]$. For this annuity the payments are made at times $1/p, 2/p, 3/p, \ldots, n$ and the amount of each payment is $1/p$.

It is a simple matter to derive an expression for $a_{\overline{n}|}^{(p)}$ from first principles. However, the following argument, possibly less immediately obvious, is an important illustration of a kind of reasoning which has widespread application.

By definition, a series of p payments, each of amount $i^{(p)}/p$ in arrear at $p$thly subintervals over any unit time interval, has the same value as a single payment of amount i at the end of the interval. By proportion, p payments, each of amount $1/p$ in arrear at $p$thly subintervals over any unit time interval, have the same value as a single payment of amount $i/i^{(p)}$ at the end of the interval.

Consider now that annuity for which the present value is $a_{\overline{n}|}^{(p)}$. The remarks in the preceding paragraph show that the p payments after time $r-1$ and not later than time r have the same value as a single payment of amount $i/i^{(p)}$ at time r. This is true for $r = 1, 2, \ldots, n$, so the annuity has the same value as a series of n payments, each of amount $i/i^{(p)}$, at times $1, 2, \ldots, n$. This means that

$$a_{\overline{n}|}^{(p)} = \frac{i}{i^{(p)}} a_{\overline{n}|} \qquad (4.2.1)$$

The alternative approach, from first principles, is to write

$$a_{\overline{n}|}^{(p)} = \sum_{t=1}^{np} \frac{1}{p} v^{t/p}$$

$$= \frac{1}{p} \frac{v^{1/p}(1 - v^n)}{1 - v^{1/p}}$$

$$= \frac{1 - v^n}{p[(1 + i)^{1/p} - 1]}$$

$$= \frac{1 - v^n}{i^{(p)}} \qquad (4.2.2)$$

which confirms equation 4.2.1.

Likewise we define $\ddot{a}_{\overline{n}|}^{(p)}$ to be the present value of a level annuity-due payable $p$thly at the rate of 1 per unit time over the time interval $[0, n]$. (The annuity payments, each of amount $1/p$, are made at times $0, 1/p, 2/p, \ldots, n - (1/p)$.)

By definition, a series of p payments, each of amount $d^{(p)}/p$, in advance at $p$thly subintervals over any unit time interval has the same value as a single payment of amount i at the *end* of the interval. Hence, by proportion, p payments, each of amount $1/p$ in advance at $p$thly subintervals, have the same value as a single payment of amount $i/d^{(p)}$ at the *end* of the interval. This means (by an identical argument to that above) that

$$\ddot{a}_{\overline{n}|}^{(p)} = \frac{i}{d^{(p)}} a_{\overline{n}|} \qquad (4.2.3)$$

It is usual to include the values of $i/i^{(p)}$ and $i/d^{(p)}$ in published tables. This enables the values of $a_{\overline{n}|}^{(p)}$ and $\ddot{a}_{\overline{n}|}^{(p)}$ to be calculated easily.

Alternatively, from first principles, we may write

$$\ddot{a}_{\overline{n}|}^{(p)} = \sum_{t=1}^{np} \frac{1}{p} v^{(t-1)/p}$$

$$= \frac{1 - v^n}{d^{(p)}} \qquad (4.2.4)$$

(on simplification), which confirms equation 4.2.3. Note that

$$a_{\overline{n}|}^{(p)} = v^{1/p} \ddot{a}_{\overline{n}|}^{(p)}$$

By combining equations 4.2.1 and 4.2.3, we obtain

$$i a_{\overline{n}|} = i^{(p)} a_{\overline{n}|}^{(p)} = d^{(p)} \ddot{a}_{\overline{n}|}^{(p)} = d \ddot{a}_{\overline{n}|} = \delta \bar{a}_{\overline{n}|} \qquad (4.2.5)$$

each expression being equal to $(1 - v^n)$.

These last equations should be obvious, since each expression represents the present value of the interest payable on a loan of 1 over a time interval of length n, according to different possible arrangements for paying the interest.

Note that, since

$$\lim_{p \to \infty} i^{(p)} = \lim_{p \to \infty} d^{(p)} = \delta \qquad \text{(by formula 4.1.8)}$$

it follows immediately from equations 4.2.2 and 4.2.4 that

$$\lim_{p \to \infty} a_{\overline{n}|}^{(p)} = \lim_{p \to \infty} \ddot{a}_{\overline{n}|}^{(p)} = \bar{a}_{\overline{n}|}$$

These equations are intuitively obvious.

Similarly, we define $s_{\overline{n}|}^{(p)}$ and $\ddot{s}_{\overline{n}|}^{(p)}$ to be the accumulated amounts of the corresponding $p$thly immediate annuity and annuity-due respectively. Thus

$$s_{\overline{n}|}^{(p)} = (1 + i)^n a_{\overline{n}|}^{(p)}$$

$$= (1 + i)^n \frac{i}{i^{(p)}} a_{\overline{n}|} \qquad \text{(by 4.2.1)}$$

$$= \frac{i}{i^{(p)}} s_{\overline{n}|} \qquad (4.2.6)$$

Also

$$\ddot{s}_{\overline{n}|}^{(p)} = (1 + i)^n \ddot{a}_{\overline{n}|}^{(p)}$$

$$= (1 + i)^n \frac{i}{d^{(p)}} a_{\overline{n}|} \qquad \text{(by 4.2.3)}$$

$$= \frac{i}{d^{(p)}} s_{\overline{n}|} \qquad (4.2.7)$$

The above proportional arguments may be applied to other varying series of payments. Consider, for example, an annuity payable annually in arrear for n years, the payment in the tth year being x_t. The present value of this annuity is obviously

$$a = \sum_{t=1}^{n} x_t v^t \qquad (4.2.8)$$

Consider now a second annuity, also payable for n years with the payment in the tth year, again of amount x_t, being made in p equal instalments in arrear over that year. If $a^{(p)}$ denotes the present value of this second annuity, by replacing the p payments for year t (each of amount x_t/p) by a single equivalent payment at the end of the year of amount $x_t[i/i^{(p)}]$, we immediately obtain

$$a^{(p)} = \frac{i}{i^{(p)}} a$$

where a is given by equation 4.2.8 above.

An annuity payable $p$thly in arrear, under which the payments continue indefinitely, is called a *perpetuity* payable $p$thly. When the rate of payment is constant and equal to 1 per unit time, the present value of such a perpetuity is denoted by $a^{(p)}_{\overline{\infty}|}$. If the payments are in advance, we have a *perpetuity-due*, with the corresponding present value denoted by $\ddot{a}^{(p)}_{\overline{\infty}|}$.

Clearly

$$\ddot{a}^{(p)}_{\overline{\infty}|} = \frac{1}{p} + a^{(p)}_{\overline{\infty}|} \tag{4.2.9}$$

By letting n tend to infinity in equations 4.2.2 and 4.2.4, we obtain (if $i > 0$)

$$a^{(p)}_{\overline{\infty}|} = \frac{1}{i^{(p)}} \tag{4.2.10}$$

and

$$\ddot{a}^{(p)}_{\overline{\infty}|} = \frac{1}{d^{(p)}} \tag{4.2.11}$$

respectively.

The present values of an immediate annuity and an annuity-due, payable $p$thly at the rate of 1 per unit time for n time units and deferred for m time units, are denoted by

$$\left. \begin{array}{r} {}_{m|}a^{(p)}_{\overline{n}|} = v^m a^{(p)}_{\overline{n}|} \\ \text{and} \quad {}_{m|}\ddot{a}^{(p)}_{\overline{n}|} = v^m \ddot{a}^{(p)}_{\overline{n}|} \end{array} \right\} \tag{4.2.12}$$

respectively.

Finally, we remark that if $p = 1$, $a^{(p)}_{\overline{n}|}$, $\ddot{a}^{(p)}_{\overline{n}|}$, $s^{(p)}_{\overline{n}|}$, and $\ddot{s}^{(p)}_{\overline{n}|}$ are equal to $a_{\overline{n}|}$, $\ddot{a}_{\overline{n}|}$, $s_{\overline{n}|}$, and $\ddot{s}_{\overline{n}|}$ respectively.

4.3 Annuities payable at intervals of time r, where $r > 1$

In section 4.2 we showed how, by replacing a series of payments to be received by an equivalent series of payments of equal value, we could immediately write down an expression for the value of a pthly annuity. The same technique, i.e. the use of equivalent payments of the same value, may be used to value a series of payments of constant amount payable at intervals of time length r, where r is some integer greater than 1.

For example, suppose that k and r are integers greater than 1 and consider a series of payments, each of amount X, due at times $r, 2r, 3r, \ldots, kr$. What is the value of this series at time 0 on the basis of an interest rate i per unit time?

The situation is illustrated in Figure 4.3.1, which shows the payments of amount X due at the appropriate times. Let us 'replace' the payment of X due at time r by a series of r payments , each of amount Y, due at times 1, $2, \ldots, r$, where Y is chosen to make these r equivalent payments of the same total value as the single payment they replace. This means that

$$Ys_{\overline{r}|} = X$$

at rate i, or

$$Y = \frac{X}{s_{\overline{r}|}} \tag{4.3.1}$$

Similarly *each* payment of amount X can be replaced by r equivalent payments of amount Y of the same value (see Figure 4.3.1). Then the original series of payments of X, due every rth time interval, has the same value as a series of kr payments of $Y = X/s_{\overline{r}|}$ due at unit time intervals. Hence the value of the annuity is

$$\frac{X}{s_{\overline{r}|}} a_{\overline{kr}|} \tag{4.3.2}$$

at rate i. (This result may also be obtained from first principles simply by summing the appropriate geometric progression.)

This technique is illustrated in example 4.3.1. The reader should make no attempt to memorize this last result, which may be obtained immediately from first principles, provided that the underlying idea has been clearly understood.

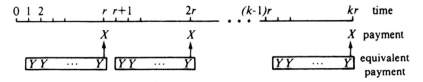

FIGURE 4.3.1   *Annuity valuation through equivalent payments*

Example 4.3.1

An investor wishes to purchase a level annuity of £120 per annum payable quarterly
in arrear for five years. Find the purchase price, given that it is calculated on the
basis of an interest rate of 12% per annum

(a) Effective
(b) Convertible half-yearly
(c) Convertible quarterly
(d) Convertible monthly.

Solution

(a) The value is

$$120a_{\overline{5}|}^{(4)} \text{ at } 12\% = 120\frac{i}{i^{(4)}}a_{\overline{5}|} = 451.583, \quad \text{say} \quad £451.58$$

(b) Since the rate of interest is a nominal one convertible half-yearly, we take the
half-year as our unit of time and 6% as our rate of interest. The annuity is
payable twice per half-year for ten half-years at the rate of £60 per half-year.
Hence its value is

$$60a_{\overline{10}|}^{(2)} \text{ at } 6\% = 60\frac{i}{i^{(2)}}a_{\overline{10}|} = 448.134, \quad \text{say} \quad £448.13$$

(c) We take the quarter-year as the unit of time and 3% as the rate of interest. The
value is thus

$$30a_{\overline{20}|} \text{ at } 3\% = 446.324, \quad \text{say} \quad £446.32$$

(d) We take the month as the unit of time and 1% as the rate of interest. The annuity
payments, of amount £30, at the end of every *third* month can be replaced by a
series of equivalent monthly payments, each of $30/s_{\overline{3}|}$ (at 1%). The value is thus

$$\frac{30}{s_{\overline{3}|}}a_{\overline{60}|} \text{ at } 1\% = 445 \cdot 084, \quad \text{say} \quad £445 \cdot 08$$

Example 4.3.2

On the basis of an effective interest rate of j per annum, building society A makes
loans which are repayable by level annuities payable annually in arrear. For any
amount of loan and term of repayment, building society B quotes the same annual
repayment as society A, but requires that the payments be made $p$thly in arrear
$(p > 1)$.

Show that, whatever the term of repayment, the effective rate of interest per
annum charged by society B is greater than

$$\hat{j} = \left(1 + \frac{j}{p}\right)^p - 1$$

In the particular case when $j = 0 \cdot 08$ and $p = 12$, find the effective rate of interest per
annum for a loan with society B when the term is (a) 10 years and (b) 25 years.

Solution

Let the term of the loan be n years and the amount lent be X. The annual repayment for either society is $X/a_{\overline{n}|j}$. Hence the effective rate of interest per annum for a loan with society B is i, where

$$X = \frac{X}{a_{\overline{n}|j}} a_{\overline{n}|i}^{(p)}$$

or

$$a_{\overline{n}|i}^{(p)} = a_{\overline{n}|j}$$

which determines i.

The left-hand side of the last equation is a monotonic decreasing function of i. Accordingly, to show that the root is greater than $\hat{j}$, it suffices to show that when $i = \hat{j}$ the left-hand side of the equation is greater than the right-hand side. Note that

$$\hat{j}^{(p)} = p[(1 + \hat{j})^{1/p} - 1]$$
$$= j \qquad \text{(by the definition of } \hat{j}\text{)}$$

Therefore

$$a_{\overline{n}|\hat{j}}^{(p)} = \frac{1 - (1 + \hat{j})^{-n}}{\hat{j}^{(p)}}$$

$$= \frac{1 - [(1 + \hat{j})^{1/p}]^{-np}}{\hat{j}^{(p)}}$$

$$= \frac{1 - \left(1 + \dfrac{j}{p}\right)^{-np}}{j} \qquad \text{(by definition of } \hat{j}\text{)}$$

$$> \frac{1 - (1 + j)^{-n}}{j} \qquad \left(\text{since}\left(1 + \frac{j}{p}\right)^{np} > (1 + j)^n\right)$$

$$= a_{\overline{n}|j}$$

as required, so $i > \hat{j}$.

In particular this shows that if $j = 0.08$ and $p = 12$, the effective rate of interest per annum for a loan with society B is always greater than $8.3\%_0$.

Since, at 8%, $a_{\overline{10}|} = 6.7101$ and $a_{\overline{25}|} = 10.6748$, the effective rates of interest per annum for loans over 10 years and 25 years with society B are i_1 and i_2 respectively where

$$a_{\overline{10}|}^{(12)} = 6.7101 \qquad \text{at rate } i_1$$

and

$$a_{\overline{25}|}^{(12)} = 10.6748 \qquad \text{at rate } i_2$$

The reader should show that $i_1 = 0.088\,806$, say 8.88%, and $i_2 = 0.084\,433$, say 8.44%. (See appendix 2 for possible methods of solving the equations.)

Note The above values are particular illustrations of a more general result. As the term of the loan increases, the effective rate of interest per annum for a loan with society B decreases.

4.4 Definition of $a_{\overline{n}|}^{(p)}$ for non-integer values of n

Let p be a positive integer. Until now the symbol $a_{\overline{n}|}^{(p)}$ has been defined only when n is a positive integer. For certain non-integral values of n the symbol $a_{\overline{n}|}^{(p)}$ has an intuitively obvious interpretation. For example, it is not clear what meaning, if any, may be given to $a_{\overline{23\cdot5}|}$, but the symbol $a_{\overline{23\cdot5}|}^{(4)}$ ought to represent the present value of an immediate annuity of 1 per annum payable quarterly in arrear for 23·5 years (i.e. a total of 94 quarterly payments, each of amount 0·25). On the other hand, $a_{\overline{23\cdot25}|}^{(2)}$ has no obvious meaning.

Suppose that n is an integer multiple of $1/p$, say $n = r/p$, where r is an integer. In this case we define $a_{\overline{n}|}^{(p)}$ to be the value at time 0 of a series of r payments, each of amount $1/p$, at times $1/p, 2/p, 3/p, \ldots, r/p = n$. If $i = 0$, then clearly $a_{\overline{n}|}^{(p)} = n$. If $i \neq 0$, then

$$a_{\overline{n}|}^{(p)} = \frac{1}{p}(v^{1/p} + v^{2/p} + v^{3/p} + \cdots + v^{r/p}) \tag{4.4.1}$$

$$= \frac{1}{p}v^{1/p}\left(\frac{1 - v^{r/p}}{1 - v^{1/p}}\right)$$

$$= \frac{1}{p}\left[\frac{1 - v^{r/p}}{(1 + i)^{1/p} - 1}\right]$$

Thus

$$a_{\overline{n}|}^{(p)} = \begin{cases} \dfrac{1 - v^n}{i^{(p)}} & \text{if } i \neq 0 \\[2mm] n & \text{if } i = 0 \end{cases} \tag{4.4.2}$$

Note that, by working in terms of a new time unit equal to $1/p$ times the original time unit and with the equivalent effective interest rate of $i^{(p)}/p$ per new time unit, we see that

$$a_{\overline{n}|}^{(p)} \text{ at rate } i = \frac{1}{p}a_{\overline{np}|} \text{ at rate } i^{(p)}/p \tag{4.4.3}$$

This formula is useful when $i^{(p)}/p$ is a tabulated rate of interest. Note that the definition of $a_{\overline{n}|}^{(p)}$ given by equation 4.4.2 is mathematically meaningful for all non-negative values of n. For our present purpose, therefore, it is convenient to adopt equation 4.4.2 as a definition of $a_{\overline{n}|}^{(p)}$ for all n. If n is not an integer multiple of $1/p$, there is no universally recognised definition of $a_{\overline{n}|}^{(p)}$. For example, if $n = n_1 + f$, where n_1 is an integer multiple of $1/p$ and

$0 < f < 1/p$, some writers define $a_{\overline{n}|}^{(p)}$ as

$$a_{\overline{n}|}^{(p)} + fv^n.$$

With this alternative definition

$$a_{\overline{23\cdot75}|}^{(2)} = a_{\overline{23\cdot5}|}^{(2)} + \tfrac{1}{4}v^{23\cdot75}$$

which is the present value of an annuity of 1 per annum, payable half-yearly in arrear for 23·5 years, together with a final payment of 0·25 after 23·75 years. Note that this is *not* equal to the value obtained from definition 4.4.2 (see example 4.4.1 below).

If $i \neq 0$, we define for all non-negative n

$$\left.\begin{array}{l}
\ddot{a}_{\overline{n}|}^{(p)} = (1 + i)^{1/p} a_{\overline{n}|}^{(p)} = \dfrac{1 - v^n}{d^{(p)}} \\[3mm]
s_{\overline{n}|}^{(p)} = (1 + i)^n a_{\overline{n}|}^{(p)} = \dfrac{(1 + i)^n - 1}{i^{(p)}} \\[3mm]
\ddot{s}_{\overline{n}|}^{(p)} = (1 + i)^n \ddot{a}_{\overline{n}|}^{(p)} = \dfrac{(1 + i)^n - 1}{d^{(p)}}
\end{array}\right\} \qquad (4.4.4)$$

where $i^{(p)}$ and $d^{(p)}$ are defined by equations 4.1.2 and 4.1.5 respectively. If $i = 0$, each of these last three functions is defined to equal n.

The reader should verify that, whenever n is an integer multiple of $1/p$, say $n = r/p$, then $\ddot{a}_{\overline{n}|}^{(p)}$, $s_{\overline{n}|}^{(p)}$, $\ddot{s}_{\overline{n}|}^{(p)}$ are the values at different times of an annuity-certain of r payments, each of amount $1/p$, at intervals of $1/p$ time unit.

As before, we use the simpler notations $a_{\overline{n}|}$, $\ddot{a}_{\overline{n}|}$, $s_{\overline{n}|}$, and $\ddot{s}_{\overline{n}|}$ to denote $a_{\overline{n}|}^{(1)}$, $\ddot{a}_{\overline{n}|}^{(1)}$, $s_{\overline{n}|}^{(1)}$, and $\ddot{s}_{\overline{n}|}^{(1)}$ respectively, thus extending the definition of $a_{\overline{n}|}$ etc. given in chapter 3 to all non-negative values of n. It is a trivial consequence of our definitions that the formulae

$$\left.\begin{array}{l}
a_{\overline{n}|}^{(p)} = \dfrac{i}{i^{(p)}} a_{\overline{n}|} \\[3mm]
\ddot{a}_{\overline{n}|}^{(p)} = \dfrac{i}{d^{(p)}} a_{\overline{n}|} \\[3mm]
s_{\overline{n}|}^{(p)} = \dfrac{i}{i^{(p)}} s_{\overline{n}|} \\[3mm]
\ddot{s}_{\overline{n}|}^{(p)} = \dfrac{i}{d^{(p)}} s_{\overline{n}|}
\end{array}\right\} \qquad (4.4.5)$$

(valid when $i \neq 0$) now hold for *all* values of n.

We may also extend the definitions of $_m|a_{\overline{n}|}^{(p)}$ and $_m|\ddot{a}_{\overline{n}|}^{(p)}$ to all values of n by the formulae

$$\left.\begin{array}{l} _m|a_{\overline{n}|}^{(p)} = v^m a_{\overline{n}|}^{(p)} \\ _m|\ddot{a}_{\overline{n}|}^{(p)} = v^m \ddot{a}_{\overline{n}|}^{(p)} \end{array}\right\} \tag{4.4.6}$$

The reader should verify that these definitions imply that

$$\left.\begin{array}{l} _m|a_{\overline{n}|}^{(p)} = a_{\overline{n+m}|}^{(p)} - a_{\overline{m}|}^{(p)} \\ _m|\ddot{a}_{\overline{n}|}^{(p)} = \ddot{a}_{\overline{n+m}|}^{(p)} - \ddot{a}_{\overline{m}|}^{(p)} \end{array}\right\} \tag{4.4.7}$$

Example 4.4.1

Given that $i = 0\cdot03$, evaluate (a) $a_{\overline{23\cdot5}|}^{(4)}$, (b) $\ddot{a}_{\overline{23\cdot75}|}^{(4)}$, (c) $_{1\cdot5}|a_{\overline{5\cdot25}|}^{(4)}$, (d) $\ddot{s}_{\overline{6\cdot5}|}^{(2)}$, and (e) $a_{\overline{23\cdot75}|}^{(2)}$.

Note The last value must be calculated from the definition 4.4.2. Compare your answer with that obtained from the alternative definition described above.

Solution

(a)
$$a_{\overline{23\cdot5}|}^{(4)} = \frac{1 - v^{23\cdot5}}{i^{(4)}} = 16\cdot8780$$

(b)
$$\ddot{a}_{\overline{23\cdot75}|}^{(4)} = \tfrac{1}{4} + a_{\overline{23\cdot5}|}^{(4)} = 17\cdot1280$$

This value may also be obtained from equation 4.4.4.

(c)
$$_{1\cdot5}|a_{\overline{5\cdot25}|}^{(4)} = v^{1\cdot5} a_{\overline{5\cdot25}|}^{(4)}$$
$$= v^{1\cdot5}\left(\frac{1 - v^{5\cdot25}}{i^{(4)}}\right) = 4\cdot6349$$

(d)
$$\ddot{s}_{\overline{6\cdot5}|}^{(2)} = \frac{(1 + i)^{6\cdot5} - 1}{d^{(2)}} = 7\cdot2195$$

(e)
$$a_{\overline{23\cdot75}|}^{(2)} = \frac{1 - v^{23\cdot75}}{i^{(2)}} = 16\cdot9391$$

The alternative definition gives

$$a_{\overline{23\cdot75}|}^{(2)} = a_{\overline{23\cdot5}|}^{(2)} + \tfrac{1}{4}v^{23\cdot75} = 16\cdot9395$$

4.5 The loan schedule for a *p*thly annuity

No new principles are involved, since this is simply a particular example of the general schedule discussed in section 3.7. For a loan repayable by a level annuity payable *p*thly in arrear over n time units and based on an interest rate i per unit time, the schedule is best derived by working with an interest rate of $i^{(p)}/p$ per time interval of length $1/p$. Thus the interest due at time r/p $(r = 1, 2, \ldots, np)$ is $i^{(p)}/p$ times the loan outstanding at time $(r - 1)/p$ (immediately after the repayment then due has been received).

For example, in relation to a loan of $a_{\overline{n}|}^{(p)}$ (at rate i) it is simple to show that the capital repaid in the rth annuity payment $(r = 1, 2, \ldots, np)$ is $(1/p)v^{n-(r-1)/p}$ and that the loan outstanding immediately after the rth payment has been received is $a^{(p)}_{\overline{n-r/p}|}$ (at rate i). (This is simply the value of the outstanding payments.)

Example 4.5.1

A loan of £1000 is repayable by a level annuity payable half-yearly in arrear for three years and calculated on the basis of an interest rate·of 15% per annum effective.

Construct the lender's schedule showing the subdivision of each payment into capital and interest and the loan outstanding after each repayment.

Solution

At 15%, $i^{(2)}/2 = 0.072\,381$, so the interest due at the end of each half-year is 7·2381% of the loan outstanding at the start of the half-year. The amount of the annual repayment is $1000/a_{\overline{3}|}^{(2)}$ at 15%, i.e. 422·68. The half-yearly payment is thus £211·34. The schedule is easily drawn up as in table 4.5.1.

Table 4.5.1   *Repayment schedule*

(1)	(2)	(3)	(4)
	Loan outstanding at start of nth	Interest due at end of nth half-year (£)	Capital repaid at end of nth half-year (£)
n	half-year (£)	$[0.072\,381 \times (2)]$	$[211.34 - (3)]$
1	1000·00	72·38	138·96
2	861·04	62·32	149·02
3	712·02	51·54	159·80
4	552·22	39·97	171·37
5	380·85	27·57	183·77
6	197·08	14·26	197·08

Exercises

4.1 (a) Given that $i = 0.0625$, find the values of $i^{(4)}$, δ, and $d^{(2)}$.
  (b) Given that $i^{(2)} = 0.0625$, find the values of $i^{(12)}$, δ, and $d^{(4)}$.
  (c) Given that $d^{(12)} = 0.0625$, find the values of $i^{(2)}$, δ, and d.
  (d) Given that $\delta = 0.0625$, find the values of $i^{(4)}$ and $d^{(2)}$.

4.2 Find, on the basis of an effective interest rate of 4% per unit time, the values of

$$a_{\overline{6}|}^{(4)}, \quad \ddot{s}_{\overline{18}|}^{(12)}, \quad {}_{14|}\ddot{a}_{\overline{10}|}^{(2)}, \quad \bar{s}_{\overline{56}|}, \quad a_{\overline{16\cdot5}|}^{(4)}, \quad \ddot{s}_{\overline{15\cdot25}|}^{(12)}, \quad {}_{4\cdot25|}a_{\overline{3\cdot75}|}^{(4)}, \quad \bar{a}_{\overline{26\cdot3}|}.$$

4.3 (a) Every three years £100 is paid into an account which earns interest at a constant rate. Find (to the nearest pound) the accumulated amount of the account immediately before the sixth payment is made, given that the interest rate is (i) 10% per annum effective, (ii) 10% per annum convertible half-yearly.

(b) Sixteen payments, each of amount £240, will be made at three-yearly intervals, the first payment being made one year from now. Find (to the nearest pound) the present value of this series of payments on the basis of an interest rate of 8% per annum effective.

4.4 (a) By showing that

$$f(x) = x(e^{\delta/x} - 1) \qquad \delta \neq 0$$

is a decreasing function on the interval $0 < x < \infty$, or otherwise, prove that

$$i^{(m+1)} < i^{(m)}$$

for any given rate of interest i.

(b) By considering the function

$$g(x) = x(1 - e^{\delta/x}) \qquad \delta \neq 0$$

or otherwise, show that

$$d^{(m+1)} > d^{(m)}$$

for any rate of interest i.

(Note that the above results show that the sequences $\{i^{(m)}\}$ and $\{d^{(m)}\}$ $(m = 1, 2, 3, \ldots)$ tend monotonically to the common limit δ from above and from below respectively. See section 4.1.)

(c) Show that the value of

$$\frac{1}{d^{(m)}} - \frac{1}{i^{(m)}}$$

does not depend on the rate of interest.

4.5 (a) On the basis of an interest rate of 12% per annum effective find (to the nearest pound) the present value of an annuity of £600 p.a. for 20 years payable
   (i) Annually in arrear;
   (ii) Quarterly in arrear;
   (iii) Monthly in arrear;
   (iv) Continuously.

(b) Find (again to the nearest pound) the present values of the annuities described in (a) on the basis of an interest rate of 12% per annum convertible quarterly.

4.6 Find the present value of a perpetuity of 1 per annum payable half-yearly

   (i) Immediately after a payment has been made;
   (ii) Three months before the next payment is made

on the basis of an interest rate of

(a) 12% per annum effective;
(b) 12% per annum convertible half-yearly;
(c) 12% per annum convertible quarterly.

4.7 An annuity is payable in arrear for 15 years. The annuity is payable half-yearly for the first five years, quarterly for the next five years, and monthly for the final five years. The annual amount of the annuity is doubled after each five-year period. On the basis of an interest rate of 8% per annum convertible quarterly for the first four years, 8% per annum convertible half-yearly for the next eight years, and 8% per annum effective for the final three years, the present value of the annuity is £2049.

Find the initial annual amount of the annuity. (The value of $i/i^{(6)}$ at 4% is 1·016 540.)

4.8 An annuity is payable for 20 years. The amount of the annuity in the tth year is £t^2. On the basis of an effective rate of interest of 5% per annum, find the present value of the annuity, assuming that it is payable

(a) Annually in advance;
(b) Quarterly in advance, the payments for each year being made in four equal instalments;
(c) Half-yearly in arrear, the payments for each year being made in two equal instalments; and
(d) Continuously, the rate of payment being constant over each year.

Hint For (b), (c), and (d) adjust the answer to (a) appropriately.

4.9 An investor effects a contract under which he pays £50 to a savings account on 1 July 1986, and at three-monthly intervals thereafter, the final payment being made on 1 October 1999. On 1 January 2000 the investor will be paid the accumulated amount of the account.

Calculate how much the investor will receive if the account earns interest at the rate of

(a) 12% per annum effective;
(b) 12% per annum convertible half-yearly;
(c) 12% per annum convertible quarterly;
(d) 12% per annum convertible monthly.

4.10 On 1 November 1985 a man was in receipt of the following three annuities, all payable by the same insurance company:

(a) £200 p.a. payable annually on 1 February each year, the final payment being on 1 February 2007;
(b) £320 p.a. payable quarterly on 1 January, 1 April, 1 July, and 1 October each year, the final payment being on 1 January 2002;
(c) £180 p.a. payable monthly on the first day of each month, the final payment being on 1 August 2004.

Immediately after receiving the monthly payment due on 1 November 1985, the man requested that these three annuities be combined into a single annuity payable half-yearly on 1 February and 1 August in each subsequent year, the final payment being made on 1 February 2007. The man's request was granted.

Find the amount of the revised annuity, given that it was calculated on the basis of an interest rate of 8% per annum effective, all months being regarded as of equal length.

4.11 (a) On 1 January and 1 July each year for the next 20 years a company will pay
a premium of £200 into an investment account. In return the company will
receive a level monthly annuity for 15 years, the first annuity payment
being made on 1 January following payment of the last premium.

Find (to the nearest pound) the amount of the monthly annuity
payment, given that it is determined on the basis of an interest rate of

  (i) 12% per annum effective;
  (ii) 12% per annum convertible half-yearly,
  (iii) 12% per annum convertible monthly.

(b) Find the monthly annuity payment as in (a), except that the first payment
of the annuity to the company is made one month after payment of the last
premium.

Hint Adjust appropriately the answers from part (a).

4.12 A loan of £9880 was granted on 10 July 1978. The loan is repayable by a level
annuity payable monthly in arrear (on the 10th of each month) for 25 years and
calculated on the basis of an interest rate of 7% per annum effective. Find

  (a) The monthly repayment;
  (b) The loan outstanding immediately after the repayment on 10 March 1992
  (c) The capital to be repaid on 10 October 1989;
  (d) (i) The total capital to be repaid, and (ii) the total amount of interest to be
  paid, in the monthly instalments due between 10 April 1996 and 10 March
  1997 (both dates inclusive);
  (e) The month when the capital to be repaid first exceeds one-half of the
  interest payment.

4.13 A loan of £19 750 was repayable by a level annuity payable monthly in
arrear for 20 years and calculated on the basis of an interest rate of 9%
per annum effective. The lender had the right to alter the conditions of the
loan at any time and, immediately after the 87th monthly repayment had been
made, the effective annual rate of interest was increased to 10%. The borrower
was given the option of either increasing the amount of his level monthly
repayment or extending the term of the loan (the monthly repayment
remaining unchanged).

  (a) Show that, if the borrower had opted to pay a higher monthly
  instalment, the monthly repayment would have been increased by £8·45.
  (b) Assume that the borrower elected to continue with the monthly repayment
  unchanged. Find the revised term of the loan and, to the nearest pound, the
  (reduced) amount of the final monthly repayment.

4.14 A loan of £11 820 was repayable by an annuity payable quarterly in arrear for
15 years. The repayment terms provided that at the end of each five-year
period the amount of the quarterly repayment would be increased by £40. The
amount of the annuity was calculated on the basis of an effective rate of interest
of 12% per annum.

  (a) Find the initial amount of the quarterly repayment.
  (b) On the basis of the lender's original schedule find the amount of principal
  repaid in (i) the third year and (ii) the thirteenth year.

(c) Immediately after paying the 33rd quarterly instalment the borrower requested that in future the repayments be of a fixed amount for the entire outstanding duration of the loan. The request was granted and the revised quarterly repayment was calculated on the original interest basis.

Find the amount of the revised quarterly repayment.

4.15 A loan is repayable over ten years by a special decreasing annuity, calculated on the basis of an effective interest rate of 10% per annum. The annuity payment each year is divided into an interest payment (equal to 10% of the loan outstanding at the start of the year) and a capital payment, which is used to reduce the amount of the loan outstanding.

The annuity decreases in such a way that, if income tax of 30% of the interest content of each annuity payment were to be deducted from each payment, the *net* amount of the payment (i.e. the capital payment plus the interest payment less tax) would be £5000 each year.

An investor who is not liable to tax will in fact receive the *gross* amount of each annuity payment (i.e. the payment without any deduction). What price should such an investor pay for the annuity to achieve an effective yield of 8% per annum?

4.16 An annuity-certain was purchased on 1 November 1985 to provide 15 instalments, payable on 1 September 1986 and thereafter on 1 January, 1 May and 1 September until 1 May 1991 inclusive. The amount of the first instalment was £1000 and each subsequent instalment is 5% greater than its predecessor.

(a) Calculate the purchase price of the annuity-certain on the basis of an effective interest rate of 6% per annum.
(b) Calculate the interest content of (i) the seventh and (ii) the fifteenth annuity instalment on this basis.

CHAPTER FIVE

DISCOUNTED CASH FLOW

5.1 Net cash flows

This chapter is largely concerned with some applications of compound interest theory to the financial assessment of investments and business ventures. These matters are of course considered by accountants, economists, and others as well as by actuaries. Some writers use terminology and symbols which differ from those usually employed by actuaries, but there are no differences of principle. (See, for example, references [3], [5], [10], [27], [30], [32], [39].)

Suppose that an investor (who may be a private individual or a corporate body) is considering the merits of an investment or business project. The investment or project will normally require an initial outlay and possibly other outlays in future, which will be followed by receipts, although in some cases the pattern of income and outgo is more complicated. The cash flows associated with the investment or business venture may be completely fixed (as in the case of a secure fixed-interest security maturing at a given date) or they may have to be estimated. The estimation of the cash inflows and outflows associated with a business project usually requires considerable experience and judgement and all the relevant factors (such as taxation and investment grants) should be considered. It is often prudent to perform calculations on more than one set of assumptions, e.g. on the basis of 'optimistic', 'average', and 'pessimistic' forecasts respectively; more complicated techniques (using statistical theory) are available to deal with this kind of uncertainty. Precision is not attainable in the estimation of cash flows for many business projects and hence extreme accuracy is out of place in many calculations.

We recall from section 3.2 that the net cash flow c_t at time t (measured in suitable time units) is

$$c_t = \text{cash inflow at time } t - \text{cash outflow at time } t \qquad (5.1.1)$$

If any payments may be regarded as continuous then $\rho(t)$, the net rate of cash flow per unit time at time t, is defined (see chapter 3) as

$$\rho(t) = \rho_1(t) - \rho_2(t) \qquad (5.1.2)$$

where $\rho_1(t)$ and $\rho_2(t)$ denote the rates of inflow and outflow at time t respectively.

86

Example 5.1.1

A businessman is considering a certain project, which involves setting up a shop. He estimates that the venture will require an initial outlay of £20 000 and a further outlay of £10 000 after one year. There will be an estimated inflow of £3000 per annum payable continuously for ten years beginning in three years' time and a final inflow of £6000 when the project ends in thirteen years' time. Measuring time in years, describe the net cash flows associated with this venture and illustrate the position on a diagram.

Solution

We have the net cash flows:

$$
\begin{aligned}
c_0 &= -20\,000 \\
c_1 &= -10\,000 \\
c_{13} &= +6000 \\
\rho(t) &= +3000 \qquad \text{for } 3 \leqslant t \leqslant 13
\end{aligned}
$$

Figure 5.1.1 illustrates these cash flows.

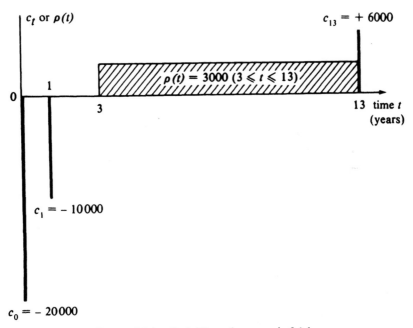

FIGURE 5.1.1 *Cash Flows for example 5.1.1*

5.2 Net present values and yields

Having ascertained or estimated the net cash flows of the investment or project under scrutiny, the investor will wish to measure its profitability in relation to other possible investments or projects. In particular, he may wish to determine whether or not it is prudent to borrow money to finance the venture.

Assume for the moment that the investor may borrow or lend money at a fixed rate of interest i per unit time. The investor could accumulate the net cash flows connected with the project in a separate account in which interest is payable or credited at this fixed rate. By the time the project ends (at time T, say), the balance in this account will be

$$\sum c_t(1 + i)^{T-t} + \int_0^T \rho(t)(1 + i)^{T-t}dt \qquad (5.2.1)$$

where the summation extends over all t such that $c_t \neq 0$. The present value at rate of interest i of the net cash flows is called the *net present value* at rate of interest i of the investment or business project, and is usually denoted by $NPV(i)$. Hence

$$NPV(i) = \sum c_t(1 + i)^{-t} + \int_0^T \rho(t)(1 + i)^{-t}dt \qquad (5.2.2)$$

(If the project continues indefinitely, the accumulation 5.2.1 is not defined, but the net present value may be defined by equation 5.2.2 with $T = \infty$.) If $\rho(t) = 0$, we obtain the simpler formula

$$NPV(i) = \sum c_t v^t \qquad (5.2.3)$$

where $v = (1 + i)^{-1}$.

Since the equation

$$NPV(i) = 0 \qquad (5.2.4)$$

is the equation of value for the project at the present time, the yield i_0 on the transaction is the solution of this equation, provided that a unique solution exists. Conditions under which the yield exists, and numerical methods for solving equation 5.2.4, have been discussed in section 3.2.

It may readily be shown that $NPV(i)$ is a smooth function of the rate of interest i and that $NPV(i) \to c_0$ as $i \to \infty$.

In economics and accountancy the yield per annum is often referred to as the *internal rate of return* (IRR) or the *yield to redemption*. The latter term is frequently used when dealing with fixed-interest securities, for which the 'running' (or 'flat') yield is also considered (see chapter 7).

Example 5.2.1

Find the net present value function $NPV(i)$ and the yield for the business venture described in example 5.1.1. Plot the graph of $NPV(i)$ for $0 \leqslant i \leqslant 0.05$.

Solution

By equation 5.2.2,

$$NPV(i) = -20\,000 - 10\,000v + 3000(\bar{a}_{\overline{13}|} - \bar{a}_{\overline{5}|}) + 6000v^{13} \qquad \text{at rate } i$$

The graph of $NPV(i)$ is as in Figure 5.2.1 (for $0 \leqslant i \leqslant 0.05$). We see from the graph that the yield i_0 (which must exist since the cash flow changes sign only once) is rather more than 2%, and by interpolation between the rates of interest 2% and $2\frac{1}{2}$% we obtain $i_0 \simeq 2.2$%. (A more accurate value is 2.197%, but this degree of accuracy is not usually necessary.)

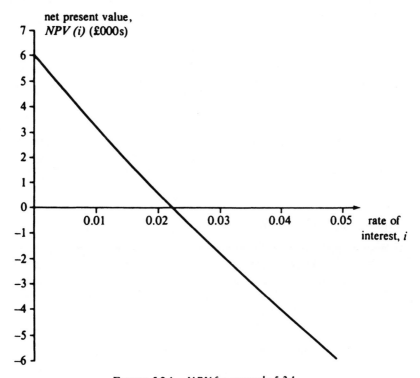

FIGURE 5.2.1   *NPV for example 5.2.1*

The practical interpretation of the net present value function $NPV(i)$ and the yield is as follows. Suppose that the investor may lend or borrow money at a fixed rate of interest i_1. Since, from equation 5.2.2, $NPV(i_1)$ is the present value at rate of interest i_1 of the net cash flows associated with the project, we conclude that the project will be profitable if and only if

$$NPV(i_1) > 0 \qquad (5.2.5)$$

Also, if the project ends at time T, then the profit (or, if negative, loss) at that time is (by expression 5.2.1)

$$NPV(i_1)(1 + i_1)^T \qquad (5.2.6)$$

Let us now assume that, as is usually the case in practice, the yield i_0 exists and $NPV(i)$ changes from positive to negative when $i = i_0$. Under these conditions it is clear that the project is profitable if and only if

$$i_1 < i_0 \qquad (5.2.7)$$

i.e. the yield exceeds the rate of interest at which the investor may lend or borrow money.

Example 5.2.2

Assume that the businessman of examples 5.1.1 and 5.2.1 may borrow or lend money at 2% per annum. Determine whether or not the business venture of example 5.2.1 is profitable, and find the profit or loss when the project ends in 13 years' time.

Solution

The net cash flow changes sign only once, so the yield i_0 exists, and it is clear that $NPV(i)$ changes sign from positive to negative at i_0. Since $i_0 \simeq 2 \cdot 2\%$ exceeds $i_1 = 2\%$, the project is profitable. By expression 5.2.6 the expected profit in 13 years' time is

$$NPV(0 \cdot 02)(1 \cdot 02)^{13} = 481(1 \cdot 02)^{13} = £622$$

'Profit-testing' of life assurance business

If random fluctuations in the numbers of death claims are ignored, the net cash flows for a large block of life assurance policies may be estimated on the basis of various assumptions about future interest rates, expenses, etc.

For example, suppose that an office has issued a number of ten-year policies with the following estimated net cash flows:

Time t (years)	Net cash flow at time t (£)
0	2 176
1	2 303
2	2 218
3	2 128
4	2 030
5	1 926
6	1 816
7	1 699
8	1 576
9	1 447
10	− 24 493

For $0 \leqslant t \leqslant 9$ the positive net cash flow at time t represents the amount by which the premium income due at time t exceeds the expenses incurred and the claims payable at that time. When $t = 10$, there is a relatively large negative cash flow, since the office has to pay the sum assured to all the policyholders who survive to that time. The life office may wish to determine the smallest rate of interest per annum that it must earn in order that this block of business should show a profit after ten years. This may be found by calculating the internal rate of return or, equivalently, by finding the rate of interest per annum i which will make the accumulated fund after ten years equal to zero. It is easily found that this value of i is 0·04. The accumulation of the fund over the next ten years at this rate of interest is as follows:

Time t (years)	Accumulated fund (after the transactions at time t years) (£)
0	2 176
1	4 566
2	6 967
3	9 373
4	11 778
5	14 176
6	16 559
7	18 921
8	21 253
9	23 551
10	0

(The reader should verify the above figures.)

Similar calculations may be carried out, perhaps with the aid of a computer, to determine the profitability of the block of business on a variety of assumptions. A more detailed discussion of this subject lies beyond the scope of this book.

5.3 The comparison of two investment projects

Suppose now that an investor is comparing the merits of two possible investments or business ventures, which we call projects A and B respectively. We assume that the borrowing powers of the investor are not limited.

Let $NPV_A(i)$ and $NPV_B(i)$ denote the respective net present value functions and let i_A and i_B denote the yields (which we shall assume to exist). It might be thought that the investor should always select the project with the higher yield, but this is not invariably the best policy. A better criterion to use is *the profit at time T* (the date when the later of the two projects ends) or, equivalently, the *net present value*, calculated at the rate of interest i_1 at which the investor may lend or borrow money. This is because A is the more profitable venture if

$$NPV_A(i_1) > NPV_B(i_1) \qquad (5.3.1)$$

The fact that $i_A > i_B$ may not imply that $NPV_A(i_1) > NPV_B(i_1)$ is illustrated in Figure 5.3.1. Although i_A is larger than i_B, the $NPV(i)$ functions 'cross over' at i'. It follows that $NPV_B(i_1) > NPV_A(i_1)$ for any $i_1 < i'$, where i' is the *cross-over rate*. There may even be more than one cross-over point, in which case the range of interest rates for which project A is more profitable than project B is more complicated.

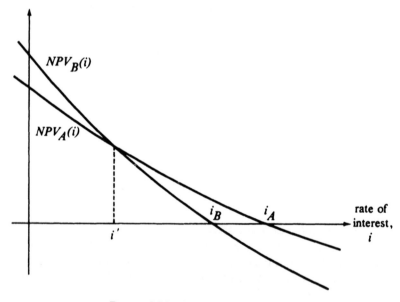

FIGURE 5.3.1   *Investment comparison*

Example 5.3.1

An investor is considering whether to invest in either or both of the following loans:

Loan A   For a purchase price of £10 000, the investor will receive £1000 per annum
         payable quarterly in arrear for 15 years.
Loan B   For a purchase price of £11 000, the investor will receive an income of £605
         per annum, payable annually in arrear for 18 years, and a return of his
         outlay at the end of this period.

The investor may lend or borrow money at 4% per annum. Would you advise him to invest in either loan, and, if so, which would be the more profitable?

Solution

We first consider loan A:

$$NPV_A(i) = -10\,000 + 1000a_{\overline{13}|}^{(4)}$$

and the yield is found by solving the equation $NPV_A(i) = 0$, or $a_{\overline{13}|}^{(4)} = 10$, which gives $i_A \simeq 5.88\%$.

For loan B we have

$$NPV_B(i) = -11\,000 + 605a_{\overline{18}|} + 11\,000v^{18}$$

and the yield (i.e. the solution of $NPV_B(i) = 0$) is $i_B = 5.5\%$. The rate of interest at which the investor may lend or borrow money is 4% per annum, which is less than both i_A and i_B, so we compare NPV_A (0.04) and $NPV_B(0.04)$.

Now $NPV_A(0.04) = £1284$ and $NPV_B(0.04) = £2089$, so it follows that, although the yield on loan B is less than on loan A, the investor will make a larger profit from loan B. We should therefore advise him that an investment in either loan would be profitable, but that, if only one of them is to be chosen, then loan B will give the higher profit.

The above example illustrates the fact that the choice of investment depends very much on the rate of interest i_1 at which the investor may lend or borrow money. If this rate of interest were $5\frac{3}{4}\%$, say, then loan B would produce a loss to the investor, while loan A would give a profit.

5.4 Different interest rates for lending and borrowing

We have assumed so far that the investor may borrow or lend money at the same rate of interest i_1. In practice, however, he will probably have to pay a higher rate of interest (j_1, say) on borrowings than the rate (j_2, say) he receives on investments. The difference $j_1 - j_2$ between these rates of interest depends on various factors, including the credit-worthiness of the investor and the expense of raising a loan.

The concepts of net present value and yield are in general no longer meaningful in these circumstances. We must calculate the accumulation of the net cash flows from first principles, the rate of interest depending on whether or not the investor's account is in credit. In many practical problems the balance in the investor's account (i.e. the accumulation of net cash flows) will be negative until a certain time t_1 and positive afterwards, except perhaps when the project ends. In order to determine this time t_1 we must solve one or more equations, as in the following example.

Example 5.4.1

A mining company is considering an open-cast coal project. It is estimated that the open-cast site will produce 10 000 tonnes of coal per annum continuously for ten years, after which period there will be an outlay of £300 000 to restore the land. The purchase price of the mining rights will be £1 000 000 and mining operations will cost £200 000 per annum, payable continuously.

The company has insufficient funds to finance this venture, but can borrow the initial outlay of £1 000 000 from a bank, which will charge interest at 12% per annum effective; this loan is not for a fixed term, but may be reduced by repayments at any time. When the mining company has funds to invest, it will receive interest calculated at 10% per annum effective on its deposits.

On the assumption that the price of coal is such that this project will just break even, determine (to the nearest month) how long the mining company will take to repay its bank indebtedness and hence calculate this minimum coal price.

Solution

Let P be the break-even price per tonne of coal. The net cash flow of the open-cast project to the mining company is as follows:

$$c_0 = -1\,000\,000$$
$$c_{10} = -300\,000$$
$$\rho(t) = k = 10\,000P - 200\,000 \qquad \text{for } 0 \leqslant t \leqslant 10$$

Since the rates of interest for borrowing and lending (12% and 10% per annum respectively) are different, we must find the time t_1 years when the mining company will have repaid its bank indebtedness. This will occur when

$$1\,000\,000(1\cdot12)^{t_1} = k\bar{s}_{\overline{t_1}|} \qquad \text{at } 12\%$$

or

$$1\,000\,000 = k\bar{a}_{\overline{t_1}|} \qquad \text{at } 12\% \qquad (1)$$

From time t_1 onwards the net cash flow may be accumulated at 10% per annum interest and, since the balance at the end of the project is to be zero, we have

$$k\bar{s}_{\overline{10-t_1}|} - 300\,000 = 0 \qquad \text{at } 10\% \qquad (2)$$

We now solve equations 1 and 2 for the two unknowns t_1 and k (from which we can find P). Eliminating k from these equations, we obtain

$$0\cdot3\bar{a}_{\overline{t_1}|0\cdot12} = \bar{s}_{\overline{10-t_1}|0\cdot10}$$

This equation must be solved numerically for t_1. By trial and interpolation we obtain $t_1 = 8\cdot481$ (to three decimal places) and hence $k = 183\,515$. Hence

$$P = \frac{200\,000 + k}{10\,000}$$

$$= £38\cdot35$$

is the minimum price per tonne of coal which would make the project profitable.

In some cases the investor must finance his investment or business project by means of a fixed-term loan without an early repayment option. In these circumstances he cannot use a positive cash flow to repay the loan gradually, but must accumulate this money at the rate of interest applicable on lending, i.e. j_2. We now reconsider the coal-mining venture of example 5.4.1 under these conditions.

Example 5.4.2

Consider again the open-cast coal project of example 5.4.1, but now suppose that the bank loan of £1 000 000 is for a fixed term of ten years with no early repayment options and that interest is payable continuously. What is the minimum price per tonne of coal which would make the project viable?

Solution

Let P' be the break-even price per tonne of coal under these conditions. Note that the annual rate of payment of loan interest is $1\,000\,000\,\delta_{0.12}$, i.e. 113 329. Hence, after paying loan interest, the mining company will have a continuous cash inflow at the rate of

$$k' = 10\,000\,P' - 200\,000 - 113\,329$$
$$= 10\,000\,P' - 313\,329 \tag{1}$$

per annum for ten years, at which time it must repay the bank loan and the cost of restoring the land. Hence, accumulating the cash inflow at 10% per annum, we must have

$$k'\bar{s}_{\overline{10}|} = 1\,300\,000 \qquad \text{at } 10\%$$

from which we obtain $k' = 77\,744$ and (from equation 1) $P' = 39 \cdot 11$. The minimum viable coal price is thus £39·11 per tonne.

As is clear by general reasoning, the new condition on the bank loan (i.e. that it cannot be repaid early) is disadvantageous to the mining company because it must accumulate cash at the (lower) rate of interest applicable to lending rather than pay off bank indebtedness (on which interest is chargeable at 12% per annum).

In many practical problems the net cash flow changes sign only once, this change being from negative to positive. In these circumstances the balance in the investor's account will change from negative to positive at a unique time t_1, or it will always be negative, in which case the project is not viable. If this time t_1 exists, it is referred to as the *discounted payback period* (DPP). It is the smallest value of t such that $A(t) \geqslant 0$, where

$$A(t) = \sum_{s \leqslant t} c_s (1 + j_1)^{t-s} + \int_0^t \rho(s)(1 + j_1)^{t-s} ds \tag{5.4.1}$$

Note that t_1 does not depend on j_2 but only on j_1, the rate of interest applicable to the investor's borrowings. Suppose that the project ends at time T. If $A(T) < 0$ (or, equivalently, if $NPV(j_1) < 0$) the project has no discounted payback period and is not profitable. If the project is viable (i.e.

there is a discounted payback period t_1) the *accumulated profit* when the project ends at time T is

$$P = A(t_1)(1 + j_2)^{T - t_1} + \sum_{t > t_1} c_t(1 + j_2)^{T - t}$$
$$+ \int_{t_1}^{T} \rho(t)(1 + j_2)^{T - t} dt \qquad (5.4.2)$$

This follows since the net cash flow is accumulated at rate j_2 after the discounted payback period has elapsed.

If interest is ignored in formula 5.4.1 (i.e. if we put $j_1 = 0$), the resulting period is called the *payback period*. As is shown in example 5.4.3, however, its use instead of the discounted payback period often leads to erroneous results and is therefore not to be recommended.

The discounted payback period is often employed when considering a single investment of C, say, in return for a series of payments each of R, say, payable annually in arrear for n years. The discounted payback period t_1 years is clearly the smallest integer t such that $A^*(t) \geq 0$, where

$$A^*(t) = -C(1 + j_1)^t + Rs_{\overline{t}} \qquad \text{at rate } j_1 \qquad (5.4.3)$$

i.e. the smallest integer t such that

$$Ra_{\overline{t}} \geq C \qquad \text{at rate } j_1 \qquad (5.4.4)$$

The project is therefore viable if $t_1 \leq n$, in which case the accumulated profit after n years is clearly

$$P = A^*(t_1)(1 + j_2)^{n - t_1} + Rs_{\overline{n - t_1}} \qquad \text{at rate } j_2 \qquad (5.4.5)$$

Example 5.4.3

An investment of £100 000 will produce an annuity of £10 500 annually in arrear for 25 years. Find the discounted payback period when the interest rate on borrowed money is 9% per annum. Find also the accumulated profit after 25 years if money may be invested at 7% per annum.

Solution

By condition 5.4.4, the discounted payback period is the smallest integer t such that

$$10\,500a_{\overline{t}} \geq 100\,000 \qquad \text{at } 9\%$$

From compound interest tables we see that the discounted payback period is 23 years. The accumulated profit after 25 years is (from equations 5.4.3 and 5.4.5)

$$P = [-100\,000(1 \cdot 09)^{23} + 10\,500s_{\overline{23}|0 \cdot 09}](1 \cdot 07)^2 + 10\,500s_{\overline{2}|0 \cdot 07}$$
$$= 26\,656$$

Note If interest is ignored, the payback period (calculated from formula 5.4.3 with $j_1 = 0$) is ten years. This is much less than the true discounted period.

Likewise £(25 × 10 500 − 100 000), i.e. £162 500, is far too high an estimate of the final profit.

If the rates (or forces) of interest on borrowing and/or lending are assumed to vary with time, one may find the accumulation of the net cash flow by formulae given in chapter 2 or by techniques to be developed in chapter 6. The determination of the net cash flow and its accumulation at any future time may be facilitated by the use of a computer. It is usual in many computer-assisted calculations to consider the net cash flow and its accumulation on a yearly basis. The resulting analysis may be easily understood and interpreted by those responsible for making investment decisions.

5.5 The effects of inflation

Consider the simplest situation, in which an investor can lend and borrow money at the same rate of interest i_1. In certain economic conditions the investor may assume that some or all elements of the future cash flows should incorporate allowances for inflation (i.e. increases in prices and wages). The extent to which the various items in the cash flow are subject to inflation may differ. For example, wages may increase more rapidly than the prices of certain goods, or vice versa, and some items (such as the income from rent-controlled property) may not rise at all, even in highly inflationary conditions.

The case when *all* items of cash flow are subject to the same rate of escalation e per time unit is of special interest. In this case we find or estimate c_t^e and $\rho^e(t)$, the net cash flow and the net rate of cash flow allowing for escalation at rate e per unit time, by the formulae

$$c_t^e = (1 + e)^t c_t, \tag{5.5.1}$$

$$\rho^e(t) = (1 + e)^t \rho(t) \tag{5.5.2}$$

where c_t and $\rho(t)$ are estimates of the net cash flow and the net rate of cash flow respectively at time t without any allowance for inflation. It follows that, with allowance for inflation at rate e per unit time, the net present value of the investment or business project at rate of interest i is

$$NPV_e(i) = \sum c_t(1 + e)^t(1 + i)^{-t} + \int_0^\infty \rho(t)(1 + e)^t(1 + i)^{-t}dt$$

$$= \sum c_t(1 + j)^{-t} + \int_0^\infty \rho(t)(1 + j)^{-t}dt \tag{5.5.3}$$

where

$$1 + j = \frac{1 + i}{1 + e}$$

or

$$j = \frac{i - e}{1 + e} \qquad (5.5.4)$$

If e is not too large, one sometimes uses the approximation

$$j \simeq i - e \qquad (5.5.5)$$

Combining equations 5.5.3 and 5.5.4, we have

$$NPV_e(i) = NPV_0 \left(\frac{i - e}{1 + e} \right) \qquad (5.5.6)$$

where NPV_0 is the net present value function with no allowance for inflation. It follows that, with inflation at rate e per unit time, the yield (or internal rate of return) i_0^e of the project is such that

$$\frac{i_0^e - e}{1 + e} = i_0$$

where i_0 is the corresponding yield if there were no inflation (see section 7.11 where *real* yields on investments are discussed). This means that

$$i_0^e = i_0(1 + e) + e \qquad (5.5.7)$$

or, if e is small,

$$i_0^e \simeq i_0 + e \qquad (5.5.8)$$

These results are of considerable practical importance, because projects which are apparently unprofitable when rates of interest are high may become highly profitable when even a modest allowance is made for inflation. It is, however, true that in many ventures the positive cash flow generated in the early years of the venture is insufficient to pay bank interest, so recourse must be had to further borrowing (unless the investor has adequate funds of his own). This in itself does not undermine the profitability of the project, but the investor would require the agreement of his lending institution before further loans could be obtained and this might cause difficulties in practice.

Calculations of this kind underlie 'low-start' mortgage schemes, which are discussed in example 5.5.2.

Example 5.5.1

A sheep farmer is considering increasing the size of his flock from 3000 to 4500 head. According to his calculations (which are all based on 1985 prices) this would give an additional annual profit, after allowing for replacement of sheep but not for financing costs, of £1·91 per extra sheep. The initial costs associated with increase of size of the flock are as follows:

Purchase price of 1500 sheep	4 292
Fencing	1 850
Reseeding 400 acres	8 000
Total	£14 142

In addition there is an annual cost of £3·27 per reseeded acre for fertilizer, i.e. £1308 per annum. The fencing may be assumed to last for 20 years and the resale value (after 20 years) of the extra sheep is taken as zero.

(a) Assuming that there will be no inflation and that the net profit each year will be received at the end of the year, find the internal rate of return for the project.
(b) What uniform annual rate of inflation will make the project viable, if the farmer may borrow and invest money at 10% per annum interest?

Solution

(a) With no inflation, the net income each year from the project is £(1·91 × 1500 − 1308) = £1557, so that the net cash flows associated with the project are

$$c_0 = -14\,142$$

Hence

$$c_1 = c_2 = \cdots = c_{20} = 1557$$

$$NPV_0(i) = -14\,142 + 1557 a_{\overline{20}|}$$

from which it follows that the internal rate of return is

$$i_0 \simeq 9.07\%$$

(b) Now suppose that the sheep farmer may borrow and lend money at the fixed rate of 10% per annum. Assuming that all prices and costs escalate at a compound annual rate e, we obtain the net present value function

$$NPV_e(i) = NPV_0\left(\frac{i-e}{1+e}\right)$$

and the internal rate of return per annum becomes (see equation 5.5.7)

$$i_0^* = 0.0907(1+e) + e$$

which is greater than 0·1 if $e > 0.0085$. It follows that, if the annual rate of inflation is greater than 0·85%, i_0^* will exceed 10% and the venture will be profitable.

Note that in this example the interest on the initial loan of £14 142 will be

£1 414·20 in the first year. Since this is less than £1557 $(1 + e)$, the sheep farmer will not have to borrow more money at any time. This is not always the case, as shown in the following example.

Example 5.5.2

In view of recent high interest rates a building society has decided to offer borrowers a 'low-start' repayment plan. The borrower of an initial sum of L_0 will pay $P(1 + e)^{t-1}$ at the end of year t $(t = 1, 2, 3, \ldots, n)$, where n is the term of the loan and e is a fixed rate of escalation. The payments are used to pay interest at rate i on the outstanding loan and, if any money remains, to reduce the loan. If the interest on the outstanding loan exceeds the year's payment the excess is added to the capital outstanding. The building society makes no allowance for expenses or for taxation.

(a) Show that, if L_t denotes the amount of loan outstanding immediately after the repayment due at time t has been made, then

$$L_t = L_0(1 + i)^t - P(1 + i)^{t-1}\ddot{a}_{\overline{t}|j}$$

where

$$j = \frac{i - e}{1 + e}$$

Hence show that

$$P = \frac{(1 + i)L_0}{\ddot{a}_{\overline{n}|j}}$$

(b) Given that $i = 0·12$ and $e = 0·1$, find, in respect of an initial loan of £10 000 repayable over 20 years,
   (i) The payment P due at the end of the first year;
   (ii) The maximum capital outstanding during the course of the loan and the year in which the capital outstanding begins to decrease.

Solution

(a) The loan outstanding at the end of a given year is simply the accumulation (at rate i) of the amount lent less the accumulation of the repayments made (see equation 3.7.4). Hence

$$L_t = L_0(1 + i)^t - \sum_{r=1}^{t} P(1 + e)^{r-1}(1 + i)^{t-r}$$

$$= L_0(1 + i)^t - P(1 + i)^{t-1} \sum_{r=1}^{t} \left(\frac{1 + i}{1 + e}\right)^{-(r-1)}$$

$$= L_0(1 + i)^t - P(1 + i)^{t-1} \sum_{r=1}^{t} (1 + j)^{-(r-1)}$$

$$= L_0(1 + i)^t - P(1 + i)^{t-1}\ddot{a}_{\overline{t}|j}$$

as required.
Since the term of the loan is n years, we must have $L_n = 0$. Hence

$$L_0(1 + i)^n - P(1 + i)^{n-1}\ddot{a}_{\overline{n}|j} = 0$$

so

$$P = \frac{L_0(1 + i)}{\ddot{a}_{\overline{n}| j}}$$

as required.

(b) (i) Using the above results, we have

$$P = \frac{10\,000 \times 1\cdot12}{\ddot{a}_{\overline{20}| j}} \qquad \text{(where } j = 0\cdot02/1\cdot1\text{)}$$

$$= £660\cdot98$$

(ii) To determine the maximum loan outstanding we compute L_t $(t = 1, 2, \ldots, 20)$ by the recurrence relation

$$L_{t+1} = 1\cdot12 L_t - 660\cdot98 \times 1\cdot1^t$$

Since $L_0 = 10\,000$, this gives the schedule in table 5.5.1, in which the payments due at the end of each year are shown. We see that L_t reaches a maximum of £14 134 at time 10, and hence the year in which the capital outstanding begins to decrease is the 11th.

Table 5.5.1 *Payment schedule*

Time t (years)	Payment at time t (£)	Loan outstanding after payment (£)
1	660·98	10 539·02
2	727·08	11 076·63
3	799·78	11 606·04
4	879·76	12 119·00
5	967·74	12 605·55
6	1064·51	13 053·70
7	1170·96	13 449·18
8	1288·06	13 775·02
9	1416·87	14 011·16
10	1558·55	14 133·95
11	1714·41	14 115·61
12	1885·85	13 923·64
13	2074·43	13 520·04
14	2281·88	12 860·57
15	2510·06	11 893·77
16	2761·07	10 559·96
17	3037·18	8 789·98
18	3340·89	6 503·88
19	3674·98	3 609·36
20	4042·48	0·00

5.6 The yield on a fund

Consider a financial institution (for example, a life office or pension fund) which had funds valued at $F(t)$ at time t years $(t_1 \leqslant t \leqslant t_2)$. The yield per annum obtained by the financial institution in the period from time t_1 to time t_2 may be found by solving the equation of value at time t_2, namely

$$F(t_1)(1 + i)^{t_2 - t_1} + \sum c_t(1 + i)^{t_2 - t} + \int_{t_1}^{t_2} \rho(t)(1 + i)^{t_2 - t} dt = F(t_2) \quad (5.6.1)$$

where c_t and $\rho(t)$ denote the net cash flow and the rate of net cash flow in respect of *new money* at time t $(t_1 \leqslant t \leqslant t_2)$. We first consider a tax-free fund, leaving the treatment of taxed funds until later. If c_t and $\rho(t)$ are known, equation 5.6.1 may be solved numerically by the usual methods.

We shall assume for the rest of this section that $t_2 = t_1 + 1$. (Thus the period with which we are concerned is one year.) In many practical applications the only information which is readily available consists of the values of $F(t_1)$, $F(t_2)$, and I, the interest income and capital gains received during the year. We shall find an approximate solution of equation 5.6.1 in terms of these quantities.

Clearly,

$$F(t_2) = F(t_1) + I + N \tag{5.6.2}$$

where N is the new money received during the year, i.e. the excess of income, *excluding interest income and capital gains*, over outgo. But N is also equal to the total net cash flow during the year, i.e.

$$N = \sum c_t + \int_{t_1}^{t_2} \rho(t) dt \tag{5.6.3}$$

If we assume that, for $t_1 \leqslant t \leqslant t_2$, $c_t = 0$ and $\rho(t) = \rho$ (a constant), then the new money is received at a constant rate over the year. This means that $N = \rho$ and hence (from equation 5.6.2) that

$$\rho = F(t_2) - F(t_1) - I$$

Equation 5.6.1 now becomes

$$F(t_1)(1 + i) + [F(t_2) - F(t_1) - I]\bar{s}_{\overline{1}|} = F(t_2) \qquad \text{at rate } i \quad (5.6.4)$$

Since $\bar{s}_{\overline{1}|} \simeq 1 + (i/2)$ for small values of i, we have

$$F(t_1)(1 + i) + [F(t_2) - F(t_1) - I]\left[1 + \frac{i}{2}\right] \simeq F(t_2)$$

and hence

$$i \simeq \frac{2I}{F(t_1) + F(t_2) - I} \tag{5.6.5}$$

This well-known formula was derived by G. F. Hardy in an article in the *Transactions of the Actuarial Society of Edinburgh*, December 1890, reprinted in *Transactions of the Faculty of Actuaries*, Vol. 9, pp. 61–2. The reader should show that formula 5.6.5 holds exactly if new money is received in two equal instalments at the beginning and end of the year.

Example 5.6.1

A friendly society, which is exempt from taxation, had the following revenue account for 1982:

	£		£
Funds at 1 January 1982	501 050	Benefit payments	31 459
Investment income	41 838	Expenses	5 541
Contribution income	19 250	Funds at 31 December 1982	525 138
	562 138		562 138

Estimate the yield per annum obtained by the society on its investments in 1982.

Solution

$I = 41\ 838$ (there being no capital gains or losses) and the yield per annum is

$$i \simeq \frac{2 \times 41\ 838}{501\ 050 + 525\ 138 - 41\ 838} = 0.0850 \quad \text{or} \quad 8.5\%$$

An alternative approach to the determination of the yield on a fund is the following, which is also due to G. F. Hardy.

Let $\delta(t)$ denote the force of interest on the fund at time $t (t_1 \leqslant t \leqslant t_2$, where $t_2 = t_1 + 1)$. By a limiting argument similar to that of section 2.9,

$$I = \int_{t_1}^{t_2} F(t)\delta(t)\mathrm{d}t \tag{5.6.6}$$

The mean force of interest obtained by the fund in the period from time t_1 to time t_2 may be defined as the weighted average

$$\delta = \frac{\displaystyle\int_{t_1}^{t_2} F(t)\delta(t)\mathrm{d}t}{\displaystyle\int_{t_1}^{t_2} F(t)\mathrm{d}t} = \frac{I}{\displaystyle\int_{t_1}^{t_2} F(t)\mathrm{d}t} \tag{5.6.7}$$

and the yield on the fund is

$$i = e^{\delta} - 1 \tag{5.6.8}$$

Now note that for small δ

$$e^{\delta} - 1 \simeq \frac{\delta}{1 - (\delta/2)}$$

(as may be shown by expanding each side in Maclaurin series), so

$$i \simeq \frac{\delta}{1 - (\delta/2)} = \frac{2I}{2\displaystyle\int_{t_1}^{t_2} F(t)\,dt - I} \tag{5.6.9}$$

Using the trapezium rule to estimate the value of the integral in the last equation, we obtain

$$i \simeq \frac{2I}{F(t_1) + F(t_2) - I}$$

as in equation 5.6.5.

A more accurate value of δ, and hence i, may be obtained if more is known about $F(t)$. For example, if $F(t)$ is known for $t = t_1, t_1 + \frac{1}{2}$ and $t_2 (= t_1 + 1)$, we may estimate $\int_{t_1}^{t_2} F(t)\,dt$ by Simpson's rule to obtain, from equation 5.6.7,

$$\delta \simeq \frac{6I}{F(t_1) + 4F\left[\dfrac{t_1 + t_2}{2}\right] + F(t_2)}$$

from which i may be found as $e^{\delta} - 1$.

If the value of the fund $F(t)$ has large discontinuities, it is preferable to divide the range of integration of $F(t)$ into sections before applying approximate integration formulae. This point is illustrated in the following example.

Example 5.6.2

A large pension fund had the following revenue account for 1980. The fund is not subject to tax and expenses of management are paid by the employer.

	£million		£million
Fund at 1 January 1980	150·8	Benefit payments	13·2
Contribution income	47·1		
Interest income	12·6		
Capital gains	nil	Fund at 1 January 1981	197·3
	210·5		210·5

The benefit payments and contribution income were fairly uniformly spread over the year except that on 30 November 1980 the employer made a special contribution to the fund of £25 million to provide certain additional benefits. Estimate the yield per annum obtained by the pension fund in 1980.

Solution

Measure time in years from 1 January 1980. An application of formula 5.6.5 gives

$$i \simeq \frac{2 \times 12 \cdot 6}{150 \cdot 8 + 197 \cdot 3 - 12 \cdot 6} = 7 \cdot 51\%$$

This formula assumes that $F(t)$ progresses fairly smoothly for $0 \leqslant t \leqslant 1$. But this is not the case here: there is a large inflow at time $t = 11/12$. A more accurate answer may be obtained by dividing the range $[0, 1]$ into two sections.

To do this we assume that $F(t)$ is linear (i.e. with a constant derivative) except for the 'jump' at $t = 11/12$. The progress of $F(t)$ is illustrated in Figure 5.6.1. Linear interpolation between $F(0)$ and $F(1) - 25$ shows that $F(11/12) = 170 \cdot 51$ (before receipt of the special payment) and hence

$$\int_0^1 F(t)dt = \int_0^{11/12} F(t)dt + \int_{11/12}^1 F(t)dt$$

$$\simeq \frac{11}{12}\left(\frac{150 \cdot 8 + 170 \cdot 51}{2}\right) + \frac{1}{12}\left(\frac{195 \cdot 51 + 197 \cdot 3}{2}\right)$$

$$= 163 \cdot 6$$

Thus $\delta \simeq 12 \cdot 6/163 \cdot 6 = 0 \cdot 07\,702$ and $i = e^\delta - 1 \simeq 0 \cdot 0801$. The estimated yield per annum is thus $8 \cdot 01\%$.

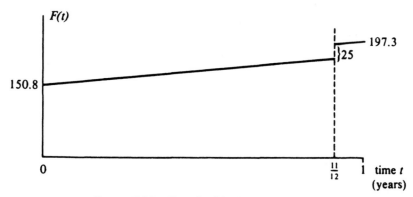

FIGURE 5.6.1   *Growth of fund in example 5.6.2*

Now consider the position of funds which are liable to tax. If tax is charged only on income and/or capital gains, it is perhaps desirable to exclude tax payments from the cash flow, to consider I as 'interest income and capital gains less taxes', and to proceed as before. The yield obtained in this manner is referred to as a *net* yield.

If, however, taxation is based on other measures (e.g. profits, or income less expenses) it is probably better to consider tax payments as 'expenses'. The yield so obtained is called a *gross* yield. In view of their rather complicated tax position, UK life funds usually calculate gross yields. Slightly different formulae may be used, depending on whether one assumes that taxes are paid (a) at the end of the year or (b) uniformly over the year. Let I be the gross interest income and capital gains received during the year, and let T be the tax payments made. In case (a), $F(t)$ falls abruptly by T at the end of the year, so

$$\int_{t_1}^{t_2} F(t)\,dt \simeq \frac{F(t_1) + [F(t_2) + T]}{2}$$

Formula 5.6.9 then gives

$$i \simeq \frac{2I}{F(t_1) + F(t_2) - I'} \tag{5.6.10}$$

where $I' = I - T$. If, on the other hand, case (b) applies, we have

$$\int_{t_1}^{t_2} F(t)\,dt \simeq \frac{F(t_1) + F(t_2)}{2}$$

and formula 5.6.9 gives

$$i \simeq \frac{2I}{F(t_1) + F(t_2) - I} \tag{5.6.11}$$

It makes little difference in practice whether one uses formula 5.6.10 or 5.6.11 for the gross yield. This is illustrated by the following example.

Example 5.6.3

A life office had the following revenue account for 1981:

	£ million		£ million
Funds at 1 January 1981	100	Benefits paid	8
Premium income	15	Expenses	2
Interest income	6	Taxation	1
Capital appreciation	3	Funds at 31 December 1981	113
	124		124

Estimate the gross yield per annum on the office's funds in 1981.

Solution

Measuring time in years from 1 January 1981, we have

$$F(0) = 100, \quad F(1) = 113, \quad I = 6 + 3 = 9, \quad T = 1$$

Hence, by formula 5.6.10,

$$i \simeq \frac{2 \times 9}{100 + 113 - 8} = 0{\cdot}0878 \quad \text{or} \quad 8{\cdot}78\%$$

and, by formula 5.6.11,

$$i \simeq \frac{2 \times 9}{100 + 113 - 9} = 0{\cdot}0882 \quad \text{or} \quad 8{\cdot}82\%$$

Note The size of any capital gains or losses displayed in the accounts may depend on the accounting practice of the institution. We shall not go further into this matter here.

5.7 Measurement of investment performance

It is frequently the case in practice that one wishes to measure the investment performance of a given fund (e.g. a pension fund or life assurance fund) over a period of several years. Suppose that the period in question is from time t_0 to time t_n (where time is measured in years). Suppose further that

$$t_0 < t_1 < t_2 < \cdots < t_n$$

so that the overall period $[t_0, t_n]$ may be divided into n specified subintervals $[t_0, t_1]$, $[t_1, t_2], \ldots, [t_{n-1}, t_n]$. (In practice these subintervals are often periods of length one year, but this is not always so.)

Using the methods of section 5.6, we may calculate the annual yield obtained on the fund in each of these subintervals. For $r = 1, 2, \ldots, n$, let i_r denote the yield per annum on the fund over the period $[t_{r-1}, t_r]$.

Consider now an investment of 1 at time t_0. If, for $r = 1, 2, \ldots, n$, this investment had earned interest at an effective annual rate i_r over the period $[t_{r-1}, t_r]$, the accumulated amount of the investment at time t_n would have been

$$(1 + i_1)^{t_1 - t_0}(1 + i_2)^{t_2 - t_1} \cdots (1 + i_n)^{t_n - t_{n-1}}$$

The constant effective annual rate of interest which would have given the same accumulation over the period t_0 to t_n is i, where

$$(1 + i)^{t_n - t_0} = (1 + i_1)^{t_1 - t_0}(1 + i_2)^{t_2 - t_1} \cdots (1 + i_n)^{t_n - t_{n-1}}$$

from which it follows that

$$i = [(1 + i_1)^{t_1 - t_0}(1 + i_2)^{t_2 - t_1} \cdots (1 + i_n)^{t_n - t_{n-1}}]^{1/(t_n - t_0)} - 1 \quad (5.7.1)$$

This value of i is generally known as the *linked internal rate of return* per annum on the fund for the period t_0 to t_n (with reference to the subdivisions defined by the intermediate times $t_1, t_2, \ldots, t_{n-1}$). In an obvious sense the linked internal rate of return is an average of the yields obtained in each subinterval. Note, however, that this average takes no account of changes in the size of the fund from period to period. It also depends on the particular subdivision of the entire period.

To allow for changes in the size of the fund with time we may also calculate the yield per annum on the fund over the entire period t_0 to t_n, using the equation of value 5.6.1. This yield is generally referred to as the *money-weighted rate of return* per annum on the fund for the period. It does not depend upon any particular subdivision of the entire period and generally gives greater weight to the yields pertaining to the times when the fund is largest.

A third index, which is often used in practice, is the *time-weighted rate of return* per annum over a given period. The time-weighted rate of return does not depend upon any particular subdivision of the period, but does require that the precise times of all cash flows of *new money* into and out of the fund over the period be known. Suppose that the cash flows of *new money*, strictly within the period, occur at times t_1, $t_2,\ldots$, t_{n-1}, where $t_1 < \ldots < t_{n-1}$. Let the given period be $[t_0, t_n]$. There may also be cash flows of new money at times t_0 and t_n. For $0 \leqslant r \leqslant n$, let V_r denote the amount of the fund at time t_r, *after* receipt of all interest and capital gains due up to and including time t_r, but *before* payment to or withdrawal from the fund of any new money at that time. Let c_r $(0 \leqslant r \leqslant n)$ denote the *net* inflow of new money to the fund at time t_r. (Thus c_r is the amount of new investment in the fund at time t_r less the amount of money withdrawn from the fund at that time.) Hence $V_r + c_r$ is the amount of the fund at time t_r *after* new money has been received or paid at that time.

For $0 \leqslant r < n$, define

$$R_r = \frac{V_{r+1}}{V_r + c_r}. \tag{5.7.2}$$

That is, R_r is the 'accumulation factor' for the fund over the time interval $[t_r, t_{r+1}]$. Let

$$R = R_0 R_1 R_2 \ldots R_{n-1} \tag{5.7.3}$$

and define i by the equation

$$(1 + i)^{t_n - t_0} = R,$$

i.e.

$$i = R^{1/(t_n - t_0)} - 1. \tag{5.7.4}$$

This value of i is the *time-weighted rate of return* per annum for the period $[t_0, t_n]$.

In some cases the linked internal rate of return (with reference to a particular subdivision), the money-weighted rate of return, and the time-weighted rate of return do not differ significantly. In many situations, however (especially if the amount of the fund varies greatly over the relevant period), these rates of return may differ considerably.

Unitized funds

Unitized funds are now commonly found in practice. In describing such funds we shall initially ignore expenses (such as dealing and administration costs) and taxation, although in practice due allowance must be made for these factors.

Suppose that at time t_0 a number of individuals subscribe various amounts totalling £$N(t_0)$ to a fund. This money is promptly invested in securities (ordinary shares, loan stocks, deposits, etc.). At time t_0, the value of the invested amounts is £$V(t_0) = $ £$N(t_0)$. The fund may be considered to consist of $N(t_0)$ 'units', each of value £$u(t_0) = $ £1. The number of units owned by any one subscriber is equal to the amount of his subscription.

At some later time, at time t_1 say, the value of the invested assets (including any income and capital gains which have been received since time t_0) is £$V(t_1)$. If this amount is divided among the $N(t_0)$ units, the value of each unit at time t_1 may be calculated as

$$u(t_1) = \frac{V(t_1)}{N(t_0)}.$$

Suppose now that at this time t_1, one of the original subscribers, who owns n units, wishes to withdraw his investment. The fraction of the total fund owned by this investor is $n/(N(t_0))$, and the value of his share at time t_1 is therefore

$$\frac{n}{N(t_0)} V(t_1) = nu(t_1).$$

If he 'surrenders' his n units, he will receive a cash payment of amount $nu(t_1)$. (The fund must realize sufficient investments at their market values to meet this amount, unless new subscribers can be found.)

There are now fewer units 'in force', the new number being

$$N(t_1) = N(t_0) - n.$$

The cash value of the fund after the surrender is

$$V(t_1) - \frac{n}{N(t_0)} V(t_1) = [N(t_0) - n] \frac{V(t_1)}{N(t_0)}$$

$$= N(t_1)u(t_1).$$

Assume also that, at time t_1, another subscriber wishes to add to his investment in the fund and that the amount he has available to invest is £X. In order to maintain the unit value at this time as £$u(t_1)$ it is convenient to 'create' and issue to this subscriber a further m units, where

$$m = \frac{X}{u(t_1)}.$$

This new money is, of course, invested immediately or, perhaps, used to pay those who surrender their units. After the surrender of units and the issue of further units to the new subscriber, the number of units in force is

$$N(t_1) + m = N(t_0) - n + m$$

and the total amount of the fund is

$$V(t_1) - \frac{n}{N(t_0)} V(t_1) + X = [N(t_0) - n] \frac{V(t_1)}{N(t_0)} + X$$

$$= [N(t_0) - n]u(t_1) + mu(t_1)$$

$$= [N(t_0) - n + m]u(t_1).$$

The above remarks are readily extended to further surrenders and purchases of units at subsequent times. Suppose that transactions are to take place at time t. Let $V(t_-)$ and $N(t_-)$ be the total market value of the fund and the total number of units in force respectively at time t, just *before* any transactions at that time. The value of the unit at time t is therefore $u(t) = V(t_-)/N(t_-)$.

Let $V(t_+)$ and $N(t_+)$ be the total market value of the fund and the total number of units in force respectively at time t, just *after* the transactions at that time.

Let

$$m(t) = N(t_+) - N(t_-).$$

Thus $m(t)$ is the net number of units created at time t (i.e. the number of new units created less the number of units surrendered).

Note that the net increase in the amount of the fund as a result of the dealings at time t is

$$V(t_+) - V(t_-) = N(t_+)u(t) - N(t_-)u(t)$$
$$= [N(t_+) - N(t_-)]u(t)$$
$$= m(t)u(t).$$

In practice, the managers of unit trusts and similar unitized funds calculate unit prices no more frequently than once every working day, or in some cases only weekly or even monthly, allowing purchases and sales of units only on 'subscription days' (the days when the unit price is calculated.) Many unit trusts accumulate income and capital gains received on investments in a cash account and distribute the income (in proportion to the number of units held) every six months or so. At such distribution dates the unit price falls by the amount of the distribution per unit. Other funds automatically reinvest income and make no distribution to unit holders. These two types of fund are generally known as *distribution funds* and *accumulation funds* respectively. Some funds have both types of unit, and maintain two unit prices, which bear a constant ratio to each other between distribution dates. The ratio is equitably adjusted when a distribution occurs.

In practice, two unit prices are maintained in every fund. The 'offer' price is used when a subscriber wishes to buy units and the (lower) 'bid' price is used when a subscriber wishes to sell units. The difference between the prices reflects an allowance for dealing expenses and management costs.

The return on a unit during a given time period (t_0, t_1) is $u(t_1)/u(t_0)$, and this is also the return on the whole fund if the number of units does not change during the period. The time-weighted rate of return per annum, i, over a series of consecutive time intervals (t_r, t_{r+1}), $r = 0, 1, \ldots, n-1$, within each of which the number of units does not change, is thus found from the equation

$$(1+i)^{t_n - t_0} = \frac{u(t_1)}{u(t_0)} \frac{u(t_2)}{u(t_1)} \cdots \frac{u(t_n)}{u(t_{n-1})} = \frac{u(t_n)}{u(t_0)}. \qquad (5.7.5)$$

(In this last equation the factor $u(t_{r+1})/u(t_r)$ corresponds to the factor R_r in equation 5.7.3). Thus the time-weighted rate of return per annum on the fund, or on the investment of any subscriber, over any period can be calculated from the unit prices at the beginning and end of the period.

The money-weighted rate of return for any subscriber depends, of course, on the number of units he purchased and sold over the period and on the times at which his transactions took place.

Example 5.7.1

The accounts of a certain friendly society, which is not subject to tax, show the following information:

Calendar year	Fund on 1 January	Interest income received in year
1980	86 932	7703
1981	91 781	8189
1982	96 316	8613
1983	100 837	9256
1984	105 054	9371
1985	109 688	

Assuming that the accounts do not reflect any capital gains or losses, and that over each calendar year benefits were paid and new money was received at a constant rate, find (a) the linked internal rate of return per annum for the five-year period 1 January 1980 to 1 January 1985 (based on calendar year subdivisions of the quinquennium) and (b) the money-weighted rate of return per annum for the period.

Solution

Since for each calendar year we are assuming a constant rate of payment of new money and benefits, we may estimate the yield on the fund each year by equation 5.6.5. Thus, for example, the annual yield for 1980 is approximately

$$\frac{2 \times 7703}{86\,932 + 91\,781 - 7703} = 0.090\,09 \quad \text{or} \quad 9.009\%$$

Let N be the net annual rate of inflow of new money over the year 1980. Then clearly

$$86\,932 + 7703 + N = 91\,781$$

from which it follows that $N = -2854$.

More generally, for $r = 1, 2, \ldots, 5$ let i_r be the annual yield on the fund for the calendar year $(1979 + r)$ and let N_r be the net annual rate of inflow of new money received for the same year. The reader should verify the figures in the following table:

r	Calendar year	Percentage yield per annum $(100i_r)$	Net annual rate of inflow of money $N_r(£)$
1	1980	9.009	−2854
2	1981	9.104	−3654
3	1982	9.137	−4092
4	1983	9.414	−5039
5	1984	9.126	−4737

From equation 5.7.1, we calculate the linked internal rate of return per annum as

$$\left\{ \prod_{r=1}^{5} (1 + i_r) \right\}^{1/5} - 1 = 0.091\,58$$

or, say, 9·16% per annum.

The money-weighted rate of return is found by solving the equation of value 5.6.1, which in this case is

$$86\,932(1 + i)^5 - [2854(1 + i)^4 + 3654(1 + i)^3 + 4092(1 + i)^2$$
$$+ 5039(1 + i) + 4737]\bar{s}_{\overline{1}|} - 109\,688 = 0$$

This gives $i = 0.091\,50$ or, say, 9.15% per annum. It will be seen that in this case the linked internal and money-weighted rates of return are almost the same.

Example 5.7.2

On 15 November in each of the years 1980 to 1984 a man invested £1000 in a unitized fund. Income is retained in the fund and used to increase the value of units. The man sold his entire holding on 15 November 1985.

Ignoring expenses and taxation and given the information in the following table relating to the value of the unit, find the time-weighted rate of return per annum on the fund over the period from 15 November 1980 to 15 November 1985. Find also the investor's money-weighted rate of return per annum. (Assume that investors may buy fractions of a unit.)

Year	*1980*	*1981*	*1982*	*1983*	*1984*	*1985*
Value of unit on 15 November (£)	1·90	2·25	2·72	2·68	2·96	3·20

Solution

Measuring time in years from 15 November 1980, we let U_r be the value of the unit at time r.

By equation 5.7.5 the time-weighted rate of return per annum, i, is given by the equation

$$(1 + i)^5 = \frac{U_5}{U_0} = \frac{3·20}{1·90}$$

from which it follows that

$$i = 0.1099, \quad \text{or} \quad 10.99\% \text{ per annum.}$$

It should be noted that the time-weighted rate of return depends only on the value of the unit at the start and end of the five-year period. It does not depend on the unit value at intermediate times.

To find the money-weighted rate of return, we must solve the appropriate equation of value. Note that the £1000 invested at time r ($r = 0, 1, \ldots, 4$) purchased $1000/U_r$ units. The total number of units purchased was thus $\sum_{r=0}^{4} 1000/U_r$, and the value of the investor's holding at the time of sale was

$$\left(\sum_{r=0}^{4} \frac{1000}{U_r} \right) U_5 = \text{£}6558·01$$

The equation of value is thus

$$1000\bar{s}_{\overline{5}|} = 6558·01 \quad \text{at rate } i,$$

from which it follows that $i = 0.0918$. The money-weighted rate of return is thus 9.18% per annum.

Exercises

5.1   A company is considering two capital investment projects. Project *A* requires an immediate expenditure of £1 000 000 and will produce returns of £270 000 at the end of each of the next eight years. Project *B* requires an immediate investment of £1 200 000 together with further expenditure of £20 000 at the end of each of the first three years, and will produce returns of £1 350 000 at the end of each of the sixth, seventh, and eighth years.

(a) Calculate (to the nearest 0·1%) the internal rate of return per annum for each project.
(b) Find the net present value of each project on the basis of an effective annual interest rate of 15%.

Comment briefly on your answers.

5.2   A businessman is considering two projects. Each project requires an immediate initial outlay and a further outlay in one year's time. For each project, the return will be a level income, payable annually for seven years and commencing in five years' time. The projects are described in the following table:

Project	Initial outlay	Outlay after one year	Annual income
A	£160 000	£80 000	£60 000
B	£193 000	£80 000	£70 000

(a) Find the internal rate of return per annum for each project.
(b) As an alternative to undertaking project *B*, the businessman may undertake project *A* and, at the same time as making the initial investment for project *A*, the businessman could also purchase, with a single premium of £33 000, a level annuity payable annually for seven years and commencing in five years' time.
   Given that the amount of the annuity would be calculated on the basis of a fixed annual rate of interest, determine how large the rate must be in order that this combined transaction be more advantageous to the businessman than project *B*.

5.3   A business venture requires an initial investment of £10 000 and a further investment of £3000 in a year's time. The venture will produce an income of £500 in two years' time, £1000 in three years' time, £1500 in four years' time, and so on, the final income being £4000 in nine years' time.

(a) Find the internal rate of return for this project.
(b) An investor has no spare cash, but may borrow money at any time at the fixed interest rate of 5% per annum, all loans being repayable in whole or in part at any time at the borrower's discretion. The investor is considering borrowing to finance the venture described above. Should he undertake the venture and, if so, what will his profit be in nine years' time on the completed transaction?

5.4   You have recently inherited a small island, which may be used for sheep rearing, goat breeding, or forestry, and have calculated that the cash flows associated with these three projects will be as follows:

Sheep rearing   Initial cost: £20 000.
Annual income: £1100, payable annually in arrear for 20 years.
Sale price after 20 years: £20 000.

Goat breeding   Initial cost: £20 000.
Annual income: £900, payable annually in arrear for 20 years.
Sale price after 20 years: £25 000.

Forestry   Planting cost: £20 000.
Sales of mature trees after 20 years: £57 300.

(a) Calculate the internal rates of return of each of these projects (to the nearest 0·1%).

(b) You do not have the capital for any of these ventures, but you have been offered a bank loan of £20 000 at 5% per annum interest payable annually in arrear. This loan is repayable in 20 years' time, there being no early repayment option. Should you require further loans, they will be granted by the bank on the same interest basis and will be repayable at the same date as the original loan. If you have any money to invest after paying bank interest, you may invest it to secure interest at 4% per annum effective.

Which of the three projects will give the largest profit in 20 years' time?

5.5   An oil company has just purchased, for £2·7 million, an oil rig and drilling rights for three years in a section of the North Sea. The rig will be operated for three years at an annual cost, payable at the end of each year, of £200 000 and at the end of the third year the rig will be sold for £100 000. At most one oil strike can occur in any one year and the probabilities of striking oil in years one, two, and three are 0·25, 0·2, and 0·1 respectively, independent of any previous strikes. The company assumes that there will be a continuous output from each strike at the rate of 50 000 barrels per year for ten years, commencing at the end of the year of the strike.

(a) Assuming a constant oil price of £20 per barrel, find the effective annual rate of interest at which the present value of the company's outlays equals the expected value of the present value of its income from the project.

(b) Assuming that oil is struck for the first time in the third year and that the price of oil will be a constant £17 per barrel, find (to the nearest 0·1%) the internal rate of return for the project.

5.6   Two business projects, each of which takes two years to complete, produce the following income and expenditure:

Project A   Initial income of £1000.
After one year, expenditure of £2000.
After two years, income of £2000.

Project B   Initial expenditure of £4000.
After one year, income of £7000.
After two years, expenditure of £1500.

(a) For each project find the rates of interest, if any, which make the present value of the income equal to the present value of the expenditure.

(b) For what range of positive interest rates does the net present value of project *A* exceed the net present value of project *B*?

(c) An investor, who has no spare cash, can borrow money at any time at a fixed annual rate of interest for any desired term. He may also lend money for any desired term at this same rate.

    Which of these projects will be the more profitable, if the fixed annual rate of interest is (i) 20%, (ii) 25%? Calculate the accumulated profit after two years from each project in both cases.

5.7 (a) In return for an immediate outlay of £10 000 an investor will receive £6000 in one year's time and £6600 in two years' time. Find, to the nearest 0·1%, the internal rate of return on this investment.

(b) A person who has no spare cash may make the investment described in (a) by borrowing the initial outlay from a bank.

    (i) If bank loans are granted on the basis of an interest rate of 16% per annum and may be partially repaid at any time, should the person make the investment and, if so, what profit will he have made at the end of the completed transaction?

    (ii) Suppose instead that the bank requires that its loan be for a fixed two-year period and that interest of 16% per annum be paid annually in arrear. If the person will be able to earn interest of 13% per annum on any spare funds, should he make the investment and, if so, what profit will he have made when the transaction is completed?

5.8 A chemical company has agreed to supply a quantity of a certain compound for each of the next seven years. The company estimates that the cash flows associated with this project will be as follows:

Outlays   Initial outlay (cost of building plant): £500 000, payable immediately. Manufacturing costs: £400 000 p.a. payable annually for seven years, the first payment being made in one year's time. Disposal of wastes and demolition of plant: £250 000 p.a. payable annually for three years, the first payment being made in eight years' time.

Income   Sales of compound: £600 000 p.a. payable annually for seven years, the first income being received in one year's time.

The company proposes to finance the project by borrowing the entire initial cost from a bank, which charges interest on loans at 15% per annum. Partial repayment of the loan will be allowed at any time and the loan will be repaid by instalments as soon as possible from the profits on the manufacture and sale of the compound.

(a) After how many years will the bank loan be paid off?

(b) After repayment of the bank loan the company will accumulate its profits in a deposit account at the bank, which pays 12% per annum interest on deposits. The costs of disposal of chemical wastes and demolition of the plant will be met from this account. Calculate the anticipated balance in the company's deposit account when the project ends in ten years' time.

5.9 (Discounted payback period)
A businessman has decided to purchase a leasehold property for £80 000, with a further payment of £5000 for repairs in one year's time. The income associated with letting the property will be £10 000 per annum, payable continuously for 20 years commencing in two years' time.

(a) (i) Given that the venture will be financed by bank loans on the basis of an effective annual interest rate of 7% and that the loans may be repaid continuously, find the discounted payback period (DPP) for the project.
(ii) Given, further, that after the loans have been repaid the businessman will deposit all the available income in an account which will earn interest at 6% per annum effective, find the accumulated amount of the account in 22 years' time.

(b) Suppose that the bank loans may be repaid partially, but only at the end of each *complete* year, interest being paid annually in arrear, and that the businessman may still deposit money at any time for any term at an annual rate of interest of 6% effective. Find (i) the discounted payback period for the project, and (ii) the accumulated amount in the businessman's account in 22 years' time.

5.10 In return for an outlay of £1000 now and a further outlay of £600 after four years a business venture will provide a single receipt of £C in one year's time ($C > 0$).

(a) Measuring time in years, write down the equation

$$NPV(i) = 0$$

for the transaction.
(b) Letting $\alpha = 400(20/3)^{1/2} 5^{1/4} (\simeq 1544.39)$, show that
(i) If $C \geqslant 1600$ the equation has one positive root;
(ii) If $\alpha < C < 1600$ the equation has two positive roots;
(iii) If $C = \alpha$ the equation has one positive root;
(iv) If $C < \alpha$ the equation has no positive roots.
(c) Find all the positive roots of the equation when
(i) $C = 1600$;
(ii) $C = 1550$;
(iii) $C = \alpha$.

5.11 Over the period 1 January 1984 to 1 January 1985 the unit prices of two accumulation funds, a 'property fund' and an 'equity fund', took the values given in the following table:

Unit prices (£)
1984

Fund	1 Jan.	1 April	1 July	1 Oct.	1985 1 Jan.
Property	1·24	1·31	1·48	1·58	1·64
Equity	1·21	0·92	1·03	1·31	1·55

(a) Find the time-weighted rate of return for each fund for the year 1984.
(b) An investor bought units in the property fund on 1 January, 1 April,

1 July, and 1 October 1984, and sold his units on 1 January 1985. Find his yield on the completed transaction if
  (i) He bought the same number of units on each date;
  (ii) He invested the same sum of money on each date.
(c) As in (b), except that the investments were made in the equity fund.

Comment on your answers.
  (Ignore expenses and taxation. You should assume that investors may buy fractional parts of a unit.)

5.12 In a particular accumulation fund income is retained and used to increase the value of the fund unit. The 'middle price' of the unit on 1 April in each of the years 1979 to 1985 is given in the following table:

Year	1979	1980	1981	1982	1983	1984	1985
Middle price of unit on 1 April (£)	1·86	2·11	2·55	2·49	2·88	3·18	3·52

(a) On the basis of the above prices and ignoring taxation and expenses, find
  (i) The time-weighted rate of return for the fund over the period 1 April 1979 to 1 April 1985.
  (ii) The yield obtained by an investor who purchased 200 units on 1 April in each year from 1979 to 1984 inclusive and who sold his holding on 1 April 1985.
  (iii) The yield obtained by a person who invested £500 in the fund on 1 April each year from 1979 to 1984 inclusive and who sold back his holding to the fund managers on 1 April 1985. (You should assume that investors may purchase fractional parts of a unit.)
(b) Suppose that, in order to allow for expenses, the fund's managers sell units at a price which is 2% above the published middle price and that they buy back units at a price which is 2% below the middle price. On this basis find revised answers to (ii) and (iii) of (a).

CHAPTER SIX

CAPITAL REDEMPTION POLICIES

6.1 Introduction and premium calculations

A *capital redemption policy* is a contract which, in return for a single payment or a series of payments of stated amount, provides a specified sum of money at the end of a fixed period. The payments made by the policyholder are known as *premiums* and the sum payable at the end of the fixed period is called the *sum assured*. The date on which the sum assured is payable is called the *maturity date*. (At one time such policies were fairly common, but they are now somewhat rare in the United Kingdom. Nevertheless, a study of these policies is a useful introduction to some of the concepts of life assurance policies.)

The premiums paid by the policyholder are invested by the insurance company to produce the sum assured at the maturity date. In calculating the premiums to be charged, the company must make (a) appropriate assumptions about the rate (or rates) of interest (net of tax) at which it will be able to invest the premiums received over the duration of the policy and (b) due allowance for the expenses which it expects to incur in relation to the policy. These two requirements are obviously of paramount importance. For example, if the term of the policy is long and interest rates are relatively high at the time the policy is effected, it may be prudent for the company to assume that there will be a fall in interest rates over the policy's duration. In sections 6.5–6.7 we discuss three possible ways in which allowance may be made for variations in the rate of interest over the term of a policy. Even if it is not thought necessary to anticipate a change in interest rates, it may be desirable to allow for a gradual increase in expenses as a result of inflation.

The essential requirement is that at the maturity date the accumulated amount of the premiums received, less any expenses incurred, should suffice to pay the sum assured.

An explicit profit margin may also be required, but it is more usual for the issuing office to charge *implicit* profit margins, by making rather conservative assumptions about interest and expenses. If there is no explicit profit margin, the *equation of value* for the policy is

$$\begin{bmatrix} \text{value of premiums to be} \\ \text{received by the company} \end{bmatrix} = \begin{bmatrix} \text{value of benefits} \\ \text{payable by the} \\ \text{company} \end{bmatrix}$$

$$+ \begin{bmatrix} \text{value of expenses,} \\ \text{associated with the} \\ \text{policy, incurred by} \\ \text{the company} \end{bmatrix} \quad (6.1.1)$$

The equation of value may, of course, be taken at any time, but the most convenient date is usually either the date of issue or the date of maturity of the policy. It is often convenient to write the equation of value 6.1.1 in the form

Value of premiums *less* value of expenses = value of benefits   (6.1.2)

Expenses may generally be classified as *initial* or *renewal*. Initial expenses are incurred only at the issue date. Renewal expenses are incurred on the payment of premiums after the first (or, in some cases, including the first). The following examples illustrate the above discussion.

Example 6.1.1

A capital redemption policy with sum assured £10 000 and term 15 years has level annual premiums, payable in advance throughout the duration of the policy. The company issuing the policy calculated the annual premium on the basis of an interest rate of 8% per annum with allowance for (a) initial expenses of £100 plus 10% of the first annual premium and (b) renewal expenses of 4% of the second and each subsequent annual premium.
    Find the annual premium for the policy.

Solution

Let P' be the annual premium. After meeting its expenses the company invests $(0.9P' - 100)$ from the first premium and $0.96P'$ from each subsequent premium. Thus, to provide the sum assured at the end of 15 years, we must have

$$(0.9P' - 100)(1 + i)^{15} + \sum_{t=1}^{14} 0.96P'(1 + i)^{15-t} = 10\,000 \qquad \text{at } 8\%$$

This equation may be expressed in terms of standard functions in several ways. For example, we may write

$$0.9P'(1 + i)^{15} + 0.96P'\ddot{s}_{\overline{14}|} = 10\,000 + 100(1 + i)^{15}$$

$$0.9P'\ddot{s}_{\overline{15}|} + 0.06P'\ddot{s}_{\overline{14}|} = 10\,000 + 100(1 + i)^{15}$$

or $$0.96P'\ddot{s}_{\overline{15}|} - 0.06P'(1 + i)^{15} = 10\,000 + 100(1 + i)^{15}$$

(The reader should be able to express the original equation of value in terms of standard functions, choosing that particular form which occurs most easily to him.)
    Using the fact that $\ddot{s}_{\overline{15}|} = s_{\overline{16}|} - 1$, from the above equations we obtain the

premium in terms of tabulated functions as

$$P' = \frac{10\,000 + 100(1 + i)^{15}}{0.96(s_{\overline{16}} - 1) - 0.06(1 + i)^{15}} \qquad \text{at } 8\%$$

$$= £368.99$$

Our next example illustrates how to deal with variations in renewal expenses.

Example 6.1.2

Consider the capital redemption policy described in the previous example. Suppose, however, that, in order to allow for an increase in expenses over the duration of the policy, the company calculated the annual premium on the basis of the same interest rate and initial expenses as above but allowed for renewal expenses (associated with the second and each subsequent premium) of (a) £5 plus (b) an increasing percentage of each premium – the amount increasing linearly from $2\frac{1}{2}\%$ of the second premium to 5% of the final premium.

Find the annual premium.

Solution

For $t = 1, 2, \ldots, 14$, let λ_t denote the increasing percentage of the premium due at time t which is absorbed by expenses. Then

$$\lambda_t - 2.5 = \frac{5 - 2.5}{14 - 1}(t - 1)$$

or

$$\lambda_t = 2.3077 + 0.1923t$$

Let P' be the annual premium. The equation of value for the policy may be expressed as

$$(0.9P' - 100)(1 + i)^{15} + \sum_{t=1}^{14}\left(P' - 5 - \frac{2.3077 + 0.1923t}{100}P'\right)(1 + i)^{15-t} = 10\,000$$

Thus

$$0.9P'(1 + i)^{15} + 0.976\,923P'\ddot{s}_{\overline{14}} - 0.001\,923P'\sum_{t=1}^{14}t(1 + i)^{15-t}$$

$$= 10\,000 + 5\ddot{s}_{\overline{14}} + 100(1 + i)^{15}$$

The summation on the left-hand side of the last equation is simply $(I\ddot{s})_{\overline{14}}$, which equals

$$\frac{\ddot{s}_{\overline{14}} - 14}{d}$$

Hence the equation for P' is

$$P'\left(0.9(1 + i)^{15} + 0.976\,923\ddot{s}_{\overline{14}} - 0.001\,923\frac{\ddot{s}_{\overline{14}} - 14}{d}\right)$$

$$= 10\,000 + 5\ddot{s}_{\overline{14}} + 100(1 + i)^{15} \quad \text{at } 8\%$$

from which we obtain

$$P' = 10\,447 \cdot 977/28 \cdot 0881 = \pounds 371 \cdot 97$$

(This premium is slightly greater than that found in example 6.1.1. The renewal expenses increase from £14·30 to £23·60, in comparison with the level renewal expenses of £14·76 for first example.)

6.2 Policy values

The particular situation in which the rate of interest is constant and there are no expenses is of some theoretical interest and worth further study. In this case the single premium and level annual premium (payable in advance) for a capital redemption policy with term n years and sum assured 1 are denoted by $A_{\overline{n}|}$ and $P_{\overline{n}|}$ respectively. These premiums are called *net* premiums (since they contain no allowance for expenses) and, at the appropriate rate of interest, are given by the equation of value

$$\text{value of net premiums} = \text{value of benefits} \tag{6.2.1}$$

Thus

$$A_{\overline{n}|} = v^n = 1 - d\ddot{a}_{\overline{n}|} \tag{6.2.2}$$

and

$$P_{\overline{n}|} = \frac{1}{\ddot{s}_{\overline{n}|}} = \frac{1}{\ddot{a}_{\overline{n}|}} - d \tag{6.2.3}$$

If there is any possibility of confusion between net and *gross* (or *office*) annual premiums (i.e. those allowing for expenses), it is customary to use the notation P for the net annual premium and P' or P'' for the gross annual premium. The notations P, P', and P'' may be used even if the sum assured is not 1, but $P_{\overline{n}|}$, $P'_{\overline{n}|}$, and $P''_{\overline{n}|}$ always refer to a sum assured of 1.

Any insurance company issuing a capital redemption policy must accumulate the premiums it receives in order to pay the sum assured at the maturity date. For a policy with level annual premiums, with term n and with sum assured 1, the symbol $_tV_{\overline{n}|}$ (where $0 \leqslant t \leqslant n$) is used to denote the accumulated amount at time t of the premiums paid *before* that time. It is usual to call $_tV_{\overline{n}|}$ the *policy value* or *reserve* at duration t. It should be noted that the value of any premium payable at time t is excluded from $_tV_{\overline{n}|}$.

The definition of $_tV_{\overline{n}|}$ implies that, for integer values of t,

$$_tV_{\overline{n}|} = P_{\overline{n}|}\ddot{s}_{\overline{t}|}$$
$$= \ddot{s}_{\overline{t}|}/\ddot{s}_{\overline{n}|} \tag{6.2.4}$$

$$= \frac{(1+i)^t - 1}{(1+i)^n - 1} \tag{6.2.5}$$

Multiplying both the numerator and denominator of equation 6.2.5 by v^n, we obtain

$$_tV_{\overline{n}|} = \frac{v^{n-t} - v^n}{1 - v^n}$$

$$= 1 - \frac{1 - v^{n-t}}{1 - v^n}$$

$$= 1 - (\ddot{a}_{\overline{n-t}|}/\ddot{a}_{\overline{n}|}) \tag{6.2.6}$$

Alternatively, we may write equation 6.2.5 in the form

$$_tV_{\overline{n}|} = v^{n-t} - \frac{1 - v^{n-t}}{(1+i)^n - 1}$$

$$= v^{n-t} - (\ddot{a}_{\overline{n-t}|}/\ddot{s}_{\overline{n}|})$$

$$= v^{n-t} - P_{\overline{n}|}\ddot{a}_{\overline{n-t}|} \tag{6.2.7}$$

Expression 6.2.7 shows that $_tV_{\overline{n}|}$ is the value at time t of the sum assured (which is payable at time n) minus the value at time t of the premiums due at and after that time. Equations 6.2.4 and 6.2.7 give alternative expressions for the policy value at time t, the former being derived *retrospectively* (i.e. by accumulating the premiums received) and the latter *prospectively* (as the value of the sum assured minus the value of the remaining premiums).

If t is not an integer we may write $t = r + f$, where r ($0 \leqslant r < n$) is an integer and $0 < f < 1$. By definition, the value of the premium payable *at* time r is *not* included in $_rV_{\overline{n}|}$. Hence, for the accumulation to time $r + f$, we have

$$_{r+f}V_{\overline{n}|} = (_rV_{\overline{n}|} + P_{\overline{n}|})(1+i)^f \tag{6.2.8}$$

where $_rV_{\overline{n}|}$ is given by the formula for the reserve at integer durations.

The prospective expression for the reserve in this case is

$$_{r+f}V_{\overline{n}|} = v^{n-r-f} - P_{\overline{n}|}(_{1-f}|\ddot{a}_{\overline{n-r-1}|})$$

$$= v^{n-r-f} - v^{1-f}P_{\overline{n}|}\ddot{a}_{\overline{n-r-1}|} \tag{6.2.9}$$

It is left as a simple exercise for the reader to verify that expressions 6.2.8 and 6.2.9 are equal.

The reader should verify that equation 6.2.8 is valid also when $f = 1$, in which case the equation indicates the manner in which the reserve accumulates from one policy anniversary to the next.

Example 6.2.1 (The effect on the reserve of a change in the rate of interest)

Let t and n be integers with $0 < t < n$. Show that $_tV_{\overline{n}|}$ is a decreasing function of the rate of interest i (where $i \geqslant 0$).

Solution

Note first (see equation 6.2.4) that

$$_tV_{\overline{x}|} = \ddot{s}_{\overline{t}|}/\ddot{s}_{\overline{t}|}$$
$$= s_{\overline{t}|}/s_{\overline{t}|}$$
$$= 1/s_{\overline{t}|}$$

As i increases $s_{\overline{t}|}$ increases, so, for any integer k, $_tV_{\overline{k}|}$ is a decreasing function of i. Also, from equation 6.2.6,

$$1 - {}_tV_{\overline{n}|} = \ddot{a}_{\overline{n-t}|}/\ddot{a}_{\overline{n}|}$$

$$= \frac{\ddot{a}_{\overline{n-t}|}}{\ddot{a}_{\overline{n-t+1}|}} \frac{\ddot{a}_{\overline{n-t+1}|}}{\ddot{a}_{\overline{n-t+2}|}} \cdots \frac{\ddot{a}_{\overline{n-1}|}}{\ddot{a}_{\overline{n}|}}$$

$$= (1 - {}_tV_{\overline{n-t+1}|})(1 - {}_tV_{\overline{n-t+2}|})\cdots(1 - {}_tV_{\overline{n}|})$$

(again by equation 6.2.6). Our remarks above show that, as i increases, each of the terms $_tV_{\overline{n-t+1}|}, {}_tV_{\overline{n-t+2}|}, \ldots, {}_tV_{\overline{n}|}$ decreases. This in turn implies that each bracket on the right-hand side of the last equation increases. Thus, as i increases, $1 - {}_tV_{\overline{n}|}$ increases, and hence $_tV_{\overline{n}|}$ decreases.

The Zillmerized reserve

Suppose that, in respect of an *n*-year capital redemption policy with sum assured 1 and annual premiums, the office premium is calculated by allowing for level annual expenses of e associated with the payment of each premium (*including* the first), and additional initial expenses of I. Note that I and e may be defined as, or may include, percentages of the office premium or the net premium.

The annual office premium $P'_{\overline{n}|}$ is given by the equation

$$P'_{\overline{n}|} \ddot{a}_{\overline{n}|} = v^n + I + e\ddot{a}_{\overline{n}|}$$

so

$$P'_{\overline{n}|} = \frac{1}{\ddot{s}_{\overline{n}|}} + \frac{I}{\ddot{a}_{\overline{n}|}} + e$$

or

$$P'_{\overline{n}|} = P_{\overline{n}|} + \frac{I}{\ddot{a}_{\overline{n}|}} + e \qquad (6.2.10)$$

The reserve at duration t (where t is an integer less than or equal to n) may be obtained *prospectively* as the value of the sum assured and expenses still to be incurred less the value of future office premiums. Thus if we denote the reserve by $_tV^*_{\overline{n}|}$, then

$$_tV^*_{\overline{n}|} = v^{n-t} + e\ddot{a}_{\overline{n-t}|} - P'_{\overline{n}|}\ddot{a}_{\overline{n-t}|}$$

$$= v^{n-t} + e\ddot{a}_{\overline{n-t}|} - \left(P_{\overline{n}|} + \frac{I}{\ddot{a}_{\overline{n}|}} + e\right)\ddot{a}_{\overline{n-t}|} \qquad \text{(by equation 6.2.10)}$$

$$= v^{n-t} - P_{\overline{n}|}\ddot{a}_{\overline{n-t}|} - I\frac{\ddot{a}_{\overline{n-t}|}}{\ddot{a}_{\overline{n}|}}$$

$$= {}_tV_{\overline{n}|} - I\frac{\ddot{a}_{\overline{n-t}|}}{\ddot{a}_{\overline{n}|}} \qquad (6.2.11)$$

$$= (1 + I)_tV_{\overline{n}|} - I \qquad (6.2.12)$$

$_tV^*_{\overline{n}|}$ is known as a *Zillmerized reserve* (after the German actuary August Zillmer, 1831–93).

Note that

$$_0V^*_{\overline{n}|} = -I$$

since it is assumed that the reserve at time 0 is taken after payment of expenses of I but before receipt of the first premium.

Alternatively, the value of $_tV^*_{\overline{n}|}$ may be found retrospectively as the accumulation of the office premiums received less the accumulation of the expenses incurred. Thus

$$_tV^*_{\overline{n}|} = P'_{\overline{n}|}\ddot{s}_{\overline{t}|} - I(1 + i)^t - e\ddot{s}_{\overline{t}|}$$

On substitution for $P'_{\overline{n}|}$ (from equation 6.2.10), this retrospective expression for $_tV^*_{\overline{n}|}$ is easily seen to equal either of the prospective expressions 6.2.11 or 6.2.12.

6.3 Policy values when premiums are payable *p*thly

When the premium is payable p times per annum the corresponding notations for the net annual premium and the policy value, or reserve, at duration t (for a policy with term n years and sum assured 1) are $P^{(p)}_{\overline{n}|}$ and $_tV^{(p)}_{\overline{n}|}$ respectively. Clearly,

$$P^{(p)}_{\overline{n}|}\ddot{s}^{(p)}_{\overline{n}|} = 1$$

so

$$P^{(p)}_{\overline{n}|} = \frac{1}{\ddot{s}^{(p)}_{\overline{n}|}} = \frac{1}{\ddot{a}^{(p)}_{\overline{n}|}} - d^{(p)} \qquad (6.3.1)$$

and, if t is an integer,

$$_tV_{\overline{n}|}^{(p)} = P_{\overline{n}|}^{(p)} \ddot{s}_{\overline{t}|}^{(p)}$$
$$= \ddot{s}_{\overline{t}|}^{(p)}/\ddot{s}_{\overline{n}|}^{(p)}$$
$$= \ddot{s}_{\overline{t}|}/\ddot{s}_{\overline{n}|}$$
$$= {}_tV_{\overline{n}|} \tag{6.3.2}$$

If t is not an integer, then we may write either (a) $t = r + (k/p)$ with $r(0 \leqslant r < n)$ and $k(0 < k < p)$ integers, or (b) $t = r + (k/p) + f$ with r and k as above and $0 < f < 1/p$. We then have

$$_{r+(k/p)}V_{\overline{n}|}^{(p)} = P_{\overline{n}|}^{(p)} \ddot{s}_{\overline{r+(k/p)}|}^{(p)} \tag{6.3.3}$$

and

$$_{r+(k/p)+f}V_{\overline{n}|}^{(p)} = [_{r+(k/p)}V_{\overline{n}|}^{(p)} + (P_{\overline{n}|}^{(p)}/p)](1+i)^f \tag{6.3.4}$$

These formulae should not be memorized. In any practical situation the policy value at duration t is best found from first principles.

6.4 Surrender values, paid-up policy values and policy alterations

The *policy value* or *reserve* of a capital redemption policy has already been defined when expenses are ignored and when the rate of interest used is the same in all calculations. For example, the policy value or reserve for an n-year annual premium policy with sum assured 1, just before payment of the premium due at time t, is $_tV_{\overline{n}|}$. If expenses are taken into account, we obtain a reserve (or policy value) allowing for expenses, such as the Zillmerized reserve $_tV_{\overline{n}|}^{*}$ given by formulae 6.2.11 and 6.2.12. ($_tV_{\overline{n}|}$ is often referred to as the *net premium reserve* in order to distinguish it from other 'reserve values' calculated on different assumptions.)

Suppose that the policyholder wishes to *surrender* his policy. This means that he wishes to terminate the contract in return for a lump sum payment, known as a *surrender value*. The office will in general grant a surrender value roughly equal to the Zillmerized reserve, or some other reserve allowing for expenses, although the position may be complicated by departures from the interest and expenses assumptions used to calculate the premiums. A full discussion of this point is beyond the scope of this book.

If, at time t, the policyholder wishes not to surrender the policy but instead to make some alteration (such as a change in the premiums payable in the future or the maturity date), the necessary calculations are carried out by means of an equation of value in which the policyholder is assumed to apply the reserve as a 'special' single premium towards the new policy. The reserve granted may be rather more generous than the surrender value, since the policyholder maintains his connection with the office. There may, however, be a charge for the cost of making the alteration. The equation of value thus takes the form

$$\begin{bmatrix} \text{value of} \\ \text{reserve for} \\ \text{the existing} \\ \text{policy} \end{bmatrix} + \begin{bmatrix} \text{value of} \\ \text{premiums for} \\ \text{the altered} \\ \text{policy} \end{bmatrix} = \begin{bmatrix} \text{value of} \\ \text{benefits of} \\ \text{the altered} \\ \text{policy} \end{bmatrix} + \begin{bmatrix} \text{value of} \\ \text{expenses,} \\ \text{including expenses} \\ \text{of alteration} \end{bmatrix}$$

(6.4.1)

This equation is usually most conveniently taken at the date of alteration. For example, suppose that the holder of an annual-premium policy with term n and sum assured 1 requests that, just before the premium payable at time t becomes due, all future premiums should be of amount P^*. If expenses are ignored throughout, the revised sum assured S^* is given by the equation of value

$$S^* v^{n-t} = {}_t V_{\overline{n}|} + P^* \ddot{a}_{\overline{n-t}|}$$

(6.4.2)

from which we obtain

$$S^* = {}_t V_{\overline{n}|} (1 + i)^{n-t} + P^* \ddot{s}_{\overline{n-t}|}$$

More complicated examples may involve expenses (see exercises).

In the particular case when $P^* = 0$ all future premiums are discontinued just before payment of the premium due at time t and the policy is made *paid-up*. When there are no expenses, the revised sum assured for an annual premium policy of original term n years is known as a *paid-up policy value* and is denoted by ${}_t W_{\overline{n}|}$. From equation 6.4.2 (with $P^* = 0$) we obtain the formula

$$\begin{aligned} {}_t W_{\overline{n}|} &= {}_t V_{\overline{n}|} (1 + i)^{n-t} \\[2mm] &= \frac{\ddot{s}_{\overline{t}|}}{\ddot{s}_{\overline{n}|}} (1 + i)^{n-t} \qquad \text{(by equation 6.2.4)} \\[2mm] &= \frac{\ddot{s}_{\overline{t}|} (1 + i)^{-t}}{\ddot{s}_{\overline{n}|} (1 + i)^{-n}} \\[2mm] &= \frac{\ddot{a}_{\overline{t}|}}{\ddot{a}_{\overline{n}|}} \end{aligned}$$

(6.4.3)

In more general circumstances, when an allowance for expenses may have to be made, the sum assured granted by the office is called the *paid-up sum assured*. This may be found by solving the equation of value 6.4.1. In this case the term 'value of premiums for the altered policy' is of course zero, so we obtain the equation

$$\begin{bmatrix} \text{value of} \\ \text{reserve for} \\ \text{the existing} \\ \text{policy} \end{bmatrix} = \begin{bmatrix} \text{value of paid-up} \\ \text{sum assured} \end{bmatrix} + \begin{bmatrix} \text{value of expenses} \\ \text{of alteration} \end{bmatrix} \quad (6.4.4)$$

A rule sometimes used in practice to calculate the paid-up sum assured is the *proportionate rule*, i.e.

$$\text{paid-up sum assured} = \frac{t}{n} \times \text{original sum assured} \qquad (6.4.5)$$

where t is the number of premiums which have been paid. It is a simple exercise (see exercise 6.5) to show that, at a positive rate of interest, $_tW_{\overline{n}|} > t/n$, so that the paid-up policy value at duration t is greater than that given by the proportionate rule. Note, however, that the paid-up policy value $_tW_{\overline{n}|}$ takes no account of expenses, and the paid-up sum assured granted in practice is usually rather less than $_tW_{\overline{n}|}$.

The calculation of paid-up sums assured, altered premiums, etc. may also be affected by departures from the assumptions concerning interest and expenses used to calculate the original premiums. A full discussion of this point is beyond the scope of this book.

Example 6.4.1

On 1 January 1964 a capital redemption policy was effected with sum assured £100 000 and term 30 years. The policy had level premiums payable quarterly in advance on 1 January, 1 April, 1 July, and 1 October each year.

On 30 September 1979 all future quarterly premiums were reduced to £100, a revised sum assured then being quoted. On 1 May 1983 the policyholder asked the insurance company to quote (i) a surrender value and (ii) a paid-up policy value.

Assuming that all the calculations were carried out on a net premium basis at 8% per annum interest, find

(a) The reduced sum assured quoted on 30 September 1979;
(b) The surrender value and paid-up policy value quoted on 1 May 1983.

Solution

(a) The original net annual premium for the policy was

$$100\,000\, P^{(4)}_{\overline{30}|} = 100\,000 \bigg/ \left(\frac{i}{d^{(4)}} s_{\overline{30}|} \right) \quad \text{at } 8\% \quad = 841{\cdot}09$$

The reserve at 30 September 1979 (i.e. at duration 15·75 years) was

$$100\,000\,{}_{15\,75}V^{(4)}_{\overline{30}|} = 841{\cdot}09 s^{(4)}_{\overline{15\,75}|} = 26\,048{\cdot}18$$

Since the revised annual premium from 1 October 1979 was £400, the revised sum assured then quoted was

$$26\,048{\cdot}18(1+i)^{14{\cdot}25} + 400 s^{(4)}_{\overline{14\,25}|} = \pounds 88\,459{\cdot}88, \quad \text{say} \quad \pounds 88\,459$$

Note: In order to facilitate further calculations, we have found the reserve at the date of alteration. A shorter method of finding the revised sum assured is to express it as the original sum assured minus the value of the premiums *not* to be received as a result of the alteration. This gives the revised sum assured directly as

$$100\,000 - (841\cdot09 - 400)\ddot{s}^{(4)}_{14\cdot25\rceil} = 88\,459$$

as before.

(b) The reserve at 1 April 1983 (just before payment of the premium then due) may be obtained as the accumulated amount of the reserve at the date of the first alteration plus the premiums subsequently received. Thus the reserve at 1 April 1983 was

$$26\,048\cdot18(1+i)^{3\cdot5} + 400\ddot{s}^{(4)}_{3\rceil} = 34\,100\cdot48 + 1622\cdot19 = 35\,722\cdot67$$

The surrender value quoted on 1 May 1983 was the reserve at that time, i.e.

$$(35\,722\cdot67 + 100)(1+i)^{1/12} = 36\,053\cdot15, \quad \text{say} \quad £36\,053$$

Since the outstanding term at this date was $10\frac{2}{3}$ years, the paid-up policy value quoted was

$$36\,053(1+i)^{32/3} = £81\,933$$

6.5 Variations in interest rates: piecewise constant i

We have already remarked that it may be desirable to allow for changes in the rate of interest over the duration of a long-term capital redemption policy. One approach, which may suffice for practical purposes in certain situations, is obtained by assuming that over successive periods the rate of interest is constant – different constant values being used for each period.

For example, in relation to an n-year capital redemption policy, if we assume an annual rate of interest of i_1 for the first n_1 years, i_2 for a further n_2 years, and i_3 thereafter (where $n_1 + n_2 < n$), and premiums are accumulated on this basis from one year to the next, then P^*, the net annual premium per unit sum assured, is given by the equation

$$P^*[\ddot{s}_{\overline{n_1}\rceil i_1}(1+i_2)^{n_2}(1+i_3)^{n-n_1-n_2} + \ddot{s}_{\overline{n_2}\rceil i_2}(1+i_3)^{n-n_1-n_2} + \ddot{s}_{\overline{n-n_1-n_2}\rceil i_3}] = 1$$

$$(6.5.1)$$

Example 6.5.1

On 1 April 1973 an office issued a 30-year capital redemption policy with annual premiums and a sum assured of £60 000. The office premium was calculated on the basis of an annual rate of interest of 8% for the first five years, 7% for the next ten years, and 6% thereafter. The office assumed that all its investments would be in securities with one-year terms. Allowance was made for initial expenses of 50% of the first office annual premium and for renewal expenses of 2% of the second and each subsequent office annual premium.

(a) Find the office annual premium.

(b) On 1 April 1985, instead of paying the premium then due, the policyholder asked for the policy to be made paid-up. The paid-up sum assured granted was the greater of

(i) The original sum assured multiplied by t/n, where n was the original term of the policy and t was the number of premiums which had been paid; and

(ii) The accumulation until the end of the original term, at $5\frac{1}{2}\%$ per annum interest, of the policy's surrender value, the surrender value being the accumulation, at $5\frac{1}{2}\%$ per annum interest, of half the first office annual premium and 97% of all subsequent premiums received.

Calculate the paid-up sum assured granted by the office. Find also the profit or loss which will be made by the office on 1 April 2003, if the office's experience follows the premium basis, the policyholder does not surrender his paid-up policy, and the cost of maintaining the paid-up policy is negligible.

Solution

(a) Let the office annual premium be P'. Then

$$60\,000 = 0.98P'[\ddot{s}_{\overline{5}|0.08}(1.07)^{10}(1.06)^{15} + \ddot{s}_{\overline{10}|0.07}(1.06)^{15} + \ddot{s}_{\overline{15}|0.06}]$$
$$- 0.48P'(1.08)^5(1.07)^{10}(1.06)^{15}$$

Hence

$$P' = 60\,000/[(0.98 \times 89.9723) - (0.48 \times 6.9270)] = £707.15$$

(b) Since 12 premiums had been paid, the paid-up sum assured is the greater of

(i) $\frac{12}{30} \times 60\,000 = 24\,000$

and

(ii) $1.055^{18}[0.5P'(1.055)^{12} + 0.97P'\ddot{s}_{\overline{11}|0.055}] = 29\,427.82$

Hence the paid-up sum assured is £29 427·82. (Note that is not necessary to evaluate the surrender value itself.)

If the office's experience over the term of the policy turns out to be as allowed for in the premium basis and there are no expenses after the policy is made paid-up, then the office's accumulation on 1 April 2003 will be

$$P'\{0.98[\ddot{s}_{\overline{5}|0.08}(1.07)^{10}(1.06)^{15} + \ddot{s}_{\overline{7}|0.07}(1.07)^3(1.06)^{15}]$$
$$- 0.48(1.08)^5(1.07)^{10}(1.06)^{15}\} = £37\,188.53$$

Hence the office's profit on 1 April 2003 will be

$$37\,188.53 - 29\,427.82, \quad \text{i.e.} \quad £7760.71$$

The use of a piecewise constant interest rate may be inappropriate in certain circumstances. Accordingly, we consider below two further methods of allowing for varying interest rates. These alternative approaches make very different assumptions about future investment. The first is an elaboration of the method described above and uses a continuous mathematical function to model the future force of interest. The second approach is fundamentally different and depends crucially upon the idea of *reinvestment rates* (see section 6.7).

6.6 Stoodley's logistic model for the force of interest

The piecewise constant model for $\delta(t)$, described in the previous section, is obviously somewhat simplistic. A more realistic approach may be obtained by using a continuous function to model future changes in interest rates. One such model, due to C. L. Stoodley (references [31], [47]), is worthy of special mention, since it possesses a particularly convenient property (see also section 2.6).

If we assume a monotonic form for $\delta(t)$, the force of interest at time t (and in particular if $\delta(t)$ is a decreasing function), we may try to model the trend in interest rates by a logistic function. If we assume that

$$\delta(t) = p + \frac{s}{1 + re^{st}} \tag{6.6.1}$$

and let

$$v(t) = \exp\left[- \int_0^t \delta(y)\,dy \right]$$

then (see equations 2.6.2, 2.6.3) it follows that

$$v(t) = \frac{1}{1 + r}e^{-(p+s)t} + \frac{r}{1 + r}e^{-pt} \tag{6.6.2}$$

$$= \frac{1}{1 + r}v_1^t + \frac{r}{1 + r}v_2^t \tag{6.6.3}$$

where v_1 and v_2 are calculated at forces of interest $(p + s)$ and p respectively.

If we denote by $a^*_{\overline{n}|}$ the value of an n-year immediate annuity under the given model, it follows directly from equation 6.6.3 that

$$a^*_{\overline{n}|} = \frac{1}{1 + r}a'_{\overline{n}|} + \frac{r}{1 + r}a''_{\overline{n}|} \tag{6.6.4}$$

where $a'_{\overline{n}|}$ and $a''_{\overline{n}|}$ are calculated at forces of interest $(p + s)$ and p respectively. Formulae similar to 6.6.4 may be obtained for continuously payable and deferred annuities and for annuities payable several times per annum.

Equations 6.6.3 and 6.6.4 indicate the considerable practical advantage of the model. The function $v(t)$ and annuity values under the varying force of interest represented by the model are obtained simply as a weighted average of the corresponding values at the constant forces of interest $(p + s)$ and p. In particular, if we are able to model the future trend of interest rates by equation 6.6.1 with values of p and s such that $(p + s)$ and p equal the force of interest at tabulated rates, then the calculation of annuity values under this model is especially simple.

The determination of the parameters p, r, and s of equation 6.6.1 is of some mathematical interest. One possible approach is to assign values to $\delta(0)$, $\delta(t_1)$ – where t_1 is some specified future time – and $\lim_{t \to \infty} \delta(t)$. These values must form a monotonic sequence – this is inherent in the use of a logistic function – and together with the value of t_1 will determine the constants p, r, and s. In practice, having found the values of p, r, and s to fit precisely a specified trend in future interest rates, we may use very slightly different parameters, which do not affect the general trend to be modelled but which ensure that tabulated rates are used for interpolation.

One procedure for determining the constants p, r, and s of equation 6.6.1 is described in reference [31].

Example 6.6.1

Suppose that we wish to model a fall in annual interest rates from a current level of around 11% to a *long-term* level of about 8%. We consider it appropriate to use a value of approximately 10% for the rate of interest four years from now.

Note that rates of interest of 11%, 10%, and 8% correspond to forces of interest of $\log 1 \cdot 11 \simeq 0 \cdot 104$, $\log 1 \cdot 1 \simeq 0 \cdot 095$, and $\log 1 \cdot 08 \simeq 0 \cdot 077$ respectively. Accordingly as a first step we consider equation 6.6.1, assuming that $\delta(0) = 0 \cdot 104$, $\delta(4) = 0 \cdot 095$ and $\delta(\infty) = 0 \cdot 077$. Using the method of reference [31], we find that $p = 0 \cdot 077$, $r = 3 \cdot 578 \, 88$, and $s = 0 \cdot 123 \, 630$. These values for p, r, and s are regarded as preliminary values. From these, after some brief experiment, we base our model on the revised parameters $p = \log(1 \cdot 08) = 0 \cdot 076 \, 961$, $r = 4$, and $s = \log(1 \cdot 25) - \log(1 \cdot 08) = 0 \cdot 146 \, 183$. We note that these new parameters imply that $\delta(0) = 0 \cdot 1062$, $\delta(4) = 0 \cdot 0948$ and $\delta(\infty) = 0 \cdot 0770$. Moreover p and $(p + s)$ are the forces of interest at rates of interest 8% and 25% respectively.

It follows from equation 6.6.3 that with these parameters

$$v(t) = \tfrac{1}{5} v^t_{0 \cdot 25} + \tfrac{4}{5} v^t_{0 \cdot 08}$$

so, for example,

$$a^*_{\overline{n}|} = \tfrac{1}{5} a_{\overline{n}| 0 \cdot 25} + \tfrac{4}{5} a_{\overline{n}| 0 \cdot 08}$$

These equations are particularly simple to apply in practice. As a further example, we leave it as an exercise for the reader to show that, under this particular model of falling interest rates,

$$_{10}|a^*_{\overline{15}|} = \tfrac{1}{5}(_{10}|a_{\overline{15}| 0 \cdot 25}) + \tfrac{4}{5}(_{10}|a_{\overline{15}| 0 \cdot 08}) = 3 \cdot 2546$$

Example 6.6.2

On the basis of Stoodley's model with the values of the parameters as in the last example (i.e. $p = 0 \cdot 076 \, 961$, $r = 4$, and $s = 0 \cdot 146 \, 183$) find the net annual premium, payable quarterly in advance, for a capital redemption policy with sum assured £100 000 and term 25 years. Find also the policy reserve and the paid-up policy value at the end of 20 years on this net premium basis.

Solution

The particularly simple expression for $v(t)$, as given by equation 6.6.3, makes it more appropriate to work with present values rather than accumulated values.

Let P be the annual premium, payable quarterly. Then, if an asterisk is used to denote an annuity value with the falling interest rates,

$$P\ddot{a}_{\overline{25|}}^{(4)*} = 100\,000\,v(25)$$

i.e.

$$P[\tfrac{1}{3}\ddot{a}_{\overline{25|}\,0\cdot25}^{(4)} + \tfrac{4}{3}\ddot{a}_{\overline{25|}\,0\cdot08}^{(4)}] = 100\,000(\tfrac{1}{3}v_{0\cdot25}^{25} + \tfrac{4}{3}v_{0\cdot08}^{25})$$

Hence

$$P = \frac{11\,757\cdot20}{9\cdot8808} = \pounds1\,189\cdot90$$

Let V and W be the reserve and paid-up policy value respectively at duration 20. Then, still working with present values, we have

$$P\ddot{a}_{\overline{20|}}^{(4)*} = V[v(20)] = W[v(25)]$$

Thus

$$P[\tfrac{1}{3}\ddot{a}_{\overline{20|}\,0\cdot25}^{(4)} + \tfrac{4}{3}\ddot{a}_{\overline{20|}\,0\cdot08}^{(4)}] = V(\tfrac{1}{3}v_{0\cdot25}^{20} + \tfrac{4}{3}v_{0\cdot08}^{20})$$
$$= W(\tfrac{1}{3}v_{0\cdot25}^{25} + \tfrac{4}{3}v_{0\cdot08}^{25})$$

Hence

$$V = P\left(\frac{9\cdot1543}{0\cdot173\,95}\right) = 62\,619\cdot72, \quad \text{or} \quad \pounds62\,620$$

$$W = P\left(\frac{9\cdot1543}{0\cdot111\,57}\right) = 92\,648\cdot65, \quad \text{or} \quad \pounds92\,648$$

6.7 Reinvestment rates

Yet another approach to variations in the rate of interest is provided by consideration of the reinvestment rates which apply over the duration of a policy (see reference [43]). Suppose that n is a positive integer and that, for $t = 0, 1, 2, \ldots, n - 1$, it will be possible at time t to make an investment of any amount such that the sum invested will be repaid *at time n* and will generate income of i_t times the sum invested, payable at times $t + 1, t + 2, \ldots, n$.

Consider, for example, an investment of 1 at time 0. This will be repaid (as 1) at time n and will generate income of i_0, payable at times $1, 2, \ldots, n$. At time 1 we may invest the income of i_0 then received. This investment will be repaid (as i_0) at time n and will generate further income of $i_1 i_0$, payable at times $2, 3, \ldots, n$. At time 2 the total income received will thus be $(i_0 + i_1 i_0)$, arising from the two previous investments. We may invest this income to be repaid (as $i_0 + i_1 i_0$) at time n and to generate further income of $i_2(i_0 + i_1 i_0)$, payable at times $3, 4, \ldots, n$.

We may continue in this manner. Thus, for $t = 1, \ldots, (n - 1)$, the income received at time t from *all* previous investments will itself then be invested for repayment at time n and to generate further income of i_t times the sum invested. At time n the investor will receive the income then due from all the previous investments in addition to the redemption proceeds of these investments.

How much will the investor receive at time n if he carries out the reinvestment procedures described above? More generally, we may consider the situation in which the investor makes a series of payments of different amounts at times $0, 1, \ldots, n - 1$. If each payment is invested in the manner described above and all resulting income is also reinvested immediately on receipt for the balance of the term (i.e. for repayment at time n), how much will the investor eventually receive?

We have already remarked that for financial contracts which relate to long-term investment it may be prudent, when current interest rates are high, to allow for a reduction in interest rates with the passing of time. Provided that the appropriate investment strategy is being followed, one way of making such an allowance is to use the model described above with a decreasing sequence $\{i_t\}$. Possibly one might assume a (relatively low) constant value of i_t for sufficiently large t.

Consider now this investment procedure in more detail, firstly in relation to a series of regular investments of equal amount.

For $t = 0, 1, \ldots, (n - 1)$ let A_t denote the total proceeds which will be received at time n by an investor who makes a series of investments, each of amount 1, at times $t, t + 1, \ldots, n - 1$ and reinvests the income in the manner described above. Clearly,

$$A_{n-1} = 1 + i_{n-1} \tag{6.7.1}$$

We now establish the following more general result.

Theorem 6.7.1

For $0 \leqslant t < n - 1$,

$$A_t = (1 + i_t) + (1 + i_t)(1 + i_{t+1}) + (1 + i_t)(1 + i_{t+1})(1 + i_{t+2}) + \cdots$$
$$\cdots + (1 + i_t)(1 + i_{t+1})\cdots(1 + i_{n-1}) \tag{6.7.2}$$

Proof

The result is established quite simply by means of a recurrence relation. The value of A_{n-1} is clearly given by equation 6.7.1. We therefore assume that $0 \leqslant t < n - 1$. Subject to the reinvestment procedure described above, the proceeds at time n of a series of investments of 1 at times $t, t + 1, \ldots, n - 1$ must equal

   1 (i.e. the repayment at time n of the initial investment of 1 at time t)
  + the proceeds of the reinvested income, i_t, received at times $t + 1, t + 2, \ldots$
    $\cdots(n - 1)$, arising from the initial investment at time t
  + i_t (i.e. the income received at time n from the initial investment at time t)
  + the proceeds of the investments of 1 at times $t + 1, t + 2, \ldots, n - 1$

Expressed algebraically, this equation is

$$A_t = 1 + i_t A_{t+1} + i_t + A_{t+1} \qquad (0 \leqslant t < n - 1)$$

or

$$A_t = (1 + i_t)(1 + A_{t+1}) \qquad (0 \leqslant t < n - 1) \qquad (6.7.3)$$

This recurrence relation and the initial value A_{n-1} enable A_t to be found for all values of t. For example, putting $t = n - 2$ in equation 6.7.3, we obtain

$$\begin{aligned} A_{n-2} &= (1 + i_{n-2})(1 + A_{n-1}) \\ &= (1 + i_{n-2})[1 + (1 + i_{n-1})] \qquad \text{(by equation 6.7.1)} \\ &= (1 + i_{n-2}) + (1 + i_{n-2})(1 + i_{n-1}) \end{aligned}$$

which establishes equation 6.7.2 for $t = n - 2$.

The general result follows immediately by induction. Assume that $1 < k < n - 1$. Put $t = k - 1$ in equation 6.7.3 to show that, if equation 6.7.2 is valid for $t = k$, then it is also valid for $t = k - 1$. Since equation 6.7.2 is valid for $t = n - 2$, it follows that it is valid for $t = n - 3, n - 4, \ldots, 1$ i.e. for all values of t.

Equation 6.7.2 is illustrated in table 6.7.1.

Consider now the investment of a single amount, rather than a series of regular investments as above. For $t = 0, 1, \ldots, n - 1$, let S_t denote the total proceeds which will be received at time n by an investor who makes a single payment of 1 at time t and reinvests the income as described above. Clearly,

$$S_{n-1} = 1 + i_{n-1} \qquad (6.7.4)$$

We may easily prove the following more general result.

Theorem 6.7.2

For $0 \leqslant t < n - 1$,

$$S_t = 1 + i_t[1 + (1 + i_{t+1}) + (1 + i_{t+1})(1 + i_{t+2}) + \cdots + (1 + i_{t+1})\cdots(1 + i_{n-1})] \qquad (6.7.5)$$

Proof

The result may be established directly (see exercise 6.14), although the proof is slightly more difficult than that of theorem 6.7.1. Accordingly, it is simpler to use an alternative argument, which makes use of the result of the previous theorem.

Subject to the reinvestment procedure described above, the proceeds at time n of a single investment of 1 at time t must equal

1 (i.e. the repayment at time n of the initial investment of 1 at time t)
+ i_t (i.e. the income received at time n from the initial investment at time t)
+ the proceeds of the reinvested income i_t, received at times $t + 1, t + 2, \ldots, n - 1$, arising from the initial investment at time t.

Table 6.7.1 *The value of A_t*

Time of investment	t	$t+1$	$t+2$	$t+3$	$\cdots$	$n-1$	n
Annual premium	1	1	1	1	$\cdots$	1	0
Interest		i_t	i_t	i_t	$\cdots$	i_t	i_t
Interest			$(1+i_t)i_{t+1}$	$(1+i_t)i_{t+1}$	$\cdots$	$(1+i_t)i_{t+1}$	$(1+i_t)i_{t+1}$
Interest				$(1+i_t)(1+i_{t+1})i_{t+2}$	$\cdots$	$(1+i_t)(1+i_{t+1})i_{t+2}$	$(1+i_t)(1+i_{t+1})i_{t+2}$
$\vdots$						$\vdots$	$\vdots$
Interest						$(1+i_t)\cdots(1+i_{n-3})i_{n-2}$	$(1+i_t)\cdots(1+i_{n-3})i_{n-2}$
Interest							$(1+i_t)\cdots(1+i_{n-2})i_{n-1}$
Total	1	$1+i_t$	$(1+i_t)(1+i_{t+1})$	$(1+i_t)(1+i_{t+1})(1+i_{t+2})$	$\cdots$	$(1+i_t)\cdots(1+i_{n-2})$	$(1+i_t)\cdots(1+i_{n-1})-1$

The first annual premium (at time t) is invested to produce interest i_t at times $t+1, t+2,...,n$ (and a return of 1 at time n). At time $t+1$ the second annual premium and interest (i.e. a total amount $1+i_t$) are invested to produce further interest of $(1+i_t)i_{t+1}$ at times $t+2$, $t+3,...,n$ (and a return of $(1+i_t)$ at time n). By continuing this argument one obtains all the terms of the array in the table and, on summing the terms of the array (by columns), one obtains equation 6.7.2.

Expressed algebraically, this equation is

$$S_t = 1 + i_t + i_t A_{t+1}$$
$$= 1 + i_t(1 + A_{t+1})$$

Substitution of the value of A_{t+1} (from equation 6.7.2) in the last equation immediately gives the required result.

Finally, we remark that, if an investor makes a series of investments c_0, $c_1, \ldots, c_{n-1}$ (with the payment c_t being made at time t), subject to the above reinvestment procedure, then his total proceeds at time n will be

$$\sum_{t=0}^{n-1} c_t S_t$$

Since the values of $S_0, S_1, \ldots, S_{n-1}$ are known (by equations 6.7.4 and 6.7.5), the proceeds of such a general series of payments can be found fairly simply.

Example 6.7.1

A special savings plan is designed to produce a capital sum at time n (measured in years). Any investment made in this plan will be repaid at its purchase price at time n. An investment made at time t ($t = 0, 1, 2, \ldots, n-1$) will produce income, payable annually in arrear until time n, of amount i_t times the sum invested. Each year the income arising from all earlier investments is automatically reinvested in the plan.

An investor is considering a special savings plan of the type described above for which $i_0 = 0\cdot1$, $i_1 = 0\cdot09$, and $i_t = 0\cdot08$ for $t \geqslant 2$. His alternative investment is a deposit account in which interest is paid annually in arrear at the constant rate of 9% per annum. He wishes to invest for a term of n years.

Find the ranges of values of n for which the special savings plan will be more advantageous to him

(a) If he intends to make a single capital investment now; and
(b) If he intends to invest equal capital amounts at the start of each of the next n years.

Solution

For a savings plan of duration n years, let $MPS(n)$ denote the final maturity proceeds when the investor makes a single payment to the plan of amount 1 at time 0, and let $MPA(n)$ denote the final maturity proceeds when the investor makes a series of annual payments to the plan of amount 1 at times $0, 1, \ldots, n-1$. Then clearly,

$$MPA(n) = A_0$$

and

$$MPS(n) = S_0$$

where A_0 and S_0 are defined by equations 6.7.2 and 6.7.5 respectively. Since $i_0 = 0\cdot1$,

$i_1 = 0.09$, and $i_t = 0.08$ for $t \geqslant 2$, it follows that, if $n \geqslant 3$,

$$MPA(n) = 1.1 + (1.1)(1.09) + (1.1)(1.09)(1.08) + \cdots + (1.1)(1.09)(1.08)^{n-2}$$
$$= 1.1 + (1.1)(1.09)[1 + 1.08 + \cdots + (1.08)^{n-2}]$$
$$= 1.1 + 1.199 s_{\overline{n-1}|0.08}$$

The investor's alternative is a deposit account for which interest is at the constant rate of 9% per annum. The maturity proceeds from a corresponding regular series of payments to the deposit account are clearly $\ddot{s}_{\overline{n}|0.09}$. Hence for a term of n years with level annual investments, the special savings plan will be more attractive if and only if

$$1.1 + 1.199 s_{\overline{n-1}|0.08} > \ddot{s}_{\overline{n}|0.09}$$

This inequality will hold only for sufficiently small values of n. By trial and error we find that when $n = 4$ the inequality holds (since $4.992 > 4.985$), but that when $n = 5$ the condition is not satisfied (since $6.503 < 6.523$). Hence for regular annual payments the special savings plan is more attractive than the deposit account when the term is four years or less.

Consider now the case when a single payment is to be invested. Note that, for $n \geqslant 3$,

$$S_0 = 1 + 0.1[1 + (1.09) + (1.09)(1.08) + \cdots + (1.09)(1.08)^{n-2}]$$
$$= 1.1 + 0.109[1 + 1.08 + \cdots + (1.08)^{n-2}]$$
$$= 1.1 + 0.109 s_{\overline{n-1}|0.08}$$

Since in this case the deposit account will produce maturity proceeds of 1.09^n, the special plan will be more attractive if and only if

$$1.1 + 0.109 s_{\overline{n-1}|0.08} > 1.09^n$$

This inequality will hold only for sufficiently small values of n. Again by trial and error we find that the condition holds when $n = 20$ (since $5.618 > 5.604$), but that when $n = 21$ the condition does not hold (since $6.088 < 6.109$). Hence, for single premium investment, the special plan is more attractive than the deposit account when the term is 20 years or less.

Note the greater range of term for which the single premium investment is more attractive. This arises because the higher value of i_0 is relatively more important for a single premium investment than for an annual premium investment.

Exercises

6.1 An office issues 20-year capital redemption policies with sum assured £10 000 and premiums payable annually in advance. The amount of the premium and any policy alterations are calculated on the basis of an effective annual interest rate of 6% with no allowance for expenses. Find

(a) The annual premium;
(b) The policy value immediately before the seventh premium is paid;
(c) The paid-up sum assured which will be granted, if premiums cease just before the seventh premium is due.

6.2 The annual premium for a 40-year capital redemption policy with sum assured £10000 is P' for the first ten years, $P'/2$ for the next ten years, and $P'/4$ for the final twenty years. The premium is calculated on the basis of an effective annual rate of interest of 5% with allowance for initial expenses of £50 plus 10% of the first premium and for renewal expenses of 4% of the second and each subsequent premium.

Find the value of P'.

6.3 (Varying interest rates)
The level annual premium for a 20-year capital redemption policy with sum assured £26 000 is payable in advance and determined on the basis of an effective annual interest rate of 8% for the first five years, 7% for the next five years, and 6% thereafter. Allowance is made for initial expenses of £130 and for renewal expenses (associated with the payment of each premium except the first) of £15 plus 2% of the level annual premium. Find

(a) The level annual premium for the policy;
(b) The yield per annum obtained by the policyholder over the completed transaction.

6.4 (Varying interest rates)
A man effects a capital redemption policy with a sum assured of £20 000 and a term of 30 years. The level annual premium for the policy is calculated on the basis of an effective interest rate of 4% per annum for the first ten years and 3% per annum thereafter. The initial expenses are £250 and the renewal expenses are 7·5% of the second and subsequent premiums. Assuming that investment is on a year-to-year basis, calculate the annual premium for this policy.

6.5 (a) Show that

$$_1W_{\overline{n}|} = \frac{1}{\ddot{a}_{\overline{n}|}} \tag{1}$$

(b) Show further that, if $1 \leqslant t \leqslant n-1$,

$$1 - {}_tW_{\overline{n}|} = s_{\overline{n-t}|}/s_{\overline{n}|}$$

and deduce that

$$1 - {}_tW_{\overline{n}|} = (1 - {}_1W_{\overline{n-t+1}|})(1 - {}_1W_{\overline{n-t+2}|})\cdots(1 - {}_1W_{\overline{n}|}) \tag{2}$$

(c) By using equations 1 and 2, show that at non-negative rates of interest $_tW_{\overline{n}|}$ is an increasing function of the rate of interest. Deduce that, at any positive rate of interest,

$$_tW_{\overline{n}|} > \frac{t}{n} \qquad 1 \leqslant t \leqslant n-1$$

6.6 (Surrender and paid-up policy values)
On 1 January 1970 a life office issued a 30-year annual premium capital redemption policy with a sum assured of £100 000. The office calculated its premiums using annual rates of interest of 8% for the first five years, 7% for the next ten years and 6% thereafter. The office assumed that all investment would be in securities with one-year terms, and added loadings for initial expenses of

50% of the first office annual premium and for renewal expenses of 2% of all office annual premiums after the first.

(a) Calculate the office annual premium.
(b) Instead of paying the premium due on 1 January 1982 the policyholder asked for the policy to be made paid-up. The paid-up sum assured granted by the office was the greater of
  (i) The original sum assured multiplied by t/n, where n was the original term of the policy and t was the number of premiums which had been paid; and
  (ii) The accumulation, at 6% per annum interest, of the policy's surrender value until the end of the original term, the surrender value being the accumulation, at 6% per annum interest, of half the first office annual premium and 95% of all subsequent premiums received.
Calculate the paid-up sum assured granted by the office. Find also the profit or loss which will be made by the office on 1 January 2000 if the office's experience follows the premium basis, the policyholder does not surrender his paid-up policy, and the cost of maintaining the paid-up policy is negligible.

6.7 On 1 April 1960 Mr X effected a 25-year capital redemption policy with sum assured £36 000 and with premiums payable half-yearly in advance throughout the duration of the policy. The premium was calculated on the basis of an interest rate of 6% per annum convertible quarterly with allowance for expenses of 10% of the gross premiums in the first year and of 3% of all subsequent gross premiums.
   On 1 October 1974, immediately before paying the premium then due, Mr X surrendered the policy for an amount equal to the accumulated value, at an interest rate of 5% per annum effective, of 95% of all premiums paid in the second and subsequent years of the policy. At the same time Mr X invested the surrender value in a deposit account to accumulate at a constant annual compound rate of interest. On 1 April 1985 Mr X withdrew the accumulated deposit.
   Given that the effective annual yield obtained by Mr X on the completed transaction was 3·886%, find the rate of interest paid in the deposit account.

6.8 Ten years ago Mr A purchased a ground rent of £1000 per annum payable annually in arrear for 25 years. At the same time he effected a 25-year capital redemption policy with sum assured equal to his purchase price and with premiums payable annually in advance for 20 years. The annual premium for the policy was calculated on the basis of an interest rate of 3% per annum effective with allowance for initial expenses of £15 plus 10% of the first annual premium and for renewal expenses (in the second to twentieth years inclusive) of £2 plus 2% of each annual premium.
   Immediately after receiving the seventh instalment of the ground rent, but before paying the annual premium then due on the capital redemption policy, Mr A made the policy paid-up and at the same time sold the remaining instalments of the ground rent at a price to give the purchaser a yield of 6% per annum effective. The paid-up sum assured was calculated as the accumulated

amount to the maturity date of the policy, at 3% per annum effective, of the premiums received less the expenses (as allowed for in the premium basis) up to the date of alteration.

Mr A now surrenders the policy for an amount equal to the discounted value, at 4% per annum effective, of the paid-up sum assured.

Given that, over the completed transaction, Mr A has obtained a yield of 4% per annum effective, find the price he paid for the ground rent.

6.9 On 1 January 1977 Mr X effected a 20-year capital redemption policy with premiums payable half-yearly in advance throughout the duration of the policy. Two offices were asked to quote a premium. Office A calculated the premium on the basis of an effective interest rate of 7% per annum for the first ten years and 6% per annum thereafter, with allowance for expenses of 5% of each gross premium and additional initial expenses of £20. Office B calculated the premium on the basis of an effective interest rate of 7% per annum throughout, with allowance for initial expenses of £30 and expenses in policy year t ($t = 1, 2, \ldots, 20$) of $(4 + 0.5t)\%$ of each gross premium received in that year.

Mr X effected the policy with office B, since the premium per half-year quoted by that office was £0·67 less than the corresponding premium quoted by office A. The sum assured was a multiple of £100.

(a) Find the sum assured and the annual premium for the policy.
(b) On 1 July 1985, instead of paying the premium then due, Mr X surrendered the policy for a sum equal to $k\%$ of the premiums paid accumulated at an effective annual interest rate of 4%. Given that the yield obtained by Mr X over the entire transaction was 3% per annum convertible half-yearly, find the value of k.

6.10 (Surrender and paid-up policy values)
A ten-year capital redemption policy with sum assured £25 000 has premiums payable quarterly in advance. The office premium is calculated at an interest rate of 8% per annum convertible half-yearly with allowance for expenses of 3·5% of all office premiums plus a further 2% of the office premiums payable in the first year.

(a) Find the quarterly gross premium.
(b) Immediately before payment of the 17th quarterly premium the policy-holder surrenders the policy for an amount equal to the greater of
  (i) The accumulation of 94% of the office premiums paid at 6% per annum effective interest; and
  (ii) The present value at 10% per annum effective interest of the proportional paid-up sum assured. (The proportional paid-up sum assured equals the original sum assured multiplied by t/n, where n is the total number of quarterly premiums due under the policy and t is the number actually paid.)
Find the surrender value received by the policyholder and the effective annual yield obtained by him over the complete transaction.

6.11 (Zillmerized reserves)
A life office issued a capital redemption policy to secure £10 000 after 15 years.

Premiums were payable annually in advance. The office used the following basis for its calculations: interest at 4% per annum; expenses of 3% of all office premiums, with additional initial expenses of 1% of the sum assured.

(b) Find the office annual premium.
(b) Find the Zillmerized reserve immediately *after* payment of the fifth annual premium.

6.12 On 1 January 1980 an office issued a capital redemption policy under which the sum assured is payable in instalments of £2000 each due on 30 June and 31 December in each of the years 1990 to 1994 inclusive. Level premiums were payable monthly in advance, the final premium being due on 1 December 1989. The office calculated the premium assuming an effective rate of interest of 10% per annum during the first three years of the policy term, 8% per annum during the next five years and 5% per annum thereafter. The office allowed for initial expenses of £80 on the issue of the policy, together with annual expenses incurred at the beginning of each year for which premiums are payable (including the first) of £5 plus 5% of the year's premiums. Allowance was also made for an expense of £20 on the payment of each instalment of the sum assured.

(a) Calculate the monthly premium.
(b) On 31 December 1985 the policyholder requested that the policy be made paid-up and the benefits altered to a single lump sum payable on 1 January 1990. The office effected the alteration using its original assumptions to calculate the reserve and allowing for an expense of £30 on payment of the single lump sum and an expense of £50 to effect the alteration. Calculate the revised sum assured.

6.13 (Premium calculations allowing for lapses)
A life office has just issued a large number of 20-year annual premium capital redemption policies, each for a sum assured of £1000. The office calculated its premiums assuming 5% per annum interest, expenses of 2·5% of all office premiums plus additional initial expenses of 2% of the sum assured, and made an allowance for lapses on all policy anniversaries. The office assumed that on each policy anniversary 1·869% of the policies in force in the preceding year would lapse due to non-payment of premium. When a policy lapses the gross premiums paid are returned immediately without interest.

Calculate the office annual premium for each of these policies and the corresponding office annual premium calculated without allowance for lapses.

6.14 (Reinvestment rates; alternative derivation of formula for proceeds of a single premium)
Let t and n be positive integers with $0 \leqslant t < n$ and let S_t be defined as in section 6.7; thus S_t is the total accumulated proceeds at time n arising from a single payment of 1 at time t.

Write down the value of S_{n-1} and explain why the following recurrence relationship holds:

$$S_t = 1 + i_t + i_t(S_{t+1} + S_{t+2} + \cdots + S_{n-1}) \qquad 0 \leqslant t < n - 1$$

Hence derive equation 6.7.5.

6.15 (Varying reinvestment rates)

Ten years ago an investor purchased a ground rent of £2000 per annum payable quarterly in arrear for 15 years. At the same time he took out a 15-year capital redemption policy with sum assured equal to his purchase price. Level premiums were payable annually in advance and were calculated on the following basis:

(a) Investment of the first premium, less expenses, would produce an income each year of 10% of the amount invested;
(b) Subsequent premiums less expenses and all interest, including that arising from the initial investment, would produce an income each year of 8% of the amount invested;
(c) Initial expenses would be £100 plus 12% of the first premium;
(d) Renewal expenses would be 8% of subsequent premiums.

The policy is made paid-up immediately before payment of the 11th premium, the office allowing a proportionate paid-up sum assured.

The investor calculates that his return on the whole transaction will be 10% per annum convertible half-yearly. What price did he pay for the ground rent?

6.16 (Varying reinvestment rates)

A 20-year capital redemption policy is issued with premiums payable annually in advance. The amount of the annual premium increases by £50 each year.

In calculating the premiums for this policy the life office assumes that

(a) The initial premium will be invested to produce an income of 10% per annum;
(b) Subsequent premiums and all amounts available for reinvestment up to and including the date of payment of the 11th premium will be invested to produce an income of 8% per annum; and
(c) Premiums from the 12th and all amounts available for reinvestment on or after the date of payment of the 12th premium will be invested to produce an income of 6% per annum.

Given that the sum assured is £63 386, find the first premium payable under the policy. (Ignore expenses.)

6.17 For both single and annual premium capital redemption policies with terms of up to 25 years, an office formerly used an interest rate of 10% per annum effective. However, because of currently available higher interest rates, the office is now calculating premiums on the basis of the reinvestment model of section 6.7 with $i_0 = 13\%$ and $i_t = 10\%$ for $t \geqslant 1$. Annual premiums are payable in advance.

The expenses allowed for in the premium bases are as follows:

Annual premium policies Initial expenses of £80 plus 5% of the first premium; renewal expenses of $2\frac{1}{2}\%$ of the second and each subsequent premium.

Single premium policies Initial expenses of £80 plus 7·5% of the single premium; no further expenses.

On both (a) the former basis and (b) the current basis, find the single premium and the level annual premium for a capital redemption policy with sum assured £10 000 and term 25 years.

Find also the corresponding premiums for terms five and fifteen years. Comment on your answers.

6.18 On the basis of an interest rate of 10%, find the values of

(a)
$$_{15}V_{\overline{20}|}, \quad _{16}V_{\overline{20}|}, \quad _{15\cdot5}V_{\overline{20}|}$$

(b)
$$_{15\cdot5}V^{(4)}_{\overline{20}|}, \quad _{15\cdot75}V^{(4)}_{\overline{20}|}, \quad _{15\cdot65}V^{(4)}_{\overline{20}|}$$

For both (a) and (b), compare the accurate third answer with the value obtained by linear interpolation between the first answer plus the appropriate premium and the second answer.

CHAPTER SEVEN

THE VALUATION OF SECURITIES

7.1 Introduction

One of the most important areas of practical application of compound interest theory is in the valuation of stock market securities and the determination of their yields. Accordingly, we shall consider this subject in some detail. In the present chapter we first give a description of some stock market terminology, particularly that used in connection with fixed-interest securities. The reader requiring a fuller treatment of this subject should refer to a book on investments, such as that by Day and Jamieson (reference [11]). Formulae for calculating the yields obtainable on fixed-interest securities are given below, and we also consider Makeham's formula. Some of the more complicated topics (e.g. cumulative sinking funds and capital gains tax) will be considered in later chapters.

7.2 Fixed-interest securities

A government, local authority, private company or other body may raise money by floating a loan on a stock exchange. The terms of the issue are set out by the borrower and investors may be invited to subscribe to the loan at a given price (called the *issue price*), or the issue may be by tender, in which case investors are invited to nominate the price that they are prepared to pay and the loan is then issued to the highest bidders, subject to certain rules of allocation. In either case the loan may be underwritten by a financial institution, which thereby agrees to purchase, at a certain price, any of the issue which is not subscribed to by other investors.

Fixed-interest securities normally include in their title the rate of interest payable, e.g. $2\frac{1}{2}\%$ Consols, a British government stock. The annual interest payable to each holder, which is often but not invariably payable half-yearly, is found by multiplying the *nominal amount* of his holding N by the rate of interest per annum D, which is generally called the *coupon rate*. (For example, in the case of $2\frac{1}{2}\%$ Consols, $D = 0.025$.) If an investor is liable to income tax at rate t_1 on the interest payments, his annual income after tax will be $(1 - t_1)DN$.

The money payable at redemption is calculated by multiplying the nominal amount held N by the *redemption price* R per unit nominal (which is often quoted 'per cent' in practice). If $R = 1$ the stock is said to be redeemable *at par*; if $R > 1$ the stock is said to be *redeemable above par* or *at a premium*; and if $R < 1$ the stock is said to be *redeemable below par* or *at a discount*. Some securities have varying coupon rates D or varying redemption prices R; these will be discussed later. The *redemption date* is the date on which the redemption money is due to be paid. Some bonds have variable redemption dates, in which case the redemption date may be chosen by the borrower (or perhaps the lender) as any interest date within a certain period, or any interest date on or after a given date. In the latter case the stock is said to have no final redemption date, or to be *undated*. We consider as illustrations the following British government stocks:

(a) *9% Treasury 1994* This stock, which was issued in 1969, bears interest at 9% per annum, payable half-yearly on 17 May and 17 November. The stock is redeemable at par on 17 November 1994.

(b) *3% British Gas 1990–5* This stock was issued in 1949 following the nationalization of the gas industry. It bears interest at 3% per annum, payable half-yearly on 1 May and 1 November, and is redeemable at par on any interest date between 1 May 1990 and 1 November 1995 (inclusive) at the option of the government. Since the precise redemption date is not predetermined, but may be chosen between certain limits by the borrower, this stock is said to have 'optional redemption dates' (see section 7.8).

(c) *3½% War Loan* This stock was issued in 1932 as a conversion of an earlier stock, issued during the 1914–1918 war. Interest is at 3½% per annum, payable half-yearly on 1 June and 1 December. This stock is now redeemable at par on any interest date the government chooses, there being no final redemption date. The stock may therefore be considered as having optional redemption dates, the second of them being infinity, and may be valued by methods discussed in section 7.8.

Note that in all three examples $R = 1$ and the frequency of interest payments per annum (which we shall denote by p) is 2. In fact, all British government stocks are redeemable at par (although, as in example (c), there may be no final redemption date), and all pay interest half-yearly except 2½% Consols (which pays interest quarterly on 5 January, 5 April, 5 July, and 5 October), 2½% Annuities and 2¾% Annuities.

The issue price and subsequent market prices of any stock are usually quoted in terms of a certain nominal amount, e.g. £100 or £1 nominal. A loan may thus be considered to consist of, say, £10 000 000 nominal divided into 100 000 bonds each of £100 nominal. The statement that an investor owns a bond of £100 nominal does not in general imply that his holding of the stock in question is worth £100. If it is worth £105, say, it is said to be *above par* or *at a premium*; if it is worth £90, say, it is said to be *below par* or *at a discount*; and if it is worth £100 the stock is said to be *at par*. We use the symbol P to denote the price per unit nominal and the symbol A for the price of nominal amount N of the stock. Thus

$$A = NP \qquad (7.2.1)$$

In practice stocks are generally quoted 'per cent', i.e. per £100 nominal. We also use the symbol C, where

$$C = NR \qquad (7.2.2)$$

to denote the *cash* received on redemption in respect of a nominal amount N of the stock.

The coupon rate, redemption price and term to redemption of a fixed-interest security serve to define the cash payments promised to a tax-free investor in return for his purchase price. If the investor is subject to taxation, appropriate deductions from the cash flow must of course be made. The value of fixed-interest stocks at a given rate of interest and the determination of their yields at a given price may be found as for any other investment or business project (see chapter 5), but, in view of the practical importance of fixed-interest securities and the special terminology used, we discuss them in detail in this chapter.

Fixed-interest securities may be classified as follows:

(a) *British government stocks, also known as gilt-edged stocks or gilts* These securities, which exist in large volumes, may be considered to offer perfect security against default by the borrower and may be bought and sold with relatively little expense. Gilts may be classified according to the term to redemption, namely Treasury Bills (three months), short-dated (up to five years), medium-dated, long-dated and undated (e.g. $3\frac{1}{2}\%$ War Loan). Since February 1986 almost all fixed-interest securities have been dealt with on the basis that the purchaser pays and the seller receives not only the quoted market price but also *accrued interest* (see equation 7.10.3) calculated from the last interest date. In recent years *index-linked* government securities have been issued in the UK. These securities, which carry interest and redemption payments linked to the Retail Prices Index, will be discussed in section 7.11.

(b) *Securities issued by British local authorities, public boards and nationalized industries* These stocks may in general be considered to have little risk of default, although the example of the Mersey Docks and Harbour Board (which went bankrupt in 1970) shows that caution should be exercised in considering the security of some of these loans.

(c) *Loans issued by overseas governments* An investor considering such securities must consider questions of currency appreciation or depreciation (unless the loan is in sterling), exchange control, taxation agreements and, in some cases, the possibility of default.

(d) *Loans issued by overseas provinces, municipalities, etc.* These stocks should of course be subjected to closer scrutiny than those of the country itself. In the United States, for example, many local stocks went into default in the 1930s.

(e) *Debentures and unsecured loan stocks* These are issued by various British and overseas companies. Even a very secure debenture will be less marketable than a government stock of similar term and have greater dealing costs. Accordingly, the yield required by investors will be higher than for a comparable government stock. In the United States such stocks are known as bonds and are classified according to their credit ratings; yields are, of course, higher for the lower-rated stocks. Yield calculations allowing for the possibility of default may be carried out, but generally we shall ignore this possibility in the valuation of securities.

There are also stocks with a coupon rate which varies according to changes in a standard rate of interest such as the rate on Treasury Bills. These stocks, and the index-linked securities mentioned above, are not fixed-interest securities in the strict sense, but have more in common with them than with ordinary shares (see section 7.3).

7.3 Ordinary shares

In addition to dealing in fixed-interest securities (which are known as 'bonds' in the United States and some other countries), the stock market deals in ordinary shares or equities (known as 'common stocks' in the United States).

Ordinary shares or *equities* are securities, issued by commercial undertakings and other bodies, which entitle their holders to receive all the net profits of the company after interest on loans and fixed-interest stocks has been paid. The cash paid out each year is called the dividend, the remaining profits (if any) being retained as reserves or to finance the company's activities. The holders of ordinary shares are proprietors, not creditors, of the company, and they usually have voting rights at the company's Annual General Meeting, although non-voting shares can be issued in the UK (but not the USA). We shall not be much concerned with equities in this book, although in example 7.3.1 we consider the 'compound interest' valuation of ordinary shares.

Reference should also be made to *preference shares*, which, despite their name, are fixed-interest securities issued by commercial companies. Interest (or, as they are usually called, dividend) payments are met only after the interest due on any bank loans and debentures has been paid, but before anything is paid to the ordinary shareholders. If there are insufficient divisible profits in any year to meet the dividends for the preference shareholders, the dividend payments are reduced or, in extreme cases, passed. In the case of cumulative preference shares, all arrears of dividend are carried forward until they are paid off (although the arrears themselves do not earn interest).

If a security is bought *ex dividend* (x.d.), the seller, not the buyer, will receive the next interest or dividend payment. If it is bought *cum dividend*, then the buyer will receive the next interest or dividend payment. (As mentioned above, most fixed-interest stocks are traded on the basis that the purchaser makes a payment to the seller for accrued interest, in proportion to the number of days between the date of purchase and the date of the last interest payment.)

Example 7.3.1 (Compound interest valuation of ordinary shares)

A pension fund, which is not subject to taxation, has a large portfolio of UK ordinary shares with a current market value of £14 700 000. The current rate of dividend payment from the portfolio is £620 000 per annum. The pension fund wishes to value its holding, at an interest rate of 6% per annum effective, on the assumption that (a) both dividend income from the shares and their market value will increase continuously at the rate of 2% per annum, and (b) the shares will be sold in 30 years' time.

What value should the pension fund place on the shares?

Solution

The value to be placed on the holding is (see section 5.5)

$$620\,000 \int_0^{30} (1{\cdot}02v)^t dt + 14\,700\,000 (1{\cdot}02v)^{30} \qquad \text{at } 6°_0$$

$$= 620\,000 \left[\frac{1 - (1{\cdot}02v)^{30}}{-\log(1{\cdot}02v)} \right] + 14\,700\,000 (1{\cdot}02v)^{30}$$

$$= (620\,000 \times 17{\cdot}798) + 4\,636\,000 = £15\,671\,000$$

Note This method gives an answer rather different from the market value of £14 700 000, and hence may be used only under certain circumstances, a discussion of which is beyond the scope of this book. (The difference between the two values lies, of course, in the fact that the market is making different assumptions concerning the future.) The method also ignores the variability of share prices and dividends and is therefore more suitable for valuing a large, diverse portfolio than a single holding. More sophisticated methods for valuing equities, using certain statistical techniques, have been developed but we do not discuss them here.

7.4 Prices and yields

We now return to the discussion of fixed-interest stocks. As in other compound interest problems, one of two questions may be asked:

(a) What price A, or P per unit nominal, should be paid by an investor to secure a net yield of i per annum?
(b) Given that the investor pays a price A, or P per unit nominal, what net yield per annum will he obtain?

To answer question (a), we set A equal to the present value, at rate of interest i per annum, of the interest and capital payments, less any taxes payable by the investor. That is,

$$A = \begin{pmatrix} \text{present value, at rate} \\ \text{of interest } i \text{ per annum,} \\ \text{of } net \text{ interest payments} \end{pmatrix} + \begin{pmatrix} \text{present value, at rate} \\ \text{of interest } i \text{ per annum,} \\ \text{of } net \text{ capital payments} \end{pmatrix} \quad (7.4.1)$$

The price per unit nominal is of course $P = A/N$, where N is the nominal amount of stock to which the payments relate.

To answer question (b), we set A in equation 7.4.1 equal to the purchase price and solve for the net yield i. If the investor is not subject to taxation the yield i is referred to as a *gross* yield. The yield quoted in the press for a fixed-interest security is often the gross nominal yield per annum, convertible half-yearly (see section 7.10). If the investor sells his holding before redemption, or if he is subject to taxation, his actual yield will in general be different from that quoted.

The yield on a security is sometimes referred to as the *yield to redemption* or the *redemption yield* to distinguish it from the *flat* (or *running*) yield, which is defined as D/P, the ratio of the coupon rate to the price per unit nominal of the stock.

Example 7.4.1

A certain debenture (i.e. a fixed-interest stock issued by a commercial company) was redeemable at par on 1 October 1937. The stock bore interest at 6% per annum, payable half-yearly on 1 April and 1 October.

(a) What price per cent should have been offered for this stock on 1 August 1915 to secure a yield of 5% per annum for a tax-free investor?
(b) What yield per annum did this stock offer to a tax-free investor who bought it at 117% on 1 August 1915?

Solution

(a) In this example we have $R = 1$, $N = 100$, $C = 100$, $D = 0.06$, and $p = 2$. The price A which should be offered on 1 August 1915 to secure a yield of 5% per annum is, by equation 7.4.1,

A = present value at 5% of interest payments

      + present value at 5% of capital payment

$$= v^{1/6}[3 + 6a_{\overline{22}|}^{(2)} + 100v^{22}] \qquad \text{at } 5\%$$

$$= 116 \cdot 19$$

(b) We now solve the equation of value

$$117 = v^{1/6}[3 + 6a_{\overline{22}|}^{(2)} + 100v^{22}] \qquad (1)$$

for the rate of interest i. By part (a), when $i = 5\%$ the right-hand side of equation 1 is 116·19, so the yield is rather below 5% per annum. Further trials and interpolation give $i \simeq 4 \cdot 94\%$ per annum.

Note   One could also work with a period of half a year; the corresponding equation of value would then be

$$117 = v^{1/3}(3 + 3a_{\overline{44}|} + 100v^{44}) \qquad \text{at rate } i'$$

which has approximate solution $i' \simeq 0 \cdot 0244$, so the effective yield per annum is

$$i \simeq (1 \cdot 0244)^2 - 1 \qquad \text{or} \qquad 4 \cdot 94\%, \text{ as before}$$

When one has to solve an equation of value by interpolation (to find a yield), it is convenient to have a rough idea of the order of magnitude of the required solution. In most situations upper and lower bounds may be found quite simply, as the following discussion indicates.

Consider a loan which will be redeemed after n years at a redemption price of R per unit nominal. Suppose that the loan bears interest, payable annually in arrear at a coupon rate of D per annum, and that an investor who is liable to income tax at rate t_1 buys the loan at a price P per unit nominal. What can be said about the magnitude of i, the investor's net annual yield?

In return for a payment of P the investor receives net interest each year of $D(1 - t_1)$ and redemption proceeds of R. His net yield i is thus that interest rate for which

$$P = D(1 - t_1)a_{\overline{n}|} + Rv^n \qquad (7.4.2)$$

If $R = P$, then clearly

$$i = \frac{D(1 - t_1)}{P}$$

If $R > P$, there is a gain on redemption, and therefore

$$i > \frac{D(1 - t_1)}{P}$$

In this case the gain on redemption is $(R - P)$. If the investor were to receive this gain in equal instalments each year over the n years rather than as a lump sum after n years, he would clearly be in a more advantageous position. In this case each year he would receive $D(1 - t_1) + (R - P)/n$ as income (and P as redemption proceeds), so his net annual yield would be $[D(1 - t_1) + (R - P)/n]/P$. This overstates i, so

$$\frac{D(1 - t_1)}{P} < i < \frac{D(1 - t_1) + (R - P)/n}{P}$$

If $R < P$ there is a loss on redemption, and hence

$$i < \frac{D(1 - t_1)}{P}$$

The loss on redemption is $(P - R)$. If the investor had to bear this loss in equal instalments each year over the n years rather than as a lump sum after n years, he would clearly be in a less advantageous position. In this case each year he would receive $D(1 - t_1) - (P - R)/n$ as income (and P as redemption proceeds) so his net annual yield would be $[D(1 - t_1) - (P - R)/n]/P$. This understates i, so

$$\frac{D(1 - t_1)}{P} > i > \frac{D(1 - t_1) - (P - R)/n}{P} = \frac{D(1 - t_1) + (R - P)/n}{P}$$

Thus in all cases i lies between $D(1 - t_1)/P$ and $[D(1 - t_1) + (R - P)/n]/P$. For most practical purposes these bounds are sufficient to indicate suitable values to be used for interpolation.

Example 7.4.2

A stock bears interest at $7\frac{1}{2}\%$ per annum, payable annually in arrear, and is redeemable at par in 20 years' time. Assuming that any interest now due will not be received by a purchaser, find the net yield per annum to an investor, liable to income tax at $33\frac{1}{3}\%$, who buys a quantity of this stock at 80%.

Solution

Note that, since the net annual interest payment is £5 on an outlay of £80 (i.e. $6\frac{1}{4}\%$) and the stock is redeemed for £100, the net yield will certainly exceed $6\frac{1}{4}\%$ per annum. The gain on redemption is £20 per £100 nominal. If this gain were paid in equal annual instalments (each of amount £1), an outlay of £80 would provide net income each year of £6, or $7\frac{1}{2}\%$. This would be a more advantageous investment than is actually available. The net annual yield is thus less than $7\frac{1}{2}\%$. We have coupon rate $D = 0.075$, price paid per unit nominal $P = 0.8$, redemption price per unit nominal $R = 1$, rate of income tax $t_1 = \frac{1}{3}$, and term to redemption $n = 20$. The equation of value is

$$P = D(1 - t_1)a_{\overline{n}|} + Rv^n \qquad \text{at rate } i$$

i.e.

$$0.8 = 0.05a_{\overline{20}|} + v^{20}$$

The above remarks indicate that i lies between 0·0625 and 0·075. When $i = 0·065$ the right-hand side of the last equation equals 0·8347, and when $i = 0·07$ the value is 0·7881. By interpolation we estimate i as 0·0687 or 6·87%. (In fact, to four decimal places, the net annual yield per cent is 6·8686. The interpolation method is thus very accurate in this case.)

An approximate solution for the rate of interest in equation 7.4.2 may be found as follows. Let $g = D/R$, so $g(1 - t_1)$ is the net interest per annum per unit redemption price. Equation 7.4.2 may now be written in the form

$$P = g(1 - t_1)Ra_{\overline{n}|} + Rv^n \qquad \text{at rate } i$$
$$= R[g(1 - t_1)a_{\overline{n}|} + (1 - ia_{\overline{n}|})]$$

from which we obtain

$$g(1 - t_1) - i - \frac{k}{a_{\overline{n}|}} = 0 \qquad (7.4.3)$$

where $a_{\overline{n}|}$ is calculated at rate i, and

$$k = (P - R)/R \qquad (7.4.4)$$

Much ingenuity has been expended on finding approximate solutions of equation 7.4.3. Here we consider only approximations based on the Maclaurin expansion of $1/a_{\overline{n}|}$, i.e.

$$\frac{1}{a_{\overline{n}|}} = \frac{i}{1 - (1 + i)^{-n}} = \frac{1}{n} + \frac{(n + 1)}{2n}i + \frac{(n^2 - 1)}{12n}i^2 + \cdots \qquad (7.4.5)$$

Ignoring powers of i above the first in equation 7.4.5, and substituting for $1/a_{\overline{n}|}$ in equation 7.4.3, we obtain

$$i = \frac{g(1 - t_1) - k/n}{1 + \left[\dfrac{n + 1}{2n}\right]k} \qquad (7.4.6)$$

This formula is generally quite accurate when n and i are not large. Greater accuracy can usually be obtained by considering powers of i up to i^2. Equations 7.4.3 and 7.4.5 then give the quadratic equation

$$\frac{k(n^2 - 1)}{12n}i^2 + \left[1 + \frac{k(n + 1)}{2n}\right]i - \left[g(1 - t_1) - \frac{k}{n}\right] = 0 \qquad (7.4.7)$$

In practical circumstances only one root of this equation is appropriate, the other being remote from the value given by equation 7.4.6. Applying formulae 7.4.6 and 7.4.7 to example 7.4.2, for which $k = -0.2$, $n = 20$ and $g(1 - t_1) = 0.05$, we obtain the approximate yields 6·70% and 6·80% respectively.

7.5 Perpetuities

Certain loans, especially those made in connection with the purchase or lease of land, may continue indefinitely or for so long (e.g. 999 years) as to be considered of indefinite duration for practical purposes. Such loans are referred to as *perpetuities* or *perpetual loans*. In some cases the 'interest' on these loans represents an annual sum payable to a landowner or feudal superior in lieu of military service or other duties. An example of this kind is the *feu-duty* payable by a feuar (or lessee) of a piece of land to his feudal superior (or landlord) under Scots law. Also, if a security is undated (i.e. if it has no final redemption date), it should in some circumstances be valued as a perpetuity (see section 7.8).

Let us assume that the next interest payment is due at time t years from the present and that interest is at rate D per annum per unit nominal, payable p times per annum. The price P per unit nominal to give a net yield of i per annum, or i, the net yield per annum to an investor who pays P per unit nominal, and is liable to income tax at rate t_1, may be found from the equation

$$P = D(1 - t_1)v^t \ddot{a}^{(p)}_{\overline{\infty}|} \qquad \text{at rate } i$$

i.e.

$$P = \frac{D(1 - t_1)v^t}{d^{(p)}} \qquad \text{at rate } i \qquad (7.5.1)$$

Example 7.5.1

As stated in section 7.2, the interest on $3\frac{1}{2}$% War Loan is payable on 1 June and 1 December each year. Considering this stock to be a perpetuity, find (a) the effective yield per annum and (b) the nominal yield per annum, convertible half-yearly, to a tax-free investor on 22 August 1983, when the price was 34·875%.

Solution

In the notation of equation 7.5.1 we have $P = 0.34875$, $D = 0.035$, $p = 2$ and $t = 101/365 = 0.27671$. We therefore solve the equation

$$0.34875 = 0.035[v^{0.27671}/d^{(2)}]$$

for i. This gives $i = 10.53\%$ and hence $i^{(2)} = 2[(1 + i)^{1/2} - 1] \simeq 10.27\%$. The latter yield, $i^{(2)}$, was quoted for $3\frac{1}{2}$% War Loan in the financial press on 22 August 1983.

Example 7.5.2

A feu-duty of £30 per annum is payable half-yearly on 15 May and 15 November. Find the present value of future payments on 15 March 1984 to a tax-free investor, using an interest rate of 10% per annum convertible half-yearly.

Solution

Work in half-year periods: the present value is

$$\frac{15v^{1/3}}{d} \quad \text{at } 5\% \quad = £309\!\cdot\!92$$

Note The Land Tenure Reform (Scotland) Act 1974 gives feuars rights to 'buy out' future payments of feu-duty. The cost of doing so is calculated by multiplying the annual feu-duty by a 'feu-duty factor' which is the present value of a feu-duty payment of 1 per annum in perpetuity. Although feu-duty is payable half-yearly (on 15 May and 15 November) and the interest on $2\frac{1}{2}\%$ Consols is payable quarterly (on 5 January, 5 April, 5 July, and 5 October), these factors are calculated weekly by means of the formula

$$\text{feu-duty factor} = \frac{\text{market price per cent nominal of } 2\frac{1}{2}\% \text{ Consols}}{2\cdot5} \quad (7.5.2)$$

(This practical formula is based on simple proportion and ignores differences in the incidence of the interest on $2\frac{1}{2}\%$ Consols and feu-duty payments during the year.) Feu-duty factors are published regularly in the Scottish press.

7.6 Makeham's formula

Consider a loan, of nominal amount N, which is to be repaid after n years at a price of R per unit nominal, and let $C = NR$. Thus C is the *cash* payable on redemption. Let the coupon rate (i.e. the annual interest per unit *nominal*) be D, and assume that interest is payable $p$thly in arrear. Thus each interest payment is of amount $DN/p = gC/p$ where

$$g = \frac{DN}{C} \quad (7.6.1)$$

$$= \frac{D}{R} \quad (7.6.2)$$

Note that g is the annual rate of interest *per unit of redemption price*.

Consider an investor, liable to income tax at rate t_1, who wishes to purchase the loan at a price to provide an effective net yield of i per annum. Let the price he should pay be A. (We assume that n is an integer multiple of

1/p and that any interest now due will not be received by the purchaser.) The price is simply the present value (at rate i) of the redemption proceeds and the future net interest payments. Thus

$$A = NRv^n + (1 - t_1)DNa_{\overline{n}|}^{(p)} \qquad \text{at rate } i$$

$$= Cv^n + (1 - t_1)gCa_{\overline{n}|}^{(p)}$$

$$= Cv^n + (1 - t_1)gC\frac{1 - v^n}{i^{(p)}}$$

$$= Cv^n + \frac{g(1 - t_1)}{i^{(p)}}(C - Cv^n)$$

Hence

$$A = K + \frac{g(1 - t_1)}{i^{(p)}}(C - K) \qquad\qquad (7.6.3)$$

where $K = Cv^n$ (at rate i) is the present value of the capital repayment, and $[g(1 - t_1)/i^{(p)}](C - K)$ is the present value of the net interest payments. Equation 7.6.3 is known as *Makeham's formula* and is of great importance. Note that $g(1 - t_1)$ is the net rate of annual interest payment *per unit redemption price* or per unit 'indebtedness'. Equation 7.6.3 is valid only when

(a) g, t_1 and R are constant throughout the term of the loan; and
(b) n is an integer multiple of $1/p$.

(See section 7.9 for the necessary modifications when these conditions are not satisfied.)

Makeham's formula remains true when the loan is repayable by instalments, provided that the coupon rate D, the rate of income tax t_1, and the redemption price R per unit nominal remain constant. To show this, we consider a loan of nominal amount $N = N_1 + N_2 + \cdots + N_m$, the nominal amount to be redeemed at time n_j being $N_j (j = 1, 2, \ldots, m)$.

The cash received on repayment of part of the loan at time n_j is $C_j = RN_j$. Equation 7.6.3 implies that the value of the capital and net interest payments associated with the jth 'tranche' of the loan is

$$A_j = K_j + \frac{g(1 - t_1)}{i^{(p)}}(C_j - K_j)$$

where

$$K_j = C_j v^{n_j}$$

The value of the entire loan is clearly

$$A = \sum_{j=1}^{m} A_j$$

$$= \sum_{j=1}^{m} \left[K_j + \frac{g(1 - t_1)}{i^{(p)}} (C_j - K_j) \right]$$

$$= K + \frac{g(1 - t_1)}{i^{(p)}} (C - K)$$

where

$$K = \sum_{j=1}^{m} K_j = \sum_{j=1}^{m} C_j v^{n_j}$$

is the value of the capital payments and

$$C = \sum_{j=1}^{m} C_j = \sum_{j=1}^{m} RN_j = R \sum_{j=1}^{m} N_j = RN$$

as before. Equation 7.6.3 thus is still valid in this situation.

The present value, or price, per unit nominal is of course $P = A/N$, provided that one purchases the *entire* loan (see section 7.7). The attractiveness of Makeham's formula lies in the fact that it enables the value of the (net) interest payments and the total value of the security to be obtained quickly from the value of the capital K, even when the stock is redeemable by instalments.

Makeham's formula may also be established by general reasoning as follows. Consider a second loan of the same total nominal amount N as the loan described above. Suppose that, as before, interest is payable $p$thly in arrear and that a nominal amount N_j of this second loan will be redeemed at time n_j $(1 \leqslant j \leqslant m)$ at a price of R per unit nominal. (The capital repayments for this second loan are thus identical to those of the original loan.) Suppose, however, that for this second loan the *net* annual rate of interest *per unit of redemption price* is $i^{(p)}$. The total 'indebtedness' (i.e. capital to be repaid) for either loan is $C = NR$. Since, by hypothesis, the net annual rate of interest payment per unit indebtedness for the second loan is $i^{(p)}$, the value at rate i of this loan is clearly C. Let K be the value of the capital payments of this second loan. (Of course, K is also the value of the capital payments of the first loan.) Then the value of the net interest payments for the second loan must be $(C - K)$. The difference between the two loans lies simply in the rate of payment of net interest. The net annual rate of interest per unit redemption price is $g(1 - t_1)$ for the original loan and $i^{(p)}$ for the second loan. *By proportion*, therefore, the value of the net interest payments for the *first* loan is obviously $g(1 - t_1)/i^{(p)}$ times the value of the net interest payments for the second loan, i.e.

$$\frac{g(1-t_1)}{i^{(p)}}(C-K)$$

The value of the first loan, being the value of the capital plus the value of the net interest, is therefore

$$K + \frac{g(1-t_1)}{i^{(p)}}(C-K)$$

as given by equation 7.6.3.

A clear grasp of the above 'proportional' argument can simplify the solution of many problems.

Note that Makeham's formula may be used either to value the security at a given rate of interest i, or to find the yield when the price A is given. It is clear that A decreases as i increases, since A is the present value of a series of positive cash flows. Equivalently, as A increases the corresponding yield i decreases. It also follows directly from Makeham's formula that

$$\left.\begin{array}{lll} \text{(a)} & A = C & \text{if} \quad g(1-t_1) = i^{(p)} \\ \text{(b)} & A > C & \text{if} \quad g(1-t_1) > i^{(p)} \\ \text{(c)} & A < C & \text{if} \quad g(1-t_1) < i^{(p)} \end{array}\right\} \qquad (7.6.4)$$

These results are, of course, obvious by general reasoning.

Example 7.6.1

A loan of £75 000 is to be issued bearing interest at the rate of 8% per annum payable quarterly in arrear. The loan will be repaid at par in 15 equal annual instalments, the first instalment being repaid five years after the issue date.

Find the price to be paid on the issue date by a purchaser of the whole loan who wishes to realize a yield of (a) 10% per annum effective, and (b) 10% per annum convertible half-yearly. (Ignore taxation.)

Solution

The capital repayments are each of amount £5000. The first repayment is after five years and the final repayment is after 19 years.

(a) Choose one year as the basic unit of time. The required yield per unit time is 10% so $i = 0.10$. Using the notation above, we have $C = 75\,000$ (since redemption is at par). The value of the capital repayment is

$$K = 5000(a_{\overline{19}|} - a_{\overline{4}|}) \qquad \text{at } 10\% \qquad = 25\,975.27$$

Note that, since redemption is at par, $g = 0.08$ and interest is paid quarterly (i.e. four times per time unit) so $p = 4$. From Makeham's formula we obtain the required price as

$$25\,975.27 + \frac{0.08}{0.10^{(4)}}(75\,000 - 25\,975.27) = £66\,636.60$$

Since $66\,636\cdot60/75\,000 = 0\cdot8885$, this price may be quoted as £88·85%.
(b) Choose six months as the basic unit of time. The required yield per unit time is 5%. Thus $i = 0\cdot05$. Note now that interest is paid *twice* per time unit, so in the notation above $p = 2$. Also, per time unit the amount of interest payable is 4% of the outstanding loan, so now we have $g = 0\cdot04$. The capital repayments occur at times $10, 12, 14, \ldots, 38$, so

$$K = \frac{5000}{a_{\overline{2}|}}(a_{\overline{40}|} - a_{\overline{10}|}) \qquad \text{at } 5\% \qquad = 25\,377\cdot27$$

Hence the value of the entire loan is

$$25\,377\cdot27 + \frac{0\cdot04}{0\cdot05^{(2)}}(75\,000 - 25\,377\cdot27)$$

$$= £65\,565\cdot63 \qquad \text{or} \qquad £87\cdot42\%$$

Note that this price is lower than that in (a).

The reader who chooses to value the interest payments from first principles will rapidly be convinced of the value of Makeham's formula.

Example 7.6.2

In relation to the loan described in the previous example, find the price to be paid on the issue date by a purchaser of the entire loan who is liable to income tax at the rate of 40% and wishes to realize a net yield of 7% per annum effective.

Solution

The capital payments have value

$$K = 5000(a_{\overline{19}|} - a_{\overline{4}|}) \qquad \text{at } 7\% \qquad = 34\,741\cdot92$$

Hence the price to provide a net yield of 7% per annum effective is

$$34\,741\cdot92 + \frac{0\cdot08(1 - 0\cdot4)}{0\cdot07^{(4)}}(75\,000 - 34\,741\cdot92)$$

$$= £63\,061\cdot89 \qquad \text{or} \qquad £84\cdot08\%$$

Example 7.6.3

A loan of nominal amount £80 000 is redeemable at 105% in four equal instalments at the end of 5, 10, 15, and 20 years. The loan bears interest at the rate of 10% per annum payable half-yearly.

An investor, liable to income tax at the rate of 30%, purchased the entire loan on the issue date at a price to obtain a net yield of 8% per annum effective. What price did he pay?

Solution

Note that the total indebtedness C is $80\,000 \times 1\cdot05$, i.e. £84 000. Each year the total interest payable is 10% of the outstanding nominal loan, so that the interest payable each year is g times the outstanding indebtedness, where $g = 0\cdot1/1\cdot05$.

Choose one year as the unit of time. Then $i = 0.08$ and at the issue date the capital payments have value

$$K = 20\,000 \times 1.05(v^5 + v^{10} + v^{15} + v^{20}) \qquad \text{at } 8\%$$

$$= 21\,000\frac{a_{\overline{20}|}}{s_{\overline{5}|}} \qquad \text{at } 8\% \qquad = 35\,144.90$$

Using the value of g described above, we obtain the price paid by the investor as

$$35\,144.90 + \frac{0.1}{1.05}\frac{(1-0.3)}{0.08^{(2)}}(84\,000 - 35\,144.90) = 76\,656.07$$

Note that the price 'per cent' is the price per £100 *nominal* amount of loan i.e. $(76\,656.07/80\,000) \times 100 = £95.82$.

Example 7.6.4

A loan of nominal amount £1 200 000 is to be issued bearing interest of 11% per annum payable half-yearly. At the end of each year part of the loan will be redeemed at 105%. The nominal amount redeemed at the end of the first year will be £10 000 and each year thereafter the nominal amount redeemed will increase by £10 000 until the loan is finally repaid. The issue price of the loan is £98.80%.

Find the net effective annual yield to an investor, liable to income tax at 40%, who purchases the entire loan on the issue date.

Solution

The term of the loan is n years, where

$$1\,200\,000 = 10\,000(1 + 2 + \cdots + n)$$
$$= 5000n(n + 1)$$

from which it follows that $n = 15$.

Note that, since the redemption price is 105%, the total indebtedness C is $1\,200\,000 \times 1.05 = 1\,260\,000$ and

$$g = \frac{0.11}{1.05} \tag{1}$$

Our unit of time is one year and $p = 2$.

At rate of interest i the capital repayments have value

$$K = 10\,000 \times 1.05 \times (Ia)_{\overline{15}|}$$

so that

$$K = 10\,500(Ia)_{\overline{15}|} \qquad \text{at rate } i \tag{2}$$

The value of the loan to provide the investor with a net yield of i per annum is thus (from equation 1)

$$A = K + \frac{0.11(1-0.4)}{1.05}\frac{}{i^{(2)}}(1\,260\,000 - K) \tag{3}$$

where K is given by equation 2.

Since the issue price is £98.80 per £100 *nominal*, the price paid by the investor was

0·988 × 1 200 000, i.e. £1 185 600. We require to find the value of i such that A (as given by equations 2 and 3) equals this figure.

Note that each £98·80 invested generates net income of £6·60 per annum and is repaid as £105. The net yield will thus be somewhat greater than 6·60/98·80 = 0·0668 or 6·68%. As a first step, therefore, we value the loan at 7%. It is left for the reader to verify that, when $i = 0·07$, $A = 1 206 860$ or 100·57%. The net yield is thus greater than 7% per annum. The reader should confirm that, when $i = 0·08$, $A = 1 127 286$ or £93·94%. By linear interpolation we estimate the net yield as

$$0·07 + \frac{(1\ 206\ 860 - 1\ 185\ 600)}{(1\ 206\ 860 - 1\ 127\ 286)} \times 0·01 = 0·0727 \quad \text{or} \quad 7·27\%$$

Example 7.6.5

Ten years ago a loan was issued bearing interest payable annually in arrear at the rate of 8% per annum. The terms of issue provided that the loan would be repaid by a level annuity of £1000 over 25 years.

An annuity payment has just been made and an investor is considering the purchase of the remaining instalments. The investor will be liable to income tax at the rate of 40% on the interest content (according to the original loan schedule) of each payment. What price should the investor pay to obtain a net yield of 10% per annum effective?

Solution

We consider four alternative solutions. Each is correct, but the relative lengths of the solutions should be noted. Our last solution makes use of a technique known as the 'indirect valuation of the capital', which is most useful in certain problems.

(a) (Solution from first principles without using Makeham's formula) The amount of the original loan was $1000a_{\overline{25}|}$ at 8%. The rth annuity payment consists of a capital payment of $1000v_{0·08}^{26-r}$ and an interest payment of $1000(1 - v_{0·08}^{26-r})$ (see chapter 3).

Fifteen payments remain to be made, the tth of which is the $(10 + t)$th overall payment. Hence the tth remaining payment consists of capital $1000v_{0·08}^{16-t}$ and interest $1000(1 - v_{0·08}^{16-t})$.

After tax the investor receives only 60% of the interest content of each payment. Accordingly, to obtain a net yield of 10% per annum he should pay

$$\sum_{t=1}^{15} (1·1)^{-t} [1000v_{0·08}^{16-t} + (1 - 0·4)1000(1 - v_{0·08}^{16-t})]$$

$$= \sum_{t=1}^{15} (1·1)^{-t} (600 + v_{0·08}^{16} \times 400 \times v_{0·08}^{-t})$$

$$= 600a_{\overline{15}|0·1} + 1·08^{-15} \times 400 \frac{1 - \left(\dfrac{1·08}{1·1}\right)^{15}}{1·1 - 1·08}$$

$$= £6080·64$$

(b) (Solution using Makeham's formula)

Note that the redemption price and rate of payment of interest are constant, so that Makeham's formula may be applied. The outstanding loan C is simply the value (on the *original* interest basis) of the remaining instalments. Thus $C = 1000a_{\overline{15}|}$ at $8\% = £8559\cdot48$. The value at 10% of the remaining capital payments is

$$K = \sum_{t=1}^{15} (1\cdot1)^{-t} \times 1000 \times v_{0\cdot08}^{16-t}$$

$$= 1000v_{0\cdot08}^{15} \frac{1 - \left(\dfrac{1\cdot08}{1\cdot1}\right)^{15}}{1\cdot1 - 1\cdot08} = 3792\cdot48$$

Hence, from Makeham's formula (with $g = 0\cdot08$), the price to be paid for a net yield of 10% is

$$3792\cdot48 + \frac{0\cdot08(1 - 0\cdot4)}{0\cdot1}(8559\cdot48 - 3792\cdot48) \qquad = £6080\cdot64 \qquad \text{as before}$$

(c) (Solution using the value of the capital, but not requiring Makeham's formula)
The value of the entire outstanding loan (with no taxation) is

$$A = 1000a_{\overline{15}|} \qquad \text{at } 10\% \qquad = 7606\cdot08$$

The value of the capital payments K has been found above to be $3792\cdot48$. The value of the *gross* interest payments I is simply the value of the entire loan less the value of the capital payments. Thus

$$I = A - K \qquad = 3813\cdot60$$

The value of the net interest payments is $0\cdot6I$, so that the price to be paid by the investor is

$$K + 0\cdot6I = 3792\cdot48 + (0\cdot6 \times 3813\cdot60) \qquad = 6080\cdot64 \qquad \text{as before}$$

(d) (Solution using indirect valuation of the capital)
Ignoring tax, we may value the loan as simply the value of the remaining instalments. Thus to an investor who is *not* liable to tax the value of the loan to give a yield of 10% per annum is

$$A = 1000a_{\overline{15}|} \qquad \text{at } 10\% \qquad = 7606\cdot08 \qquad (1)$$

as in (c).
However, if K denotes the value at 10% of the remaining capital payments, by Makeham's formula the value of the loan to an investor who is not liable to tax is

$$K + \frac{0\cdot08}{0\cdot1}(8559\cdot48 - K) \qquad (2)$$

since £8559·48 (i.e. $1000a_{\overline{15}|0\cdot08}$) is the outstanding loan. By equating these two expressions for the price, we obtain

$$7606\cdot08 = K + \frac{0\cdot08}{0\cdot1}(8559\cdot48 - K)$$

from which it follows that $K = 3792\cdot48$. Note now that the value of the *capital* payments to an investor who pays no tax is the same as the value of the capital payments to an investor who is liable only to *income* tax. (The reader should study the above argument carefully. It is important to realize that, instead of finding the value of K and then using this value to find the price, we have found the price (ignoring tax) directly and then used Makeham's formula 'backwards' to obtain the value of K.)

Having found the value of K, we may use Makeham's formula to find the price to be paid by the investor who is liable to income tax as

$$3792 \cdot 48 + \frac{0 \cdot 08(1 - 0 \cdot 4)}{0 \cdot 1}(8559 \cdot 48 - 3792 \cdot 48)$$

$$= £6080 \cdot 64$$

The arithmetical simplicity of this solution should be noted. The indirect valuation avoids the need to determine K as the sum of a geometric series.

Note that the indirect determination of K could be used in conjunction with the solution (c) above.

7.7 The effect of the term to redemption on the yield

Consider first a loan of nominal amount N which has interest payable $p$thly at the annual rate of D per unit nominal. Suppose that the loan is redeemable after n years at a price of R per unit nominal. An investor, liable to income tax at rate t_1, wishes to purchase the loan at a price to obtain a net effective annual yield of i.

As before, let $g = D/R$ and $C = NR$, so that $gC = DN$. The price to be paid by the investor is

$$A(n,i) = \begin{cases} (1 - t_1)DNa_{\overline{n}|}^{(p)} + Cv^n \\ (1 - t_1)gCa_{\overline{n}|}^{(p)} + C[1 - i^{(p)}a_{\overline{n}|}^{(p)}] \qquad \text{at rate } i \\ C + [(1 - t_1)g - i^{(p)}]Ca_{\overline{n}|}^{(p)} \end{cases} \quad (7.7.1)$$

(The last equation is obvious by general reasoning. If the net annual rate of interest per unit indebtedness were $i^{(p)}$, the value of the loan would be C. In fact the net annual rate of interest per unit indebtedness is $(1 - t_1)g$ and the second term in the right-hand side of the last equation is the value of net interest in excess of the rate $i^{(p)}$.)

The following are immediate consequences of equations 7.7.1:

(a) If $i^{(p)} = (1 - t_1)g$, then, for any value of n, $A(n, i) = C$.
(b) If $i^{(p)} < (1 - t_1)g$, then, regarded as a function of n, $A(n, i)$ is an increasing function (i.e. if $n_2 > n_1$, then $A(n_2, i) > A(n_1, i)$).
(c) If $i^{(p)} > (1 - t_1)g$, then, regarded as a function of n, $A(n, i)$ is a decreasing function.
(d) For any fixed n, $A(n, i)$ is a decreasing function of i.

$\left.\rule{0pt}{6em}\right\} (7.7.2)$

These simple observations are not without significance. One very important application of the above is to be found in relation to loans which have optional redemption dates (see section 7.8). It is, however, convenient first to discuss some related matters.

Consider two loans, each of which is as described in the first paragraph of this section except that the first loan is redeemable after n_1 years and the second loan after n_2 years, where $n_1 < n_2$. Suppose that an investor, liable to income tax at a fixed rate, wishes to purchase one of the loans for a price B. Then

(a) If $B < C$ (where $C = NR$), the investor will obtain a higher net yield by purchasing the first loan (i.e. the loan which is repaid earlier).

(b) If $B > C$, the investor will obtain a higher net yield by purchasing the second loan (i.e. the loan which is repaid later).

(c) If $B = C$, the net yield will be the same for either loan.

$\left.\begin{array}{r}\\\\\\\\\\\\\end{array}\right\}(7.7.3)$

This is easily seen as follows. If $B = C$, then remark 7.7.2(a) shows that for either loan the net annual yield is i, where $i^{(p)} = (1 - t_1)g$. This establishes 7.7.3(c). Suppose now that $B < C$ and let i_1 be the net annual yield obtainable from the first loan. Clearly $i_1^{(p)} > (1 - t_1)g$. Let B^* be the price which should be paid to provide a net annual yield of i_1 from the *second* loan. Then

$$B^* = A(n_2, i_1) \qquad \text{(by definition of } B^*)$$

$$< A(n_1, i_1) \qquad \text{(by 7.7.2(c), since } i_1^{(p)} > (1 - t_1)g \text{ and } n_2 > n_1)$$

$$= B \qquad \text{(by definition of } i_1)$$

Thus a purchaser of the second loan who wishes to obtain a net annual yield of i_1 should pay *less* than B. If in fact he pays B, then obviously (or by 7.7.2(d)) his net yield will be less than i_1. This establishes 7.7.3(a). The proof of 7.7.3(b) follows similarly.

The above results are intuitively obvious. If $B < C$, the purchaser will receive a capital gain when either loan is redeemed. From the investor's viewpoint, the sooner he receives this capital gain the better. He will therefore obtain the greater yield on the loan which is redeemed first. On the other hand, if $B > C$ there will be a capital loss when either loan is redeemed. The investor will wish to defer this loss as long as possible. He will therefore obtain the greater yield on the loan which is redeemed later.

When an investor purchases *part* of a loan redeemable by instalments, the yield he will obtain depends on the actual date (or dates) at which his holding is chosen for redemption. In relation to such a loan, issued in bonds of equal nominal amount, suppose that a nominal amount N_r is redeemable at time n_r $(r = 1, 2, \ldots, k)$ where $n_1 < n_2 < \cdots < n_k$. Suppose that (for each bond) the purchase price per unit nominal and redemption price per unit nominal are P and R respectively. Consider a purchaser of one bond, subject to income tax at rate t_1, and let the net yield per annum which he will obtain if his bond is redeemed at time n_r be denoted by $i_r (r = 1, 2, \ldots, k)$. It follows by remarks 7.7.3 that, if P is less than R, i_r *decreases* as r (or n_r) increases. Similarly, if P is greater than R, i_r *increases* as r (or n_r) increases. If the bonds redeemed at any one time are drawn by lot, the probability of obtaining a particular yield i_r is of course equal to

$$p_r = N_r \bigg/ \sum_{r=1}^{k} N_r$$

and the expected value of the yield, in the probabilistic sense, is therefore

$$i^* = \sum_{r=1}^{k} p_r i_r = \sum_{r=1}^{k} N_r i_r \bigg/ \sum_{r=1}^{k} N_r \qquad (7.7.4)$$

It should be noted that this quantity is *not* in general equal to the net yield i on the whole loan, but in most practical cases i^* and i will be quite close to each other. It is also clear that both i^* and i will lie somewhere between i_1 and i_k. The probability that the yield obtained by the purchaser of one bond will exceed a particular value i' may easily be found from the fact that the yields i_r increase or decrease with term (see examples 7.7.1 and 7.7.2).

Example 7.7.1

Consider the loan described in example 7.6.3, and suppose that an investor, who is subject to income tax at 30%, purchases one bond of £100 nominal on the issue date for £95·82. Find the net yield per annum he will obtain, assuming redemption after 5, 10, 15, and 20 years, and plot these net yields on a graph. Find also the probability that the net yield will exceed 9% p.a. Find the expected value of his yield and show that it is not identical with the net yield which he would obtain if he purchased the entire issue at the same price.

Solution

Let the net yield per annum be i_r, if redemption of the bond occurs at time $5r$ $(r = 1, 2, 3, 4)$. The equation of value for finding i_r is

$$95·82 = K + \frac{0·1(0·7)}{1·05 i_r^{(2)}}(105 - K)$$

where $K = 105 v^{5r}$ at rate i_r.

The solutions (expressed as percentages, correct to three decimal places) are as follows:

$$i_1 = 9·066\%, \quad i_2 = 8·109\%, \quad i_3 = 7·803\%, \quad i_4 = 7·660\%$$

The probability that he will obtain a yield of at least 9% is therefore 0·25. The expected value of his yield is $i^* = (i_1 + i_2 + i_3 + i_4)/4 = 8·16\%$, whereas (see example 7.6.3) the net yield if the investor purchased the entire loan would be 8% per annum. These results are illustrated in Figure 7.7.1.

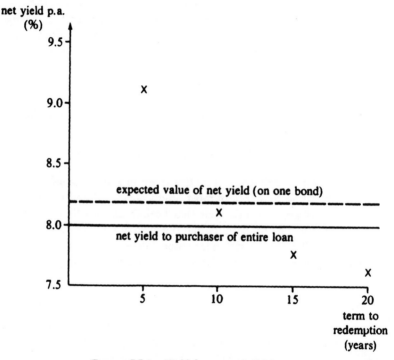

FIGURE 7.7.1   *Yield for example 7.7.1*

Example 7.7.2

A loan of nominal amount £100 000 is to be issued in bonds of nominal amount £100 bearing interest of 6% per annum payable half-yearly in arrear. The loan will be repaid over 20 years, 50 bonds being redeemed at the end of each year at a price of £120%. The bonds redeemed in any one year will be drawn by lot. The issue price of the loan is £94·32%.

An investor, liable to income tax at the rate of 25%, is considering the purchase of all or part of the loan.

(a) Show that if he purchases the entire loan his net annual yield on the transaction will be 7%.

(b) Show that if he purchases only one bond his net annual yield could be as high as 32·36% or as low as 5·61% and find the probability that he will achieve a net annual yield of (i) at least 8%, and (ii) between 6% and 8%.

Solution

(a) Note that £5000 nominal is redeemed each year at 120%. In our usual notation we have $g = 0.06/1.2 = 0.05$, $p = 2$, $t_1 = 0.25$, and $C = 120\,000$ (for the entire loan). The value of the entire loan to provide a net annual yield of i is thus, by Makeham's formula,

$$A = K + \frac{(1 - 0.25)0.05}{i^{(2)}} (120\,000 - K)$$

where

$$K = 6000a_{\overline{20}|} \qquad \text{at rate } i$$

The reader should verify that, when $i = 0.07$, the value of A is £94 318. For practical purposes this is the same as £94 320, which is the issue price of the entire loan, so that, if the investor purchases the entire loan, he will obtain a net annual yield of 7%.

(b) Now assume that the investor purchases a single bond for £94·32. This purchase price is less than £120, the redemption price of the bond. Our remarks above indicate that the sooner the bond is redeemed the greater will be the yield to the investor. If the bond is redeemed after n years, the net annual yield is that rate of interest for which

$$94.32 = (1 - 0.25)6a_{\overline{n}|}^{(2)} + 120v^n$$

The reader should verify from this equation that the values of i corresponding to $n = 1$ and $n = 20$ are 0·3236 and 0·0561 respectively. This establishes the required result.

(i) The above discussion shows that there will be a 'critical' term t such that the investor's net annual yield will be at least 8% if and only if his bond is redeemed within t years. In order to determine t, we consider formally the equation

$$94.32 = (1 - 0.25)6a_{\overline{t}|}^{(2)} + 120v^t \qquad \text{at 8\%} \qquad (1)$$

Since at 8%

$$a_{\overline{t}|}^{(2)} = \frac{1 - v^t}{0.08^{(2)}}$$

equation 1 is equivalent to

$$94.32 = 0.75 \times 6 \times \frac{1 - v^t}{0.08^{(2)}} + 120v^t \qquad \text{at 8\%}$$

from which it follows that

$$v^t = 0.5901 \qquad \text{at 8\%}$$

There is, of course, no *integer* value of t which satisfies this last equation. Since, at 8%,

$$v^6 > 0.5901 > v^7$$

the investor's net annual yield will be at least 8% if and only if his bond is redeemed within six years. The required probability is thus $6/20 = 0.3$.

(The reader may verify that, if the bond is redeemed after six years, the net annual yield will be 8·53% and that, if it is redeemed after seven years, the net annual yield will be 7·92%.)

(ii) Using the same method the reader should verify that the net annual yield will be at least 6% if and only if the bond is redeemed within 15 years. This means that the net annual yield will be between 6% and 8% if and only if the bond is redeemed after 7, 8,..., 14, or 15 years. There are nine possible redemption dates, so that the required probability is 9/20 = 0·45.

Note A more detailed analysis of this problem shows that the net annual percentage yield to the purchaser of one bond has mean 9·05 and standard deviation 6·07.

7.8 Optional redemption dates

Sometimes a security is issued without a fixed redemption date. In such cases the terms of issue may provide that the borrower can redeem the security *at his option* at any interest date on or after some specified date. Alternatively, the issue terms may allow the borrower to redeem the security at his option at any interest date on or between two specified dates (or possibly on any one of a series of dates between two specified dates).

An example of the first situation is provided by $2\frac{1}{2}$% Consols. When this stock was issued, the government reserved the right to repay the stock at par on 5 April 1923 or at any later interest date, so it may now be repaid at any interest date the government chooses. An illustration of a range of redemption dates is given by 12% Exchequer 2013–17: this stock may be redeemed at par on any 12 June or 12 December between 12 December 2013 and 12 December 2017 inclusive.

The latest possible redemption date is called the *final redemption date* of the stock, and if there is no such date (as in the case of $2\frac{1}{2}$% Consols), then the stock is said to be *undated*. It is also possible for a loan to be redeemable between two specified interest dates, or on or after a specified interest date, at the option of the lender, but this arrangement is less common than when the borrower chooses the redemption date.

An investor who wishes to purchase a loan with redemption dates at the option of the borrower cannot, at the time of purchase, know how the market will move in the future and hence when the borrower will repay the loan. He thus cannot know the net yield which he will obtain. However, by using equation 7.7.2 or equation 7.7.3 he can determine either:

(1) The maximum price to be paid, if his net yield is to be *at least* some specified value; or

(2) The minimum net yield he will obtain, if the price is some specified value.

We shall now show how he may do so. Suppose that the outstanding term of the loan, n years, may be chosen by the borrower subject to the restriction that $n_1 \leqslant n \leqslant n_2$. (We assume that n_1 is an integer multiple of $1/p$ and that n_2 is an integer multiple of $1/p$ or is infinite.) Using the notation of section 7.7, we let $A(n, i)$ be the price to provide a net annual yield of i, if the loan is redeemed at time n.

Suppose that the investor wishes to achieve a net annual yield of at least i. It follows from 7.7.2 that

(a) If $i^{(p)} < (1 - t_1)g$, then $A(n_1, i) < A(n, i)$ for *any* value of n such that $n_1 < n \leqslant n_2$. In this case, therefore, the investor should value the loan on the assumption that redemption will take place at the *earliest* possible date. If this does in fact occur, his net annual yield will be i. If redemption occurs at a later date, the net annual yield will exceed i.

(b) If $i^{(p)} > (1 - t_1)g$, then $A(n_2, i) < A(n, i)$ for *any* value of n such that $n_1 \leqslant n < n_2$. In this case, therefore, the investor should value the loan on the assumption that redemption will occur at the *latest* possible date. (If $n_2 = \infty$, the loan should be valued as a perpetuity.) This strategy will ensure that his net annual yield will be at least i.

(c) If $i^{(p)} = (1 - t_1)g$, the net annual yield will be i irrespective of the actual redemption date chosen.

Suppose, alternatively, that the price of the loan is given. What can be said about the yield which the investor will obtain? As before, let P and R be the purchase price per unit nominal and the redemption price per unit nominal respectively. The minimum net annual yield is obtained by solving an appropriate equation of value. The following are immediate consequences of equation 7.7.3:

(a) If $P < R$, the investor should determine the net annual yield on the assumption that the loan will be repaid at the *latest* possible date. If this does in fact occur, his net annual yield will be that calculated. If redemption takes place at an earlier date, the net annual yield will be greater than that calculated.

(b) If $P > R$, the investor should determine the net annual yield on the assumption that the loan will be repaid at the *earliest* possible date. The actual yield obtained will be at least the value calculated on this basis.

(c) If $P = R$, the net annual yield is i, where $i^{(p)} = g(1 - t_1)$, irrespective of the actual redemption date chosen.

Example 7.8.1

A fixed-interest security bears interest at 7% p.a. payable half-yearly and is to be redeemed at 105% at an interest date n years from the present, where $10 \leqslant n \leqslant 15$ and n is to be determined by the borrower. What price per cent should be paid for this stock by

(a) A tax-free investor who requires a net yield of at least 5% p.a.; and
(b) An investor subject to income tax at 40% who also requires a net yield of at least 5% p.a.?

Solution

(a) We have

$$g = \frac{0 \cdot 07}{1 \cdot 05} = 0 \cdot 066\,67, \quad t_1 = 0, \quad i^{(p)} = 0 \cdot 05^{(2)} = 0 \cdot 049\,39$$

so

$$g(1 - t_1) > i^{(p)}$$

and we assume redemption as early as possible, i.e. at time 10 years. By Makeham's formula, the required price per unit nominal is

$$K + \frac{g}{i^{(p)}}(C - K) = 1 \cdot 05v^{10} + \frac{0 \cdot 066\,67}{0 \cdot 049\,39}(1 \cdot 05 - 1 \cdot 05v^{10}) \qquad \text{at } 5\%$$

$$= 1 \cdot 1918$$

Therefore the maximum price is £119·18%.
(If the investor pays this price and the loan is redeemed after 15 years, his net annual yield will be 5·42%.)

(b) We now have

$$t_1 = 0 \cdot 4 \quad \text{and} \quad g(1 - t_1) = 0 \cdot 04$$

so

$$g(1 - t_1) < i^{(2)}$$

Hence we assume redemption after 15 years (as late as possible). By Makeham's formula, the required price per unit nominal is

$$1 \cdot 05v^{15} + \frac{0 \cdot 04}{0 \cdot 049\,39}(1 \cdot 05 - 1 \cdot 05v^{15}) \qquad \text{at } 5\%$$

$$= 0 \cdot 9464$$

Therefore the maximum price is £94·64%.
(If the investor pays this price and the loan is redeemed after ten years, his net annual yield will be 5·35%.)

Example 7.8.2

A loan bearing interest of 12% per annum payable quarterly is to be issued at a price of £92%. The entire loan will be redeemed at par on any interest date chosen by the borrower between 20 and 25 years (inclusive) from the issue date. Find the minimum net annual yield which will be obtained by an investor, liable to income tax at 30%, who purchases part of the loan.

Solution

Since the issue price is less than the redemption price, our remarks above indicate that the investor should value the loan on the assumption that it will be redeemed at the latest possible date. Net interest each year is £8·40 per £100 nominal, so that the appropriate equation of value is

$$92 = 8·40a_{\overline{23}|}^{(4)} + 100v^{25} \qquad \text{at rate } i$$

The yield is clearly somewhat greater than 9%. By interpolation we find that $i = 0·095\,46$; hence the net annual yield will be at least 9·546%. (If the loan is in fact repaid after 20 years, the net annual yield will be 9·61%.)

Note If the stock is being valued between two interest dates (or between the issue date and the first interest date), the above rules concerning prices and yields may not be correct, as the assumption that n_1 is an integer multiple of $1/p$ and n_2 is an integer multiple of $1/p$, or is infinite, does not hold (see exercise 7.19). In this case the term which gives the greatest, or least, price or yield should be determined by further analysis or by numerical calculations. See also section 7.9.

7.9 Valuation between two interest dates: more complicated examples

In the preceding sections we have generally assumed that a fixed-interest security was being valued at the issue date or just after receipt of an interest payment. If one is valuing a stock at an interest date immediately *before* receipt of the interest payment, the value to an investor who pays income tax at rate t_1 is clearly

$$A^* = A + (1 - t_1)DN/p = A + (1 - t_1)gC/p \qquad (7.9.1)$$

where A denotes the corresponding 'ex dividend' value, which may be found by Makeham's formula or otherwise.

When valuing between two interest dates, or between the date of issue and the first interest date, we may use one of the following methods, (a) and (b). Let the valuation date be a fraction m of a year $(0 < m < 1/p)$ after the preceding interest date or the date of issue, and hence a fraction $m'(= 1/p - m)$ of a year from the next interest date.

Method (a) Let A denote the value of the security, at rate of interest i per annum, just *after* the last interest payment was made, or at the issue date if there has been no such payment; A may be found by Makeham's formula or by other methods. The value a fraction m of a year later is thus

$$A' = (1 + i)^m A \qquad (7.9.2)$$

Method (b) Instead of working from the preceding interest date (or issue date), let us work from the next interest date. Let A^* be the value of the stock, at rate of interest i per annum, just *before* the next interest payment is made; A^* may be found by formula 7.9.1. It follows that the present value of this stock is

$$A' = (1 + i)^{-m'} A^* \qquad (7.9.3)$$

Our first example illustrates the use of both these methods. In the subsequent examples we consider how to deal with the valuation of securities when the redemption price, coupon rate or rate of income tax is not constant. In some cases it is simplest to proceed from first principles, while in others Makeham's formula may be used to value the loan in sections, or in other ways. Our final example, the solution of which may at first seem difficult, illustrates how a clear grasp of the principles underlying Makeham's formula enables an indirect valuation of the capital to shorten the solution considerably.

Example 7.9.1

The middle market price (i.e. the average of the buying and selling prices) quoted on the London Stock Exchange on 15 December 1976 for Southampton 6% stock 1981 was £66%. This stock paid interest half-yearly on 1 April and 1 October, and was redeemable at par at the end of September 1981. Find the gross yield per annum obtainable on a purchase of this stock at the above price.

Solution

Method (a) The value per £100 nominal of this stock on 1 October 1976 *after* the payment of the interest then due is, by Makeham's formula,

$$A = K + \frac{0 \cdot 06}{i^{(2)}}(100 - K)$$

where $K = 100v^5$ at rate i (the gross annual yield). Hence the value on 15 December 1976 is

$$A' = (1 + i)^{75/365}A$$

and we now solve the equation $A' = 66$ to give $i \simeq 0 \cdot 1776$ so that the gross annual yield is 17·76%.

Method (b) The value per £100 nominal just after receipt of the interest payment on 1 April 1977 is, by Makeham's formula,

$$A = K + \frac{0 \cdot 06}{i^{(2)}}(100 - K)$$

where $K = 100v^{4 \cdot 5}$ at rate i, the gross annual yield. The value just *before* receipt of this interest payment is $(A + 3)$ and the value on 15 December 1976 is

$$A' = (1 + i)^{-107/365}(A + 3)$$

We then find the gross yield per annum by solving $A' = 66$. This gives $i \simeq 0 \cdot 1776$ as before.

Example 7.9.2 (Varying redemption price)
An issue of 1000 debentures each of £100 nominal bears interest at $1\frac{1}{2}$% per annum, payable yearly, and is redeemable as follows:

250 bonds after 30 years at £110%
250 bonds after 40 years at £120%

250 bonds after 50 years at £130%
250 bonds after 60 years at £140%

Find the price payable for the issue by an investor requiring a yield of $2\frac{1}{2}\%$ per annum. (Ignore taxation.)

Solution

If the bonds were redeemed at 60%, their total value would be £60 000 (since they each carry an interest payment of £1·50 per annum on an outlay of £60, with redemption at this price). The value of the 'extra' redemption money is

$$250(50v^{30} + 60v^{40} + 70v^{50} + 80v^{60}) \quad \text{at } 2\tfrac{1}{2}\% \quad = \text{£}21\,183$$

So the total price payable is £81 183.

In more complicated problems involving varying redemption prices it is often helpful to use one of the following approaches:

Method (a)   Consider a fixed redemption price, value using Makeham's formula, and then adjust the answer to take 'excess' or 'shortfall' in the redemption money into account.

Method (b)   Split the loan into sections, for each of which the redemption price is constant, and value each section by Makeham's formula.

Method (b) is generally more suitable when the redemption price R has only two or three values, whereas method (a) may be more suitable when R takes many values.

Similar methods apply when the coupon rate D or the rate of income tax t_1 varies. One may use first principles or either of the following methods:

Method (a)   Consider the net rate of interest to be constant, value by Makeham's formula and adjust the answer to allow for the actual net interest rates.

Method (b)   Split the loan into sections, assume that for each the net rate of interest is constant, and value each section by Makeham's formula. (Note, however, that adjustments will generally have to be made, since the net interest rate for some sections may not be constant (see example 7.9.5).)

In certain rather complicated examples the redemption price and the net rate of interest may vary. Such problems should be tackled by a combination of the methods described above.

Example 7.9.3 (Varying coupon rate)

A loan stock of £1 000 000 nominal is redeemable at 110% by ten equal annual instalments of capital, the first due in one year's time. Interest is payable half-yearly in arrear at the rate of $(4\cdot5 + t)\%$ per annum in year t ($t = 1, 2, \ldots, 10$). What price should be paid for the stock to yield 8% per annum convertible half-yearly? Ignore taxation.

Solution

Work in half-yearly periods. The value of the capital repayments is

$$K = 110\,000(v^2 + v^4 + \cdots + v^{20}) \qquad \text{at } 4\%$$

$$= \frac{110\,000a_{\overline{20}|}}{s_{\overline{2}|}} \qquad \text{at } 4\% \qquad = 732\,810$$

Assume firstly that interest is constant at 5·5% per annum, or 2·75% per period. By Makeham's formula, the value of the loan is

$$K + \frac{0\cdot0275}{1\cdot1 \times 0\cdot04}(1\,100\,000 - K) = 962\,300$$

To this must be added the value of the extra interest payments, which is

$$\tfrac{1}{2}[9000(v^3 + v^4) + 2 \times 8000(v^5 + v^6) + 3 \times 7000(v^7 + v^8)$$

$$+ \cdots + 9 \times 1000(v^{19} + v^{20})]$$

$$= \tfrac{1000}{2}(v^3 + v^4)[9(1 + v^{16}) + 16(v^2 + v^{14})$$

$$+ 21(v^4 + v^{12}) + 24(v^6 + v^{10}) + 25v^8]$$

$$= 106\,680$$

Hence the total value of the loan is £1 068 980.

Example 7.9.4 (Varying redemption price and varying rate of income tax)

A loan of nominal amount £10 000 is issued in bonds of £100 nominal bearing interest at 9% per annum payable annually in arrear. Five bonds will be redeemed at the end of each of the first ten years and ten bonds at the end of each of the next five years. During the first ten years redemption will be at £112·50 per bond and thereafter it will be at £120 per bond.

An investor has been offered the entire issue of bonds for £10 190. At present the investor pays tax on income at the rate of 40% but a date has been fixed after which the rate will be 30%.

Given that these terms offer the investor a yield of 7% per annum net effective, find how many interest payments will be subject to the higher rate of tax.

Solution

Assume first that the redemption price is constant at 112·5%. The value of the capital repayments is

$$K = 562\cdot5a_{\overline{10}|} + 1125v^{10}a_{\overline{5}|} \quad \text{at } 7\% \quad = 6295\cdot63$$

and the value of the issue, assuming income tax to be at 30%, is

$$K + \frac{0\cdot08 \times 0\cdot7}{0\cdot07}(11\,250 - K) = 10\,259$$

The addition for the higher redemption price in the last five years is

$$1000 \times 0.075(a_{\overline{15}|} - a_{\overline{10}|}) \quad \text{at } 7\% \quad = 156$$

Hence the total value of the loan, assuming that income tax is constant at 30%, is £10 415. From this must be deducted the value of the additional income tax on the first n interest payments, where n is such that the value of the loan is reduced to £10 190. We draw up the following schedule:

(1)	(2) Nominal amount outstanding at beginning of	(3) Gross interest at time t	(4) Value of 10% of gross interest at time t	(5) Cumulative total
Year (t)	year t	$0.09 \times (2)$	$0.1 \times (3) \times v_{0.07}^{t}$	of (4)
1	10 000	900	84.1	84.1
2	9 500	855	74.7	158.8
3	9 000	810	66.1	224.9
4	8 500	765	58.4	

Since the actual price is £225 less than £10 415, we have $n = 3$, i.e. the 40% rate of income tax applies to the interest payments made at the end of the first three years but not to those made later.

Example 7.9.5 (Varying coupon rate)

A loan of £3 000 000 nominal bears half-yearly interest at $3\frac{1}{2}\%$ per annum for the first ten years and at 4% per annum thereafter. It is redeemable at par by equal annual instalments of capital over the next 30 years. Find the net yield per annum to a financial institution, subject to income tax at the rate of 37.5%, which bought the loan at par.

Solution

We split the loan into two sections: the first section, of nominal amount £1 000 000, consists of that part of the loan which is redeemable in the first ten years, and the second section, of nominal amount £2 000 000, is the remainder of the loan. Makeham's formula may be applied to value each section, but an adjustment must be made to allow for the fact that interest on the second section is at $3\frac{1}{2}\%$, not 4%, during the first ten years. Let the unknown yield be i; we have (in an obvious notation)

$$K_1 = 100\,000 a_{\overline{10}|} \qquad \text{at rate } i$$
$$K_2 = 100\,000(a_{\overline{30}|} - a_{\overline{10}|}) \qquad \text{at rate } i$$

The value of the first section is

$$A_1 = K_1 + \frac{0.035 \times 0.625}{i^{(2)}}(1\,000\,000 - K_1)$$

We now value the second section. If interest on this section were always at 4%, its value would be

$$A_2 = K_2 + \frac{0.04 \times 0.625}{i^{(2)}}(2\,000\,000 - K_2)$$

During the first ten years interest is actually at $3\frac{1}{2}\%$ p.a., so we must deduct the value of the 'excess' net interest, i.e.

$$2\,000\,000 \times 0.005 \times 0.625 \times a^{(2)}_{\overline{10}|} \qquad \text{at rate } i$$

We therefore solve the equation

$$3\,000\,000 = A_1 + A_2 - 6250a^{(2)}_{\overline{10}|}$$

A rough solution is $3.75\% \times 0.625 \simeq 2.3\%$, and evaluation of $A_1 + A_2 - 6250a^{(2)}_{\overline{10}|}$ shows that this is correct (to the nearest 0.1%).

Example 7.9.6

A loan of £1 000 000 is to be repaid over 40 years by a level annuity payable monthly in arrear. The amount of the annuity is determined on the basis of an interest rate of 8% per annum effective.

   An investor, liable to income tax at the rate of 40% on the interest content (according to the original schedule) of each annuity payment, wishes to purchase the entire loan on the issue date at a price to obtain a net yield of 6% per annum convertible half-yearly. What price should he pay?

Solution

Note first that the amount of the repayment annuity per annum is £X, where

$$Xa^{(12)}_{\overline{40}|} = 1\,000\,000 \qquad \text{at } 8\%$$

This gives

$$X = 80\,933.82$$

The required net yield is 6% per annum convertible half-yearly. Choose a half-year as the unit of time and note that the required net yield is 3% per unit time. Per half-year the annuity is of amount $X/2 = 40\,466.91$. Since the annuity is payable six times per half-year, the value of the entire loan (ignoring tax) is

$$A = 40\,466.91a^{(6)}_{\overline{80}|} \qquad \text{at } 3\% \qquad = 1\,237.320 \qquad (1)$$

(Note that $0.03^{(6)} = 0.029\,632$.)

   Let K be the value (at 3% per half-year) of the capital payments. Interest is paid each month of amount $0.08^{(12)}/12$ times the outstanding loan. If, in fact, interest were payable each month of amount $0.03^{(6)}/6$ times the outstanding loan, the value of the loan would be £1 000 000. Hence, using Makeham's formula 'backwards', we have

$$1\,237\,320 = K + \frac{0.08^{(12)}/12}{0.03^{(6)}/6}(1\,000\,000 - K)$$

from which we immediately obtain

$$K = 216\,214 \qquad (2)$$

Clearly $(A - K)$ is the value of the gross interest payments. Allowing for tax, the investor should pay

$$K + (1 - 0.4)(A - K) = A - 0.4(A - K) = 0.6A + 0.4K$$

From equations 1 and 2, this is easily calculated to be £828 878 (or, say, £82·89%).

Note: The value of K may, of course, be obtained directly as

$$K = \left[\frac{80\,933\cdot82}{12} - 1\,000\,000 \times \frac{0\cdot08^{(12)}}{12} \right] \sum_{t=1}^{480} (1\cdot08)^{(t-1)/12}(1\cdot03)^{-t/6}$$

However, the evaluation of this expression is somewhat longer, and more liable to error, than the indirect method above.

7.10 A formula for finding the yield

A method for finding the gross yield per annum, convertible half-yearly, of most fixed-interest stocks is given in reference [13]. Since these formulae are often used in practice and have been referred to in the prospectuses of many recent loans, we quote the relevant passage, slightly modified to allow for the practice (adopted in the UK in February 1986) of excluding accrued interest from quoted market prices.

Redemption yields are calculated taking accrued interest as part of the price and using a true compound interest formula, i.e. finding the value of v to give $f(v) = 0$ where

$$f(v) = v^p \left[C_1 + C \frac{(1 - v^n)v}{(1 - v)} + Rv^n \right] - P - \sum_i B_i v^{b_i} \qquad (7.10.1)$$

v	is the discounting factor per period (e.g. half-year)
R	is the redemption amount
C	is the coupon amount per period
C_1	is the actual coupon due at the next payment date (which may be zero if the stock is already quoted 'ex dividend', or may be a first fractional payment)
n	is the integral number of periods till redemption from the next payment date
p	is the fractional period till the next payment date
P	is the price actually payable (with 'accrued interest' added in)
B_1, B_2 etc.	are outstanding calls on a partly-paid stock
b_1, b_2 etc.	are the fractional periods till these calls are due.

When the root of $f(v)$ has been found, the gross yield y convertible half-yearly is obtained from

$$y = 200 \left(\frac{1}{v^{k/2}} - 1 \right) \text{per cent} \qquad (7.10.2)$$

Where the stock has a range of optional redemption dates the earliest or latest is used, whichever gives the lower redemption yield.

It is of interest to further consider equation 7.10.1. In the first instance, the unit of time for the equation of value $f(v) = 0$ is $1/k$ year, where $1/k$ year is the interval of time between any two consecutive interest dates, and $v = 1/(1 + i)$, where i is the effective yield per period of $1/k$ year. The 'fractional period until the next payment date', p, is the time until the next interest date, *measured in units of* $1/k$ *year*. Thus if the time, in years, to the next interest date is F, then $p = kF$. The factor v^p in equation 7.10.1 therefore allows for the time until the next interest date, and the term in square brackets is the value, at that time, of the interest and redemption money. Note that $(1 - v^n)v/(1 - v) = a_{\overline{n}|}$. Formula 7.10.1 with $C_1 = C_{p_1}$ applies at the issue date when the first interest payment is due at time p (measured in units of $1/k$ year) from the issue date, and the first interest payment is fractional. This is quite a common arrangement in practice (see example 7.10.1). If, on the other hand, the stock is 'ex dividend' (see section 7.2), then $C_1 = 0$, as the seller will receive the next interest payment.

The price P must include an allowance for accrued interest. This is calculated by the formula

accrued interest = annual coupon rate × nominal amount of stock × θ

where

$$\theta = \begin{cases} \dfrac{1}{365} \times \begin{pmatrix} \text{number of days from issue date} \\ \text{or last interest date up to} \\ \text{settlement date} \end{pmatrix} & \text{if the stock is \textit{not} x.d.} \\[4mm] \dfrac{-1}{365} \times \begin{pmatrix} \text{number of days from settlement} \\ \text{date up to next interest date} \end{pmatrix} & \text{if the stock is x.d.} \end{cases} \tag{7.10.3}$$

The price P payable in respect of the relevant nominal amount (e.g. per £1 nominal or per £100 nominal) equals the market price plus accrued interest.

A stock is said to be *partly-paid* if it is being paid for by instalments, which are referred to as *calls*. The amounts of any outstanding calls are denoted by $B_1, B_2, \ldots$ and are payable at times $b_1, b_2, \ldots$ respectively, where $b_1, b_2, \ldots$ are measured in time units of $1/k$ year. The final term on the right-hand side of formula 7.10.1 is therefore equal to the present value of any outstanding calls.

Equation 7.10.1 may be solved numerically. Since the solution v refers to a period of $1/k$ year (i.e. $v = 1/(1 + i)$, where i is the effective yield per period of $1/k$ year), the corresponding effective annual yield is

$$i' = (1 + i)^k - 1 = \frac{1}{v^k} - 1$$

The nominal annual yield, convertible half-yearly, is thus

$$2[(1 + i')^{1/2} - 1] = 2\left(\frac{1}{v^{k/2}} - 1\right)$$

and the quantity y is this rate expressed as a percentage.

Example 7.10.1

A debenture was issued on 1 May 1986 at a price of £95% and is redeemable at par on 1 January 1993. Interest is at 6% per annum payable quarterly on 1 January, 1 April, 1 July, and 1 October, the interest payment on 1 July 1986 being pro rata. Find the gross yield per annum convertible half-yearly (a) at the issue date and (b) on 15 June 1986, when the price of the stock, *allowing for accrued interest*, was £94 per cent ex dividend.

Solution

We first remark that the interval between any two consecutive interest dates is a quarter-year, so $k = 4$ and we begin by working in periods of a quarter-year. We work in terms of £100 nominal.

(a) In the notation of this section, we have $1/k = 1/4$ and $F = 1/6$, so $p = 2/3$. We also have $C = 3/2$ (since the annual interest of £6 is payable in four equal instalments) and $C_1 = Cp = 1$. The equation of value 7.10.1 is thus

$$f(v) = v^{2/3}(1 + \tfrac{3}{2}a_{\overline{26}|} + 100v^{26}) - 95 = 0$$

which gives $v = 0.982\,91$. The gross annual yield convertible half-yearly is then found, by equation 7.10.2, to be

$$y = 200\left(\frac{1}{v^2} - 1\right)\% = 7.02\%$$

(b) The debenture is now 'ex dividend', so $C_1 = 0$. Also $F = 1/24$, so $p = 1/6$ and equation 7.10.1 is now
$$f(v) = v^{1/6}(\tfrac{3}{2}a_{\overline{26}|} + 100v^{26}) - 94 = 0$$

The solution of this equation is $v = 0.982\,54$, whence we find that the gross annual yield convertible half-yearly is

$$y = 200\left(\frac{1}{v^2} - 1\right)\% = 7.17\%$$

Note   In all examples and exercises, the reader should assume that by the 'price' of a stock we mean the price *actually payable* (i.e. including any adjustment for accrued interest).

7.11 Real returns and index-linked stocks

Inflation may be defined as a fall in the purchasing power of money. It is usually measured with reference to an index representing the cost of certain goods and (perhaps) services. In the UK the index used most frequently is the Retail Prices Index (RPI), which is calculated monthly by the Central Statistical Office. It is the successor to various other official indices dating back many years.

Real investment returns, as opposed to the *money* returns we have so far considered, take into account changes in the value of money, as measured by the RPI or another such index. It is possible for all calculations relating to discounted cash flow, yields on investments, etc. to be carried out using units of real purchasing power rather than units of ordinary currency, provided of course that one defines the index being used.

It is a simple matter to write down the equation of value in 'real' terms for any transaction. Suppose that a transaction involves cash flows c_1, $c_2, \ldots, c_n$, the rth cash flow occurring at time t_r. (Note that the cash flows are *monetary* amounts.) If the appropriate index has value $Q(t)$ at time t, the cash flow c_r at time t_r will purchase $c_r/Q(t_r)$ 'units' of the index. By the real yield on the transaction we mean the yield calculated on the basis that the investor's receipts and outlays are measured in units of index (rather than monetary units). The real internal rate of return (or yield) on the transaction, measured in relation to the index Q, is thus that value of i for which

$$\sum_{r=1}^{n} \frac{c_r}{Q(t_r)}(1 + i)^{-t_r} = 0 \qquad (7.11.1)$$

This equation is, of course, equivalent to

$$\sum_{r=1}^{n} c_r \frac{Q(t_k)}{Q(t_r)}(1 + i)^{-t_r} = 0 \qquad (7.11.2)$$

Equation 7.11.2 may be considered as the equation of value for the transaction, measured in units of purchasing power at a particular time t_k. Similar equations may of course be developed for continuous cash flows (see chapter 5).

In the remainder of this chapter we shall, unless it is otherwise stated, calculate real yields with reference to the RPI.

Example 7.11.1

On 16 January 1980 a bank lent £25 000 to a businessman. The loan was repayable three years later, and interest was payable annually in arrear at 10% per annum. Ignoring taxation and assuming that the RPI for any month relates to the middle of that month, find the real annual rate of return, or yield, on this transaction. Values of the RPI for the relevant months are as follows:

Calendar year	1980	1981	1982	1983
Value of RPI for January	245·3	277·3	310·6	325·9

Solution

In money terms, the yield is of course 10% p.a. To obtain the annual yield in real terms we work in units of purchasing power, so that for this example the equation of value 7.11.1 becomes

$$\frac{25\,000}{Q(0)} = 2500\left[\frac{v}{Q(1)} + \frac{v^2}{Q(2)} + \frac{v^3}{Q(3)}\right] + 25\,000\frac{v^3}{Q(3)}$$

where $v = 1/(1 + i)$ and $Q(t)$ is the RPI at time t years, measured from mid-January 1980. Thus we have

$$1 = 0\cdot1\left(\frac{245\cdot3}{277\cdot3}\right)v + 0\cdot1\left(\frac{245\cdot3}{310\cdot6}\right)v^2 + 1\cdot1\left(\frac{245\cdot3}{325\cdot9}\right)v^3$$

i.e.

$$1 = 0\cdot088\,460v + 0\cdot078\,976v^2 + 0\cdot827\,953v^3$$

When $i = 0$, the right-hand side of the last equation is $0\cdot995\,389$, so the real rate of return is negative. When $i = -0\cdot005$, the right-hand side is $1\cdot009\,174$. By linear interpolation, $i \simeq -0\cdot0017$, so the real rate of return is negative and is approximately equal to $-0\cdot17\%$ per annum.

Note The RPI for any month relates to a specific applicable date in that month (usually the second or third Tuesday) and is normally published on the Friday of the second or third week of the following month. For example, the RPI for March 1983 was $327\cdot9$. This value was published on 22 April 1983 and applied to 15 March 1983 (see reference [54]).

For many purposes it is necessary to postulate a value of the RPI for a date between the applicable dates in consecutive months. Such a value may be estimated by interpolation. It is generally better to interpolate between values of $\log Q(t)$ rather than between values of $Q(t)$, since $\log Q(t)$ is usually more nearly linear.

In conditions of relatively stable money, such as existed in Britain for about 100 years before 1914, real and money rates of return are roughly equal. In modern times rates of inflation have been relatively high, and the real return on UK government stocks, even to a tax-free investor, has fluctuated between positive and negative values. The real rate of return on fixed-interest investments, after allowing for taxation, has often been negative.

Theoretically, at least, the idea of index-linking the outlays and receipts on any particular transaction is quite straightforward. In its simplest form an index-linked investment has a series of cash flows of 'nominal' amount $c_1, c_2, \dots, c_n$, the rth cash flow being due at time t_r. The actual *monetary* amount of the rth cash flow is defined to be

$$c_r \frac{R(t_r)}{R(t_0)} \tag{7.11.3}$$

where R is some specified index and t_0 is a prescribed base date. (It is, of course, essential that, for $r = 0, 1, \dots, n$, the value of $R(t_r)$ is known at time t_r.) The real rate of return per annum on such an investment, measured in relation to an index Q, is (see equation 7.11.1) that value of i for which

$$\sum_{r=1}^{n} c_r \frac{R(t_r)}{R(t_0)} \frac{1}{Q(t_r)}(1 + i)^{-t_r} = 0 \tag{7.11.4}$$

Note that if Q, the index used to measure the real rate of return, were to be identically equal to R, the index used to determine the amount of each cash flow, then equation 7.11.4 would simplify to

$$\sum_{r=1}^{n} c_r (1 + i)^{-t_r} = 0$$

so the real rate of return would be the same as the money rate of return on the corresponding 'unindexed' transaction (with cash flows of monetary amounts $c_1, c_2, \ldots, c_n$). In this special case, therefore, the real rate of return does *not* depend on the movement of the index. In general, however, the indices Q and R are distinct, and the real rate of return on any index-linked transaction does depend on the movement of the relevant indices.

In reality this simplest form of index-linking, in which the same index is used both to determine the cash flows and to measure the real rate of return, is not feasible, since there does not exist a suitable index with values which form a continuous function and are *immediately* known at *all* times. These are clearly essential requirements, at least for the index relative to which the real rate of return is calculated. We have already remarked that the RPI, for example, is known accurately on only one day in each month, its value on that day being known (somewhat misleadingly) as the value 'for the month'. Moreover the value for any one month is not known until around the middle of the following month. Thus, although real returns are, in the UK, nearly always measured in relation to the RPI (interpolation being used as necessary to estimate the index values $Q(t_r)$ in equations 7.11.1 and 7.11.4), the index R used to determine cash flows for an index-linked transaction is generally a step function, constant over any calendar month and based on the RPI with a time lag. (The use of a time lag overcomes difficulties which would otherwise arise since the index value for a given month is not published during that month.) Following reference [54], we describe below two particular examples of UK index-linked investments. For this purpose it is convenient to define the value of the RPI 'applicable' on any given date to be the value of the RPI for the calendar month which contains that date. (Note that the value applicable at any time is *not* in general the value of the index at that time.)

Index-linked National Savings Certificates

These were first issued by the UK government in June 1975. The first issue was available only to persons of retirement age, but the second issue (November 1980) was free from such restrictions. The value of a certificate held for one year or more is linked to the RPI with a time lag of two months. For the first issue there was a bonus of 4% if the certificate was held for five years and this bonus is itself subsequently indexed. (On the second issue supplements of varying amount are paid, but we ignore these for our present purposes.) The repayment value of a certificate is guaranteed never to be less than its purchase price.

Measuring time in years from some fixed origin, we define

$R(t)$ = value of RPI applicable two calendar months before time t

$$(7.11.5)$$

(Thus over any calendar month $R(t)$ is constant and equal to the published value for two months earlier.)

A certificate of the first issue purchased for £A at time t_1 and repaid at a subsequent time t_2 is redeemed for an amount £B, where

$$B = \begin{cases} A & \text{if } 0 < t_2 - t_1 < 1 \\ \max\left\{ A, A\left[\dfrac{R(t_2)}{R(t_1)} \right] \right\} & \text{if } 1 \leqslant t_2 - t_1 < 5 \\ \max\left\{ A, A\left[\dfrac{R(t_1 + 5)}{R(t_1)} + 0{\cdot}04 \right]\left[\dfrac{R(t_2)}{R(t_1 + 5)} \right] \right\} & \text{if } \quad t_2 - t_1 \geqslant 5 \end{cases}$$

$$(7.11.6)$$

The following example is given in reference [54].

Example 7.11.2

An index-linked National Savings Certificate was purchased in July 1981 for £100 and redeemed in August 1982.

(a) For what amount was the certificate redeemed?
(b) Estimate the real rate of return on the transaction if the certificate was
    (i) Bought on 1 July 1981 and sold on 1 August 1982;
    (ii) Bought on 1 July 1981 and sold on 31 August 1982;
    (iii) Bought on 31 July 1981 and sold on 1 August 1982;
    (iv) Bought on 31 July 1981 and sold on 31 August 1982.

The relevant RPI values are:

19 May 1981	294·1
16 June 1981	295·8
14 July 1981	297·1
18 August 1981	299·3
15 June 1982	322·9
13 July 1982	323·0
17 August 1982	323·1
14 September 1982	322·9

Solution

(a) Since the certificate was purchased in July 1981 and sold in August 1982, it follows from equations 7.11.6 and 7.11.5 that the redemption money (in each of the four cases) was

$$£100 \frac{\text{value of RPI for June 1982}}{\text{value of RPI for May 1981}} = £100 \frac{322 \cdot 9}{294 \cdot 1} = £109 \cdot 79$$

(b) Let $Q(t)$ denote the value of the RPI at time t. Since there are 28 days from 16 June 1981 to 14 July 1981, we estimate Q (1 July 1981) by the linear interpolation formula

$$\log Q \text{ (1 July 1981)} \simeq \tfrac{13}{28} \log Q \text{ (16 June 1981)} + \tfrac{15}{28} \log Q \text{ (14 July 1981)}$$
$$= \tfrac{13}{28} \log 295 \cdot 8 + \tfrac{15}{28} \log 297 \cdot 1$$

from which we obtain

$$Q \text{ (1 July 1981)} \simeq 296 \cdot 50$$

In a similar manner we estimate the other intermediate RPI values as

$$Q \text{ (31 July 1981)} \quad = 298 \cdot 17$$
$$Q \text{ (1 August 1982)} = 323 \cdot 05$$
$$Q \text{ (31 August 1982)} = 323 \cdot 00$$

The real rate of return per annum in each case is given (see equation 7.11.1) by the equation

$$\frac{-100}{Q(t_0)}(1+i)^{-t_0} + \frac{109 \cdot 79}{Q(t_1)}(1+i)^{-t_1} = 0$$

where t_0 and t_1 denote the times of purchase and sale respectively, measured in years from some origin.

The last equation may be written in the form

$$(1+i)^{t_1 - t_0} = \frac{109 \cdot 79}{100} \frac{Q(t_0)}{Q(t_1)} \tag{7.11.7}$$

(i) Purchase on 1 July 1981; redemption on 1 August 1982;

$$Q(t_0) = 296 \cdot 50; \quad Q(t_1) = 323 \cdot 05 \qquad \text{(see above)}$$

The time from purchase to redemption was one year and 31 days, so that

$$t_1 - t_0 = 1 \tfrac{31}{365} = 1 \cdot 084 \, 93$$

On substitution of these values in equation 7.11.7 it follows immediately that $i = 0 \cdot 0071$. The real rate of return per annum is thus $0 \cdot 71\%$.

The reader should verify that the other answers are (ii) $0 \cdot 67\%$, (iii) $1 \cdot 33\%$, and (iv) $1 \cdot 24\%$.

The range of values of the answers should be noted. The effect of the time lag is to produce different yields for cases (i) and (iv), although in each of these cases the certificate is held for the same length of time.

Index-linked government stocks

The UK government issued its first index-linked stock in March 1981. This particular stock (2% Index-Linked Treasury Stock 1996) carries a coupon of 2% per annum. Interest is payable half-yearly on 16 March and 16 September. The final interest payment will be on 16 September 1996, on which date the stock will be redeemed at par. Both interest and redemption price are indexed by the RPI with a time lag of eight months (see below).

A number of similar stocks have been issued, with different redemption dates and annual coupon rates of either 2% or 2·5%. All have had essentially the same method of indexation. Each stock is redeemable on a single date at an indexed par. The indexation is simply described, if we recall our definition above of the value of the RPI applicable at any time and let

$$R(t) = \text{value of RPI applicable eight calendar months before time } t$$

$$(7.11.8)$$

For an index-linked stock issued at time t_0 with annual coupon rate D (e.g. 0·02 or 0·025) payable half-yearly, the monetary amount of an interest payment due at any particular time t_r is

$$\frac{D}{2} 100 \frac{R(t_r)}{R(t_0)}$$

per £100 nominal stock. Similarly the redemption money per £100 nominal is

$$100 \frac{R(t_n)}{R(t_0)}$$

where t_n denotes the time of redemption.

The use of an eight-month time lag for indexation means that an investor who purchases stock at any time will, at the time of purchase, always know the monetary amount of the first interest payment he will receive. For example, an investor who purchased 2% Index-Linked Treasury Stock 1996 ex dividend on 1 March 1984 would have received his first interest on 16 September 1984, on which date the amount of interest paid per £100 nominal was

$$\frac{0 \cdot 02}{2} 100 \frac{\text{RPI value for January 1984}}{\text{RPI value for July 1980}}$$

On 1 March 1984 the value of this expression was known to be £1·27 (to two decimal places, rounded down).

The calculation of observed real returns for index-linked stocks is carried out quite simply using equation 7.11.1 (or 7.11.2). In order to estimate prospective future real yields, we must make explicit assumptions about the future movement of the RPI. In any given situation it may be desirable to estimate future real returns on the basis of alternative assumptions relating to future values of the RPI (see reference [54]).

Example 7.11.3

On 21 December 1983 an investor, who is not liable to taxation, purchased a quantity of 2% Index-Linked Treasury Stock 1996 at a price of £106·375 per cent nominal. At the time of purchase, using the latest available RPI value, he estimated his real yield on the transaction on the assumption that he would hold the stock until redemption and that the RPI would grow continuously at a constant rate of (a) 5% per annum and (b) 7% per annum.

Given the information below, find the estimates made by the investor of his real yield under each assumption.

On 21 December 1983 the latest known value of the RPI was 341·9 for November 1983. (This value was published on 16 December 1983 and related to 15 November 1983.) A description of the stock is given earlier in this section. In particular, note that the stock was issued in March 1981. The values of the RPI for July 1980 and July 1983 are 267·9 and 336·5 respectively.

Solution

This example is typical of the kind of calculation which must be carried out in practice. Although with hindsight we see that neither of the underlying assumptions has been realized, we are required to obtain the investor's original estimates. Accordingly our solution assumes that the calculation is being made at the time of purchase.

The first interest payment will be received on 16 March 1984. Interest is payable half-yearly. The final interest payment will be made on 16 September 1996, which is the redemption date of the stock.

It is convenient to label successive calendar months by an integer variable as follows: November 1983 as month 0, December 1983 as month 1, January 1984 as month 2, etc.

We measure time in *half-years* from the date of purchase and let i be the real yield per half-year. The periods from 16 September 1983 to 16 March 1984, and from 21 December 1983 (the purchase date) to 16 March 1984, consist of 182 days and 86 days respectively, so that the first interest payment will be received at time f where $f = 86/182$. Over the entire transaction the investor will receive 26 interest payments, the kth payment being at time $(k - 1 + f)$. (For practical purposes we ignore the fact that, except in leap years, the half-year from 16 March to 16 September is slightly longer than that from 16 September to 16 March.)

Let r be the assumed compound rate of growth of the RPI *per half-year.* (We shall assume that either $1 + r = 1·05^{1/2}$ or $1 + r = 1·07^{1/2}$.) Then

$$Q(t) = Q(0)(1 + r)^t$$

where $Q(t)$ is the RPI value at exact time t. The RPI value for month 0 is 341·9. We assume that, if $l \geq 0$, the RPI value for month l will be $341·9(1 + r)^{l/6}$. (As we do not know the precise day for the official calculation of the index each month, this is a reasonable hypothesis.)

We work per £100 nominal of stock. Recalling the eight-month time lag for indexation, we note that the first interest payment will be of amount

$$\frac{0 \cdot 02}{2} 100 \frac{\text{RPI July 1983}}{\text{RPI July 1980}} = \frac{336 \cdot 5}{267 \cdot 9} = 1 \cdot 25$$

The amount of this payment is known on the purchase date. The payment will be made at time f, when the value of the RPI will be $Q(f) = Q(0) (1 + r)^f$.

For $j \geqslant 2$, the investor's jth interest payment will be received in month $(6j - 2)$, and will be of amount

$$\frac{0 \cdot 02}{2} 100 \frac{\text{RPI month } (6j - 10)}{\text{RPI July 1980}}$$

We estimate this to be

$$\frac{341 \cdot 9(1 + r)^{(6j - 10)/6}}{267 \cdot 9} = \frac{341 \cdot 9}{267 \cdot 9}(1 + r)^{j - (5/3)}$$

This payment will be received at time $(j - 1 + f)$, when the value of the RPI will be $Q(0) (1 + r)^{j - 1 + f}$.

The redemption proceeds will be paid at time $(25 + f)$ with the final (26th) interest payment. The estimated redemption proceeds will thus be

$$100 \frac{341 \cdot 9}{267 \cdot 9}(1 + r)^{26 - (5/3)}$$

Combining the above results, we use equation 7.11.1 to determine the real yield per half-year. It is that value of i for which

$$0 = \frac{-106 \cdot 375}{Q(0)} + \frac{1 \cdot 25}{Q(0)(1 + r)^f}(1 + i)^{-f}$$

$$+ \sum_{j=2}^{26} \frac{341 \cdot 9}{267 \cdot 9}(1 + r)^{j - (5/3)} \frac{1}{Q(0)(1 + r)^{j - 1 + f}}(1 + i)^{-(j - 1 + f)}$$

$$+ 100 \frac{341 \cdot 9}{267 \cdot 9}(1 + r)^{26 - (5/3)} \frac{1}{Q(0)(1 + r)^{25 + f}}(1 + i)^{-(25 + f)}$$

Since the factor $Q(0)$ cancels throughout, this equation is easily seen to be equivalent to

$$106 \cdot 375 = (1 + i)^{-f} \left\{ 1 \cdot 25(1 + r)^{-f} + \frac{341 \cdot 9}{267 \cdot 9}(1 + r)^{-f - (2/3)} \right.$$

$$\left. \times \left[\sum_{j=2}^{26} (1 + i)^{1 - j} + 100(1 + i)^{-25} \right] \right\}$$

or

$$106 \cdot 375 = (1 + i)^{-f} \left\{ 1 \cdot 25(1 + r)^{-f} + \frac{341 \cdot 9}{267 \cdot 9}(1 + r)^{-f - (2/3)} \right.$$

$$\left. \times \left[\frac{1 - (1 + i)^{-25}}{i} + 100(1 + i)^{-25} \right] \right\}$$

The right-hand side of either of the last two equations is obviously a decreasing function of i. Given any value of r, it is relatively straightforward to obtain the solution for i to any desired degree of accuracy.

(a) If $1 + r = 1.05^{1/2}$, the above equation becomes (since $f = 86/182$)

$$106.375 = (1 + i)^{-86/182} \left\{ 1.235\,674 + 1.241\,244 \right.$$

$$\left. \times \left[\frac{1 - (1 + i)^{-25}}{i} + 100(1 + i)^{-25} \right] \right\}$$

When $i = 0.017$ the right-hand side equals 106.9287. When $i = 0.0175$ the value is 105.7695. By interpolation we estimate i as $0.017\,24$. The estimated real yield is thus 1.72% per half-year, or 3.44% per annum convertible half-yearly.

(b) If $1 + r = 1.07^{1/2}$, the equation is

$$106.375 = (1 + i)^{-86/182} \left\{ 1.230\,177 + 1.227\,975 \right.$$

$$\left. \times \left[\frac{1 - (1 + i)^{-25}}{i} + 100(1 + i)^{-25} \right] \right\}$$

When $i = 0.0165$ the right-hand side equals 106.9544. When $i = 0.017$ the value is 105.7932. By interpolation we estimate i as $0.016\,749$. The estimated real yield is thus 1.67% per half-year, or 3.34% per annum convertible half-yearly.

Exercises

7.1   For a certain perpetuity, interest is payable on 1 June and 1 December each year. The amount of each interest payment is £1·75 per £100 nominal.
   Find the effective annual yield obtained by an investor, not liable to taxation, who purchased a quantity of this stock on 14 August 1984, when the market price was 35·125%.

7.2   A fixed-interest loan bears interest of 10% per annum payable half-yearly in arrear. The loan will be redeemed at 105% by five instalments, of equal nominal amount, on 31 July in each of the years 2016 to 2020 inclusive.
   An investor, who is liable to income tax at the rate of 40%, purchased the entire loan on 19 June 1986, at a price to obtain a net yield of 9% per annum effective. Find the price per cent paid, assuming that the purchase was made

   (a) 'ex dividend';
   (b) 'cum dividend'

   in relation to the interest payment due on 31 July 1986.

7.3   A loan of nominal amount £300 000 in bonds of nominal amount £100 is to be repaid by 30 annual drawings, each of 100 bonds, the first drawing being one year after the issue date. Interest will be payable quarterly in arrear at the rate of 8% per annum. Redemption will be at par for the first 15 drawings and at 120% thereafter.
   An investor, who will be liable to income tax at the rate of 40%, purchases the entire loan on the issue date at a price to obtain a yield per annum of 7% net effective.
   What price does the investor pay for each bond?

7.4   A loan of nominal amount £1 000 000 is to be issued bearing interest at the rate of 8% per annum payable quarterly in arrear. At the end of the 15th and each subsequent year £75 000 nominal of the loan will be redeemed. The entire outstanding balance of the loan will be repaid at the end of the 25th year.

An investor, liable to income tax at the rate of 30%, wishes to purchase the entire loan on the issue date at a price to obtain a yield of 7% per annum net effective. Find the price he should pay, if

(a) The redemption price is constant and equal to 110%;
(b) The redemption price at the end of year t $(t = 15, 16, \ldots, 25)$ is $(125 - t)\%$.

7.5   A loan stock bears interest at the rate of 11% per annum. Interest is payable on 15 May and 15 November each year and the entire loan is redeemable at par on either of these dates in the year 2018 or in any subsequent year.

An investor, who is liable to income tax at the rate of 50%, purchased part of the loan on 15 November 1986 (just after payment of the interest then due).

(a) Find the maximum price per cent nominal he should have paid to be certain of obtaining a net effective annual yield of 4%. Assuming that he paid this price, find the maximum possible net yield he may obtain from the investment.
(b) Find the maximum price per cent nominal he should have paid to be certain of obtaining a net effective annual yield of 7%. Assuming that he paid this price, find the maximum possible net yield he may obtain from the investment.

7.6   A loan of nominal amount £10 000 is to be issued in bonds of nominal amount £100 bearing interest at $4\frac{1}{2}\%$ per annum payable quarterly in arrear. Ten bonds will be redeemed at the end of the first and of each subsequent year until the loan is repaid. For each bond redeemed at the end of the nth year the redemption price will be R_n, where

$$R_n = 100 + \frac{n^2}{10}$$

Find the price to be paid by a purchaser of the entire loan to obtain a yield of 7% per annum effective. (Ignore taxation.)

7.7   A certain foreign loan stock, of £175 million nominal, paid interest at 3% per annum, payable quarterly in arrear on 16 July, 16 October, 16 January, and 16 April. The stock was issued on 16 April 1870 and was redeemable at par according to the following schedule:

Years	Nominal amount redeemed (£ million)
1879–1907	1
1908–1925	2
1926–1938	3
1939–1945	4
1946–1950	5
1951–1953	6

Stock was drawn for redemption by lot in bonds of £100 nominal on 16 April in each of these years.

(a) Find the price per £100 nominal of the outstanding amount of the loan on 16 April 1915 on the basis of a rate of interest of $3\frac{1}{2}\%$ p.a. effective.

(b) Find the probability that a purchaser of £100 nominal at this price on 16 April 1915 would obtain a yield of at least 5% p.a. effective.

(Ignore taxation, and assume that the purchaser did not receive the interest payable on the date of purchase.)

7.8   A loan of nominal amount £1 650 000 is to be issued bearing interest of $5\frac{1}{2}\%$ per annum payable half-yearly in arrear. The loan will be in bonds of nominal amount £100.

The loan will be redeemed at 110% by annual drawings, the first repayment being made after five years. Initially 1000 bonds will be redeemed and the number of bonds redeemed at subsequent drawings will increase by 100 each year until the loan is repaid. The bonds repaid in any one year will be drawn by lot.

(a) A syndicate, liable to income tax at the rate of 25%, is considering the purchase of the entire loan on the issue date. What price should the syndicate pay in order to obtain a net yield of 4% per annum effective?

(b) An investor, liable to income tax at the rate of 30%, purchased one bond on the issue date for £107. What is the probability that he will obtain a net yield of at least 4% per annum effective?

7.9   On 7 April 1986 the government of a certain country issued two index-linked stocks, with terms of 20 years and 30 years respectively. For each stock interest is payable half-yearly in arrear and the annual coupon rate is 3%. Both interest and capital payments are indexed by reference to the country's 'cost of living index' with a time-lag of eight months.

The index value for August 1985 was 187·52, and at the issue date of the stocks the latest known value of the index was 192·10, the value for February 1986.

The issue price of each stock was such that, if the cost of living index were to increase continuously at the rate of 6% per annum effective, a purchaser of either stock would obtain a real yield on his investment of 3% per annum convertible half-yearly. (This real yield is measured in relation to the cost of living index.)

(a) Show that the issue prices of the stocks were equal and find the common issue price.

(b) Show that, if the cost of living index were to increase continuously at the rate of 4% per annum effective, the 20-year stock would provide a greater real yield than the 30-year stock but that, if the cost of living index were to increase continuously at the rate of 8% per annum effective, the opposite would be true.

7.10 On 14 August 1984 the market price of $2\frac{1}{2}\%$ Index-Linked Treasury Stock 2020 was 85·625% nominal. This stock is redeemable on 16 April 2020 and interest is payable on 16 April and 16 October each year. Interest and capital payments are indexed by reference to the RPI with a time lag of eight months. The stock was originally issued in October 1983. The value of the RPI for February 1983 (i.e. the 'base' month) was 327·3 and the value for February

1984 was 344·0. On 14 August 1984 the latest known value of the RPI was 351·9, the value for June 1984.

(a) Find the amount of the interest payment (per £100 nominal stock) on 16 October 1984.
(b) On 14 August 1984 an investor, who is not liable to tax, calculated the effective real yield he would obtain by purchasing the stock, on the assumption that the RPI would increase continuously from its latest known value at the rate of 10% per annum effective. What answer did he obtain?

7.11 A loan of nominal amount £50000 is to be issued in bonds of nominal amount £100 bearing interest at 6% per annum payable annually in arrear. At the end of each of the next 20 years ten bonds will be redeemed at par and at the end of each of the following 20 years fifteen bonds will be redeemed at 150%. The bonds to be redeemed in any one year will be chosen by lot.

On the issue date an investor buys a single bond at a price of £125.

(a) Find the probability that the effective annual yield obtained by the investor will be
(i) Between 3% and 5%;
(ii) Negative.
(b) Find the effective annual yield obtained by the investor if his bond is redeemed after four years.

(Ignore taxation.)

7.12 On 31 December 1981 a loan was issued to be repaid over 12 years by a level annuity payable quarterly in arrear on the last days of March, June, September, and December. The amount of the annuity was calculated on the basis of an interest rate of 12% per annum convertible quarterly. The total interest paid in 1985, according to the original schedule, was £6374·41.

On 31 December 1985 an investor, liable to income tax at the rate of 40% on the interest content (according to the original schedule) of each annuity instalment, purchased the annuity instalments due after that date.

Find the purchase price, if the net yield obtained by the investor was

(a) 8% per annum convertible quarterly;
(b) 8% per annum effective.

7.13 A loan of £100000 is repayable over 40 years by a level annuity payable at the end of every fourth year. The amount of the annuity is calculated on the basis of an interest rate of 4% per annum effective.

An investor, who will be liable to income tax on the interest content (according to the original schedule) of each annuity payment at the rate of 50% for 16 years and at the rate of 25% thereafter, purchases the entire loan at the issue date to obtain a yield of 4% per annum net effective. Find the price paid for the loan.

7.14 A loan of nominal amount £30000 is to be redeemed at par in three instalments, each of nominal amount £10000, at the end of 8, 16, and 24 years. Interest will be payable annually in arrear at the rate of 6% per annum for the first 8 years, 4% per annum for the next 8 years, and 2% per annum for the final 8 years.

An investor, who will be liable to income tax on the interest payments at the rate of 30% for 12 years and at the rate of 50% thereafter, pays £26 000 to purchase the entire loan on the issue date.

Calculate the net effective annual yield obtained by the investor.

7.15 A loan of £100 000 is to be repaid over 15 years by a level annuity payable monthly in arrear. The amount of the annuity is calculated on the basis of an interest rate of 16% per annum convertible quarterly.

An investor, liable to income tax at the rate of 80% on the interest content (according to the original schedule) of each annuity instalment, wishes to purchase the entire loan on the issue date and calculates that in order to obtain a yield on his investment of 5% per annum net effective he should pay £86 467 for the loan. In the event, however, he has to pay £88 000 for this investment.

Find (a) the amount of each monthly annuity repayment, and (b) the net effective annual yield that the investor will obtain.

7.16 A loan of nominal amount £8000 is to be issued bearing interest of 10% per annum payable quarterly in arrear. At the end of the second, fourth, sixth and eighth years a nominal amount of £2000 of the loan is to be redeemed at a premium which is to be proportional to the time elapsed from the issue date.

An investor, who will be liable to income tax on the interest payments at the rate of 40% for five years and at the rate of 50% thereafter, calculates that, in order to obtain a yield of 7% per annum net effective on his investment, he should offer a price of £7880·55 for the entire loan.

Find the redemption prices of the loan.

7.17 A loan of £100 000 will be repayable by a level annuity, payable annually in arrear for 15 years. The amount of the annuity is calculated on the basis of an interest rate of 8% per annum effective.

An investor, who is liable to income tax at the rate of 40% on the interest content (according to the original schedule) of each annuity payment, wishes to purchase the entire loan on the issue date.

Find the price he should pay in order to achieve a net effective annual yield of (a) 7% and (b) 8%.

7.18 A loan of nominal amount £6000 bearing interest at a rate of $8\frac{1}{2}$% per annum payable annually in arrear will be redeemed in six annual instalments of nominal amount £1000, the first at the end of the tenth year and the last at the end of the fifteenth year. At the end of the $(10 + k)$th year $(0 \leqslant k \leqslant 5)$ the redemption price will be £$[85 + (k/2)(k + 1)]$%. At the end of the tenth and of each subsequent year the borrower has the option to redeem the outstanding balance of the loan at the redemption price then ruling.

An investor who is liable to income tax at a rate of 70% wishes to buy the entire loan at the issue date. Find the maximum price he should pay in order to be certain of achieving a yield of at least 6% per annum net effective.

Note This slightly unusual example does not fit precisely into the treatment of optional redemption dates in the text. The investor must protect himself against *every* option available to the lender and calculate the price accordingly.

7.19 (Optional redemption dates; purchase date *not* the date of issue or that of an interest payment.)
A loan bears interest of 10% per annum, payable annually in arrear. The loan is redeemable at par on any one of the first five anniversaries of the issue date, the actual redemption date being chosen by the borrower.

Six months after the issue date an investor buys a quantity of the loan at a price of 102%.

Show that (in contrast to the rule on page 169, which does *not* apply as the purchase is not on the issue date or an interest date) the investor's yield will be least if the loan is redeemed at the latest possible date.

CHAPTER EIGHT

CAPITAL GAINS TAX

8.1 Introduction

Capital gains tax, which was introduced in the UK by the Finance Act 1965, is a tax levied on the difference between the sale or redemption price of a stock (or other asset) and the purchase price, if lower. In contrast to income tax, this tax is normally payable once only in respect of each disposal, at the date of sale or redemption. Certain assets, including some fixed-interest securities, may be exempt from capital gains tax; in addition, exemption from this tax may be granted after an asset has been held for a certain period. In some cases capital losses may be offset against capital gains on other assets (see section 8.6). Unless there is some form of 'indexation', capital gains tax may be criticized as unfair, in that in times of inflation it does not take into account the falling value of money: it taxes 'paper' gains as well as real ones. A system of indexation of capital gains was introduced in the UK in 1981 and subsequently modified. After assets have been held for a year the calculation of any capital gains tax liability takes into account movements in the Retail Prices Index.

In this chapter we shall be concerned with the effect of capital gains tax on the prices and yields of fixed-interest securities. In general we shall assume that the stock in question will be held to redemption. If a stock is sold before the final maturity date, the capital gains tax liability will in general be different, since it will be calculated with reference to the sale proceeds rather than the corresponding redemption money. (See examples 8.4.2 and 8.4.3 for illustrations of these calculations.) For simplicity, we shall ignore the possibility of indexation in our calculations. In theory, it is relatively simple to allow for indexation; in practice, however, changes in legislation may make some capital gains tax calculations extremely complicated.

Example 8.1.1

A loan of nominal amount £100 000 will be repaid at 110% after 15 years. The loan bears interest of 9% per annum payable annually in arrear. On the issue date the loan is purchased for £80 000 by an investor who is liable to income tax at the rate of 40% and to capital gains tax at the rate of 30%.
Find the investor's net effective annual yield for the transaction.

Solution

It is important to understand the operation of capital gains tax. Since the redemption price is 110%, the purchase price of £80 000 acquires redemption proceeds of £110 000. There is thus a capital gain of £30 000 on redemption. The capital gains tax payable is therefore £9000 (i.e. 30% of £30 000), so that after tax the investor retains £101 000 of the redemption proceeds.

Interest payable each year is £9000 before tax. The net interest received by the investor is thus £5400 per annum (after tax at 40%).

The investor's net yield per annum is therefore that rate of interest for which

$$80\,000 = 5400a_{\overline{15}|} + 101\,000v^{15}$$

The reader should verify that the solution for the rate of interest is 0·077 37. The net annual yield is thus 7·74%.

8.2 Valuing a loan with allowance for capital gains tax

Consider a loan which bears interest at a constant rate and is redeemable by instalments. Suppose that at a particular time the total nominal amount of loan outstanding is $N = (N_1 + N_2 + \cdots + N_r)$ and that a nominal amount N_j will be redeemed after a further n_j years $(1 \leqslant j \leqslant r)$, where $n_1, n_2, \ldots, n_r$ are integer multiples of $1/p$. Suppose further that the redemption price per unit nominal is constant and equal to R.

What price should be paid by an investor who wishes to purchase the entire outstanding loan to obtain a net yield of i per annum, allowing for his tax liability? Assume that interest is payable p times per annum at the rate g per annum on the capital outstanding, measured in cash, not nominal, terms. Let $C = RN$. If the investor has no tax liability, the price to be paid is of course

$$A = K + \frac{g}{i^{(p)}}(C - K) \tag{8.2.1}$$

or

$$A = K + I$$

where K and I denote the value of the gross capital and interest payments respectively.

If the investor is liable to income tax at rate t_1 but is not liable to capital gains tax, the value of the tax payable must be deducted from the price given by equation 8.2.1. Thus the price to be paid is A', where

$$A' = K + \frac{g(1 - t_1)}{i^{(p)}}(C - K) \tag{8.2.2}$$

Finally, consider the price to be paid by an investor who is liable to capital gains tax at the rate t_2 in addition to income tax at the rate t_1. If

$$g(1 - t_1) \geq i^{(p)} \tag{8.2.3}$$

then $A' \geq C$, so the price paid allowing for only income tax is not less than the total redemption monies receivable. If this price is paid, there will be a capital *loss* on redemption. In this case, therefore, the price to be paid is simply A' (as given by equation 8.2.2) and in fact no capital gains tax is payable. (We are assuming that it is not permissible to offset the capital loss against any other capital gain: see example 8.6.1.)

If, however,

$$g(1 - t_1) < i^{(p)} \tag{8.2.4}$$

then A', the price paid to allow for only income tax, is less than the total redemption monies and this will obviously be true also for the price allowing in addition for capital gains tax. In this case let A'' be the price to be paid allowing for both income tax and capital gains tax. At time n_j, when a nominal amount N_j is repaid, the redemption money received is $N_j R$. If the price paid for the entire outstanding loan is A'', under current UK practice the tranche redeemed at time n_j is considered (for tax purposes) to have cost $(N_j/N)A''$ (i.e. for capital gains tax calculations, the total purchase price paid is divided among the different tranches in proportion to the *nominal* amounts redeemed). The capital gain at time n_j is thus deemed to be $N_j R - (N_j/N)A''$ and the total value of the capital gains tax payable is thus

$$\sum_{j=1}^{r} t_2 \left(N_j R - \frac{N_j}{N} A'' \right) v^{n_j} = t_2 \frac{NR - A''}{NR} \sum_{j=1}^{r} N_j R v^{n_j}$$

$$= t_2 \frac{C - A''}{C} K \tag{8.2.5}$$

$$= t_2 \frac{R - P''}{R} K \tag{8.2.6}$$

where P'' is the price per unit nominal.

It is essential to understand the derivation of the last expression. It may be recalled simply by the following argument, which depends crucially on the fact that a constant proportion of each loan repayment is absorbed by capital gains tax. The total capital repayment (i.e. the redemption monies actually paid) is C and this has value K. The total capital gains tax payable is $t_2(C - A'')$. By proportion, therefore, the value of the capital gains tax is $t_2(C - A'')(K/C)$, as given by equation 8.2.5.

The price A'' is the value of the net proceeds, after allowance for all tax liability. This implies that

$$A'' = K + \frac{g(1 - t_1)}{i^{(p)}}(C - K) - t_2 \frac{C - A''}{C} K \qquad (8.2.7)$$

from which we immediately obtain

$$A'' = \frac{(1 - t_2)K + (1 - t_1)(g/i^{(p)})(C - K)}{1 - t_2 K/C} \qquad (8.2.8)$$

If the principles underlying equations 8.2.5 and 8.2.6 are fully understood, the valuation of a stock with allowance for capital gains tax often involves little more work than when this form of taxation is ignored.

For future reference, it is convenient to summarize the above results as follows.

If $A''(i)$ denotes the price to provide a net annual yield of i, and K and I are the values (at rate of interest i) of the *gross* capital and interest payments respectively, we have

$$A''(i) = \begin{cases} K + (1 - t_1)I & \text{if } i^{(p)} \leqslant g(1 - t_1) \\ \dfrac{(1 - t_2)K + (1 - t_1)I}{1 - t_2 K/C} & \text{if } i^{(p)} > g(1 - t_1) \end{cases} \qquad (8.2.9)$$

where $I = (g/i^{(p)})(C - K)$. Note that the condition $i^{(p)} > g(1 - t_1)$ is equivalent to

$$i > \left[1 + \frac{g(1 - t_1)}{p} \right]^p - 1$$

As i increases $A''(i)$ decreases. (Both K and I are decreasing functions of i, while $1 - t_2 K/C$ is an increasing function of i.) When $i^{(p)} = g(1 - t_1)$, each of the expressions on the right-hand side of equation 8.2.9 is equal to C.

Example 8.2.1

A loan of nominal amount £500 000 was issued bearing interest of 8% per annum payable quarterly in arrear. The loan will be repaid at £105% by 20 annual instalments, each of nominal amount £25 000, the first repayment being ten years after the issue date.

An investor, liable to both income tax and capital gains tax, purchased the entire loan on the issue date at a price to obtain a net effective annual yield of 6%. Find the price paid, given that his rates of taxation for income and capital gains are

(a) 40% and 30% respectively;
(b) 20% and 30% respectively.

Solution

Note that $C = 500\,000 \times 1{\cdot}05 = 525\,000$.

Also,

$$K = 25\,000 \times 1{\cdot}05(a_{\overline{29}|} - a_{\overline{9}|}) \qquad \text{at } 6\% \qquad = 178\,211$$

and

$$I = \frac{0{\cdot}08}{1{\cdot}05}\frac{1}{0{\cdot}06^{(4)}}(C - K) = 450\,158$$

The price payable if there had been no tax liability is

$$A = K + I = 628\,368$$

(a) $t_1 = 0{\cdot}4$ and $t_2 = 0{\cdot}3$.

In this case the price allowing for income tax only is

$$A - t_1(A - K) = K + (1 - t_1)I = 448\,306$$

which is less than $525\,000$ (i.e. C). Thus capital gains tax is payable and the price actually paid by the investor is given by equation 8.2.7 as

$$A'' = 448\,306 - t_2\frac{C - A''}{C}K$$

Thus

$$A'' = 448\,306 - 0{\cdot}3\left(1 - \frac{A''}{525\,000}\right)178\,211$$

so that

$$A'' = \frac{448\,306 - (0{\cdot}3 \times 178\,211)}{1 - (0{\cdot}3 \times 178\,211/525\,000)} = 439\,610$$

The price paid was thus £439 610 or £87·922%.

(b) $t_1 = 0{\cdot}2$ and $t_2 = 0{\cdot}3$.

In this case the price allowing for income tax only is

$$178\,211 + (1 - 0{\cdot}2) \times 450\,158 = 538\,337$$

Since this exceeds £525 000, the price paid by the investor was £538 337 and in fact he has no liability for capital gains tax.

8.3 Capital gains tax when the redemption price or the rate of tax is not constant

When (in relation to a loan repayable by instalments) either the redemption price or the rate of capital gains tax is not constant, it is important to fall back on first principles in order to value the capital gains tax. As we have already remarked, it is vital to appreciate that for taxation purposes the 'price' deemed to have been paid for a part of the loan redeemed at any one time is calculated from the total price paid in proportion to the *nominal* amount redeemed.

Suppose, as before, that when the loan is purchased the total nominal amount outstanding is $N = (N_1 + N_2 + \cdots + N_r)$. Suppose further that a nominal amount N_j will be redeemed after n_j years, at which time the redemption price per unit nominal will be R_j and the rate of capital gains tax t_j^*.

If the price paid for the entire loan is A'', the tranche redeemed at time n_j for an amount $N_j R_j$ is considered as having cost $(N_j/N)A''$. This means that the total capital gains tax payable has value

$$\sum_{j=1}^{r} t_j^* \left(N_j R_j - \frac{N_j}{N} A'' \right) v^{n_j} \tag{8.3.1}$$

$$= \sum_{j=1}^{r} t_j^* \left(\frac{R_j - P''}{R_j} \right) K_j \tag{8.3.2}$$

where $K_j = N_j R_j v^{n_j}$ is the value of the capital repayment for the jth tranche and P'' is the price per unit nominal.

In practice, either of these summations is evaluated quite simply by grouping together those terms for which the tax rate and the redemption price are constant (see example 8.3.1).

If the value obtained for P'' by using the above expression for the value of the capital gains tax payable is such that, for certain values of j, $R_j < P''$, there will be no tax liability in respect of these tranches. In this case P'' must be recalculated by omitting the appropriate terms from the summations 8.3.1 and 8.3.2. If these capital losses can be offset against capital gains on other tranches, more detailed calculations are required (see section 8.6).

Example 8.3.1

Fifteen years ago a loan of nominal amount £300 000 was issued bearing interest of 8% per annum payable annually in arrear. The loan was to be repaid over 30 years, a nominal amount of £10 000 being repayable at the end of each year. The redemption price is par for the first ten years, 105% for the next ten years, and 110% for the final ten years.

The 15th capital repayment has just been made and an investor wishes to purchase the entire outstanding loan. The investor is not liable for income tax but will be liable for capital gains tax at the rate of 35% for the next ten years and at the rate of 30% thereafter. What price should the investor pay, if he wishes to realize a net effective annual yield of 10% on his investment?

Solution

The loan has an outstanding term of 15 years, during which three distinct situations will pertain:

(a) First five years: redemption price 105%,   CGT rate 35%;
(b) Second five years: redemption price 110%,   CGT rate 35%;
(c) Final five years: redemption price 110%,   CGT rate 30%.

Ignoring the premiums on redemption, the investor may value the future capital repayments, as $(K_1 + K_2 + K_3)$, where $K_1 = 10\,000 a_{\overline{5}|}$, $K_2 = 10\,000\,_5|a_{\overline{5}|}$ and $K_3 = 10\,000\,_{10}|a_{\overline{5}|}$ (at 10%). Thus $K_1 = 37907\!\cdot\!87$, $K_2 = 23\,537\!\cdot\!80$, $K_3 = 14\,615\!\cdot\!12$, and $K_1 + K_2 + K_3 = 76\,060\!\cdot\!79$.

The nominal amount of loan outstanding at the time of purchase is £150 000. Hence the value of the loan without any allowance for tax (but with allowance for the redemption premiums) is

$$76\,060\!\cdot\!79 + \frac{0\!\cdot\!08}{0\!\cdot\!10}(150\,000 - 76\,060\!\cdot\!79) + (0\!\cdot\!05 \times 37\,907\!\cdot\!87)$$

$$+ \, 0\!\cdot\!1(23\,537\!\cdot\!80 + 14\,615\!\cdot\!12) = 140\,922\!\cdot\!84$$

We must now value the capital gains tax. Let A'' be the price paid by the investor. Since equal nominal amounts are redeemed each year, each outstanding tranche is considered as having cost $A''/15$, so that the capital gains tax has value

$$0\!\cdot\!35 \sum_{t=1}^{5}\left[(10\,000 \times 1\!\cdot\!05) - \frac{A''}{15}\right]v^t$$

$$+ \, 0\!\cdot\!35 \sum_{t=6}^{10}\left[(10\,000 \times 1\!\cdot\!1) - \frac{A''}{15}\right]v^t + 0\!\cdot\!3\sum_{t=11}^{15}\left[(10\,000 \times 1\!\cdot\!1) - \frac{A''}{15}\right]v^t$$

or

$$(0\!\cdot\!35 \times 1\!\cdot\!05 \times K_1) + (0\!\cdot\!35 \times 1\!\cdot\!1 \times K_2) + (0\!\cdot\!3 \times 1\!\cdot\!1 \times K_3)$$

$$- \frac{A''}{15}(0\!\cdot\!35 a_{\overline{10}|} + 0\!\cdot\!3\,_{10}|a_{\overline{5}|})$$

which equals

$$27\,816\!\cdot\!18 - 0\!\cdot\!172\,603\,A''$$

Hence A'' is given by the equation

$$A'' = 140\,922\!\cdot\!84 - (27\,816\!\cdot\!18 - 0\!\cdot\!172\,603\,A'')$$

from which it follows that $A'' = 136\,701\!\cdot\!80$.

Note Since the outstanding nominal loan is £150 000, this price could be quoted as £91·13%.

8.4 Finding the yield when there is capital gains tax

An investor who is liable for capital gains tax may wish to determine the net yield on a particular transaction in which he has purchased a loan at a given price.

One possible approach is to determine the price on two different net yield bases and then estimate the actual yield by interpolation. This approach, which is obviously comparable with that described in the previous chapter, is not always the quickest method. Since the purchase price is known, so too is the amount of the capital gains tax, and the net receipts for the investment are thus known. In this situation one may more easily write down an equation of value which will provide a simpler basis for interpolation, as illustrated by our next example.

Example 8.4.1

A loan of £1000 bears interest of 6% per annum payable yearly and will be redeemed at par after ten years. An investor, liable to income tax and capital gains tax at the rates of 40% and 30% respectively, buys the loan for £800. What is his net effective annual yield?

Solution

Note that the net income each year of £36 is 4·5% of the purchase price. Since there is a gain on redemption, the net yield is clearly greater than 4·5%.

The gain on redemption is £200, so that the capital gains tax payable will be £60 and the net redemption proceeds will be £940. The net effective yield p.a. is thus that value of i for which

$$800 = 36a_{\overline{10}|} + 940v^{10}$$

If the net gain on redemption (i.e. £140) were to be paid in equal instalments over the ten-year duration of the loan rather than as a lump sum, the net receipts each year would be £50 (i.e. £36 + £14). Since £50 is 6·25% of £800, the net yield actually achieved is less than 6·25%. When $i = 0.055$, the right-hand side of the above equation takes the value 821·66, and when $i = 0.06$ the value is 789·85. By interpolation, we estimate the net yield as

$$i = 0.055 + \frac{821.66 - 800}{821.66 - 789.85} 0.005 = 0.0584$$

The net yield is thus 5·84% per annum.

Alternatively, using equation 8.2.7 or equation 8.2.8, we may find the prices to give net yields of 5·5% and 6% per annum. These prices are £826·27 and £787·81, respectively. (The reader should verify these values.) The yield may then be obtained by interpolation. However, this alternative approach is somewhat longer than the first method.

Example 8.4.2

Assume that in respect of the loan described in example 8.2.1 the investor's tax rates for income and capital gains were 40% and 30% respectively, and that he purchased the entire loan on the issue date at a price to obtain a net yield of 6% per annum effective. Assume further that eight years after purchasing the loan (immediately after receiving the interest payment then due) the investor sold the entire loan to another investor who was liable to the same rates of income tax and capital gains tax. The price paid by the second investor was such as to provide him with a net effective yield of 6% per annum and the original purchaser paid capital gains tax on the proceeds of the sale.

Find the net effective annual yield obtained on the completed transaction by the first investor.

Solution

The price paid by the first investor was £439 610 (see the solution to example 8.2.1). At the time he sells the loan it has an outstanding term of 21 years and the first payment of capital will occur after two years. The second investor values the gross capital payments as

$$K = (25\,000 \times 1 \cdot 05)(a_{\overline{21}|} - a_{\overline{1}|}) \qquad \text{at } 6\% \qquad = 284\,043$$

and the gross income payments as

$$I = \frac{0 \cdot 08}{1 \cdot 05} \frac{1}{0 \cdot 06^{(4)}}(C - K) \qquad \text{(where } C = 525\,000)$$

$$= 312\,778$$

For the second investor the value of the gross receipts less income tax (only) is

$$K + (1 - 0 \cdot 4)I = 471\,710$$

which is less than £525 000. Hence capital gains tax will be payable by the second investor and the price he actually paid is A'', where

$$A'' = 471\,710 - \frac{0 \cdot 3(525\,000 - A'')}{525\,000} 284\,043$$

from which it follows that $A'' = 461\,384$.

The capital gain realized by the sale for the first investor is thus £21 774 (i.e. £461 384 − £439 610). The capital gains tax payable is therefore £6532·2 and the net sale proceeds are £454 851·8. The net yield per annum for the first investor is thus the value of i for which

$$439\,610 = (1 - 0 \cdot 4)40\,000 a_{\overline{8}|}^{(4)} + 454\,851 \cdot 8 v^8$$

The reader should verify that $i = 0 \cdot 0593$ or $5 \cdot 93\%$.

Note Because of the earlier incidence of capital gains tax, the net yield is less than 6%, although this rate was used as the net yield basis for both purchasers. This is in contrast to the situation which applies when only income tax is involved.

Example 8.4.3

A certain irredeemable stock bears interest at $5\frac{1}{2}\%$ per annum, payable quarterly on 31 March, 30 June, 30 September, and 31 December. On 31 August of a certain year, an investor bought a quantity of the stock at a price of £49·50%; he sold it exactly one year later at a price of £57·71%.

Given that the investor was liable to capital gains tax at the rate of 30% and income tax at the rate of 40%, find the net yield per annum he obtained.

Solution

Consider the purchase of £100 nominal of the stock. This cost £49·50 and provided net annual income of £3·30 (i.e. 0·6 × £5·50). The net sale proceeds, after payment of capital gains tax, were

$$57\cdot71 - 0\cdot3(57\cdot71 - 49\cdot50) = 55\cdot25$$

The net annual yield is that rate of interest for which

$$49\cdot50 = 3\cdot3v^{30/365}\ddot{a}_{\overline{1}|}^{(4)} + 55\cdot25v$$

A rough solution is

$$i \simeq \frac{3\cdot3 + 55\cdot25 - 49\cdot50}{49\cdot50} = 0\cdot183$$

and by interpolation we obtain $i = 0\cdot1895$, or $18\cdot95\%$.

8.5 Optional redemption dates

In section 7.8 we considered the consequences for a lender, subject only to income tax, of the borrower having a choice of redemption dates. In fact, even with the additional complication of capital gains tax a lender should adopt the same strategy when valuing a loan for which the borrower has optional redemption dates. In the latter situation, however, a little more care is needed with the argument to see that this is indeed the case.

Suppose then that a person, liable to both income and capital gains tax, wishes to realize a net annual yield of at least i on a given loan. Suppose further that, subject to specified conditions, the borrower has a choice as to when he repays the loan.

What price should the lender offer to achieve a net annual yield of at least i?

As before, we assume that interest is payable p times per annum at the rate g per annum on the capital outstanding (measured by cash, rather than nominal, amount). Note first that, if $i^{(p)} = g(1 - t_1)$, the lender should pay C. In this case, whenever the loan is repaid the lender will achieve a net annual yield of i.

If $i^{(p)} < g(1 - t_1)$, it follows from equation 8.2.9 that the price to provide a net yield of i is

$$A^* = K + (1 - t_1)I$$

$$= K + \frac{g(1 - t_1)}{i^{(p)}}(C - K)$$

$$= \frac{g(1 - t_1)C}{i^{(p)}} - \left[\frac{g(1 - t_1)}{i^{(p)}} - 1\right]K$$

Since by hypothesis $i^{(p)} < g(1 - t_1)$, it follows that the least value of A^* will occur when K takes its greatest value. Thus the lender should value the loan on the basis that the borrower will choose that option for which the capital payments have the *greatest* possible value. (In particular, if the entire loan must be repaid at one time, the lender should assume that the loan will be repaid as soon as possible.) If this option is in fact chosen, the lender's net yield will be i; otherwise it will exceed i.

If $i^{(p)} > g(1 - t_1)$, the price to provide a net yield of i is (see equation 8.2.9) $f(K)$, where K is the value at rate i of the gross capital payments and

$$f(x) = \frac{(1 - t_2)x + [g(1 - t_1)/i^{(p)}](C - x)}{1 - t_2 x/C} \qquad (8.5.1)$$

It is readily verified from this last equation that

$$f'(x) = (1 - t_2)\left[1 - \frac{g(1 - t_1)}{i^{(p)}}\right] / \left(\frac{x}{C}\right)^2$$

Since by hypothesis $i^{(p)} > g(1 - t_1)$, it follows that $f'(x) > 0$, so f is an increasing function. Hence, the smaller the value of K, the smaller the value of $f(K)$. Thus, if $i^{(p)} > g(1 - t_1)$, the lender should value the loan on the basis that the borrower will choose that option for which the capital payments have the *least* possible value. (In particular, if the entire loan must be repaid at one time, the lender should assume that the loan will be repaid as late as possible.) This strategy will ensure that the actual net yield to the lender is at least i.

Example 8.5.1

A loan of nominal amount £100 000 is to be issued bearing interest payable half-yearly in arrear at the rate of 8% per annum. The terms of the issue provide that the borrower shall repay the loan (at par) in ten consecutive annual instalments each of nominal amount £10 000, the first repayment being made any time (at the borrower's option) between 10 and 25 years from the issue date.

An investor, liable to income tax at the rate of 40% and to capital gains tax at the rate of 30%, wishes to purchase the entire loan on the issue date at a price to guarantee him a net yield of at least 7% per annum.

(a) What price should he pay?
(b) Given that he paid the price as determined above and that the first capital repayment will actually be made after 14 years, find the net yield which the lender will in fact achieve on this transaction.

Solution

(a) Using the standard notation we have $C = 100\,000$, $p = 2$, $g = 0.08$, $t_1 = 0.4$, $t_2 = 0.3$, and $i = 0.07$. Since $g(1 - t_1) = 0.048$, which is clearly less than $0.07^{(2)}$, the lender should assume that the capital repayments have the least possible value. Since the repayments must be made in ten consecutive years, the lender should assume that redemption occurs as late as possible, i.e. the first repayment will be after 25 years.

On this basis

$$K = 10\,000\,_{24|}a_{\overline{10|}} \qquad \text{at } 7\%$$
$$= 10\,000(a_{\overline{34|}} - a_{\overline{24|}})$$
$$= 13\,846{\cdot}75$$

The value of the net interest payments is thus (by Makeham's formula)

$$\frac{0{\cdot}08(1 - 0{\cdot}4)}{0{\cdot}07^{(2)}}(100\,000 - 13\,846{\cdot}75) = 60\,092{\cdot}87$$

Hence the price to be paid is A'', where

$$A'' = 13\,846{\cdot}75 + 60\,092{\cdot}87 - \frac{0{\cdot}3(100\,000 - A'')}{100\,000}13\,846{\cdot}75,$$

from which we obtain

$$A'' = 72\,810{\cdot}14$$

Thus he should pay £72 810·14 for the loan. If the borrower delays repaying the loan until the latest permitted date, the lender's net annual yield will be 7%. If redemption occurs earlier than this, the net annual yield to the lender will exceed 7%.

(b) Now suppose that the price paid for the entire loan was £72 810·14 and that the first capital repayment will be made after 14 years. Each tranche of the loan is deemed to have cost £7281·01, so that the capital gains tax payable with each redemption is thus 0·3 (10 000 − 7281·01), i.e. £815·70. The net proceeds of each redemption payment are thus £9184·30. Note that the value on the issue date (at rate i) of the gross redemption monies is

$$10\,000(a_{\overline{23|}} - a_{\overline{13|}}) \qquad \text{at rate } i$$

Hence, using Makeham's formula, we obtain the net yield to the lender as that interest rate for which

$$72\,810{\cdot}14 = 9184{\cdot}30(a_{\overline{23|}} - a_{\overline{13|}}) + \frac{0{\cdot}08(1 - 0{\cdot}4)}{i^{(2)}}[100\,000 - 10\,000(a_{\overline{23|}} - a_{\overline{13|}})]$$

The reader should verify that the yield is 7·44%.

8.6 Offsetting capital losses against capital gains

Until now we have considered the effects of capital gains tax on the basis that it is *not* permitted to offset capital gains by capital losses. In some situations, however, it may be permitted to do so. This may mean that an investor, when calculating his liability for capital gains tax in any year, is allowed to deduct from his total capital gains for the year the total of his capital losses (if any). If the total capital losses exceed the total capital gains, no 'credit' will generally be given for the overall net loss, but no capital gains tax will be payable.

A detailed treatment of this topic is beyond the scope of this book, but the following elementary example indicates the kind of situation which may arise.

Example 8.6.1

Two government stocks each have an outstanding term of four years. Redemption will be at par for both stocks. Interest is payable annually in arrear at the annual rate of 15% for the first stock and 8% for the second stock. Interest payments have just been made and the prices of the stocks are £105·80% and £85·34% respectively.

(a) Verify that an investor, liable for income tax at the rate of 35% and capital gains tax at the rate of 50%, who purchases either of these stocks (but *not* both) will obtain a net yield on his transaction of 8% per annum.
(b) Assume now that the investor is allowed to offset capital gains by capital losses. Show that, if the proportion of his available funds invested in the 8% stock is such that the overall capital gain is zero, he will achieve a net yield on the combined transaction of 8·46% per annum.

Solution

(a) We omit details, since this is perfectly straightforward. (See section 8.4.)
(b) Assume that the investor applies a fraction λ of his funds to purchase the 8% stock (and, thus, a fraction $1 - \lambda$ to the 15% stock). We shall work on the basis of a total investment of £1000. The *nominal* amounts of each stock purchased are thus

$$\frac{1000\lambda}{85\cdot34}100 \qquad \text{of the 8\% stock}$$

$$\frac{1000(1 - \lambda)}{105\cdot80}100 \qquad \text{of the 15\% stock}$$

The net income received each year is

$$0\cdot65\left[\frac{1000\lambda}{85\cdot34}8 + \frac{1000(1 - \lambda)}{105\cdot80}15\right] = 92\cdot16 - 31\cdot22\lambda$$

The total *gross* redemption proceeds are

$$\frac{1000\lambda}{85\cdot34}100 + \frac{1000(1 - \lambda)}{105\cdot80}100 = 945\cdot18 + 226\cdot60\lambda$$

If $\lambda = (1000 - 945\cdot18)/226\cdot60$ (i.e. 0·2419), the total gross redemption proceeds will be 1000. In this case the capital gain on the 8% stock will be exactly offset by the capital loss on the 15% stock. In this situation no capital gains tax will be payable and the net redemption proceeds will be £1000 (i.e. the amount originally invested). Since, for this value λ, the combined net annual income from the two stocks is

$$92\cdot16 - (31\cdot22 \times 0\cdot2419) = 84\cdot61$$

the net yield on the combined transaction is 8·461%, as required.

The above example indicates how the investor may exploit the offsetting of gains by losses to obtain a greater net yield than he could obtain from either stock alone.

Exercises

8.1 A loan is to be issued bearing interest of 9% per annum payable half-yearly in arrear. The loan will be redeemed after 15 years at 110%.

   An investor, liable to income tax at the rate of 45% and to capital gains tax at the rate of 30%, is considering the purchase of part of the loan on the issue date.

   (a) What price (per cent nominal) should he pay in order to achieve a net effective annual yield of 8%?
   (b) Given that the price actually payable by the investor will be £80%, find his net effective annual yield.

8.2 A loan is to be redeemed in 15 annual instalments of equal nominal amount, the first instalment being paid five years after the issue date. The redemption price is 120% and the loan will bear interest of 8·4% per annum payable half-yearly *in advance*.

   An investor, liable to income tax at the rate of 30% and to capital gains tax at the rate of 25%, purchased the entire loan on the issue date at a price to obtain a net effective annual yield of 7%.

   What price per cent nominal did the investor pay?

8.3 Investors A and B are both liable to capital gains tax at the rate of 40%, but neither is liable to income tax.

   Investor A bought a bond of nominal amount £1000 bearing interest of 6% per annum payable half-yearly in arrear. The bond was to be redeemed at par ten years after the date of purchase, and the price paid by A was such that, if he had held the bond until it was redeemed, he would have obtained a net yield on his investment of 10% per annum.

   Five years after purchasing the bond (and immediately after receiving the interest then due) A sold it to B, paying capital gains tax on the excess of his selling price over his original price. The bond was held by B until it was redeemed.

   (a) If the purchase price paid by B were such that he obtained a net yield on his investment of 10% per annum, find the net annual yield obtained by A over his completed transaction.
   (b) If the purchase price paid by B were such that A obtained a net yield over his completed transaction of 10% per annum, find the net annual yield obtained by B on his investment.

8.4 A zero-coupon bond (i.e. a bond bearing no interest) was purchased m years ago by investor A who is liable to capital gains tax at rate t. At the time of purchase the outstanding term of the bond was n years ($n > m$). The price paid by A will provide him with a net effective annual yield of i if he holds the bond until it is redeemed.

   Investor A now wishes to sell the bond. He will be liable to capital gains tax on the excess of his selling price over his purchase price.

   (a) Derive an expression in terms of t, n, and i for the purchase price (per unit redemption money) paid by A.
   (b) Derive also an expression in terms of t, n, m, and i for the price (per unit

redemption money) at which A should now sell the bond in order to obtain a net annual yield of i on the completed transaction.

(c) Assume that in fact the bond is sold by A to a second investor, who is also liable to capital gains tax at rate t, at a price which will provide the *new* purchaser with a net annual yield of i, if he holds the bond until it is redeemed.

Derive an equation from which can be found the value of j, the net annual yield obtained by A on the completed transaction.

Find the value of j when $n = 10$, $m = 5$, $t = 0.4$, and $i = 0.1$.

8.5 A loan of nominal amount £100 000 is redeemable at 115% by triennial payments of capital. The cash amount of the tth repayment is £$(15 000 + 1000t(t - 1))$ and the first repayment is due ten years after the loan is issued. Interest on the loan is payable half-yearly in arrear at the rate of 6% per annum. Find the value of the whole loan to a lender, liable to tax on income at 30% and to tax on capital gains at 40%, who wishes to obtain a net yield of 7% per annum on his investment.

8.6 A redeemable stock was issued on 1 January 1965 in bonds of £100 nominal. Interest was payable half-yearly in arrear at 4% per annum for the first ten years and at 3% per annum for the next ten years. Redemption was at par after 20 years. On 1 July 1969 the stock was quoted at a price which would have given a tax-free investor a yield of 6% per annum. A bond was bought on that date by an investor who was subject at all times to income tax at 40% and to capital gains tax at 20%. This investor held the bond until redemption. What net annual yield did he obtain?

8.7 A loan of £1 000 000 nominal has just been issued in bonds of £100 nominal bearing interest at 4% per annum payable half-yearly in arrear. The loan is redeemable by drawings at the end of each year of 400 bonds per annum for the first seven years and 600 bonds per annum thereafter. Redemption is at par during the first seven years and at 115% thereafter. The issue price is such as to give the purchaser of the entire loan, which is a life office subject to income tax at 35% and to capital gains tax at 30%, a net yield of 6% per annum.

(a) Calculate the price per bond paid by the life office.
(b) Determine the number of individual bonds which will give the life office a net yield of less than 6% per annum.

8.8 A loan of nominal amount £500 000 is issued to be repaid by drawings of equal nominal amount at the end of each year for 20 years. Interest is payable monthly in arrear at the rate of 12% per annum and redemption is at par for the first ten drawings and at 110% thereafter.

An investor is liable to income tax at the constant rate of 40%, and to capital gains tax at the rate of 40% until the first five drawings have been made and at the rate of 30% thereafter. He purchases the entire loan on the issue date at a price to obtain a net yield of 8% per annum.

What price does he pay?

8.9 A loan of nominal amount £100 000 is to be repaid at par by 20 instalments each of nominal amount £5000, the first instalment being payable at the end of ten years and subsequent instalments at intervals of two years thereafter. Interest is payable annually in arrear at the rate of 6% per annum during the first 30 years and at 7% per annum thereafter.

Find the issue price paid to earn 8% per annum net by a purchaser of the whole loan who is subject to income tax at 35% and capital gains tax at 30%.

8.10 An insurance company has just purchased, at a price of 94·5%, a loan of £880 000 nominal bearing interest at $5\frac{1}{2}$% per annum payable half-yearly in arrear. The loan is repayable at par by annual instalments, the first being in five years' time. The first instalment is £50 000, and the amount of each subsequent instalment is £6000 more than the preceding one.

Find the net annual yield which will be obtained by the insurance company, given that it pays income tax at 37·5% and capital gains tax at 30%.

8.11 A loan of nominal amount £300 000 in bonds of nominal amount £100 is to be repaid by 30 annual drawings, each of 100 bonds, the first drawing being one year after the date of issue. Interest will be payable quarterly in arrear at the rate of 8% per annum. Redemption will be at par for the first 15 drawings and at 120% thereafter.

An investor is liable to income tax at the constant rate of 40%, and to capital gains tax at the rate of 40% for ten years and 25% thereafter. He purchases the entire loan on the issue date at a price to obtain a net yield of 7% per annum.

What price does the investor pay for each bond?

8.12 On 31 December 1974 company A issued a loan stock bearing interest at 7% per annum payable half-yearly in arrear and repayable at par by 15 annual instalments commencing on 31 December 1980. The first instalment was £15 000 and the instalments were to increase by £5000 per annum. An investor B, subject only to income tax at the rate of 33%, purchased the entire loan on the issue date at a price to yield him 10% per annum net on his investment.

(a) What price did B pay?
(b) On 1 January 1985 legislation was introduced which altered the taxation position of both A and B. From that date B became subject to tax on income at the rate of 50% and on future capital gains (in excess of the purchase price) at the rate of 10%. Company A proposed that the rate of interest from 1 January 1985 be reduced to 5% per annum payable half-yearly in arrear and that each of the remaining repayment instalments be increased by a constant percentage λ. If B was still to obtain a net yield of 10% per annum on the whole transaction, find the value of λ.

Hint Consider the relation between the value of the net income given up and the value of the net redemption premiums receivable under the revised arrangement.

8.13 10 000 bonds each of £100 nominal bear interest at 4% per annum payable half-yearly, and are to be repaid according to the following schedule:

At the end of year	Number of bonds redeemed	Redemption price per bond
10	1000	80
13	1500	80
15	2000	100
18	2500	100
20	2000	120
23	1000	120

Find the net yield per annum obtained by a purchaser of the entire issue at a price of 85% if he is subject to income tax at 35% and to capital gains tax at 30%. (Capital losses may not be offset against capital gains for tax purposes.)

CHAPTER NINE

CUMULATIVE SINKING FUNDS

9.1 Introduction

A loan is said to be repayable by a *cumulative sinking fund* when a fixed sum is applied periodically to repay the loan as follows:

(a) Interest is paid at a stated rate on the loan outstanding at the start of each period. This interest may be paid at the end of each period or it may be paid in equal amounts at regular intervals throughout each period.

(b) At the end of each period, after the interest for the period has been paid, the balance of the fixed sum is used to redeem part of the outstanding loan at a stated price.

The portion of the loan redeemed at the end of any one period is normally selected by lot. (A loan which is repayable by a cumulative sinking fund usually consists of a series of numbered bonds from which drawings are made periodically.) The fixed periodical sum is called the *service* of the loan. In certain cases the rate of interest or the redemption price may not be constant. We shall consider below the possible complications which are caused by changes in these factors.

It is obvious that, if the rate of interest is constant, the amount required for payment of interest decreases from one period to the next as more and more of the loan is redeemed, so that an ever-increasing part of the service is available to repay the loan at the end of each period. We consider this point further in section 9.2, where we also give an example in which the redemption price of the loan is not constant.

Since the redemption date of any one bond is determined by chance, the value on any given basis of *part* of a loan redeemable by a cumulative sinking fund cannot be determined precisely. However, if an investor buys the *whole* outstanding loan, the value of the entire loan can be calculated accurately. In general it is unusual for an investor to buy the entire loan, but, if he purchases a substantial part of the loan, he may reasonably assume that the experience (in relation to redemption) of that part will be similar to that of the whole loan and value his purchase accordingly. It is important to realize that some such assumption is implicit in the valuation of a loan which is redeemable by a cumulative sinking fund. At one time loans repayable by a cumulative sinking fund were quite common in the UK and many such loans are still outstanding. They are now issued much less frequently, although examples do still occur from time to time.

It is also important to realize that a loan repayable by a cumulative sinking fund is simply a particular kind of fixed-interest security, and may therefore be valued by the methods described in chapters 7 and 8. In this chapter we describe certain techniques which may be useful in relation to such loans, and give several examples to illustrate our remarks.

9.2 The relationship between successive capital repayments

Consider a loan of nominal amount N, bearing interest at the constant rate D per unit nominal per annum, which is to be redeemed by a cumulative sinking fund which operates by drawings at the end of each year. Let the annual service of the loan be S, and assume that interest is payable in arrear p times per annum.

At the end of each year, after the interest for the year has been paid, the balance of the service is used to redeem part of the outstanding loan. Let the redemption price per unit nominal applicable at time t (measured in years from the issue date) be R_t ($t = 1, 2, \ldots$). (In practice the redemption price is often constant, so that R_t does not depend on t, but it is convenient to consider the more general case here.) Let n_t be the nominal amount of loan which is repaid at time t, and let N_t be the nominal amount of loan outstanding after the repayment at time t has been made.

Consider the period $(t - 1, t)$. The loan outstanding at the start of the period is N_{t-1}, so that the amount of interest paid during the period is DN_{t-1}. (This interest will be paid at time t if $p = 1$; otherwise it will be paid in p equal instalments over the period.) The balance of the service, available for drawings at time t, is $(S - DN_{t-1})$. Since the redemption price per unit nominal at this time is R_t, it follows that

$$n_t R_t = S - DN_{t-1}$$

or

$$S = DN_{t-1} + n_t R_t \tag{9.2.1}$$

Similarly, by considering the period $(t, t + 1)$, we have

$$S = DN_t + n_{t+1} R_{t+1}$$

Equating these last two expressions for S, we obtain

$$n_{t+1} R_{t+1} = D(N_{t-1} - N_t) + n_t R_t$$

Since

$$N_{t-1} - N_t = n_t$$

it follows that

$$n_{t+1} R_{t+1} = n_t(D + R_t)$$

or

$$n_{t+1} R_{t+1} = n_t R_t \left(1 + \frac{D}{R_t}\right) \qquad (9.2.2)$$

Now let x_t denote the amount of money which is applied to drawings at time t. Then

$$x_t = n_t R_t$$

and equation 9.2.2 may be written in the form

$$x_{t+1} = \left(1 + \frac{D}{R_t}\right) x_t \qquad (9.2.3)$$

A clear understanding of the above discussion is essential for the solution of most problems relating to cumulative sinking funds. In particular, it should be noted that none of the above results depends on p, so that they are valid whether interest is payable with the same frequency as, or more frequently than, the sinking fund drawings.

The solution of the following example makes use of the above results.

Example 9.2.1

A loan of nominal amount £500 000, bearing interest of 5% per annum payable half-yearly in arrear, is to be repaid over 45 years by a cumulative sinking fund operating by annual drawings. The redemption price of the loan will be 115% for the first five years, 110% for the next ten years, and par thereafter.

The annual service of the loan is of a constant amount such that the money available for the sinking fund at the final drawing is exactly sufficient to repay the loan outstanding at that time. The entire loan is purchased on the issue date at a price of 105% by an investor who is liable to income tax at the rate of 40%.

(a) Find the amount of the annual service of the loan.
(b) Assuming that the investor holds the loan until final redemption, find the net annual yield which he will obtain on the completed transaction.

Solution

(a) The interest required in the first year is £25 000. For $1 \leqslant t \leqslant 45$ let R_t and x_t denote respectively the redemption price per unit nominal at the end of year t and the cash applied to the sinking fund drawing at that time. Note that the amount of the annual service is $(25\,000 + x_1)$.

It follows from equation 9.2.3 that

$$x_{t+1} = \left(1 + \frac{0\cdot05}{R_t}\right)x_t \qquad 1 \leqslant t < 45 \tag{1}$$

Since

$$R_t = \begin{cases} 1\cdot15 & \text{for } 1 \leqslant t \leqslant 5 \\ 1\cdot1 & \text{for } 6 \leqslant t \leqslant 15 \\ 1\cdot0 & \text{for } 16 \leqslant t \leqslant 45 \end{cases}$$

equation 1 implies that, if we define

$$g_1 = \frac{0\cdot05}{1\cdot15}, \qquad g_2 = \frac{0\cdot05}{1\cdot1}, \qquad g_3 = 0\cdot05$$

then

$$x_t = \begin{cases} x_1(1 + g_1)^{t-1} & 1 \leqslant t \leqslant 5 \\ x_1(1 + g_1)^5(1 + g_2)^{t-6} & 6 \leqslant t \leqslant 15 \\ x_1(1 + g_1)^5(1 + g_2)^{10}(1 + g_3)^{t-16} & 16 \leqslant t \leqslant 45 \end{cases} \tag{2}$$

In the final year redemption is at par. This means that x_{45} must equal the nominal amount of loan outstanding immediately after the penultimate sinking fund drawing. In the final year the total amount required as service (to pay capital *and* interest) is $(1 + g_3)x_{45}$. We have also remarked that the annual service of the loan is $(25\,000 + x_1)$. Hence

$$(1 + g_3)x_{45} = 25\,000 + x_1$$

Combining this equation with equation 2, we obtain

$$x_1(1 + g_1)^5(1 + g_2)^{10}(1 + g_3)^{30} = 25\,000 + x_1$$

from which it follows that

$$x_1 = 3406\cdot16$$

The annual service of the loan is therefore £28 406·16.

(b) On the issue date, *if interest were payable annually in arrear*, the value at rate i of the loan to an investor who is not liable to tax would be

$$28\,406\cdot16a_{\overline{45}|i}$$

At the same time the value of the capital payments is

$$K = \sum_{t=1}^{45} x_t v^t$$

Using equation 2, we may express this in the form

$$K(i) = x_1 \left[\sum_{t=1}^{5} v^t(1 + g_1)^{t-1} + (1 + g_1)^5 \sum_{t=6}^{15} v^t(1 + g_2)^{t-6} \right.$$
$$\left. + (1 + g_1)^5(1 + g_2)^{10} \sum_{t=16}^{45} v^t(1 + g_3)^{t-16} \right]$$

On simplification, this gives

$$K(i) = x_1 \left[\frac{1 - \left(\frac{1+g_1}{1+i}\right)^5}{i-g_1} + \left(\frac{1+g_1}{1+i}\right)^5 \frac{1 - \left(\frac{1+g_2}{1+i}\right)^{10}}{i-g_2} \right.$$
$$\left. + \left(\frac{1+g_1}{1+i}\right)^5 \left(\frac{1+g_2}{1+i}\right)^{10} \frac{1 - \left(\frac{1+g_3}{1+i}\right)^{30}}{i-g_3} \right] \qquad (3)$$

When allowance is made for the half-yearly payment of interest the value of the loan at the issue date to a purchaser who is liable to income tax at rate 40% is simply

$$A(i) = K(i) + 0.6 \frac{i}{i^{(2)}} [28\,406 \cdot 16 a_{\overline{25}|} - K(i)]$$

where $K(i)$ is given by equation 3.

Since the price paid for the entire loan was £525 000, we require to find that value of i for which $A(i) = 525\,000$. By trial and interpolation we find $i = 0.028\,24$, so that the investor's net annual yield is 2·82%.

In certain other problems, the above ideas are still most useful, since it is often possible to divide the term of the loan into intervals over each of which the interest rate is constant. We may then apply these results over each such interval. Note also that in deriving equation 9.2.3 we have used only the fact that the service of the loan and the rate of payment of interest are the same in each of the two consecutive periods $(t-1, t)$ and $(t, t+1)$. This property is sometimes useful in problems when the amount of the service or the rate of interest changes from time to time (see example 9.4.1).

9.3 The term of the loan when the redemption price is constant

In most practical situations the amount of the periodic service of the loan is specified. The determination of the term of the loan is then one of the first calculations which an investor has to make. In this situation, however, there may be a minor complication in relation to the final period. For this reason, therefore, we prefer to consider first the equivalent problem of finding the amount of the service sufficient to repay the loan over a specified term. We use the notation of section 9.2.

Suppose that the redemption price is constant, so $R_t = R$ for all values of t. Let $C = NR$ and let $g = D/R$. Equation 9.2.3 then gives

$$x_{t+1} = (1+g)x_t, \qquad (9.3.1)$$

from which it immediately follows that

$$x_t = (1+g)^{t-1}x_1 \qquad t \geqslant 1$$

and that, if m does not exceed the term of the loan,

$$\sum_{t=1}^{m} x_t = x_1 s_{\overline{m}|} \qquad \text{at rate } g \qquad (9.3.2)$$

Note now that, since the interest paid in the first year is $DN = gC$,

$$x_1 = S - gC \qquad (9.3.3)$$

Let the term of the loan be n years. Since the loan will be repaid when the monies applied to drawings have totalled $RN = C$, it follows from equations 9.3.2 and 9.3.3 that

$$(S - gC)s_{\overline{n}|} = C \qquad \text{at rate } g \qquad (9.3.4)$$

Hence

$$S = gC + \frac{C}{s_{\overline{n}|}} \qquad (9.3.5)$$

$$= gC + C\left(\frac{1}{a_{\overline{n}|}} - g\right)$$

$$= \frac{C}{a_{\overline{n}|}} \qquad \text{at rate } g \qquad (9.3.6)$$

Either of equations 9.3.5 and 9.3.6 gives the amount of the annual service. Equation 9.3.6 may be written as

$$C = Sa_{\overline{n}|} \qquad \text{at rate } g \qquad (9.3.7)$$

which shows that the total 'indebtedness' C (in cash, rather than nominal, terms) is the value *at rate g* of an annuity of the service, payable annually in arrear throughout the duration of the loan. (Note that equation 9.3.7 does *not* involve p, the frequency of interest payments.)

An alternative approach to the above discussion is provided by defining z, *the initial rate of sinking fund*, to be the ratio of the *nominal* amount of loan repaid after one year to the total *nominal* amount of loan. Since (see equation 9.3.3) the capital payment at the end of the first year is $(S - gC)$, it follows by definition that

$$z = \frac{(S - gC)}{R} \frac{1}{N}$$

$$= \frac{S - gC}{C} \qquad (9.3.8)$$

and hence that

$$S = (g + z)C \qquad (9.3.9)$$

Since

$$x_1 = S - gC$$

$$= zC \qquad \text{(by equation 9.3.9)}$$

it follows from the remarks above that

$$x_t = zC(1 + g)^{t-1}$$

and that the nominal amount of loan repaid at the end of year t is

$$n_t = zN(1 + g)^{t-1} \qquad (9.3.10)$$

Note finally that, since $(S - gC) = zC$, equation 9.3.4 may be written in the form

$$z s_{\overline{n}|} = 1 \qquad \text{at rate } g \qquad (9.3.11)$$

This equation determines the initial rate of sinking fund necessary to repay the loan over n years.

The above discussion indicates how to find the constant annual service required to repay a loan over a given term. In practice, however, the amount of the service is often specified and it is the term of the loan which has to be found. In this case we again consider equation 9.3.7, but now regard it as determining n, since C, S, and g are known.

Often there will be no integer n satisfying equation 9.3.7. More generally, therefore, suppose that n is the unique integer for which

$$S a_{\overline{n-1}|} < C \leqslant S a_{\overline{n}|} \qquad \text{at rate } g$$

or, equivalently (see equation 3.3.10),

$$(S - gC) s_{\overline{n-1}|} < C \leqslant (S - gC) s_{\overline{n}|} \qquad \text{at rate } g \qquad (9.3.12)$$

It follows by equations 9.3.2 and 9.3.3 that the total cash indebtedness extinguished after $n - 1$ years is $(S - gC)s_{\overline{n-1}|}$ at rate g, and that in year n the loan will be completely repaid. The term of the loan is therefore n years. The redemption money required in the final year is

$$C - (S - gC)s_{\overline{n-1}|} \qquad \text{at rate } g \qquad (9.3.13)$$

and, on adding interest (which is calculated by multiplying the outstanding *cash* indebtedness by g), the service in the final year is

$$(1 + g)[C - (S - gC)s_{\overline{n-1}|}] \qquad \text{at rate } g \qquad (9.3.14)$$

It may easily be shown that this is less than the normal service, S, except when $(S - gC)s_{\overline{n}|}$ at rate $g = C$ (in which case the service in the final year is S).

The above discussion is illustrated by example 9.3.2. We also remark that the inequalities 9.3.12 are equivalent to

$$zs_{\overline{n-1}|} < 1 \leqslant zs_{\overline{n}|} \qquad \text{at rate } g \qquad (9.3.15)$$

One final practical point should be noted. A loan will generally be issued in bonds of some specified nominal amount (e.g. £100, £10, or £1). It is not possible to redeem a fractional part of a bond, so that when sinking fund drawings are made a very small part of the money available for redemption may not be used. The precise manner in which such 'unused' money is treated will be specified in the conditions of the loan. (It may simply be ignored or carried forward to the next year's drawings, with or without interest.) For practical purposes the effect is of virtually no significance, and may be ignored by an investor when valuing the loan.

Example 9.3.1

A loan of nominal amount £100 000 is to be issued bearing interest of 6% per annum. The loan will be redeemed over 15 years at a price of 120% by a cumulative sinking fund which operates by annual drawings. The annual service of the loan is of a constant amount, such that the money available for the sinking fund at the final drawing is exactly sufficient to repay the loan outstanding at that time.

An investor, who is liable to income tax at the rate of 30%, wishes to purchase the entire loan on the issue date at a price to obtain an effective net annual yield of 8%. What price should he pay, if interest on the loan is payable (a) annually in arrear, (b) quarterly in arrear?

Solution

Using the notation above, we note that $R = 1\cdot2$, $g = 0\cdot06/1\cdot2 = 0\cdot05$, $C = 100\,000 \times 1\cdot2 = 120\,000$, and $n = 15$. Let the annual service be S. Then (by equation 9.3.6)

$$S = \frac{120\,000}{a_{\overline{15}|}} \qquad \text{at } 5\% \qquad = 11\,561\cdot07 \qquad \text{or, say,} \qquad £11\,561$$

Note that this value of S applies whatever the frequency of interest payments.

(a) Assume that interest is payable annually in arrear. The investor is simply purchasing a level annuity for 15 years, tax being payable on the interest content of the annuity each year. The value of the investor's gross receipts is $11\,561a_{\overline{15}|}$ at 8%, i.e. £98 956.

The interest required in the first year is £6000, so the capital payment at the end of that year is £5561. From equation 9.3.1 it follows that the value (at 8%) of the capital payments is

$$K = \sum_{t=1}^{15} 5561(1\cdot05)^{t-1}1\cdot08^{-t}$$

$$= 5561\,\frac{1 - (1\cdot05/1\cdot08)^{15}}{0\cdot03} = 63\,883$$

The value of the gross interest payments is thus £(98 956 − 63 883), i.e. £35 073. Hence the value of the loan to provide a net annual yield of 8% is A, where

$$A = 63\,883 + (1 - 0\cdot3)35\,073 = 88\,434$$

The investor should therefore pay £88 434 for the loan.

Note Having found the value of the gross payments to be £98 956, we could find the value of the capital indirectly from the equation (based on Makeham's formula)

$$98\,956 = K + \frac{0\cdot05}{0\cdot08}(120\,000 - K)$$

This gives $K = 63\,883$, as before. This approach avoids the need to sum the geometric series above.

(b) Assume that interest is payable quarterly in arrear. In this case the *capital* payments form exactly the same series as in (a). Each year the same amount of interest is paid as in (a), but now the interest is payable in four equal instalments over each year. The value of the net interest is therefore $0\cdot08/0\cdot08^{(4)}$ times the corresponding value in (a). Hence the price which the investor should pay is

$$A = 63\,883 + (1 - 0\cdot3)\frac{0\cdot08}{0\cdot08^{(4)}}35\,073 = 89\,159$$

The investor should therefore pay £89 159 for the loan.

Note The reader should note carefully how the solution of case (b) follows almost immediately from that of case (a) by a simple adjustment for the fact that the interest is payable quarterly. This illustrates the important point that, in problems where interest is paid with a greater frequency than the sinking fund drawings, it is usually still highly relevant to first consider the corresponding problem in which the interest payments and sinking fund drawings occur with the same frequency.

Example 9.3.2

A loan of nominal amount £500 000 was issued bearing interest of 11% per annum payable half-yearly in arrear. The loan is repayable at a price of 110% by a cumulative sinking fund operating by annual drawings. Each year £65 000 is available to service the loan. The sixth sinking fund drawing has just been made and an investor, liable to income tax at the rate of 40% and to capital gains tax at the rate of 25%, wishes to purchase the entire outstanding loan at a price to obtain a net annual yield of 9%. What price should he pay?

Solution

We must first find the term of the loan. Note that (in our previous notation) $R = 1\cdot1$, $g = 0\cdot11/1\cdot1 = 0\cdot1$, $C = 550\,000$ (for the entire loan), and $S = 65\,000$.
   We obtain the term by considering the equation (see equation 9.3.7)

$$550\,000 = 65\,000 a_{\overline{n}|} \qquad \text{at } 10\%$$

Since, at 10% interest,

$$a_{\overline{19}|} < \frac{\cdot 550\,000}{65\,000} \leqslant a_{\overline{20}|}$$

it follows that the term of the loan is 20 years. The interest paid over the first year is £55 000, so that the capital payment at the end of the first year is £10 000. The total amount of money applied to the first 19 drawings is $10\,000\,s_{\overline{19}|}$ at 10%, i.e. £511 591. The outstanding indebtedness at the start of the final year is thus £38 409, so that the service required for the final year is £(38 409 × 1·1) = £42 250, say.

Now consider the value of the outstanding loan to the investor who is making a purchase six years after the issue date. We begin by assuming that interest is paid annually and subsequently adjust our calculations to allow for the true position.

If interest were payable annually, the investor would simply be buying an annuity of £65 000 for 13 years followed by a final payment of £42 250. The value (at 9%) of his *gross* receipts would therefore be

$$65\,000 a_{\overline{13}|} + 42\,250 v^{14} \qquad \text{at } 9\% \qquad = £499\,292$$

Note that the total money used for the first six drawings is $£10\,000 s_{\overline{6}|0\cdot1}$, i.e. £77 156, so that the outstanding indebtedness immediately after the sixth drawing is £472 844. Hence the value of the remaining capital payments is K, where

$$499\,292 = K + \frac{0\cdot1}{0\cdot09}(472\,844 - K)$$

from which it follows that $K = 234\,812$.

Note   The value of K could also be found as

$$\sum_{t=1}^{13} (10\,000 \times 1\cdot1^{5+t} \times 1\cdot09^{-t}) + (38\,409 \times 1\cdot09^{-14})$$

Hence the value of the gross interest payments, if they were paid annually, would be £(499 292 − 234 812) = £264 480.

Since interest is in fact paid half-yearly and the investor is liable to tax, the price he should pay is A, where

$$A = 234\,812 + \left[(1 - 0\cdot4)\frac{0\cdot09}{0\cdot09^{(2)}}264\,480 \right] - \left[0\cdot25(472\,844 - A)\frac{234\,812}{472\,844} \right]$$

(see equation 8.2.7). This gives $A = 386\,242$.

The investor should therefore pay £386 242 for the entire outstanding loan.

Note that at the time of purchase the outstanding *nominal* amount of the loan is £472 844/1·1 = £429 858. The price *per cent* (nominal) which the investor should pay is therefore

$$\frac{£386\,242}{429\,858}100 = £89\cdot85$$

9.4 Further examples

Example 9.4.1

A loan of nominal amount £300 000 is to be issued bearing interest payable quarterly in arrear at the rate of 6% per annum for ten years and at the rate of 8·4%

per annum thereafter. The loan is repayable by a cumulative sinking fund operating by annual drawings at a price of 120%. Each year £28 000 is available to service the loan. Find the net annual yield obtained by an investor, liable to income tax at the rate of 35%, who buys the entire loan on the issue date at a price of 95%.

Solution

We must first find the term of the loan. Note that $R = 1\cdot2$, $C = 360\,000$, $S = 28\,000$. During the first ten years the rate of interest is 6%, so during this period $g = 0\cdot06/1\cdot2 = 0\cdot05$.

Let £x_t be the cash applied to drawings at the end of the tth year. Clearly $x_1 = 10\,000$ and hence (from equation 9.3.2)

$$\sum_{t=1}^{10} x_t = 10\,000 s_{\overline{10}|} \quad \text{at } 5\% \quad = 125\,779$$

The outstanding indebtedness at the start of the 11th year is therefore £234 221. After ten years the rate of interest payable on the loan increases to 8·4%. Thus during the 11th and subsequent years $g = 0\cdot084/1\cdot2 = 0\cdot07$. The amount of interest paid in the 11th year is £$(0\cdot07 \times 234\,221) = £16\,395$, so that at the end of the 11th year the amount available for drawings is £11 605.

For $t \geqslant 11$, the relationship between successive capital payments (except possibly in the final year) is

$$x_{t+1} = 1\cdot07 x_t$$

so that the total sum applied to drawings in years 11, 12,..., 10 + r is $11\,605\, s_{\overline{r}|\,0\cdot07}$. To find the term of the loan we therefore consider the equation

$$11\,605 s_{\overline{n}|} = 234\,221$$

or

$$s_{\overline{n}|} = 20\cdot183 \quad \text{at } 7\%$$

Since, at 7%,

$$s_{\overline{13}|} < 20\cdot183 < s_{\overline{14}|}$$

the term of the loan is 24 years, a reduced service being applied in the final year. The total sum applied to drawings in the first 23 years is

$$125\,779 + 11\,605 s_{\overline{13}|} \quad \text{at } 7\% \quad = 359\,511$$

This means that at the end of 24 years the money applied to the final sinking fund drawing is £489. The service required in the final year is thus £$(1\cdot07 \times 489) = £523$.

The amount paid by the investor for the entire loan is £285 000. We must find the rate of interest at which this is the value of the investor's net receipts.

Firstly, ignoring taxation, we value the loan on the assumption that interest is payable annually in arrear. On this basis the value (at any given rate i) of the loan is

$$A(i) = 28\,000 a_{\overline{23}|} + 523 v^{24} \quad \text{at rate } i$$

The value of the capital payments is

$$K(i) = \sum_{t=1}^{10} 10\,000(1\cdot05)^{t-1}(1+i)^{-t}$$

$$+ \sum_{t=11}^{23} 11\,605(1\cdot07)^{t-11}(1+i)^{-t} + 489(1+i)^{-24}$$

Thus

$$K(i) = 10\,000\frac{1 - [1\cdot05/(1 + i)]^{10}}{i - 0\cdot05}$$

$$+ 11\,605\frac{1 - [1\cdot07/(1 + i)]^{13}}{(1 + i)^{10}(i - 0\cdot07)} + 489(1 + i)^{-24}$$

provided that $i \neq 0\cdot05$ or $0\cdot07$.

The price to obtain a net annual yield of i, allowing for the fact that interest is payable quarterly, is

$$A'(i) = K(i) + (1 - 0\cdot35)\frac{i}{i^{(4)}}[A(i) - K(i)]$$

and we must find the value of i for which this equals 285 000. The reader should verify the values in the following table:

i	$A(i)$	$K(i)$	$A'(i)$
0·06	344 624	174 686	287 601
0·065	329 679	165 899	274 917

By interpolation we estimate the investor's net annual yield to be

$$0\cdot06 + \frac{287\,601 - 285\,000}{287\,601 - 274\,917}0\cdot005 = 0\cdot0610, \quad \text{or} \quad 6\cdot10\%.$$

Example 9.4.2

A loan of nominal amount £1 000 000 is to be issued bearing interest at $5\frac{1}{2}\%$ per annum payable half-yearly in arrear. Each year £165 000 will be available to service the loan. This will be used firstly to pay interest and then to redeem, by drawings at the end of each year, 5% of the nominal amount of the loan outstanding at a price of 132%. The balance of the service will then be used to redeem as much as possible of the remainder of the loan by drawings at 110%.

(a) What proportion of the total nominal amount of the loan will be redeemed at 132%?

(b) An investor, liable to income tax at the rate of 37·5%, is considering the purchase of the entire loan on the issue date at a price to obtain a net annual yield of 6%. What price should he pay?

Solution

(a) We must first find the term of the loan. Because of the somewhat unusual conditions of the loan, it is best to fall back on first principles. Let $F(t)$ denote the *nominal* amount of loan outstanding immediately after all the drawings have been made at the end of the tth year ($t \geqslant 1$), and let $F(0) = 1\,000\,000$.

Consider the application of the service during the tth year. The amount paid in interest during the year is $0\cdot055F(t - 1)$. At the end of the year a nominal amount $0\cdot05F(t - 1)$ is redeemed at 132%. This requires a cash payment of $0\cdot066F(t - 1)$. The balance of the service is available for drawings at 110%. This means that (except for the final year, when a reduced service will apply)

$$F(t) = F(t - 1) - 0\cdot05F(t - 1) - [165\,000 - 0\cdot055F(t - 1) - 0\cdot066F(t - 1)]/1\cdot1$$

so

$$F(t) = 1.06F(t-1) - 150\,000 \qquad (1)$$

Since $F(0) = 1\,000\,000$, equation 1 implies that

$$F(t) = 1.06^t \times 1\,000\,000 - 150\,000s_{\overline{t}|} \qquad \text{at } 6\%$$
$$= 2\,500\,000 - 1\,500\,000 \times 1.06^t \qquad (2)$$

Since

$$1.06^8 < \frac{2.5}{1.5} < 1.06^9$$

the term of the loan is nine years. After eight years the nominal amount of loan outstanding is $F(8)$, i.e. £109 228.

The total nominal amount of loan redeemed at 132% is

$$\sum_{t=0}^{8} 0.05F(t) = 0.05 \sum_{t=0}^{8} [2\,500\,000 - 1\,500\,000 \times 1.06^t]$$
$$= 1\,125\,000 - 75\,000s_{\overline{9}|} \qquad \text{at } 6\% \qquad = 263\,151$$

The proportion of the total nominal amount which is redeemed at 132% is thus 0.263 151.

(b) The service required for the final year consists of interest of

$$£(0.055 \times 109\,228) = £6008$$

and a capital payment of

$$£\{[(0.05 \times 1.32) + (0.95 \times 1.1)]109\,228\} = £121\,352$$

The total amount of the service in the final year is thus £127 360.

If interest were payable annually, the value at 6% of the loan ignoring taxation would be

$$A = 165\,000a_{\overline{8}|} + 127\,360v^9 \qquad \text{at } 6\% \qquad = £1\,100\,000$$

For $1 \leqslant t \leqslant 8$ the capital payment at time t is

$$165\,000 - 0.055F(t-1) = 27\,500 + 82\,500 \times 1.06^{t-1}$$

so the value at 6% of the capital payment is

$$K = \sum_{t=1}^{8} [27\,500 + 82\,500 \times 1.06^{t-1}]1.06^{-t} + 121\,352 \times 1.06^{-9}$$
$$= 27\,500a_{\overline{8}|} + (8 \times 82\,500 \times 1.06^{-1}) + (121\,352 \times 1.06^{-9}) \qquad \text{at } 6\%$$

$$= 865\,239$$

Hence, allowing for half-yearly payment of interest and for income tax, the investor should pay

$$A' = K + \frac{0.06}{0.06^{(2)}}(1 - 0.375)(A - K)$$

$$= 865\,239 + 148\,894 = 1\,014\,133$$

or, say, £101.41%.

Example 9.4.3

Some years ago a loan of nominal amount £100 000 was issued bearing interest at $5\frac{1}{2}\%$ per annum payable quarterly in arrear. Each year £8000 is available to service the loan. The annual service is used first to pay the interest due and secondly to redeem at the end of each year as much as possible of the outstanding loan, the redemption price being par for the first 15 years and 110% thereafter. A redemption of part of the loan has just been made and the nominal amount now outstanding is £67 812.

An investor, who is liable to income tax at the rate of 50% and to capital gains tax at the rate of 40%, now wishes to purchase the entire outstanding loan at a price to obtain a net yield of 7% per annum. What price per cent should he pay?

Solution

We work with a unit of issued nominal amount £100 (rather than £100 000), so that the annual service of the loan is £8, except possibly in the final year.

Note that, in the notation of section 9.2, $D = 0.055$. Also $R_t = 1$ for $1 \leqslant t \leqslant 15$ and $R_t = 1.1$ for $t \geqslant 16$. As before, let x_t denote the capital payment at the end of the tth year. It follows from equation 9.2.3 that

$$x_{t+1} = \begin{cases} 1.055x_t, & \text{for } 1 \leqslant t \leqslant 15 \\ 1.05x_t, & \text{for } t \geqslant 16 \qquad \text{(except in the final year)} \end{cases} \tag{1}$$

During the first year £5.5 of the service is required for interest, so that £2.5 is the capital payment at the end of the first year. Thus $x_1 = 2.5$. It follows from equation (1) that

$$\sum_{t=1}^{15} x_t = 2.5 s_{\overline{15}|} \qquad \text{at } 5\tfrac{1}{2}\%, \qquad \text{i.e. } 56.021\,66$$

and that

$$x_{16} = 2.5(1.055)^{15} = 5.581\,19$$

Since redemption is at par for the first 15 years, 56.021 66 is the nominal amount of the loan repaid during this period and at the end of 15 years the nominal loan outstanding is 43.987 34. After 15 years redemption is at 110%, so the loan will be finally repaid when the capital payments in the 16th and subsequent years total $1.1(43.978\,34) = 48.376\,17$. Hence, if the loan is repaid after $(15 + n)$ years, it follows from equation 1 (since $x_{16} = 5.581\,19$) that

$$5.581\,19 s_{\overline{n-1}|} < 48.376\,17 \leqslant 5.581\,19 s_{\overline{n}|} \qquad \text{at } 5\%$$

This implies that $n = 8$ and that the term of the loan is 23 years.

For the first 22 years the annual service is £8. During years 16 to 22 inclusive the total capital payments amount to $5.581\,19\,s_{\overline{7}|}$ at 5% i.e. 45.442 10, so that during this period the nominal amount of loan redeemed is 45.442 10/1.1, i.e. 41.311 00. Thus the nominal amount outstanding at the start of the final year is $(43.978\,34 - 41.311\,00) = 2.667\,34$ and the service required for the final year is $2.667\,34(1.1 + 0.055) = 2.934\,07 + 0.146\,70 = 3.080\,77$. Note that the final capital payment is of amount 2.934 07.

We must now determine the number of capital repayments which have been made at the time when the investor wishes to purchase the outstanding loan. Clearly, less than 15 payments have been made and the most recent capital payment is the tth, where

$$100 - 2 \cdot 5 s_{\overline{t}|} = 67 \cdot 812 \qquad \text{at } 5\tfrac{1}{2}\%$$

which implies that $t = 10$.

In order to determine the value of the net income and gross capital payments, we value the outstanding loan initially on the basis of annual interest payments. We then value the capital directly and hence find the value of the net interest payable quarterly. Finally, having calculated the value of the capital gains tax payable (in terms of the purchase price), we obtain an equation from which the price can be found.

On the basis of annual interest, the value of the gross payments outstanding at the purchase date is

$$8 a_{\overline{12}|} + 3 \cdot 080 \, 77 v^{13} \qquad \text{at } 7\% \qquad = 64 \cdot 819 \, 90 \tag{2}$$

The value of the gross capital payments outstanding is $K = K_0 + K_1$, where K_0 denotes the value of capital payments made at par and K_1 denotes the value of capital payments made at 110%. Now

$$K_0 = \sum_{t=1}^{5} 2 \cdot 5 (1 \cdot 055)^{9+t} (1 \cdot 07)^{-t}$$

$$= 2 \cdot 5 (1 \cdot 055)^{10} \left[\frac{1 - \left(\dfrac{1 \cdot 055}{1 \cdot 07} \right)^5}{1 \cdot 07 - 1 \cdot 055} \right]$$

$$= 19 \cdot 403 \, 28 \tag{3}$$

and

$$K_1 = \sum_{t=6}^{12} 5 \cdot 581 \, 19 (1 \cdot 05)^{t-6} (1 \cdot 07)^{-t} + 2 \cdot 934 \, 07 (1 \cdot 07)^{-13}$$

$$= 5 \cdot 581 \, 19 (1 \cdot 07)^{-5} \left[\frac{1 - \left(\dfrac{1 \cdot 05}{1 \cdot 07} \right)^7}{1 \cdot 07 - 1 \cdot 05} \right] + 1 \cdot 217 \, 54$$

$$= 24 \cdot 617 \, 73 + 1 \cdot 217 \, 54 = 25 \cdot 835 \, 27 \tag{4}$$

Hence

$$K = 45 \cdot 238 \, 55. \tag{5}$$

From equations 2 and 5 we see that the value of gross interest, assumed to be payable annually, is $(64 \cdot 819 \, 90 - 45 \cdot 238 \, 55)$ i.e. $19 \cdot 581 \, 35$. (This is obtained as the value of *all* future payments less the value of future *capital payments*.)

Hence the value of the *net* interest, payable quarterly, is

$$(1 - 0 \cdot 5) \frac{0 \cdot 07}{0 \cdot 07^{(4)}} \, 19 \cdot 581 \, 35 = 10 \cdot 044 \, 06 \tag{6}$$

Now let A be the price to be paid for the entire outstanding loan (per £100 nominal of the *original* loan). We must find the value of the capital gains tax payable. Recall (see section 8.3) that for taxation purposes the total purchase price is apportioned between the different tranches of the loan in proportion to the *nominal* amounts redeemed at the end of each year. The nominal amount redeemed at the end of the tth year is x_t if $1 \leqslant t \leqslant 15$, and $x_t/1\cdot1$ if $t \geqslant 16$. At the time of purchase the nominal amount of loan outstanding is £67·812. Hence at the purchase date the value of the capital gains tax payable (at the rate of 40%) is

$$0\cdot4\left[\sum_{t=11}^{15}\left(x_t-\frac{x_t}{67\cdot812}A\right)v^{t-10}\right.$$

$$\left.+\sum_{t=16}^{23}\left(x_t-\frac{x_t/1\cdot1}{67\cdot812}A\right)v^{t-10}\right] \qquad \text{at } 7\%$$

$$=0\cdot4\left[\left(1-\frac{A}{67\cdot812}\right)\sum_{t=11}^{15}x_tv^{t-10}\right.$$

$$\left.+\left(1-\frac{A}{1\cdot1\times67\cdot812}\right)\sum_{t=16}^{23}x_tv^{t-10}\right]$$

$$=0\cdot4\left[\left(1-\frac{A}{67\cdot812}\right)19\cdot40328\right.$$

$$\left.+\left(1-\frac{A}{74\cdot593}\right)25\cdot83527\right] \qquad \text{(by equations 3 and 4)}$$

$$=18\cdot0954-0\cdot25299A \tag{7}$$

Since the purchase price is simply the value of the net receipts, by combining equations 5, 6, and 7 we obtain

$$A=45\cdot238\,55+10\cdot044\,06-(18\cdot09\,54-0\cdot252\,99A)$$

so

$$A=(45\cdot238\,55+10\cdot044\,06-18\cdot0954)/(1-0\cdot252\,99)$$
$$=49\cdot781$$

The investor should therefore pay £49 781 for the entire outstanding loan. Since the nominal outstanding is £67 812 the price per cent to be paid is

$$£\left(\frac{49\,781}{67\,812}\right)100, \quad \text{i.e. £73·41}$$

Exercises

9.1 (Finding the term)

A loan of nominal amount £1 000 000 bears interest at 4% per annum and is redeemable at 105% by a cumulative sinking fund operating annually. The initial rate of sinking fund is 3%.

(a) Find the annual service of the loan and the term of the loan.
(b) Find the reduced annual service required in the final year.
(c) After how many years will more than 50% of the nominal amount of the loan have been redeemed?

9.2 Consider the loan described in exercise 9.1, and suppose that it is to be valued at the date of issue at a rate of interest of 5% per annum. Find the present value of the loan, assuming that

(a) Interest is payable annually and the investor is not subject to taxation.
(b) Interest is payable half-yearly and the investor is not subject to taxation.
(c) Interest is payable half-yearly and the investor is subject to income tax at 37·5% but not to capital gains tax.

9.3 A loan bears interest at $5\frac{1}{2}$% per annum payable half-yearly in arrear. The loan is redeemable at £110 per cent by a cumulative sinking fund with a total annual service of £9·90 per £100 nominal of initial loan. After payment of interest each year the balance of the annual service is applied at the end of the year to redeem part of the loan.
   (a) Find the value on the issue date per £100 nominal of loan of
      (i) The capital payments, and
      (ii) The interest payments,
      assuming an effective annual rate of interest of 6%.
   (b) Hence or otherwise find the price per £100 nominal which should be paid by a purchaser of the entire loan who requires a net annual yield of 6% and who is subject to income tax at 35% and to capital gains tax at 15%.

9.4 For integer n and $0 \leqslant f < 1$, let

$$a_{\overline{n+f}|} = a_{\overline{n}|} + f(a_{\overline{n+1}|} - a_{\overline{n}|})$$

(See the discussion in section 4.4. Note that this is *not* the definition given by equation 4.4.2.)
   A loan bears interest annually in arrear at a rate of g per unit of redemption price. The loan is redeemable at C per £100 nominal by a cumulative sinking fund operating by annual drawings. The total annual service of the loan is $(g + z)C$ per £100 nominal. If appropriate, there will be a reduced service in the final year.
   Show that the present value, at a rate of interest of i per annum effective, of the above loan is exactly equal to

$$(g + z)Ca_{\overline{n+f}|i}$$

where n is an integer, $0 \leqslant f < 1$, and n and f are obtained from the equation

$$a_{\overline{n+f}|g} = \frac{1}{g + z}$$

9.5 A loan of nominal amount £1 000 000 bears interest at $5\frac{1}{4}$% per annum payable half-yearly in arrear and is redeemable by annual drawings at 105% by means of a cumulative sinking fund. The annual service of the loan is £74 500.

   (a) What price should a purchaser of the whole loan pay on the issue date in order to obtain a net yield of 6% per annum if he pays tax at 40% on interest and at 30% on capital gains?
   (b) How much of the nominal amount of the loan will have been repaid after 15 years?

9.6 A loan of nominal amount £1 000 000, bearing interest at $2\frac{1}{2}\%$ per annum payable half-yearly in arrear, is to be redeemed at 125% by a cumulative sinking fund at the initial rate of 4% per annum, operating yearly. Find the issue price to give a net yield per annum of 2%, assuming income tax to be at 37·5% and capital gains tax to be at 20% for the duration of the loan.

9.7 (Varying redemption price)
A loan of nominal amount £250 000 bears interest at the rate of 11% per annum payable annually in arrear and is to be repaid by a level annuity, also payable annually in arrear, over 20 years. Each year, after payment of interest on the loan still outstanding, the balance of the annuity payment is to be applied to repay as much as possible of the loan. The redemption price is 110% in the first ten years and 122·22% thereafter.

(a) If X is the amount of the first annuity payment which will be applied to repay the loan after interest has been paid, show that the amount of the rth annuity payment applied in repayment of the loan is

$$X(1\cdot1)^{r-1} \qquad \text{for } 1 \leqslant r \leqslant 11$$
$$X(1\cdot1)^{10}(1\cdot09)^{r-11} \qquad \text{for } 12 \leqslant r \leqslant 20$$

(b) Find the annual annuity payment necessary to repay the loan.
(c) Calculate the price to be paid for the whole loan on the issue date to yield 8% per annum net after allowing for tax on capital gains at 20% and tax on income at 30%.

9.8 A few years ago a company issued a loan bearing interest at 9% per annum payable half-yearly and redeemable at 112·5% by a cumulative sinking fund operating by yearly drawings. The annual service of the loan is now 10% of the original nominal amount of the loan. It is an anniversary of the issue date, the payment due has just been made, and 80% of the loan is now outstanding.

(a) What yield (to the nearest 0·25%) would you obtain by purchasing the whole of the outstanding loan at a price of 102%?
(b) Immediately after you have purchased this loan at 102% the company runs into financial difficulties. It suggests that it will pay only interest on the outstanding loan for the next five years. Thereafter the full service of the loan will be resumed and, to compensate for the delay in repayment, future redemptions of the loan will be at 150%. Does the suggested arrangement offer a higher yield than that originally promised?
   In your answer you should assume that any payments due under either the original or the suggested system of repayment will certainly be made. (Ignore taxation.)

9.9 On 1 January 1955 an industrial company issued a loan of £1 000 000 nominal bearing interest at 5% per annum payable yearly and redeemable at par by a cumulative sinking fund operating by annual drawings. The annual service was exactly sufficient to pay interest and to redeem the loan in 30 years. On the issue date the entire loan stock was purchased for £950 890 by a financial institution which pays income tax at 37·5%.
   The company was unable to make any sinking fund payments in the 11th to 15th years (inclusive) of the loan, interest only being paid during this period.

On 1 January 1970 the company agreed to increase the nominal amount of loan outstanding to £N, this new amount being repayable at par by a cumulative sinking fund. The new annual service was exactly sufficient to pay interest at 5% per annum, payable annually, and to repay the loan in 15 years from 1 January 1970. The value of N was calculated to be such that the net yield to the financial institution on the whole transaction was the same as it would have obtained if the original terms had been adhered to. Find N.

9.10 On 1 January 1975 a loan of £1 000 000 nominal was issued bearing interest at 7% per annum, payable half-yearly, and redeemable at par by a cumulative sinking fund operating yearly. The annual service was such that the loan would be repaid in 20 years and the entire loan was purchased at a price of 90% by a life office, which pays tax on income at 37·5% and on capital gains at 30%.

On 1 January 1985, just after making the payment then due, the borrower requested that the annual service be reduced to £60 000 per annum, with a corresponding extension of the term. Assuming that the life office values loans at a net rate of interest of 5% per annum, find the reduction in the value (on 1 January 1985) of the outstanding loan caused by the proposed change.

9.11 A loan of £500 000 nominal bears half-yearly interest at 4% p.a. and is repayable by a cumulative sinking fund operating by annual drawings. The annual service was £45 000 (with perhaps a smaller payment in the final year) and the redemption price was 115% for the first ten years and par thereafter. The loan was issued on 1 February 1981, when it was purchased in its entirety for £461 000 by a pension fund, which pays no tax.

   (a) Calculate the yield per annum obtainable by the pension fund if it held the loan until final redemption.
   (b) Just after receiving the interest and capital due on 1 February 1985, the pension fund sold its entire holding in this loan for £403 000. Find the yield per annum actually obtained by the pension fund in respect of its dealings in this loan.

9.12 A lender granted a loan bearing interest at $5\frac{1}{2}$% per annum payable yearly in arrear and repayable at £110 for each £100 lent by means of a cumulative sinking fund operating by annual drawings over a period of 20 years. The borrower met all payments in full for a period of five years but for the next three years was able to pay only interest on the outstanding loan. Thereafter the annual service was increased to complete repayment on the date originally planned.

What net yield per annum (to the nearest 0·1%) was obtained by the lender, who is subject to income tax at 40% but not to capital gains tax?

CHAPTER TEN

YIELD CURVES, DISCOUNTED MEAN TERMS, MATCHING, AND IMMUNIZATION

10.1 Introduction

In this chapter we discuss various topics relating to fixed-interest securities and the valuation of assets and liabilities. Some of these topics are of interest to private investors, life assurance offices (in connection with valuations, for example), pension funds and other financial institutions. It is sometimes convenient, in a theoretical discussion of these topics, to make certain assumptions which may not always hold in practice. Consequently some of the techniques presented in this chapter may require modification before they can be applied directly to practical problems. A description of these modifications is beyond the scope of this book.

10.2 Yield curves and related topics

Compound interest techniques play an important role in the analysis of gilt-edged and other fixed-interest securities. Many of the relevant ideas are of a relatively advanced nature, and in this section we describe only briefly the idea of a yield curve and certain related topics. The interested reader may consult Day and Jamieson (reference [11]) for a more detailed discussion and further references.

If $\delta(t)$, the force of interest per unit time at time t, is known for all values of t and the market operates on a consistent basis, then $v(t)$, the value at time 0 of a payment of 1 due at time t, is known (see equation 2.5.3).

Let us first suppose that the market operates on a consistent basis and that there is available an infinite range of fixed-interest stocks, with all possible terms to redemption and interest payable continuously. Let us also assume initially that there is no taxation. We denote by $P_g(t)$ the current value (i.e. the price at time 0) of a fixed-interest security which is redeemable for 1 at time t and bears interest, payable continuously at the rate of g per unit time until it is redeemed. (For convenience we shall call this 'the stock (g, t)'.) Then clearly

$$P_g(t) = g \int_0^t v(s)\mathrm{d}s + v(t) \tag{10.2.1}$$

We may also define $r_g(t)$, 'the term t redemption yield for coupon g', to be the yield obtained by an investor who purchases a quantity of the stock (g, t) at time 0 at the price of $P_g(t)$ and holds the stock until it is redeemed. Thus $r_g(t)$ is defined by the equation

$$P_g(t) = g\,\bar{a}_{\overline{t}|} + v^t \qquad \text{at rate } r_g(t) \qquad (10.2.2)$$

Finally, we define $y(t)$, 'the term t par yield', to be the coupon rate per unit time of the stock with term t which has price 1 (per unit redemption money). Thus $y(t)$ is defined by the equation

$$P_{y(t)}(t) = 1 \qquad (10.2.3)$$

It is clear that, regarded as functions of t, $\delta(t)$, $y(t)$, $P_g(t)$, and $r_g(t)$ are closely related. Provided that the market operates in a consistent manner (see equation 2.3.4), any one of these four functions determines each of the other three.

In practice, of course, we do not have stocks with continuously payable interest. Even if we overlook this point, we cannot escape the fact that in no real market do there exist stocks with every possible coupon rate or term to redemption. Thus, although conceptually $P_g(t)$ may be defined for all positive values of g and t, in reality the best we may hope to achieve is to know the values of $P_g(t)$ and $r_g(t)$ for a finite set of stocks $(g_1, t_1), \ldots, (g_n, t_n)$. By considering only an appropriate subset of the available stocks we may assume that the coupon rates $g_1, \ldots, g_n$ are all equal or of similar magnitude.

For a given coupon rate (or range of coupon rates) we may fit an appropriate curve to a set of observed redemption yields, regarded as a function of t, the term to redemption. The resulting curve is known as a *redemption yield curve*. In practice any acceptable method of curve-fitting may be used. For the *Financial Times – Actuaries (FTA)* Yield Indices, introduced in 1977, the curve-fitting is done by a mathematical formula. The coefficients of the formula are calculated daily by the minimization of a weighted sum of squares of the difference between the value of the yield given by the formula and the observed yield (see reference [58]).

In practice, because of the effects of taxation and other factors, the shape of the redemption yield curve varies somewhat with the coupon. Accordingly the *FTA* yield curves are found for each of three coupon bands, consisting of low-, medium-, and high-coupon stocks (the irredeemable stocks are included in each band). As an illustration, we show in figure 10.2.1 the three redemption yield curves (one for each coupon band) on 8 October 1984.

We may ascertain the force of interest $\delta(t)$ implied by the fitted yield curve and then use this to value any particular stock. Whether or not the value

yield (%)

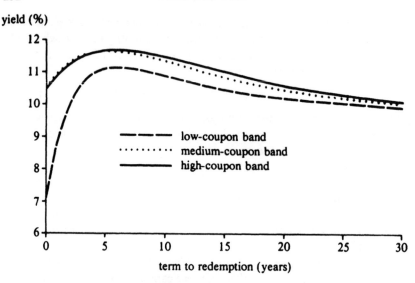

FIGURE 10.2.1 *Redemption yield curves: 8 October 1984*

thus found exceeds the actual market price of the stock may be an indication that it is relatively dear or cheap. Various other mathematical models are also used in practice (see reference [6]). The use of redemption yields alone to assess the relative values of different stocks may be fairly criticized as being too simplistic: other factors should be considered. These topics are beyond the scope of this book.

Except in the case of absolute matching, for which no assumptions about interest rates are required (see section 10.6), for the remainder of this chapter we shall make the simplifying assumptions that, at any given time, all stocks are valued at the same constant force of interest, and that this force of interest is the same for borrowers and lenders.

10.3 The discounted mean term of a project

Consider a business project or investment which will produce a series of cash flows. The net present value of the project at a force of interest δ per annum is (see section 5.2)

$$NPV(\delta) = \sum c_t e^{-\delta t} \qquad (10.3.1)$$

where the summation extends over those values of t for which c_t, the net cash flow at time t, is non-zero. At this force of interest, if $NPV(\delta) \neq 0$, the *discounted mean term* of the project, $T(\delta)$, is defined to be

$$T(\delta) = \frac{\sum t c_t e^{-\delta t}}{\sum c_t e^{-\delta t}} \qquad (10.3.2)$$

Note that the discounted mean term is a weighted average of the future times at which there are cash flows. The 'weight' associated with any given time is the present value (at force of interest δ) of the net cash flow due at that time.

In the more general situation in which there is also a continuous cash flow (see section 5.1), the net rate of payment being $\rho(t)$ per annum at time t years, the net present value of the project at force of interest δ per annum is

$$NPV(\delta) = \sum c_t e^{-\delta t} + \int_0^\infty \rho(t) e^{-\delta t} dt \qquad (10.3.3)$$

and the discounted mean term of the project is defined to be

$$T(\delta) = \frac{\sum t c_t e^{-\delta t} + \int_0^\infty t \rho(t) e^{-\delta t} dt}{\sum c_t e^{-\delta t} + \int_0^\infty \rho(t) e^{-\delta t} dt} \qquad (10.3.4)$$

provided that $NPV(\delta) \neq 0$.

Example 10.3.1

A mine owner estimates that the net cash flow from his mining operations will be as follows:

Time (years)	Net cash flow (£)
1	500 000
2	400 000
3	300 000
4	200 000
5	100 000

Calculate the discounted mean term of the project at an interest rate of 6% per annum.

Solution

By formula 10.3.2, the discounted mean term is

$$\frac{5v + (2 \times 4v^2) + (3 \times 3v^3) + (4 \times 2v^4) + 5v^5}{5v + 4v^2 + 3v^3 + 2v^4 + v^5} \qquad \text{at } 6\%$$

$$= \frac{4\cdot7170 + 7\cdot1200 + 7\cdot5566 + 6\cdot3367 + 3\cdot7363}{4\cdot7170 + 3\cdot5600 + 2\cdot5189 + 1\cdot5842 + 0\cdot7473}$$

$$= \frac{29\cdot4666}{13\cdot1274} = 2\cdot245 \text{ years}$$

Example 10.3.2

Ten years ago an investor purchased an annuity-certain payable continuously for 25 years. Throughout the tth year of payment ($1 \leqslant t \leqslant 25$) the rate of payment of the annuity is £t per annum.

    Find the discounted mean term of the *remaining* annuity payments on the basis of an effective annual interest rate of (a) 5% and (b) 10%.

Solution

Measure time in years from the present. Note that over the coming year the rate of annuity payment is £11 per annum, in the following year £12 per annum, etc. In the final year of payment (i.e. the year commencing 14 years from now) the rate of payment is £25 per annum.

    Using equation 10.3.4, we may express the discounted mean term of the future payments, at force of interest δ per annum, as

$$\frac{\displaystyle\int_0^1 t11e^{-\delta t}dt + \int_1^2 t12e^{-\delta t}dt + \int_2^3 t13e^{-\delta t}dt + \cdots + \int_{14}^{15} t25e^{-\delta t}dt}{\displaystyle\int_0^1 11e^{-\delta t}dt + \int_1^2 12e^{-\delta t}dt + \int_2^3 13e^{-\delta t}dt + \cdots + \int_{14}^{15} 25e^{-\delta t}dt}$$

The reader should verify that on simplification this expression becomes

$$\frac{\dfrac{11}{\delta^2} + \dfrac{1}{\delta}(Ia)_{\overline{14}|} + \dfrac{1}{\delta^2}a_{\overline{14}|} - 25\left(\dfrac{15v^{15}}{\delta} + \dfrac{v^{15}}{\delta^2}\right)}{10\bar{a}_{\overline{15}|} + (I\bar{a})_{\overline{15}|}}$$

which equals

$$\frac{\dfrac{11}{\delta} + (Ia)_{\overline{14}|} + \dfrac{1}{\delta}a_{\overline{14}|} - 25\left(15v^{15} + \dfrac{v^{15}}{\delta}\right)}{i[10a_{\overline{15}|} + (Ia)_{\overline{15}|}]}$$

Evaluating the last expression at rates of interest 5% and 10%, we obtain the answers (a) 7·656 years and (b) 6·809 years respectively.

10.4 Volatility

Consider an investment or business project with net cash flows as described above. Suppose that the net present value of the project is currently determined on the basis of an annual force of interest δ_0, but that this may change at very short notice to another value, δ_1 say, not very different from δ_0. If this were to happen, the *proportionate* change in the net present value of the project would be

$$\frac{\Delta NPV(\delta_0)}{NPV(\delta_0)} = \frac{NPV(\delta_1) - NPV(\delta_0)}{NPV(\delta_0)}$$

$$\simeq (\delta_1 - \delta_0)\frac{NPV'(\delta_0)}{NPV(\delta_0)} \tag{10.4.1}$$

(We use the notation $NPV'(\delta)$ to denote the derivative with respect to δ of $NPV(\delta)$. We also assume that $NPV(\delta_0) \neq 0$.)

We now define the *volatility* of the project at force of interest δ_0 to be

$$\frac{-NPV'(\delta_0)}{NPV(\delta_0)} \qquad (10.4.2)$$

With this definition, equation 10.4.1 may be expressed verbally as

$$\left(\begin{array}{l} \text{proportionate change} \\ \text{in net present value} \end{array} \right) \simeq -(\text{change in force of interest}) \times (\text{volatility})$$

$$(10.4.3)$$

Note that the definition of the volatility at force of interest δ_0 (expression 10.4.2) may be expressed as

$$\left[-\frac{d}{d\delta} \log NPV(\delta) \right]_{\delta = \delta_0} \qquad (10.4.4)$$

By combining equation 10.3.3 with our definition 10.4.2, we see that, at force of interest δ, the volatility of the project equals

$$\frac{\sum tc_i e^{-\delta t} + \int_0^\infty t\rho(t)e^{-\delta t}dt}{\sum c_i e^{-\delta t} + \int_0^\infty \rho(t)e^{-\delta t}dt} \qquad (10.4.5)$$

It is clear from equation 10.3.4 and the expression 10.4.5 that, at any force of interest, we have the identity

$$\text{discounted mean term} \equiv \text{volatility} \qquad (10.4.6)$$

We shall therefore use the symbol $T(\delta)$ to refer to either the discounted mean term or the volatility of an investment or business project.

Returning to formula 10.4.1, we see that the proportionate change in the net present value on a small immediate change in the force of interest is approximately equal to $-(\delta_1 - \delta_0)T(\delta_0)$. Thus, the proportionate profit or loss on a small change in interest rates depends more or less directly on (a) the size of the change in the force of interest, which is similar to the size of the change in the rate of interest, and (b) the volatility of the investment; if interest rates rise there will, of course, be a loss and if they fall there will be a profit to the investor. It follows that the volatility (or, equivalently, the discounted mean term) of a fixed-interest investment is of some interest to actual and prospective investors and often appears in the financial press. In practical applications the volatility of a fixed-interest stock is calculated by formula 10.3.2, although in theoretical work other formulae are also useful. In the next section we shall give expressions for the volatility of certain fixed-interest stocks, and discuss the variation of volatility with the coupon rate and the term to redemption.

10.5 The volatility of certain fixed-interest securities

Consider a unit nominal amount of a fixed-interest stock, which bears interest at D per annum and is redeemable in n years' time at price R. Let $g = D/R$ (see section 7.6). Let the annual force of interest implied by the current price of this stock be δ_0 (i.e. the current price of the stock equals the present value, at force of interest δ_0 per annum, of the future interest and capital payments). We suppose an investor has purchased the stock, so his future net cash flows are all positive.

If interest is payable annually in arrear and n is an integer, the discounted mean term of the stock at force of interest δ_0 is given by equation 10.3.2 as

$$T(\delta_0) = \frac{g(Ia)_{\overline{n}|} + ne^{-\delta n}}{ga_{\overline{n}|} + e^{-\delta n}} \qquad \text{at force of interest } \delta_0 \qquad (10.5.1)$$

In the particular case when g equals i_0, the rate of interest per annum corresponding to the annual force of interest δ_0, we obtain

$$T(\delta_0) = \frac{(\ddot{a}_{\overline{n}|} - nv^n) + nv^n}{(1 - v^n) + v^n} \qquad \text{at rate of interest } i_0$$

$$= \ddot{a}_{\overline{n}|} \qquad \text{at rate of interest } i_0 \qquad (10.5.2)$$

By letting n tend to infinity in equation 10.5.1 we obtain the volatility at force of interest δ_0 of a perpetuity as

$$T(\delta_0) = \frac{g(Ia)_{\overline{\infty}|}}{ga_{\overline{\infty}|}} \qquad \text{at force of interest } \delta_0$$

$$= \frac{1}{d} \qquad \text{at rate of interest } i_0 \qquad (10.5.3)$$

Note that the volatility does *not* depend on the coupon rate of the perpetuity.

Note also that for a *zero-coupon* bond redeemable in n years' time, it is an immediate consequence of equation 10.5.1 that

$$T(\delta_0) = n \qquad (10.5.4)$$

for *all* values of δ_0.

It is clear from equation 10.5.1 that the volatility at force of interest δ_0 of a fixed-interest stock depends on both the annual interest rate per unit indebtedness g and the term to redemption n of the stock. We wish to consider the variation of the volatility $T(\delta_0)$ with g and n. In order to simplify our calculations, we assume that interest is payable continuously. Equation 10.3.4 then implies that

$$T(\delta_0) = \frac{g(I\bar{a})_{\overline{n}|} + nv^n}{g\bar{a}_{\overline{n}|} + v^n} \qquad \text{at force of interest } \delta_0 \qquad (10.5.5)$$

We first show that, for each fixed term n years, the volatility of the stock decreases as g increases. To prove this, we note that

$$\frac{\partial T(\delta_0)}{\partial g} = \frac{(I\bar{a})_{\overline{n}|}(g\bar{a}_{\overline{n}|} + e^{-\delta_0 n}) - \bar{a}_{\overline{n}|}[g(I\bar{a})_{\overline{n}|} + ne^{-\delta_0 n}]}{(g\bar{a}_{\overline{n}|} + e^{-\delta_0 n})^2}$$

$$= \frac{e^{-\delta_0 n}[(I\bar{a})_{\overline{n}|} - n\bar{a}_{\overline{n}|}]}{(g\bar{a}_{\overline{n}|} + e^{-\delta_0 n})^2} < 0$$

since

$$(I\bar{a})_{\overline{n}|} = \int_0^n te^{-\delta_0 t}\,dt < n\int_0^n e^{-\delta_0 t}\,dt = n\bar{a}_{\overline{n}|}$$

(This result is illustrated in Figure 10.5.1.)

We now consider the variation of volatility with the term to redemption for a fixed value of g. Equation 10.5.5 implies that

$$\lim_{n\to\infty} T(\delta_0) = \frac{(I\bar{a})_{\overline{\infty}|}}{\bar{a}_{\overline{\infty}|}} \qquad \text{at force of interest } \delta_0$$

$$= \frac{1}{\delta_0} \qquad\qquad (10.5.6)$$

and also that

$$\lim_{n\to 0} T(\delta_0) = 0 \qquad\qquad (10.5.7)$$

Note that neither of these limiting values depends on the annual coupon rate. By straightforward (but somewhat tedious) calculation (see exercise 10.11) it can be shown that

$$\frac{\partial T(\delta_0)}{\partial n} = \frac{[\delta_0 + g(g - \delta_0)n - (g - \delta_0)^2\bar{a}_{\overline{n}|}]e^{-\delta_0 n}}{\delta_0(g\bar{a}_{\overline{n}|} + e^{-\delta_0 n})^2} \qquad \text{at force of interest } \delta_0$$

$$(10.5.8)$$

and hence that the sign of $\partial T(\delta_0)/\partial n$ is the same as that of

$$\delta_0 + (g - \delta_0)[g(n - \bar{a}_{\overline{n}|}) + \delta_0\bar{a}_{\overline{n}|}] \qquad \text{at force of interest } \delta_0$$

Since $n - \bar{a}_{\overline{n}|}$ is a non-negative increasing unbounded function of n, we have the following results:

(a) If $g \geqslant \delta_0$, the volatility $T(\delta_0)$ increases steadily from zero when $n = 0$ to the limiting value $1/\delta_0$ as n tends to infinity.

(b) If $g < \delta_0$, the volatility $T(\delta_0)$ increases from zero when $n = 0$ until n is such that

$$g(n - \bar{a}_{\overline{n}|}) + \delta_0 \bar{a}_{\overline{n}|} = \frac{\delta_0}{\delta_0 - g} \qquad \text{at force of interest } \delta_0 \quad (10.5.9)$$

after which term the volatility decreases steadily to the limiting value $1/\delta_0$ as n tends to infinity. (Equation 10.5.9 may be solved numerically, as in exercise 10.4.)

We therefore have the possibly unexpected fact that, at a given force of interest δ_0, the volatility of certain long-dated low-coupon redeemable stocks exceeds that of a perpetuity. The redeemable stocks in question are those such that $T(\delta_0) > (1/\delta_0)$, i.e. those for which

$$\frac{\dfrac{g}{\delta_0^2}(1 - e^{-\delta_0 n}) + \left(1 - \dfrac{g}{\delta_0}\right)n e^{-\delta_0 n}}{\dfrac{g}{\delta_0}(1 - e^{-\delta_0 n}) + e^{-\delta_0 n}} > \frac{1}{\delta_0}$$

This inequality holds if and only if

$$\frac{g}{\delta_0^2}(1 - e^{-\delta_0 n}) + \left(1 - \frac{g}{\delta_0}\right)n e^{-\delta_0 n} > \frac{g}{\delta_0^2}(1 - e^{-\delta_0 n}) + \frac{e^{-\delta_0 n}}{\delta_0}$$

i.e., if and only if

$$\delta_0 > g + \frac{1}{n} \qquad (10.5.10)$$

The last inequality may be written in either of the forms

$$n > \frac{1}{\delta_0 - g}$$

$$g < \delta_0 - \frac{1}{n}$$

Thus, if g and δ_0 are given with $0 < g < \delta_0$, a redeemable stock with interest rate g per unit indebtedness will have greater volatility (at force of interest δ_0) than a perpetuity if and only if the term to redemption exceeds $1/(\delta_0 - g)$. Alternatively, if δ_0 and n are given, a redeemable stock with term n will have greater volatility (at force of interest δ_0) than a perpetuity if and only if the interest rate per unit indebtedness is less than $\delta_0 - (1/n)$.

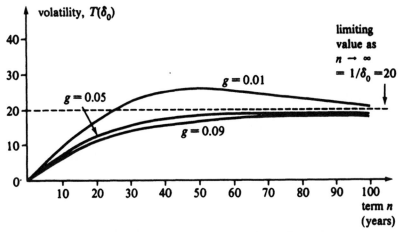

FIGURE 10.5.1   *Variation of volatility with term*

The variation of volatility with term (when $\delta_0 = 0.05$) is illustrated in Figure 10.5.1.

An explanation of the fact that certain low-coupon long-dated securities are more volatile than a perpetuity is that, in the case of the redeemable stock, the main 'weight' attaches to the repayment of capital, which is very distant, whereas for the perpetuity the main 'weights' attach to the next few years' interest payments. Hence the discounted mean term of the redeemable stock may be expected to be greater than that of the perpetuity.

Example 10.5.1

A speculator buys large quantities of a fixed-interest security when he expects interest rates to fall, with the intention of selling in a short time to realize a profit. At present he is choosing between the following two secure fixed-interest stocks:

Security 1 which bears interest at 5% per annum payable annually in arrear and is repayable at par in five years' time.
Security 2 which bears interest at 11% per annum payable annually in arrear and is repayable at par in six years' time.

These stocks always have the same gross yield per annum, at present 10%. Which should he buy in order to obtain the larger capital gain on a small fall in interest rates? (Ignore taxation.)

Solution

We calculate the volatilities of each of the securities using formula 10.5.1.
The volatility of security 1 is

$$\frac{0.05(Ia)_{\overline{5}|} + 5v^5}{0.05a_{\overline{5}|} + v^5} \qquad \text{at 10\% interest} \qquad = 4.488$$

The volatility of security 2 is

$$\frac{0 \cdot 11(Ia)_{\overline{5}|} + 6v^6}{0 \cdot 11a_{\overline{5}|} + v^6} \qquad \text{at } 10\% \text{ interest} \qquad = 4 \cdot 725$$

Hence the speculator should buy security 2.

Example 10.5.2

Suppose that the force of interest per annum for valuing secure fixed-interest stocks is δ. When $\delta = 0 \cdot 08$ the prices per unit nominal of two secure fixed-interest stocks, A and B, are equal and it is known that when $0 \cdot 05 \leqslant \delta \leqslant 0 \cdot 08$ the volatility of stock A is not less than that of stock B. Prove that, if the force of interest changes immediately from $0 \cdot 08$ per annum to $0 \cdot 05$ per annum, then the new price of stock A will not be less than the new price of stock B.

Solution

Let $V_A(\delta)$ and $V_B(\delta)$ denote the prices per unit nominal of stocks A and B respectively when the force of interest is δ per annum. We are given that

$$V_A(0 \cdot 08) = V_B(0 \cdot 08)$$

and hence

$$\log V_A(0 \cdot 08) = \log V_B(0 \cdot 08)$$

Also, by equation 10.4.4,

$$-\frac{d}{d\delta}[\log V_A(\delta) - \log V_B(\delta)] \geqslant 0 \qquad 0 \cdot 05 \leqslant \delta \leqslant 0 \cdot 08$$

It follows that if we set

$$f(\delta) = \log V_A(\delta) - \log V_B(\delta)$$

then $f(0 \cdot 08) = 0$ and $f'(\delta) \leqslant 0$ for $0 \cdot 05 \leqslant \delta \leqslant 0 \cdot 08$.

Hence, by the mean value theorem of elementary calculus, $f(0 \cdot 05) \geqslant 0$, and we have

$$\log V_A(0 \cdot 05) \geqslant \log V_B(0 \cdot 05)$$

i.e.

$$V_A(0 \cdot 05) \geqslant V_B(0 \cdot 05)$$

as required.

10.6 The matching of assets and liabilities

In a general business context, the *matching* of assets and liabilities of a company requires that the company's assets be chosen as far as possible in such a way as to make the assets and liabilities equally responsive to the influences which affect them both. In this wide sense, matching includes the matching of assets and liabilities in terms of currencies and the degree of inflation-linking. In this chapter, however, we shall consider both assets and liabilities to be in money terms. Accordingly, we shall use the word 'matching' to refer only to a suitable choice of the terms of the assets in relation to the terms of the liabilities, so as to reduce the possibility of loss arising from changes in interest rates.

The *liabilities* of a business are the sums which it has contracted to pay in the future. Let S_t denote the liability at time t (i.e. the money which must be paid out at that time). Let P_t be the money to be *received* by the company at time t from its business operation, excluding any investment proceeds due at that time. We define the *net liability-outgo* at time t to be

$$L_t = S_t - P_t \qquad (10.6.1)$$

The net liabilities of the business are the sums $\{L_t\}$. In the notation of section 5.1, the net cash flow at time t is $c_t = -L_t$ for all t. We shall assume that all the liabilities and receipts are discrete, although the arguments given here may easily be generalized to include continuous cash flows.

Example 10.6.1

Consider a life office which issued a 20-year capital redemption policy ten years ago with a sum assured of £10 000 and an annual premium of £288·02. Measuring time in years from the present and assuming that the premium now due has been paid and that expenses may be ignored, calculate the net liabilities $\{L_t\}$ in respect of this policy.

Solution

We have

$$P_t = 288\cdot02 \qquad \text{for } t = 1, 2, \ldots, 9$$
$$S_t = 10\,000 \qquad \text{when } t = 10$$

hence

$$L_t = \begin{cases} -288\cdot02 & \text{for } t = 1, 2, \ldots, 9 \\ +10\,000 & \text{for } t = 10 \end{cases}$$

The proceeds at time t of the company's investments, whether capital, interest, or both, are referred to as the *asset-proceeds* at time t years, and are denoted by A_t. The *assets* of the business are the collection of asset-proceeds (or, in ordinary usage, the securities providing them).

Suppose first that L_t is never negative and that the asset-proceeds are such that, for *all* values of t,

$$A_t = L_t \qquad (10.6.2)$$

It is clear that in this case the business will always have exactly sufficient cash to meet the net liabilities, no matter what the pattern of interest rates may be at present or in the future. The business is therefore said to be *absolutely matched*. If the net cash flow is sometimes positive and sometimes negative (as in example 10.6.1), absolute matching is not possible, as some investment must take place in the future (at unknown interest rates) to provide for the liabilities in the final stages of the project.

Assume that at present the annual force of interest for valuing assets and liabilities is δ. The present value of the net liabilities is thus

$$V_L(\delta) = \sum L_t v^t \qquad \text{at force of interest } \delta \qquad (10.6.3)$$

and the present value of the assets is

$$V_A(\delta) = \sum A_t v^t \qquad \text{at force of interest } \delta \qquad (10.6.4)$$

We shall assume for the remainder of this section that $V_L(\delta) > 0$ for all δ, or at any rate for all values of δ under consideration.

If the force of interest per annum will remain constant at δ_0, at least until all payments have been made and received, then it is clear that any set of secure assets satisfying the condition

$$V_A(\delta_0) = V_L(\delta_0) \qquad (10.6.5)$$

will be exactly sufficient to meet the net liabilities as they emerge in the future.

Consider, for example, the policy given in example 10.6.1 and assume that the force of interest per annum will remain constant at $\delta_0 = \log 1{\cdot}05$ (which corresponds to a rate of interest of 5% p.a.). If the office possesses any fixed-interest assets of value

$$V_L(\delta_0) = 10\,000 v^{10} - 288{\cdot}02 a_{\overline{9}|} \qquad \text{at 5\%}$$

$$= 6139{\cdot}13 - 2047{\cdot}19 = 4091{\cdot}94$$

then it will have exactly sufficient cash to meet the liability under the policy. (Since the annual premium was also calculated at 5% interest and expenses were ignored, this sum is the net premium reserve for the policy; see section 6.4.)

In practice, however, it cannot normally be assumed that the force of interest will remain constant until all transactions have been completed, and the company will wish to make a profit (or, at any rate, not make a loss) if interest rates change. A degree of protection against changes in interest rates may be obtained by using the concept of *immunization*, which we shall discuss in the next two sections.

It is possible, of course, that the investor or management of the business may deliberately maintain an unmatched position in order to profit from an anticipated rise, or fall, in interest rates: this policy is known as 'taking a view of the market'. Some businesses are, however, essentially custodians of money belonging to others – life assurance companies and pension funds fall into this category – and such institutions are as a rule constrained by law or considerations of solvency into a largely defensive position with respect to changes in interest rates. (The real position is, however, more complicated: 'with-profits' policies may exist, and the life office may have shareholders, so an element of risk in investment policy may be acceptable.) The theory of immunization is of interest to many financial institutions (although certain modifications may be necessary before it can be applied in practice).

10.7 Redington's theory of immunization

In 1952, F. M. Redington (see reference [38]) presented a theory of *immunization*, under which the investor (in Redington's paper a life office) is protected against small changes in the rate of interest. We now give a brief description of this theory, referring the reader to Redington's original paper for further details.

Consider a financial institution, which for definiteness we take to be a life office, which has net liability-outgo L_t and asset-proceeds A_t at time t years. We assume that the current force of interest is δ_0, and that the value at this force of interest of the net liabilities equals that of the asset-proceeds, i.e.

$$V_A(\delta_0) = V_L(\delta_0) \tag{10.7.1}$$

It is further assumed that the asset-proceeds may be altered without dealing costs, and that the following additional conditions are satisfied:

$$V'_A(\delta_0) = V'_L(\delta_0) \tag{10.7.2}$$

and

$$V''_A(\delta_0) > V''_L(\delta_0) \tag{10.7.3}$$

(The notations $V'_A(\delta)$ and $V''_A(\delta)$ are used to denote the first and second derivatives with respect to δ of $V_A(\delta)$. We define $V'_L(\delta)$ and $V''_L(\delta)$ similarly.)

If these conditions are satisfied, it follows by elementary calculus that the function

$$f(\delta) = V_A(\delta) - V_L(\delta)$$

equals zero when $\delta = \delta_0$, and has a minimum turning point there. Thus there is a neighbourhood of δ_0 such that, if δ lies therein but is not equal to δ_0, then

$$V_A(\delta) > V_L(\delta) \tag{10.7.4}$$

The position is illustrated by Figure 10.7.1.

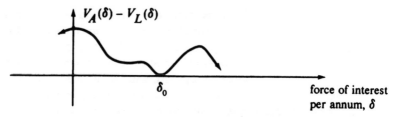

FIGURE 10.7.1   *Immunization against small changes in rate of interest*

An investor whose investments are such that conditions 10.7.1, 10.7.2, and 10.7.3 hold is thus said to be 'immunized' against small changes in the rate of interest, as it is clear that any immediate small change in the rate of interest will lead to a surplus, in the sense that the present value of the assets will exceed that of the net liabilities. We now discuss how conditions 10.7.1, 10.7.2, and 10.7.3 may be interpreted in practice.

Since

$$L_t = S_t - P_t$$

equations 10.7.1 and 10.7.2 are equivalent to

$$\sum (P_t + A_t)v^t = \sum S_t v^t \qquad \text{at force of interest } \delta_0 \qquad (10.7.5)$$

and

$$\sum t(P_t + A_t)v^t = \sum t S_t v^t \qquad \text{at force of interest } \delta_0 \qquad (10.7.6)$$

These two equations imply that

$$\frac{\sum t(P_t + A_t)v^t}{\sum (P_t + A_t)v^t} = \frac{\sum t S_t v^t}{\sum S_t v^t} \qquad \text{at force of interest } \delta_0 \qquad (10.7.7)$$

Thus, at force of interest δ_0, the discounted mean terms of the *total assets* (i.e. the receipts, P_t, plus the asset-proceeds, A_t) and of the *liabilities*, S_t, are equal. Let their common value be denoted by $T(\delta_0)$. Note that condition 10.7.3 may be expressed as

$$\sum t^2 A_t v^t > \sum t^2 L_t v^t \qquad \text{at force of interest } \delta_0$$

i.e.

$$\sum t^2 (A_t + P_t)v^t > \sum t^2 S_t v^t \qquad \text{at force of interest } \delta_0$$

Combined with equations 10.7.5 and 10.7.6, the last inequality is equivalent to the condition

$$\sum [t - T(\delta_0)]^2 (A_t + P_t)v^t > \sum [t - T(\delta_0)]^2 S_t v^t \quad \text{at force of interest } \delta_0$$
$$(10.7.8)$$

Hence the 'spread' of the total assets about their discounted mean term must, if the conditions for immunization are satisfied, exceed that of the liabilities; that is to say, the spread of the receipts and asset-proceeds about the discounted mean term must exceed that of the liabilities. Since the net liabilities $\{L_t\}$ are predetermined, the asset-proceeds $\{A_t\}$ must be chosen in such a way as to satisfy conditions 10.7.5, 10.7.7 and 10.7.8. An equivalent formulation of these conditions is easily shown to be:

$$\sum A_t v^t = \sum L_t v^t \qquad \text{at force of interest } \delta_0 \qquad (10.7.9)$$

$$\frac{\sum t A_t v^t}{\sum A_t v^t} = \frac{\sum t L_t v^t}{\sum L_t v^t} \qquad \text{at force of interest } \delta_0 \qquad (10.7.10)$$

$$\sum t^2 A_t v^t > \sum t^2 L_t v^t \qquad \text{at force of interest } \delta_0 \qquad (10.7.11)$$

Equation 10.7.9 may be expressed as 'the value of the assets must equal that of the net liabilities', and equation 10.7.10 may be expressed as 'the discounted mean term of the assets must equal that of the net liabilities'.

The arguments given here may easily be generalised to include continuous cash flows.

As an illustration of how the conditions for an immunised position may be satisfied, let us consider the capital redemption policy specified in example 10.6.1. In order that equations 10.7.9 and 10.7.10 should hold, the value of the assets, at 5% interest, must be equal to the value of the net liabilities, viz. £4 091·94 (see example 10.6.1), and the discounted mean term of the assets, at 5% interest, must be equal to

$$\frac{\sum t L_t v^t}{\sum L_t v^t} \qquad \text{at 5\% interest}$$

$$= \frac{10 \times 10\,000 v^{10} - 288·02 (Ia)_{\overline{5}|}}{10\,000 v^{10} - 288·02 a_{\overline{5}|}} \qquad \text{at 5\% interest}$$

$$= 12·66 \text{ years.}$$

Condition 10.7.11 must also be satisfied. Since equations 10.7.9 and 10.7.10 are satisfied, this last condition will be met if inequality 10.7.8 holds. This inequality states that the spread of the total assets (i.e., assets and receipts) about the common discounted mean term, $T(\delta_0)$, should be greater than the spread of the liabilities about $T(\delta_0)$. Since, in this case, all the liabilities are due at time 10 years, we have $T(\delta_0) = 10$, and the spread of the liabilities about this time is zero. It follows that any set of assets of present value £4091·94 and discounted mean term 12·66 years will be such that the office will be immunized against any immediate small change in the rate of interest. If a large variety of secure fixed-interest stocks exist, an appropriate portfolio of assets may be chosen in one of many ways. For example, if the life office is able and willing to invest in a stock redeemable at par and bearing interest at 5% per annum payable continuously, the term n years should be such that

$$12 \cdot 66 = \text{the volatility of the stock}$$

$$= \frac{0 \cdot 05(I\bar{a})_{\overline{n}|} + nv^n}{0 \cdot 05\,\bar{a}_{\overline{n}|} + v^n} \qquad \text{at an interest rate of 5\% effective}$$

from which it follows that

$$1 \cdot 05^n - 0 \cdot 003\,088n - 2 \cdot 576\,541 = 0$$

Hence $n \simeq 19 \cdot 88$ years.

Similar calculations may be performed for other coupon rates, and if a fixed term (of at least 12·66 years) is required, the appropriate coupon rate may be calculated.

As time goes by, the portfolio of investments must be varied, in theory continuously but in practice at discrete intervals, to achieve an immunized position. If the annual force of interest used to value assets and net liabilities may change only at certain defined dates, e.g. times $t_1(=0)$, $t_2,\dots$ years (as in Figure 10.7.2), then the portfolio of assets should be suitably chosen *just before* these times. Since no change in the force of interest can occur between these dates, no profit or loss may be made between the dates, and if the force of interest changes slightly at times $t_1, t_2,\dots$ then a small profit will be made at each of these times.

It is also true that, even if the above conditions hold, a loss may occur in the event of a *large* change in the rate of interest: Redington's theory covers

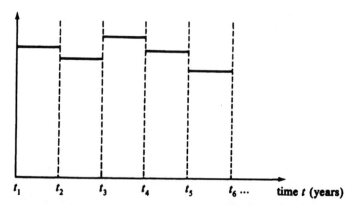

FIGURE 10.7.2 *Force of interest per annum used at time* t *to value assets and liabilities*

only small changes. There is, however, another theory of immunization – which, to avoid confusion with the theory presented by Redington, we refer to as 'full' immunization – under which the investor makes a profit on any immediate changes whatever in the rate of interest. We shall discuss this concept of 'full' immunization in the next section.

10.8 Full immunization

Consider the investor to have a liability-outgo of S due at time t_1 years and to hold total assets (in the sense defined in the preceding section) providing A at time $t_1 - a$ and B at time $t_1 + b$. It is assumed that a and b are positive and for practical use it is also necessary to assume that $a \leqslant t_1$. Suppose that exactly two of four values A, B, a, b are known. It may be shown (see exercise 10.12) that in many cases it is possible to determine the other two values uniquely in such a way that the two equations

$$Ae^{\delta_0 a} + Be^{-\delta_0 b} = S \qquad (10.8.1)$$

and

$$Aae^{\delta_0 a} = Bbe^{-\delta_0 b} \qquad (10.8.2)$$

are satisfied. There are exceptional cases when no solution exists, but if the known quantities are (a) a, b (b) B, b (c) A, a or (d) A, b then unique values of the other two quantities may certainly be found.

Let $V(\delta)$ denote the present value, at force of interest δ per annum, of the total assets less the present value of the liability, i.e. let

$$V(\delta) = e^{-\delta t_1}(Ae^{\delta a} + Be^{-\delta b} - S) \qquad (10.8.3)$$

It may easily be shown that conditions 10.8.1 and 10.8.2 are equivalent to $V(\delta_0) = 0$ and $V'(\delta_0) = 0$. Assume that these conditions are satisfied. We thus have

$$V(\delta) = e^{-\delta t_1}[Ae^{\delta_0 a}e^{(\delta - \delta_0)a} + Be^{-\delta_0 b}e^{-(\delta - \delta_0)b} - S]$$

$$= e^{-\delta t_1}Ae^{\delta_0 a}\left[e^{(\delta - \delta_0)a} + \frac{a}{b}e^{-(\delta - \delta_0)b} - \left(1 + \frac{a}{b}\right)\right] \qquad (10.8.4)$$

using equation 10.8.2 and the fact that, by equations 10.8.1 and 10.8.2,

$$Ae^{\delta_0 a}\left(1 + \frac{a}{b}\right) = Ae^{\delta_0 a}\left(1 + \frac{Be^{-\delta_0 b}}{Ae^{\delta_0 a}}\right) = S$$

Now consider the sign of the function

$$f(x) = e^{ax} + \frac{a}{b}e^{-bx} - \left(1 + \frac{a}{b}\right)$$

Note that $f(0) = 0$. Since

$$f'(x) = a(e^{ax} - e^{-bx})$$

it follows that

$$f'(x) \begin{cases} > 0 & \text{for } x > 0 \\ = 0 & \text{for } x = 0 \\ < 0 & \text{for } x < 0 \end{cases}$$

and thus that $f(x) > 0$ for all $x \neq 0$. Hence, by equation 10.8.4 (with $\delta - \delta_0 = x$),

$$V(\delta) > 0 \qquad \text{for } all \ \delta \neq \delta_0 \tag{10.8.5}$$

In order that the theory of 'full' immunization may be applied in a more general context, it is necessary to split up the liabilities in such a way that each item of liability-outgo is linked to two items of the total assets, the earlier of which may be a 'receipt' (an item of positive cash flow arising from the business itself) due *before* the item of liability-outgo. It then follows by the argument given above that the asset associated with this item of liability-outgo (or the later asset, if there are two) may be varied in amount and date so that equations 10.8.1 and 10.8.2 hold. If this procedure is followed for all items of liability-outgo, the entire business will be immunized against *any* immediate change in the rate of interest. (See exercise 10.8 for an illustration of this process.)

As in the case of Redington's theory of immunization, one must modify the assets as time goes by so as to achieve an immunized position. If the force of interest may change only at certain specified times, as in Figure 10.7.2, then suitable assets may be chosen just before these times. Various practical difficulties attend the application of this theory of immunization, notably the absence in the market of appropriate zero-coupon securities and possible variations in the force of interest with the term to redemption and the coupon rate (see section 10.2). (See, for example, the paper by A. D. Shedden, reference [45].)

Example 10.8.1

Consider the 20-year capital redemption policy described in example 10.6.1, and let the current rate of interest per annum be 5%. Show that the life office can, by suitable investment, achieve a fully immunized position (in which it will make a profit on any immediate change in the rate of interest).

Solution

As there is only one liability, and nine 'receipts' (i.e. premiums), we shall split the liability into nine equal sections, each of amount $10\,000/9 = 1111 \cdot 11$. Consider the item of liability-outgo linked to the premium due at time r years.

We have the equations:

$$288\cdot02(1\cdot05)^{10-r} + B(1\cdot05)^{-b} = 1111\cdot11$$

and

$$288\cdot02(10-r)(1\cdot05)^{10-r} = Bb(1\cdot05)^{-b}$$

where B denotes the proceeds of a zero-coupon bond due at time $10 + b$ years from the present. By direct solution, we obtain

$$b = \frac{\left(\dfrac{288\cdot02}{1111\cdot11}\right)(10-r)(1\cdot05)^{10-r}}{1 - \left(\dfrac{288\cdot02}{1111\cdot11}\right)(1\cdot05)^{10-r}}$$

and

$$B = \frac{1111\cdot11 - 288\cdot02(1\cdot05)^{10-r}}{(1\cdot05)^{-b}}$$

$$= \text{the proceeds of a zero-coupon bond due at time } 10 + b \text{ years}$$

The investments may thus consist of nine zero-coupon bonds of the following amounts maturing at the times shown:

r	Amount of zero-coupon bond (£)	Time to maturity (years)
1	892·55	16·05
2	873·52	14·97
3	858·75	14·02
4	847·40	13·19
5	838·82	12·47
6	832·50	11·84
7	828·06	11·29
8	825·17	10·80
9	823·58	10·37

The above portfolio of zero-coupon bonds is such that the office is fully immunized against any immediate change in the rate of interest. (It is not claimed that this is the only such portfolio: many others can be constructed.)

Example 10.8.2

An investor has a single liability of £1000 due in 15 years' time. The yield on zero-coupon stocks of any term is currently 4% per annum, and the investor possesses cash equal to the present value of his liability, i.e. $1000v^{15}$ at 4% = £555·26. He wishes to invest in 10-year and 20-year zero-coupon stocks in such a way that he will make a profit on any immediate change in the force of interest. How much of each security should he buy, and how large a profit will he make if the rate of interest per annum immediately becomes 0·01, 0·02, 0·03, 0·05, 0·06, 0·07, or 0·08?

Solution

Equations 10.8.1 and 10.8.2 give

$$Ae^{5\delta_0} + Be^{-5\delta_0} = 1000$$

and

$$5Ae^{5\delta_0} = 5Be^{-5\delta_0}$$

where

$$\delta_0 = \log_e(1{\cdot}04)$$

We thus have the equations

$$A(1{\cdot}04)^5 + Bv^5 = 1000 \qquad \text{at } 4\% \text{ interest}$$
$$A(1{\cdot}04)^5 = Bv^5 \qquad \text{at } 4\% \text{ interest}$$

These equations may easily be solved, giving $A = 410{\cdot}96$ and $B = 608{\cdot}32$. The quantities of zero-coupon stocks which he should buy are therefore those providing £410·96 at time 10 years and £608·32 at time 20 years. The profit on an immediate change in the rate of interest to i per annum is, by formula 10.8.3,

$$v^{15}[410{\cdot}96(1 + i)^5 + 608{\cdot}32v^5 - 1000] \qquad \text{at rate } i$$

In the following table we give the present values of the liability ($1000v^{15}$), the assets ($410{\cdot}96v^{10} + 608{\cdot}32v^{20}$) and the profit to the investor for each of the specified rates of interest:

Rate of interest per annum	Present value of liability	Present value of assets	Present value of profit
0·00	1000·00	1019·28	19·28
0·01	861·35	870·58	9·23
0·02	743·01	746·51	3·50
0·03	641·86	642·61	0·75
(0·04	555·26	555·26	0·00)
0·05	481·02	481·56	0·54
0·06	417·27	419·16	1·89
0·07	362·45	366·11	3·66
0·08	315·24	320·87	5·63

Exercises

10.1   (Discounted mean term)
   (a) Calculate, at an effective annual interest rate of (i) 5% (ii) 15%, the discounted mean term of the following set of payments:
      £100 payable now
      £230 payable in five years' time
      £600 payable in thirteen years' time.
   (b) An investor is considering the purchase of an annuity, payable annually in arrear for 20 years. The first payment is £1000. Using a rate of interest of 8% per annum, calculate the discounted mean term of the annuity when

(i) The payments remain level over the term;
(ii) The payments increase by £100 each year;
(iii) The payments increase at a rate of 8% per annum compound;
(iv) The payments increase at a rate of 10% per annum compound.

10.2   (a) At a given rate of interest a series of payments has present value V_1 and discounted mean term t_1. At the same rate of interest a second series of payments has present value V_2 and discounted mean term t_2. Show that, if $V_1 + V_2 \neq 0$, at the given rate of interest the discounted mean term of both series combined is

$$\frac{V_1 t_1 + V_2 t_2}{V_1 + V_2}$$

Generalize this result to the case of n different series of payments.
        (b) An investor is entitled to receive an annuity of £10 000 per annum, payable annually in arrear for ten years. The first payment will be made in one year's time. The investor is under an obligation to make two payments of £30 000, the first being due in five years' time and the second ten years later.
            On the basis of an interest rate of 10% per annum
            (i) Find the present value and the discounted mean term of (1) The payments to be received by the investor, (2) The payments to be made by the investor.
            (ii) Using the result of (a), or otherwise, determine the discounted mean term of the investor's net cash flow.

10.3   (a) A special annuity payable annually in arrear for n years is such that the payment at time t years is

$$(n - t + 1)t = (n + 1)t - t^2$$

Show that the present value of the annuity may be expressed as

$$(n + 1)(Ia)_{\overline{n}|} - \frac{2(Ia)_{\overline{n}|} - a_{\overline{n}|} - n^2 v^{n+1}}{1 - v}$$

Find the value of this expression when the effective rate of interest is 9% per annum and $n = 9$.
        (b) An insurance company issues only annual premium capital redemption policies with term ten years. Premiums, payable in advance, are calculated at an effective rate of interest of 9% per annum, expenses being ignored.
            (i) On 1 January 1985 the company issued a group of policies with total annual premiums of £20 000. Find, at 9% per annum effective, the discounted mean term of the *future* cash movements (inflows less outflows) for this block of business just after the receipt of the annual premiums on 1 January 1985.
            (ii) Assume that the company has issued the same volume of new business on each 1 January from 1976 to 1985 inclusive. Find, again at 9% per annum effective, the discounted mean term of the *future* cash movements for all of these policies just after receipt of the annual premiums on 1 January 1985.

10.4   (The term at which volatility reaches a maximum)
       A fixed-interest stock bears interest of 5% per annum payable continuously
       and is redeemable at par in n years' time, where n is not necessarily an integer.

   (a) On the basis of a constant force of interest per annum of 0·07, determine
       the volatility of the stock if (i) $n = 20$, (ii) $n = 60$.
   (b) The volatility of the stock, on the basis of a specified constant force of
       interest per annum δ, may be considered to be a function of n. When
       $\delta = 0·07$, for what value of n is this volatility greatest and what is its
       maximum value?

10.5   (a) A company is liable to make four payments at five-yearly intervals, the
           first payment being due five years from now. The amount of the tth
           payment is £$(1000 + 100t)$. The company values these liabilities at an
           effective rate of interest of 5% per annum. On this basis find (i) the present
           value and (ii) the discounted mean term of these liabilities.
       (b) An amount equal to the total value of the liabilities (on the basis of an
           effective annual interest rate of 5%) is immediately invested in two newly
           issued loans, one redeemable at the end of ten years and the other at the
           end of 30 years. Each loan bears interest at 5% per annum payable
           annually in arrear. Both loans are issued and redeemable at par. Given
           that, on the basis of an effective annual interest rate of 5%, the discounted
           mean term of the asset-proceeds is the same as the discounted mean term
           of the liability-outgo, determine how much is invested in each of the
           loans.

10.6   On each 1 January for more than 20 years, the government of a country has
       issued a 20-year fixed-interest stock, redeemable at par and bearing interest at
       $3\frac{1}{8}$% per annum, payable annually in arrear. On 1 January 1984, when the
       market prices of these stocks were consistent with a yield of 5% per annum
       effective, an office operating in this country issued a ten-year capital
       redemption policy with annual premiums and sum assured £100 000. The
       annual premium was calculated on the basis of an interest rate of 5% per
       annum effective with allowance for expenses of $6\frac{1}{4}$% of each office premium.

   (a) Calculate the office annual premium.
   (b) The office invested the first premium, less expenses of $6\frac{1}{4}$%, in that
       fixed-interest stock for which the discounted mean term, at 5% per
       annum interest, was closest to ten years.

   Find (i) The outstanding term of this stock, and
        (ii) The nominal amount purchased.

10.7   The market prices of all fixed-interest stocks are currently determined on the
       basis of a constant force of interest of 5% per annum. A man, owing £100 000
       due in five years' time, possesses cash equal to the present value of his liability
       on this basis.
         The man wishes to invest all his cash in a fixed-interest stock, which pays
       interest continuously at a constant annual rate until the stock is redeemed at
       par. The term of the stock may be chosen by the man and need not be an
       integer. The man wishes to make the discounted mean term of his assests

equal to the discounted mean term of his liability at a force of interest of 5% per annum.

(a) Show that, if the rate of payment of interest on the stock is k% per annum, the term of the stock must be that value of n for which

$$e^{0.05n} = \frac{25 + 15k + n(k - 5)}{15k}$$

and find the value of n when (i) $k = 5$, (ii) $k = 10$.

(b) Assume that the stock purchased has an annual coupon rate of 10% and the term found in (a). Show that the man will then be immunized against small changes in the force of interest. What would be the present value of his profit, if the basis for valuing assets and liabilities were to change immediately to a constant annual force of interest of (i) $5\frac{1}{2}$%, (ii) $4\frac{1}{2}$%?

10.8 In a certain country investors may trade in zero-coupon stocks of any term. There are no dealing expenses, and all zero-coupon stocks are redeemable at par, exempt from taxation, and priced on the basis of the same constant force of interest.

An investor, who owes £1 million due in ten years' time, possesses exactly sufficient cash to meet this debt on the basis of the current constant market force of interest per annum δ_0. Show that, by purchasing suitable quantities of five-year and fifteen-year zero-coupon bonds, the investor can be sure of making a profit on any immediate change in the constant force of interest used by the market to value zero-coupon stocks. Assuming further that $\delta_0 = 0.05$, and that the appropriate quantities of each bond are purchased, find the profit the investor will make now if this force of interest per annum changes immediately to (a) 0.07, (b) 0.03.

10.9 An insurance company has a liability of £100 000 due in eight years' time. The company, which has exactly sufficient money to cover the liability on the basis of a constant force of interest of 5% per annum (which is also the market basis for pricing all stocks), now wishes to invest this money in a mixture of the following securities:

(i) Zero-coupon bonds redeemable at par in 20 years' time
(ii) Very short-term deposits (which may be regarded as interest-bearing cash).

(a) The company requires that, on the basis of a constant force of interest of 5% per annum, the discounted mean term of the assets equal that of the liabilities. Find the amounts to be invested in each of securities (i) and (ii).

(b) Suppose that these investments are in fact made. Find the present value of the profit to the organization (i.e. the difference between the present value of the assets and that of the liabilities) on the basis of a constant force of interst per annum of (1) 3%, (2) 7%.

10.10 An insurance company has just issued a 15-year single premium capital redemption policy with sum assured £10 000. The amount of the single premium was calculated on the assumption that, for all values of $t \geqslant 0$, $\delta(t)$, the force of interest per annum in t years' time, would equal $\log(1\cdot08)$. Expenses were ignored.

The company immediately invested part of the single premium in cash and the balance in a 20-year zero-coupon stock (redeemable at par), the price of which was consistent with the constant value of $\delta(t)$ assumed above. The amount invested in cash was such that, on the basis of this constant value of $\delta(t)$, the discounted mean term of the assets equalled the discounted mean term of the liability.

(a) Calculate the single premium and the nominal amount of stock purchased.

(b) Immediately after the investment of the single premium, economic forces caused a change in market rates of interest. Calculate the present value of the profit or loss to the company on the assumption that future cash flows are discounted at the new market rates of interest for which

(i) $\delta(t) = \log(1\cdot05)$ for all t
(ii) $\delta(t) = \log(1\cdot1)$ for all t
(iii) $\delta(t)$ is a linear function of t, with $\delta(0) = \log(1\cdot05)$ and $\delta(20) = \log(1\cdot1)$.

10.11 (Variation of volatility with term)

Let $T(\delta_0)$ be the volatility, at constant force of interest δ_0 per annum, of a fixed-interest stock with interest payable continuously at the annual rate D per unit nominal and redeemable at R per unit nominal in n years' time. Let $g = D/R$. Show that

$$\frac{\partial T(\delta_0)}{\partial n} = \frac{[\delta_0 + g(g-\delta_0)n - (g-\delta_0)^2 \bar{a}_{\overline{n}|}]e^{-\delta_0 n}}{\delta_0(g\bar{a}_{\overline{n}|} + e^{-\delta_0 n})^2} \qquad \text{at force of interest } \delta_0.$$

10.12 (Full immunization)

Suppose that δ_0 and S are known positive numbers. Suppose further that *two* of the four *positive* numbers a, A, b, and B are given. Consider the system of simultaneous equations

$$\left.\begin{aligned} Ae^{\delta_0 a} + Be^{-\delta_0 b} &= S \\ Aae^{\delta_0 a} &= Bbe^{-\delta_0 b} \end{aligned}\right\}$$

for the two 'unknown' numbers from $\{a, A, b, B\}$.

(a) Show that if the given pair of values is either (i) a, b, (ii) B, b (with $Be^{-\delta_0 b} < S$), (iii) A, a (with $Ae^{\delta_0 a} < S$) or (iv) A, b (with $A < S$) then there exists a unique pair of *positive* values for the two 'unknown' numbers which satisfies the system of equations.

(b) Let $\delta_0 = 0\cdot05$ and $S = 1$. Show that, if $a = 15$ and $B = 0\cdot98$, then there are *two* distinct pairs of positive numbers (b, A) which satisfy the above equations.

CHAPTER ELEVEN

CONSUMER CREDIT

11.1 Consumer credit legislation

In recent years the governments of various countries have enacted laws aimed at making people who borrow money or buy goods or services on credit more aware of the true cost of credit and enabling them to compare the true interest rates implicit in various lending schemes. Examples of laws of this type are the Consumer Credit Act 1974 in the UK and the Consumer Credit Protection Act 1968 (which contains the 'truth in lending' provisions) in the USA.

In this chapter we give a brief outline of the general principles underlying recent UK legislation. The relevant Act and Regulations are somewhat complex and, accordingly, the reader must refer to the appropriate legislation for precise details.

Regulations made under powers in the Consumer Credit Act 1974 lay down what items should be treated as entering into the total charge for credit and how the rate of charge for credit should be calculated. The rate is known as the 'Annual Percentage Rate of Charge' (APR) and is defined in such a way as to be the effective annual rate of interest on the transaction, obtained by solving the appropriate equation of value, taking into account all the items entering into the total charge for credit. The total charge for credit and the APR have to be disclosed in advertisements and in quotations for consumer credit agreements. Provided that all the advances precede all the repayments and the total amount of the repayments is greater than the total amount of the advances, then (see theorems 3.2.1 and 3.2.2) there is only one root of the equation of value and this root is positive. However, if the repayments commence before the last advance is made, the equation of value may have more than one positive root (see exercise 11.2). In this case the APR is defined to be the *least* positive root. As from April 2000 in all cases the APR is to be quoted as a percentage rate rounded to one decimal place. The Regulations define 'rounding' in such a way that 'midway' values are rounded up. Thus, if the rate i is such that $0 \cdot 1545 \leqslant i < 0 \cdot 1555$, the quoted APR is 15·5%. (Under earlier Regulations APR results were 'truncated' to one decimal place rather than rounded. This means that the official Consumer Credit Tables published in 1977 (see reference [8]) are no longer valid.

Regulation Z of the US Consumer Credit Protection Act 1968 requires the disclosure of the 'finance charge' (defined as the excess of the total repayments over the amount lent) and the 'annual percentage rate', which is the nominal rate of interest per annum convertible as often as the repayments are made (e.g. monthly or weekly). The value quoted must be accurate to one-quarter of 1%.

11.2 Flat rates of interest

In many situations where a loan is to be repaid by level instalments at regular intervals it is a common commercial practice to calculate the amount of each repayment instalment by specifying a *flat rate* of interest for the transaction. The operation of flat rates of interest is as follows.

Consider a loan of P which is to be repayable over a certain period by n level instalments. Suppose that the flat rate of interest for the transaction is F per specified time unit. (Note that the time unit used to specify F need *not* be the time interval between repayments: in practice the time unit used to specify F is generally a year.) The total *charge for credit* for the loan is defined to be

$$D = PFk \qquad (11.2.1)$$

where k is the repayment period of the loan, measured in units of time used in the definition of F.

The total amount repaid is defined to be the amount of the loan plus the charge for credit, i.e. $(P + D)$. Each instalment is thus of amount

$$\frac{P + D}{n} \qquad (11.2.2)$$

Example 11.2.1

A loan of £4000 is to be repaid by level monthly instalments over 2 years. What is the amount of the monthly repayment if the flat rate of interest for the transaction is (a) 10% per annum (b) 1% per month? Assuming that the payments are in arrear, find the APR in each case.

Solution

(a) The period of the loan is two years, so that (from equation 11.2.1) the charge for credit for the loan is $4000 \times 0.1 \times 2 = £800$. The total repaid is thus £4800 and the amount of each monthly instalment is £4800/24 = £200.

   The APR is obtained from the equation

$$12 \times 200 a_{\overline{2}|}^{(12)} = 4000 \qquad \text{at rate } i$$

   from which $i = 0.19747$, so that the APR is 19.7%.

(b) The period of the loan is 24 months, so that (from equation 11.2.1) the charge for credit for the loan is $4000 \times 0.01 \times 24 = £960$. The total repaid is thus £4960 and the monthly instalment is thus £4960/24 = £206·67. The APR is obtained from the equation

$$12 \times 206\text{·}67 \times a_{\overline{2}|}^{(12)} = 4000$$

from which $i = 0\text{·}238\,39$. The APR is thus 23·8%.

Suppose now that a given time unit is specified (e.g. one year) and that the flat rate of interest is F per unit time. Suppose that a loan of P is to be repaid by regular level instalments (in arrear), there being m repayments per unit time. As before, let the total number of repayments be n.

The total period of the loan is thus n/m time units and the charge for credit for the loan is

$$D = PF\frac{n}{m}$$

The amount of each repayment is

$$\frac{P + D}{n} = P\left(\frac{1}{n} + \frac{F}{m}\right) \tag{11.2.3}$$

The effective rate of interest i per unit time is thus given by the equation

$$P = mP\left(\frac{1}{n} + \frac{F}{m}\right)a_{\overline{n/m}|}^{(m)}$$

or

$$1 = \left(\frac{m}{n} + F\right)a_{\overline{n/m}|}^{(m)} \qquad \text{at rate } i \tag{11.2.4}$$

Equation 11.2.4 may be written in the form

$$m[(1 + i)^{1/m} - 1] = \left(\frac{m}{n} + F\right)[1 - (1 + i)^{-n/m}]$$

If we expand the terms in this last equation by the binomial theorem and ignore powers of i higher than i^3 and terms in Fi^3 (which is of the same order of magnitude as i^4), we obtain (on simplification, including division throughout by i)

$$\frac{-nF}{m} + i\frac{(n + 1)}{2m} + \frac{n(n + m)}{2m^2}Fi - \frac{(n + 1)(n + 3m - 1)}{6m^2}i^2 = 0 \tag{11.2.5}$$

A first approximation for i is obtained from this equation by ignoring the terms in i^2 and Fi (which is the same order of magnitude as i^2). This gives

$$\frac{-nF}{m} + i\frac{(n + 1)}{2m} = 0$$

or

$$i = 2F\frac{n}{n+1} \tag{11.2.6}$$

This shows that, if n is large, a first approximation for i is $2F$. This result is intuitively obvious, since on average the outstanding loan is about half the original loan, on which 'interest' is charged at the constant rate F per unit time.

A second approximation may be obtained iteratively by writing equation 11.2.5 in the form

$$i\left[\frac{n+1}{2m} + \frac{n(n+m)}{2m^2}F - \frac{(n+1)(n+3m-1)}{6m^2}i\right] = \frac{nF}{m}$$

and replacing i inside the square bracket in this equation by the first approximation (i.e. $2Fn/(n+1)$). On simplifying the resulting equation, we obtain

$$i = \frac{2F}{\dfrac{n+1}{n} + F\left(\dfrac{n-3m+2}{3m}\right)} \tag{11.2.7}$$

Note that if the time unit for specifying F, the flat rate of interest, is equal to the interval between loan repayments, then $m = 1$. In this case equation 11.2.7 gives

$$i = \frac{2F}{\dfrac{n+1}{n} + F\left(\dfrac{n-1}{3}\right)} \tag{11.2.8}$$

as an approximation for the effective rate of interest per unit time.

Either of the approximations 11.2.6 and 11.2.7 may be used to obtain a rough estimate of the effective rate of interest in any transaction based on flat rates, but neither can be relied upon to give answers correct when rounded to the nearest one-tenth of 1%. They are therefore not suitable for use under the Consumer Credit Act 1974, but they are a very useful method of obtaining a first approximation for one of the more accurate methods (see appendix 2). See also exercise 11.9, where another approximation is given, and reference [12].

Example 11.2.2

A motoring organization offers loans to members for buying cars. The repayments are calculated on the basis of a flat rate of interest of 12% per annum.

(a) Find the monthly repayment for a loan of £3000 repayable over five years.

(b) Use equations 11.2.7 and 11.2.8 to estimate the effective rate of interest for the transaction (i) per annum (ii) per month.
(c) Find the APR for the transaction.

Solution

(a) The charge for credit for the loan is $3000 \times 0.12 \times 5 = 1800$, so that the monthly repayment is $£(3000 + 1800)/60 = £80$.
(b) (i) We estimate the effective interest rate per annum from equation 11.2.7 with $F = 0.12$, $n = 60$, and $m = 12$. This gives $i = 0.2175$ or 21.75%.
   (ii) Note that a flat rate of interest of 12% per annum is the same as a flat rate of 1% per month. We therefore use equation 11.2.8 with $F = 0.01$ and $n = 60$. This gives $i = 0.0165$. The effective rate of interest per month is thus estimated to be 1.65% (Since $1.0165^{12} = 1.2170$, the two estimates are in reasonably close agreement.)
(c) The APR is given by the equation

$$3000 = 12 \times 80 \times a_{\overline{5}|}^{(12)} \qquad \text{at rate } i$$

This gives $i = 0.22311$, so that the APR is 22.3%. (Note that neither of the above estimates gives this value.)

11.3 Early repayment terms: the rule of 78

Consider a loan of P, repayable by n level instalments each of amount $(P + D)/n$, so that the total charge for credit for the loan is D. If we choose as our unit of time the period between loan repayments, the effective rate of interest per unit time for the loan is i, where

$$\frac{P + D}{n} a_{\overline{n}|} = P \qquad \text{at rate } i \qquad (11.3.1)$$

This implies that

$$D a_{\overline{n}|} = P(n - a_{\overline{n}|})$$

and hence (by division) that

$$\frac{P + D}{n} = \frac{D}{n - a_{\overline{n}|}} \qquad \text{at rate } i \qquad (11.3.2)$$

Suppose that an instalment has just been paid and that a further t instalments remain to be paid. Under a compound interest loan schedule (see section 3.8) the outstanding loan is

$$\frac{P + D}{n} a_{\overline{t}|} \qquad \text{at rate } i$$

and the borrower would have to pay this sum to pay off the loan. The total amount of the outstanding instalments is $t(P + D)/n$, and the sum required from the borrower to repay the loan may be regarded as the total amount of the remaining instalments less an *interest rebate* – i.e. a refund of part of the total charge for credit – because of the early repayment. The appropriate fraction of the total charge for credit to be allowed as a rebate is therefore k, where

$$\frac{P+D}{n} a_{\overline{t}|} = t\frac{P+D}{n} - kD \qquad \text{at rate } i$$

Hence

$$kD = \frac{P+D}{n}(t - a_{\overline{t}|})$$

$$= \frac{D}{n - a_{\overline{n}|}}(t - a_{\overline{t}|}) \qquad \text{(by 11.3.2)}$$

Thus

$$k = \frac{t - a_{\overline{t}|}}{n - a_{\overline{n}|}} \qquad \text{at rate } i \qquad (11.3.3)$$

The above calculations determine the appropriate rebate under a conventional compound interest schedule. In relation to loans made on the basis of a flat rate of interest an alternative procedure, known as 'the rule of 78', is generally used to determine the loan outstanding at any time. For most practical purposes, this alternative method, which is simpler to apply, gives results which are similar to those of the compound interest schedule. We describe this method below.

The total charge for credit D is divided into a number of 'units'. It is assumed that there are n units in the first repayment instalment, $(n-1)$ units in the second instalment, $(n-2)$ units in the third instalment, and so on. The final instalment (i.e. the nth) contains 1 unit of the credit charge.

The total number of credit charge 'units' repaid is $n(n+1)/2$ (i.e. $1 + 2 + \cdots + n$). Since the total charge for credit is D, one unit is of amount $2D/[n(n+1)]$. The rth repayment instalment contains $(n+1-r)$ credit charge units and is therefore considered as consisting of 'interest'

$$\frac{2D(n+1-r)}{n(n+1)} \qquad (11.3.4)$$

as part of the total charge for credit, and a capital repayment of amount

$$\frac{P+D}{n} - \frac{2D(n+1-r)}{n(n+1)} \qquad (11.3.5)$$

On this basis the lender may easily draw up a schedule to indicate the progress of the loan repayment (see exercise 11.7).

If a repayment schedule is drawn up in the above manner, the amounts of successive capital repayments form an *arithmetic* progression (as opposed to the geometric progression in a compound interest loan schedule). Moreover, there is no obvious relationship between the fraction of the total charge for credit contained in any repayment and the loan outstanding after the previous repayment.

In relation to a loan repayable over one year by monthly instalments $n = 12$ and $n(n + 1)/2 = 78$. For this reason the determination of the outstanding loan by the above method is known as the 'rule of 78'. This name is used even when the repayment period is not one year.

Immediately after a repayment has been made, at a time when a further t instalments remain to be paid, the outstanding loan is simply the total of the capital repayments contained in the final t instalments. This is obviously the same as the total of the outstanding instalments less the total of the credit charge units (according to the schedule) contained in these instalments. Since these instalments contain a total of $t(t + 1)/2$ (i.e. $1 + 2 + \cdots + t$) units of credit charge, the outstanding loan according to the rule of 78 is

$$t\frac{P + D}{n} - \frac{1}{2}t(t + 1)\frac{2D}{n(n + 1)}$$

or

$$t\frac{P + D}{n} - k'D \qquad (11.3.6)$$

where

$$k' = \frac{t(t + 1)}{n(n + 1)} \qquad (11.3.7)$$

Thus, according to the rule of 78, the rebate for early repayment is a fraction k' of the total charge for credit, where k' is given by equation 11.3.7. This contrasts with an interest rebate of k times the total interest payable under a compound interest loan schedule (where k is given by equation 11.3.3). It is possible to show (see example 11.3.1) that $k' < k$, so that the rule of 78 is to the advantage of the lender; however, it should be noted that the lender has in general no option to terminate the loan if interest rates rise, whereas the borrower may generally do so if he can obtain cheaper credit elsewhere, or for other reasons. A borrower who repays his outstanding loan early (on the basis of the rule of 78) will pay a greater effective rate of interest on the transaction than he would have if he had continued to repay the loan by instalments. For practical purposes, however, the difference between the two rates of interest is generally relatively small (see example 11.3.3).

In practical situations early repayment of a loan may occur between instalment payment dates. For this reason, and also to allow for his administrative expenses, it is sometimes permissible for a lender to apply a modified version of the rule of 78, under which the fraction of the total interest charge permitted as a rebate for early repayment allows for a 'deferment period'. In this case, when t instalments are outstanding, the appropriate fraction is k'', where

$$k'' = \begin{cases} 0 & \text{if} \quad t \leqslant \alpha \\[2mm] \dfrac{(t-\alpha)(t-\alpha+1)}{n(n+1)} & \text{if } \alpha < t \leqslant n \end{cases} \qquad (11.3.8)$$

and α is generally 1, 2, or 3. (In the UK for monthly credit plans the minimum rebate permitted under the 1983 Regulations relating to the Consumer Credit Act 1974 is given by equation 11.3.8 with $\alpha = 1$ for loans repayable over more than five years and $\alpha = 2$ for loans of term five years or less. It is obvious that in certain situations this modified rule of 78 can penalize the borrower quite severely. In Australia the unmodified rule of 78 must be used for calculating interest rebates.)

Example 11.3.1

Let t and n be integers with $1 \leqslant t < n$. Show that at any positive rate of interest

$$\frac{t(t+1)}{n(n+1)} < \frac{t - a_{\overline{t}|}}{n - a_{\overline{n}|}}$$

(This shows that the interest rebate granted under the rule of 78 is less than the corresponding rebate under a compound interest loan schedule.)

Solution

Let $v = (1+i)^{-1}$, where i is positive. Note that $1, v, v^2, v^3,\ldots$ form a decreasing sequence. Hence, if $l < m$,

$$\frac{1 + v + v^2 + \cdots + v^{l-1}}{l} > \frac{1 + v + v^2 + \cdots + v^{m-1}}{m}$$

Multiplying both sides of this inequality by $(1-v)$ and rearranging the resulting terms, we immediately obtain

$$\frac{1 - v^l}{1 - v^m} > \frac{l}{m}$$

if $l < m$. In particular,

$$\frac{1-v}{1-v^{t+1}} > \frac{1}{t+1}, \quad \frac{1-v^2}{1-v^{t+1}} > \frac{2}{t+1}, \quad \ldots, \quad \frac{1-v^t}{1-v^{t+1}} > \frac{t}{t+1}$$

By adding this last set of inequalities we see that

$$\frac{t-(v+v^2+\cdots+v^t)}{1-v^{t+1}} > \frac{1+2+\cdots+t}{t+1} = \frac{t}{2}$$

Hence

$$1+\frac{1-v^{t+1}}{t-(v+v^2+\cdots+v^t)} < 1+\frac{2}{t}$$

i.e.

$$\frac{(t+1)-(v+v^2+\cdots+v^{t+1})}{t-(v+v^2+\cdots+v^t)} < \frac{(t+1)(t+2)}{t(t+1)}$$

Thus

$$\frac{t-(v+v^2+\cdots+v^t)}{t(t+1)} > \frac{(t+1)-(v+v^2+\cdots+v^{t+1})}{(t+1)(t+2)}$$

By repeated application of this last inequality it follows that, if $t < n$,

$$\frac{t-(v+v^2+\cdots+v^t)}{t(t+1)} > \frac{n-(v+v^2+\cdots+v^n)}{n(n+1)}$$

from which the required result follows immediately.

Example 11.3.2

A loan of £900 is being repaid by 18 level monthly instalments, calculated on the basis of a flat rate of interest of 10% per annum.

(a) Find
  (i) The total charge for credit for the loan;
  (ii) The amount of each monthly instalment;
  (iii) The APR for the loan.
(b) Find the loan outstanding immediately after the 12th instalment has been repaid on the basis of
  (i) A true compound interest loan schedule (based on the rate of interest implied in the original transaction);
  (ii) The rule of 78.

Solution

(a) (i) The period of the loan is 1·5 years, so that the total charge for credit is £900 × 0·1 × 1·5 = £135.
  (ii) The amount of each monthly instalment is £(900 + 135)/18 = £57·50.
  (iii) The effective rate of interest per month is given by the equation

$$900 = 57 \cdot 5 a_{\overline{18}|} \qquad \text{at rate } i$$

from which we obtain $i = 0 \cdot 015\,145$. Since $1 \cdot 015\,145^{12} = 1 \cdot 197\,67$, the APR for the loan is 19·8%.
(b) (i) On the basis of a compound interest loan schedule with an effective interest rate of $i = 1 \cdot 5145\%$ per month, the loan outstanding after one year is $57 \cdot 5 a_{\overline{6}|}$ at rate i, i.e. £327·43.

Note This may be obtained alternatively as follows. The fraction of the total charge for credit to be allowed as a rebate when there are six outstanding instalments is

$$(6 - a_{\overline{6}|})/(18 - a_{\overline{18}|}) \qquad \text{at rate } i = 0.130\,176$$

The rebate allowed is thus £135 × 0.130 176 = £17.57. Since the remaining instalments total £345, the outstanding loan is £(345 − 17.57) = £327.43, as above.

(ii) On the basis of the rule of 78 the fraction of the charge for credit allowed as rebate is $(6 \times 7)/(18 \times 19) = 0.122\,807$, so that the interest rebate is £135 × 0.122 807 = £16.58. The outstanding loan is therefore £(345 − 16.58), i.e. £328.42.

Example 11.3.3

A loan of £6000 is being repaid over four years by level monthly instalments, calculated on the basis of a flat rate of interest of 12.75% per annum.

(a) Find the APR for the loan.
(b) Immediately after paying the 24th monthly instalment the borrower wishes to repay the entire outstanding loan. Find the effective rate of interest for the transaction, if the interest rebate is calculated according to
   (i) The rule of 78 (i.e. equation 11.3.7);
   (ii) The modified rule of 78 with a two-month time lag (i.e. equation 11.3.8 with $\alpha = 2$).

Solution

(a) The charge for credit for the loan is £(6000 × 0.1275 × 4) = £3060, so the monthly instalment is £(3060 + 6000)/48 = £188.75. The effective rate of interest per month is given by the equation

$$6000 = 188.75 a_{\overline{48}|} \qquad \text{at rate } i$$

from which we obtain $i = 0.018\,260$. Since $1.018\,26^{12} = 1.242\,52$, the APR is 24.3%.

(b) (i) After two years a further 24 instalments remain to be paid. The interest rebate allowed by the rule of 78 is thus

$$\frac{24 \times 25}{48 \times 49} 3060 = £780.61$$

so the outstanding loan is $(24 \times 188.75) - 780.61 = £3749.39$.

The effective rate of interest per month for the transaction is that rate of interest for which

$$6000 = 188.75 a_{\overline{24}|} + 3749.39 v^{24}$$

This gives $i = 0.018\,973$, from which the effective annual rate of interest is obtained as 25.3%. Note that this is greater than the APR found above.

(ii) On the basis of the modified rule of 78 the interest rebate allowed after two years is

$$£\frac{(24 - 2) \times (25 - 2)}{48 \times 49} 3060 = £658.32$$

so that the outstanding loan is £3871·68. The reader should verify that this implies an effective monthly rate of interest for the completed transaction of 1·976 48%, which corresponds to an annual rate of 26·5%.

11.4 Loans repayable by a single instalment

Consider a loan of amount P, which is to be repaid by a single instalment of amount $(P + D)$ after n years (where n need not be an integer). The charge for credit is D and the effective annual rate of interest on the transaction is, of course,

$$i = \left(1 + \frac{D}{P}\right)^{1/n} - 1 \tag{11.4.1}$$

The APR on the transaction is thus $100i$ per cent (quoted to the *lower* one decimal place).

Suppose now that the loan is to be repaid early, t years before the originally specified repayment date (where t need not be an integer). On the basis of the rate of interest implicit in the original agreement, the sum required to repay the loan is

$$P(1 + i)^{n-t} = P\left(1 + \frac{D}{P}\right)^{1 - (t/n)} \tag{11.4.2}$$

In practice, however, the borrower may have the right to repay the loan early, but the lender will have no power to compel early repayment. The lender will also have administration costs associated with the early termination of the contract. For these reasons the 1983 Regulations relating to the Consumer Credit Act 1974 permit the sum required to repay the loan early to be rather larger than that given by equation 11.4.2.

In essence, the Regulations require that the interest rebate be proportional to the period from the 'settlement date' to the originally specified repayment date. Note, however, that, as in the case of the loans repayable by several instalments, a 'deferment period' is allowed, so that the settlement date may be taken as the actual date of the early repayment deferred by one month for a loan of original term greater than five years or by two months for a loan of original term of five years or less. This means that, if the period from the settlement date (deferred as described above) to the originally specified repayment date is t' years, the minimum permitted interest rebate is

$$D\frac{t'}{n} \tag{11.4.3}$$

so that the amount required to repay the loan early must not exceed

$$P + D - \frac{t'}{n}D \tag{11.4.4}$$

Note (see exercise 2.1) that

$$P\left(1 + \frac{D}{P}\right)^{1 - (t/n)} < P\left[1 + \left(1 - \frac{t}{n}\right)\frac{D}{P}\right]$$

$$= P + D - \frac{t}{n}D$$

This means that, even without deferment, proportional calculation of the interest rebate permits a larger payment for early settlement than the value given by equation 11.4.2.

The interested reader may refer to the note by Daykin (reference [12]) for a further discussion on consumer credit calculations under UK legislation. Note, however, that some Regulations have been amended since the note was written.

Example 11.4.1

On 21 December 1985 a man borrowed £160 to be repaid on 20 September 1986 by a single payment of £185. On 1 May 1986 the man repaid the loan early.

(a) Find the APR on the transaction as originally agreed.
(b) The interest rebate allowed for early repayment was calculated proportionally, but based on a two-month deferment of the settlement date. Find the sum repaid by the borrower and the APR on the completed transaction.

Solution

The original term of the loan was 273 days. The 'settlement date' is taken as 1 July 1986 (since there is a two-month deferment). The period from the settlement date to 20 September 1986 is 81 days.

(a) The effective annual yield i is given by the equation

$$160(1 + i)^{273/365} = 185$$

from which it follows that $i = 0.214\,23$. The APR is thus 21·4%.

(b) The charge for credit is £25 and the interest rebate allowed for early settlement is therefore

$$\tfrac{81}{273}25 = £7·42$$

The sum paid by the borrower to repay the loan early was thus

$$£(185 - 7·42) = £177·58$$

The loan was repaid after 131 days. By solving the equation

$$160(1 + i')^{131/365} = 177·58$$

to obtain $i' = 0.337\,04$, we calculate the APR on the completed transaction to be 33·7%.

Exercises

11.1 A television set, which costs £200, may be purchased by 52 weekly instalments of £4, payable in arrear. What is the APR on this transaction?
Note A week is taken as a 52nd part of a year.

11.2 (Multiple roots of the equation of value)
A farmer pays a deposit of £100 to a credit company on 1 April to secure the right to borrow money to finance the harvest. On 1 July he borrows £1760 and this is repayable in two instalments, £605 on 1 October and £1331 on 1 January of the following year. What is the APR under the agreement?

Note Assume that all three-month periods are of equal length.

11.3 (An account varying from credit to debit and vice versa)
A bank charges interest at $1\frac{1}{2}\%$ per month on accounts overdrawn by £1000 or more and at 1% per month on accounts overdrawn by less than £1000. Interest charges are levied at the end of each month.

On 1 January a certain customer, whose account was neither in credit nor in debit, borrowed £2000 from the bank.

(a) What monthly payment, in arrear, should he make in each of the next 12 months in order that his account should again be at zero on the following 1 January? (You should assume that he makes no other payments from, or into, this bank account during the year.)

(b) Assume that the customer decided to make these monthly payments. What is the APR on the transaction?

11.4 A loan of £1000 is repayable in 60 monthly instalments (in arrear) of £27·92. What is the APR for this loan?

11.5 A loan of £1000 is agreed to be repaid by 36 monthly instalments of £48 starting one month after the date of the loan. The consumer settles the loan at the end of the 30th month (i.e. when six monthly instalments are outstanding). Find the rebate under formula 11.3.8 with $\alpha = 2$, and the sum actually payable by the borrower to settle the loan.

11.6 The purchaser of a motor car borrows £5000, repayable by 24 monthly instalments, in arrear, calculated at a flat rate of interest of 10% per annum.

(a) Calculate his monthly repayment and the APR on the transaction.

(b) Just after making the 12th repayment, the borrower pays off the outstanding loan. The lender calculates the amount required for early settlement by using the (possibly modified) rule of 78. Find the sum payable by the borrower and the yield per annum on the completed transaction, assuming that the rule of 78 is applied as in equation 11.3.8 with (i) $\alpha = 0$, (ii) $\alpha = 2$.

11.7 A finance company lends a customer £1200 repayable by six monthly instalments, in arrear, calculated on the basis of a flat rate of interest of 2% per month.

(a) Calculate the monthly repayment and i, the effective rate of interest per month on the transaction.

(b) Construct a schedule showing the loan outstanding after each repayment, using
(i) A compound interest schedule at rate of interest i per month;
(ii) The arithmetic progression of the charge for credit in each instalment, as defined by the unmodified rule of 78 (see formula 11.3.6).

11.8 (Tax relief on interest payments in a flat interest schedule)
Consider a loan of P repayable by n level monthly instalments (in arrear), calculated on the basis of a flat rate of interest of F per annum. The amount of each monthly payment is thus

$$P\left(\frac{1}{n} + \frac{F}{12}\right)$$

On a 'flat interest schedule' each monthly payment is regarded as consisting of a capital payment of amount P/n and an 'interest' payment of amount $PF/12$.

If a borrower can obtain tax relief at rate t on the interest content of each payment, the amount of each payment is, in effect, reduced by $t(PF/12)$. Show by general reasoning that the net effective rate of interest per month when tax relief is permitted is higher on the basis of a flat interest schedule than it would be if interest payments were calculated on a compound interest schedule (see section 3.8) at the gross rate of interest per month implicit in the transaction. Illustrate this result by calculating the net effective rate of interest per month on a loan of £6000 repayable by 12 level monthly instalments if interest is at 13% per annum flat and the borrower may obtain tax relief at 30% on interest payments, and comparing it with the net effective rate of interest per month if interest were calculated on a compound interest schedule.

11.9 Consider a loan of P repayable by n level instalments, payable in arrear m times per annum. Let D be the charge for credit. Show that the APR is approximately equal to

$$\frac{2mD}{P(n + 1) + \dfrac{D}{3}(n - 3m + 2)}$$

Hint See formula 11.2.7.

11.10 On 15 April 1985 a man borrowed £2000 to be repaid one year later by a single payment of £2200. On 17 July 1985 the borrower repaid the loan early.
Find the sum paid by the borrower to terminate the contract, and the APR on the completed transaction, assuming that the interest rebate allowed for early settlement was calculated by proportion and that

(a) There was no deferment of the 'settlement date';
(b) The 'settlement date' was deferred for two months.

11.11 On 3 January 1986 a man borrowed £160 to be repaid by a single payment of £185 on 3 October 1986. On 14 May 1986 the man repaid the loan early, the interest rebate being determined on the basis of a two-month deferment of the settlement date.

(a) Find the APR for the loan as originally agreed.
(b) Find the sum paid by the borrower to repay the loan early and the APR on the completed transaction.

CHAPTER TWELVE

AN INTRODUCTION TO STOCHASTIC INTEREST RATE MODELS

12.1 Preliminary remarks

We have frequently remarked that financial contracts are often of a long-term nature. Accordingly, at the outset of many contracts there may be considerable uncertainty about the economic and investment conditions which will prevail over the duration of the contract. Thus, for example, if it is desired to determine premium rates on the basis of one fixed rate of interest, it is nearly always necessary to adopt a conservative basis for the rate to be used in any calculations.

An alternative approach to recognizing the uncertainty that in reality exists is provided by the use of a *stochastic interest rate model*. In such models no single interest rate is used. Variations in the rate of interest are allowed for by the application of probability theory. Possibly one of the simplest models is that in which each year the rate of interest obtained is independent of the rates of interest in all previous years and takes one of a finite set of values, each value having a constant probability of being the actual rate for the year. Alternatively, the rate of interest may take any value within a specified range, the actual value for the year being determined by some given probability density function.

Even simple models, such as those described above, can be of practical use. For example, they may serve to illustrate the financial consequences (sometimes considerable!) to an office of the departure of its actual investment experience from that assumed in its premium bases.

In recent years much attention has been paid to the development of stochastic interest rate models: see, for example, references [2], [35], [36], [37], [44], [52], [53], [55], [56], and [57].

The power of modern computers makes it relatively simple to study considerably more sophisticated models. For example, it is possible (and probably more realistic) to drop the independence assumption and to build in some measure of dependence between the rates of interest in successive years (see section 12.5).

For many stochastic interest rate models the attainment of theoretically exact results is difficult. However, this is not an insuperable barrier to further progress. The use of *simulation techniques* often provides revealing insights of

269

practical importance into the financial risks involved in many kinds of contract.

A discussion in depth of stochastic interest rate models is beyond the scope of this book. By studying the references given above the reader will quickly discover the complexities which may arise in the application of such models to practical problems. Although in this chapter we are able to present only a very brief introduction to the subject, it is important to recognize that, by combining elementary financial concepts with probability theory, stochastic interest rate models offer a powerful tool for the analysis of financial problems. Moreover, this tool is fundamentally different from the deterministic approach hitherto considered throughout this book.

12.2 An introductory example

At this stage we consider briefly an elementary example, which – although necessarily artificial – provides a simple introduction to the probabilistic ideas implicit in the use of stochastic interest rate models.

Suppose that a company wishes to issue now a block of single-premium capital redemption policies, each policy having a term of 15 years. The premium will be invested in a fund which grows under the action of compound interest at a constant rate throughout the term of the policy. This constant rate of interest is not known now, but will be determined immediately after the policies have been issued.

Suppose that the effective annual rate of interest for the fund will be 2%, 4%, or 6% and that each of these values is equally likely. In the probabilistic sense the expected value of the annual interest rate for the fund is

$$E[i] = (\tfrac{1}{3} \times 0.02) + (\tfrac{1}{3} \times 0.04) + (\tfrac{1}{3} \times 0.06) = 0.04 \qquad (12.2.1)$$

or 4%.

Consider now a policy with unit sum assured. Let P be the single premium charged. If i denotes the annual rate of interest for the fund in which the premiums are invested, the accumulated profit – i.e. the company's profit at the time the policy matures – will be

$$P(1 + i)^{15} - 1 \qquad (12.2.2)$$

This will equal $[P(1.02)^{15} - 1]$, $[P(1.04)^{15} - 1]$, or $[P(1.06)^{15} - 1]$, each of these values being equally likely. The expected value of the accumulated profit is therefore

$$E[\text{accumulated profit}] = \tfrac{1}{3}[P(1.02)^{15} - 1] + \tfrac{1}{3}[P(1.04)^{15} - 1]$$
$$+ \tfrac{1}{3}[P(1.06)^{15} - 1]$$
$$= 1.847\,79\,P - 1 \qquad (12.2.3)$$

Since the expected value of the annual yield is 4%, the company might choose to determine P on the basis of this rate of interest. In this case

$$P = v^{15} \quad \text{at } 4\% \quad = 0.555\,26$$

and hence (from equation 12.2.3)

$$E[\text{accumulated profit}] = 0.026\,01$$

It should be noted that in this case the use of the average rate of interest in the premium basis does *not* give zero for the expected value of the accumulated profit. (The reader should verify that the company's accumulated profit will be $0.330\,72, 0$, or $-0.252\,69$, each of these values being equally likely. Although it is possible for the company to make a loss, it is equally possible for the company to make a greater profit.)

An alternative viewpoint is provided by consideration of the net present value of the policy immediately after it is effected, i.e. the value at the outset of the policy of the company's contractual obligation less the value of the single premium paid. If i is the annual yield on the fund, the net present value is

$$(1 + i)^{-15} - P \tag{12.2.4}$$

This will equal $(1.02^{-15} - P), (1.04^{-15} - P)$, or $(1.06^{-15} - P)$, each of these values being equally likely. The expected value of the net present value is thus

$$E[\text{net present value}] = \tfrac{1}{3}(1.02^{-15} - P) + \tfrac{1}{3}(1.04^{-15} - P) + \tfrac{1}{3}(1.06^{-15} - P)$$

$$= 0.571\,88 - P \tag{12.2.5}$$

If, as before, P has been determined on the basis of the average rate of interest of 4% as $0.555\,26$, it follows that

$$E[\text{net present value}] = 0.016\,62$$

Again it should be noted that this is *not* zero.

Note also that the value of P for which the expected value of the accumulated profit is zero is *not* the value of P for which the expected net present value is zero (see exercise 12.2).

12.3 Independent annual rates of return

In our previous example the effective annual rate of interest *throughout the duration* of the policy was 2%, 4%, or 6%, each of these values being equally likely. A more flexible model is provided by assuming that over each *single year* the annual yield on invested funds will be one of a specified set of values or lie within some specified range of values, the yield in any particular year being independent of the yields in all previous years and being determined by a given probability distribution. For example, we might assume that each year the yield obtainable will be 2%, 4%, or 6%, each of these values being equally likely. Alternatively we might assume that each year all yields between 2% and 6% are equally likely (in which case the density function for i is uniform on the interval [0·02, 0·06]).

Measure time in years. Consider the time interval $[0, n]$ subdivided into successive periods $[0, 1], [1, 2], \ldots, [n-1, n]$. For $t = 1, 2, \ldots, n$ let i_t be the yield obtainable over the tth year, i.e. the period $[t-1, t]$. Assume that money is invested only at the beginning of each year. Let F_t denote the accumulated amount at time t of all money invested before time t and let P_t be the amount of money invested at time t. Then, for $t = 1, 2, 3, \ldots,$

$$F_t = (1 + i_t)(F_{t-1} + P_{t-1}) \tag{12.3.1}$$

It follows from this equation that a single investment of 1 at time 0 will accumulate at time n to

$$S_n = (1 + i_1)(1 + i_2) \cdots (1 + i_n) \tag{12.3.2}$$

Similarly a series of annual investments, each of amount 1, at times $0, 1, 2, \ldots, n-1$ will accumulate at time n to

$$
\begin{aligned}
A_n = \ & (1 + i_1)(1 + i_2)(1 + i_3) && \cdots && (1 + i_n) \\
& + (1 + i_2)(1 + i_3) && \cdots && (1 + i_n) \\
& + \ \ \vdots && \vdots && \vdots \\
& \quad + \ (1 + i_{n-1})(1 + i_n) \\
& \qquad + (1 + i_n)
\end{aligned}
\tag{12.3.3}
$$

Note that A_n and S_n are random variables, each with its own probability distribution function. For example, if the yield each year is 0·02, 0·04, or 0·06 and each value is equally likely, the value of S_n will be between 1·02^n and 1·06^n. Each of these extreme values will occur with probability $(1/3)^n$.

In general, a theoretical analysis of the distribution functions for A_n and S_n is somewhat difficult. It is often more useful to use simulation techniques in the study of practical problems (see section 12.4). However, it is perhaps worth noting that the moments of the random variables A_n and S_n can be found relatively simply in terms of the moments of the distribution for the yield each year. This may be seen as follows.

Moments of S_n

From equation 12.3.2 we obtain

$$(S_n)^k = \prod_{t=1}^{n} (1 + i_t)^k$$

and hence

$$E[S_n^k] = E\left[\prod_{t=1}^{n}(1 + i_t)^k\right]$$

$$= \prod_{t=1}^{n} E[(1 + i_t)^k] \qquad (12.3.4)$$

since (by hypothesis) $i_1, i_2, \ldots, i_n$ are independent. Using this last expression and given the moments of the annual yield distribution, we may easily find the moments of S_n.

For example, suppose that the yield each year has mean j and variance s^2. Then, letting $k = 1$ in equation 12.3.4, we have

$$E[S_n] = \prod_{t=1}^{n} E[1 + i_t]$$

$$= \prod_{t=1}^{n}(1 + E[i_t])$$

$$= (1 + j)^n \qquad (12.3.5)$$

since, for each value of t, $E[i_t] = j$.

With $k = 2$ in equation 12.3.4 we obtain

$$E[S_n^2] = \prod_{t=1}^{n} E[1 + 2i_t + i_t^2]$$

$$= \prod_{t=1}^{n}(1 + 2E[i_t] + E[i_t^2])$$

$$= (1 + 2j + j^2 + s^2)^n \qquad (12.3.6)$$

since, for each value of t,

$$E[i_t^2] = (E[i_t])^2 + \text{var}[i_t] = j^2 + s^2$$

The variance of S_n is

$$\text{var}[S_n] = E[S_n^2] - (E[S_n])^2$$

$$= (1 + 2j + j^2 + s^2)^n - (1 + j)^{2n} \qquad (12.3.7)$$

from equations 12.3.5 and 12.3.6.

These arguments are readily extended to the derivation of the higher moments of S_n in terms of the higher moments of the distribution of the annual rate of interest.

Moments of A_n

It follows from equation 12.3.3 (or from equation 12.3.1) that, for $n \geqslant 2$,

$$A_n = (1 + i_n)(1 + A_{n-1}) \qquad (12.3.8)$$

The usefulness of this equation lies in the fact that, since A_{n-1} depends only on the values $i_1, i_2, \ldots, i_{n-1}$, the random variables i_n and A_{n-1} are independent. (By assumption the yields each year are independent of one another.) Accordingly, equation 12.3.8 permits the development of a recurrence relation from which may be found the moments of A_n (see for example, reference [55]). We illustrate this approach by obtaining the mean and variance of A_n.

Let

$$\mu_n = E[A_n]$$

and let

$$m_n = E[A_n^2]$$

Since

$$A_1 = 1 + i_1$$

it follows that

$$\mu_1 = 1 + j \qquad\qquad (12.3.9)$$

and

$$m_1 = 1 + 2j + j^2 + s^2 \qquad\qquad (12.3.10)$$

where, as before, j and s^2 are the mean and variance of the yield each year.

Taking expectations of equation 12.3.8, we obtain (since i_n and A_{n-1} are independent)

$$\mu_n = (1 + j)(1 + \mu_{n-1}) \qquad n \geqslant 2 \qquad\qquad (12.3.11)$$

This equation, combined with initial value μ_1, implies that, for all values of n,

$$\mu_n = \ddot{s}_{\overline{n}|} \qquad \text{at rate } j \qquad\qquad (12.3.12)$$

Thus the expected value of A_n is simply $\ddot{s}_{\overline{n}|}$, calculated at the mean rate of interest.

Since

$$A_n^2 = (1 + 2i_n + i_n^2)(1 + 2A_{n-1} + A_{n-1}^2)$$

by taking expectations we obtain, for $n \geqslant 2$,

$$m_n = (1 + 2j + j^2 + s^2)(1 + 2\mu_{n-1} + m_{n-1}) \qquad\qquad (12.3.13)$$

As the value of μ_{n-1} is known (by equation 12.3.12), equation 12.3.13 provides a recurrence relation for the calculation successively of $m_2, m_3, m_4, \ldots$. (This equation may also be solved by the methods of difference equations: the authors wish to thank R. Bruynel, A. Boyd and C. Greeff for pointing this out to them.) The variance of A_n may be obtained as

$$\text{var}[A_n] = E[A_n^2] - (E[A_n])^2$$
$$= m_n - \mu_n^2 \tag{12.3.14}$$

In principle the above arguments are fairly readily extended to provide recurrence relations for the higher moments of A_n. (The calculations are quite trivial for a computer.) However, a knowledge of the first few moments does not define the distribution and it is generally more practicable to use simulation methods to gain further understanding of actual problems. We illustrate such methods in section 12.4.

Example 12.3.1

A company considers that on average it will earn interest on its funds at the rate of 4% p.a. However, the investment policy is such that in any one year the yield on the company's funds is equally likely to take any value between 2% and 6%.

For both single and annual premium capital redemption policies with terms of 5, 10, 15, 20, and 25 years and premium £1, find the mean accumulation and the standard deviation of the accumulation at the maturity date. (Ignore expenses.)

Solution

The annual rate of interest is uniformly distributed on the interval [0·02, 0·06]. The corresponding probability density function is constant and equal to 25 (i.e. $1/(0\cdot06 - 0\cdot02)$). The mean annual rate of interest is clearly

$$j = 0\cdot04$$

and the variance of the annual rate of interest is

$$s^2 = \int_{0\cdot02}^{0\cdot06} 25(x - 0\cdot04)^2 dx = \tfrac{4}{3}10^{-4}$$

We are required to find $E[A_n]$, $(\text{var}[A_n])^{1/2}$, $E[S_n]$, and $(\text{var}[S_n])^{1/2}$ for $n = 5, 10, 15, 20,$ and 25.

Table 12.3.1 *Accumulations for example 12.3.1*

Term (years)	Single premium £1		Annual premium £1	
	Mean accumulation (£)	Standard deviation (£)	Mean accumulation (£)	Standard deviation (£)
5	1·216 65	0·030 21	5·632 98	0·094 43
10	1·480 24	0·051 98	12·486 35	0·283 53
15	1·800 94	0·077 48	20·824 53	0·578 99
20	2·191 12	0·108 86	30·969 20	1·004 76
25	2·665 84	0·148 10	43·311 74	1·593 92

Substituting the above values of j and s^2 in equations 12.3.5 and 12.3.7, we immediately obtain the results for the single premium policies. For the annual premium policies we must use the recurrence relation 12.3.13 (with $\mu_{n-1} = \bar{s}_{\overline{n-1}|}$ at 4%) together with equation 12.3.14.

The results are summarized in table 12.3.1. It should be noted that, for both annual and single premium policies, the standard deviation of the accumulation increases rapidly with the term.

The log-normal distribution

We have already remarked that in general a theoretical analysis of the distribution functions for A_n and S_n is somewhat difficult, even in the relatively simple situation when the yields each year are independent and identically distributed. There is, however, one special case for which an exact analysis of the distribution function for S_n is particularly simple.

Suppose that the random variable $\log(1 + i_t)$ is normally distributed with mean μ and variance σ^2. In this case, the variable $(1 + i_t)$ is said to have a *log-normal* distribution with parameters μ and σ^2. A detailed discussion of this distribution, including economic arguments to support its use, is given in reference [1].

Equation 12.3.2 is equivalent to

$$\log S_n = \sum_{t=1}^{n} \log(1 + i_t)$$

The sum of a set of independent normal random variables is itself a normal random variable. Hence, when the random variables $(1 + i_t)$ $(t \geq 1)$ are independent and each has a log-normal distribution with parameters μ and σ^2, the random variable S_n has a log-normal distribution with parameters $n\mu$ and $n\sigma^2$.

Since the distribution function of a log-normal variable is readily written down in terms of its two parameters, in the particular case when the distribution function for the yield each year is log-normal we have a simple expression for the distribution function of S_n.

12.4 Simulation techniques

All modern computers are able to generate 'random numbers'. (Many pocket calculators also have this facility.) In a strict sense such numbers, when generated by computer, are 'pseudo-random', in that they are produced by some well-defined numerical algorithm, rather than truly 'random', with values which are determined purely by chance. The essential point, however, about the sequence of values produced by an efficient random number generator is that, from a statistical viewpoint, it is indistinguishable from a sequence of values produced by a genuinely random process. For practical purposes, therefore, the pseudo-random values generated by computer may be considered as if they were indeed genuinely random numbers.

There is an extensive literature concerning random number generation and related simulation problems. The interested reader may consult references [26] and [41] for an introduction to the topic.

Example 12.4.1

As a simple illustration of a simulation exercise, consider the model of independent annual rates of return (as described in section 12.3) in relation to a fund for which the yield each year will be 2%, 4%, or 6%, each of these values being equally likely. Suppose that at the start of each year for 15 years £1 is paid into the account. To how much will the account accumulate at the end of 15 years?

Solution

This question can be answered only with hindsight. At the start of the 15-year period all that can be said with certainty is that the accumulated amount will lie between £17·6393 (i.e. $\ddot{s}_{\overline{15}|}$ at 2%) and £24·6725 (i.e. $\ddot{s}_{\overline{15}|}$ at 6%). The lower value will occur only if in every year of the period the annual yield is 2%. Likewise, the higher value will occur only if the yield every year is 6%. Since each year there is a probability of 1/3 that the yield will be 2% and (by assumption) the yields in different years are independent of one another, the probability that the yield will be 2% in *every* year of the period is $(1/3)^{15}$, which is less than 10^{-7}. This shows how unlikely it is that the accumulation will be the lowest possible value.

The yield each year has expected value

$$j = E[i_t] = \tfrac{1}{3}(0{\cdot}02 + 0{\cdot}04 + 0{\cdot}06) = 0{\cdot}04 \tag{12.4.1}$$

It follows from equation 12.3.12 that the accumulated amount of the fund (after 15 years) has mean value

$$\mu_{15} = E[A_{15}] = \ddot{s}_{\overline{15}|} \quad \text{at } 4\% \quad = 20{\cdot}824\,533 \tag{12.4.2}$$

The variance of the annual yield is

$$s^2 = E[(i_t - j)^2] = \tfrac{1}{3}[(-0{\cdot}02)^2 + 0^2 + (0{\cdot}02)^2] = \tfrac{8}{3}10^{-4} \tag{12.4.3}$$

The variance of the accumulated amount may be found from equations 12.3.10, 12.3.13 (with $\mu_{n-1} = \ddot{s}_{\overline{n-1}|}$ at 4%), and 12.3.14. This gives the variance as

$$\text{var}[A_{15}] = 0{\cdot}670\,76 \tag{12.4.4}$$

The standard deviation of the accumulated amount, σ_{15} say, is

$$\sigma_{15} = (\text{var}[A_{15}])^{1/2} = 0{\cdot}819\,00 \tag{12.4.5}$$

How likely is it that the accumulated amount of the fund (after 15 years) will be less than £19, or greater than £22·50? The answers to these and similar questions are of practical importance. Theoretically exact answers are difficult to obtain, but much valuable information may be obtained by *simulation*. This is illustrated below.

Suppose that we generate a sequence of 15 random numbers from the interval

[0, 1], say $\{x_i : i = 1, 2, \ldots, 15\}$. The value of x_t is used to determine i_t, the yield on the fund over year t, by the rule

$$i_t = \begin{cases} 0\cdot02 & \text{if } 0 \leqslant x_t < \frac{1}{3} \\ 0\cdot04 & \text{if } \frac{1}{3} \leqslant x_t < \frac{2}{3} \qquad 1 \leqslant t \leqslant 15 \qquad (12.4.6) \\ 0\cdot06 & \text{if } \frac{2}{3} \leqslant x_t \leqslant 1 \end{cases}$$

Since x_t is randomly chosen from the interval [0, 1], equation 12.4.6 ensures that each possible yield has probability 1/3 of being the actual yield in year t. The accumulated amount A_{15} (calculated by equation 12.3.3), corresponding to the sequence $\{x_i : i = 1, 2, \ldots, 15\}$ is thus a fair realization of the experience of the fund over one particular period of 15 years.

We may repeat this procedure many times. For each simulation we generate a sequence of 15 random numbers from the interval [0, 1] and then calculate the accumulated amount at the end of 15 years, using equations 12.4.6 and 12.3.3. By studying the distribution of the large number of resulting accumulations we gain valuable insights.

As an illustration, 10 000 simulations were carried out for the distribution of A_{15}. The sample mean of the resulting values was 20·822 07 and the sample standard deviation was 0·815 97. (Compare these values with μ_{15} and σ_{15}, as given by equations 12.4.2 and 12.4.5.) The distribution of the values is indicated in table 12.4.1.

Table 12.4.1 *Distribution of* $(A_{15} - \mu_{15})/\sigma_{15}$ *for 10 000 simulations*

Range	< −3	−3 to −2·5	−2·5 to −2·0	−2·0 to −1·5	−1·5 to −1·0	−1·0 to −0·5	−0·5 to −0·0	0·0 to 0·5	0·5 to 1·0	1·0 to 1·5	1·5 to 2·0	2·0 to 2·5	2·5 to 3·0	⩾3
Number of values in range	1	22	165	462	930	1592	1914	1877	1463	880	446	171	59	18

Each of the intervals used to define the above grouping is closed on the left. Thus ' − 3 to − 2 5' includes − 3, but excludes − 2 5.

Thus, for example, of the 10 000 simulations, 1877 produced accumulations which lay between μ_{15} and $\mu_{15} + 0\cdot5\sigma_{15}$; 18 accumulations were greater than or equal to $\mu_{15} + 3\sigma_{15}$; only one accumulation was less than $\mu_{15} - 3\sigma_{15}$; 5086 accumulations were less than μ_{15}. The lowest accumulation was 18·1394; the greatest accumulation was 23·9657; 80 of the accumulations were less than 19; 227 of the accumulations were greater than 22.5. These figures suggest that the probability that any one 15-year period will produce an accumulated amount of less than £19 is approximately 0·008. The accuracy of this estimate is a matter of statistical inference. An approximate 95% confidence interval for the probability is (0·0063, 0·0097), from the formula $\bar{p} \pm 1\cdot96\sqrt{[\bar{p}(1 - \bar{p})/n]}$ with $\bar{p} = 0\cdot008$ and $n = 10\,000$. Similarly, the probability that the accumulated fund will exceed £22·50 is approximately 0·0227, with a corresponding approximate 95% confidence interval (0·0198, 0·0256).

The above discussion relates to one particular example only, but the underlying ideas are readily extended to more general situations. In addition more sophisticated simulation methods may be available: for example, it is possible on many computers to generate random samples drawn from a population with a normal, log-normal, or some other well-known distribution.

At this stage it is perhaps worth considering briefly how, given only the ability to generate random numbers on the interval [0, 1], we may simulate a more general random variable. Suppose that the random variable X takes values on the interval $[a, b]$. (If $a = -\infty$ or $b = +\infty$, the interval is to be regarded as open or half-open, as appropriate.) Suppose that $f(x)$, the probability density function of X, is strictly positive on the interior of the interval. Let F be the distribution function for X. Thus, for $a \leqslant x \leqslant b$,

$$F(x) = \text{probability } \{X \leqslant x\} \tag{12.4.7}$$

Clearly, $F(a) = 0$ and $F(b) = 1$. Since, by assumption, X has a strictly positive density function, F is a strictly increasing function over the interval $[a, b]$. The situation is illustrated in Figure 12.4.1.

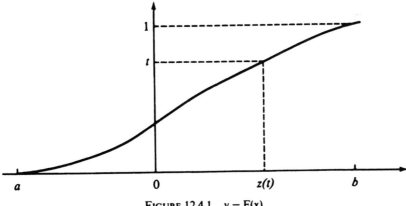

FIGURE 12.4.1 $y = F(x)$

If t is any real number in the interval [0, 1], there is a unique real number in the interval $[a, b]$, $z(t)$ say, such that $F[z(t)] = t$ (see Figure 12.4.1).

Suppose that α is a given value in the interval $[a, b]$ and that t is chosen randomly from the interval [0, 1]. What is the probability that $z(t)$ will not exceed α?

Consideration of Figure 12.4.2 shows that $z(t)$ will not exceed α if and only if $0 \leqslant t \leqslant F(\alpha)$. Since, by hypothesis, t is chosen at random from the interval [0, 1],

$$\text{probability } \{z(t) \leqslant \alpha\} = F(\alpha)$$
$$= \text{probability } \{X \leqslant \alpha\} \tag{12.4.8}$$

This equation shows that $z(t)$ may be regarded as a fair realization of the random variable X. Thus, provided that we are able to 'invert' the distribution function F (i.e. solve the equation $F(x) = t$) by generating random numbers from the interval $[0, 1]$, we may simulate the random variable with distribution function $F(x)$.

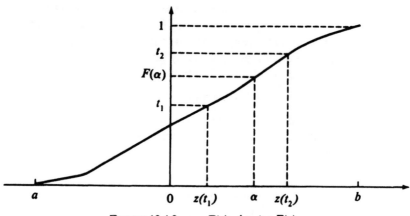

FIGURE 12.4.2 $y = F(x)$, *showing* $F(\alpha)$

Example 12.4.2

Consider a non-negative random variable with exponential probability density function

$$f(x) = 2e^{-2x} \qquad x \geqslant 0$$

Show how to simulate the drawing of a random sample from this distribution.

Solution

Since

$$F(x) = \int_0^x f(r) dr$$

$$= \int_0^x 2e^{-2r} dr$$

$$= 1 - e^{-2x}$$

it follows that, if

$$F[z(t)] = t$$

then

$$1 - e^{-2z(t)} = t$$

Hence

$$z(t) = -\tfrac{1}{2} \log(1 - t)$$

By generating a random number t from the interval $[0, 1]$ and using the last equation to find $z(t)$, we may simulate the sampling procedure very simply.

Example 12.4.3

By adopting a particular investment strategy a company expects that on average the annual yield on its funds will be 0·08 (i.e. 8%). However, the investment policy is one of comparatively high risk and it is anticipated that the standard deviation of the annual yield will be 0·07 (i.e. 7%). The yields in different years may be assumed to be independently distributed.

(a) Find the expected value and the standard deviation of the accumulated amount after 15 years of (i) a single premium of £1000, (ii) a series of 15 annual premiums, each of £1000.

Assuming further that each year $1 + i$ (where i is the annual yield on the company's funds) has a log-normal distribution,

(b) Calculate the probability that a single premium accumulation will be (i) less than 60% of its expected value, (ii) more than 150% of its expected value.
(c) Using simulation methods, estimate the corresponding probabilities for an annual premium accumulation.

Solution

(a) In the notation previously used, we are given that

$$j = E[i] = 0.08$$

and that

$$s = (\text{var}[i])^{1/2} = 0.07$$

(i) The single premium accumulation has expected value

$$1000 E[S_{15}] = 1000(1.08)^{15} = £3172.17$$

and standard deviation

$$1000(\text{var}[S_{15}])^{1/2} = £808.14$$

(These values are given by equations 12.3.5 and 12.3.7 respectively.)
(ii) The annual premium accumulation has expected value

$$1000 E[A_{15}] = 1000\ddot{s}_{\overline{15}|} \quad \text{at } 8\% \quad = £29\,324.28$$

and standard deviation

$$1000(\text{var}[A_{15}])^{1/2} = £5080.29$$

(The last value is obtained using equation 12.3.14.)
The relative magnitudes of these values should be noted. For example, in the case of a single premium accumulation, the standard deviation is some 25% of the expected value. This is a reflection of the wide range of possible yields each year and the corresponding spread of possible accumulations.
(b) We now assume that each year $\log(1 + i_t)$ has a normal distribution. We are given that $E[i_t] = 0.08$ and $\text{var}[i_t] = 0.07^2$. It is easily shown (see exercise 12.7) that this implies that the normal distribution for $\log(1 + i_t)$ has mean

$$\mu = 0.074\,865$$

and standard deviation

$$\sigma = 0{\cdot}064\,747$$

Our remarks in section 12.3 show that $\log(S_{15})$ has a normal distribution with mean $\mu' = 15\mu = 1{\cdot}122\,975$ and standard deviation $\sigma' = \sigma\sqrt{15} = 0{\cdot}250\,764$.

(i) Let Φ denote the distribution function of a standard normal random variable. Note that $0{\cdot}6E[S_{15}] = 1{\cdot}903\,30$. Since $S_{15} < 1{\cdot}903\,30$ if and only if $\log(S_{15}) < 0{\cdot}643\,589$, it follows that

$$\text{probability } (S_{15} < 0{\cdot}6E[S_{15}]) = \Phi\left(\frac{0{\cdot}643\,589 - \mu'}{\sigma'}\right)$$

$$= \Phi(-1{\cdot}911\,70)$$

$$= 0{\cdot}028$$

(ii) Similarly, since $\log(1{\cdot}5E[S_{15}]) = \log(4{\cdot}758\,25) = 1{\cdot}559\,880$, it follows that

$$\text{probability } (S_{15} > 1{\cdot}5E[S_{15}]) = 1 - \Phi\left(\frac{1{\cdot}559\,880 - \mu'}{\sigma'}\right)$$

$$= 1 - \Phi(1{\cdot}742\,29)$$

$$= 0{\cdot}041$$

(c) In order to estimate the corresponding probabilities for the annual premium accumulation we use simulation. Having found the values of μ and σ, we may readily simulate the accumulation (see reference [41] for one possible way of generating normally distributed random variables). The yield in the tth year is determined by generating a value, x_t say, drawn randomly from a normal distribution with mean μ and standard deviation σ, and letting

$$\log(1 + i_t) = x_t$$

or

$$1 + i_t = e^{x_t}$$

The outcome of the simulation of the annual premium accumulation is summarized in table 12.4.2, in which we denote the expected value and standard deviation of the accumulation (as calculated above) by μ_{15} and σ_{15} respectively.

Table 12.4.2 Distribution of $(A_{15} - \mu_{15})/\sigma_{15}$ for 10 000 annual premium accumulations

Range	< −3	−3 to −2·5	−2·5 to −2·0	−2·0 to −1·5	−1·5 to −1·0	−1·0 to −0·5	−0·5 to 0·0	0·0 to 0·5	0·5 to 1·0	1·0 to 1·5	1·5 to 2·0	2·0 to 2·5	2·5 to 3·0	⩾ 3
Number of values in range	0	4	77	401	1058	1839	2000	1797	1223	833	424	203	82	59

The spread of the distribution should be noted. A total of 5379 accumulations were less than the expected value. This is a reflection of the fact that more than one-half of a log-normal distribution lies below its mean. Thus each year the yield was more likely to be less than 8% than to exceed 8%. The smallest accumulation was 16·0604 and the greatest was 54·1707. The smallest accumulation corresponds to an effective yield over the 15-year period of only 0·85% p.a. For the largest accumulation the effective annual yield is 14·89%.

Note that $0·6E[A_{15}] = 17·5946$ and $1·5E[A_{15}] = 43·9864$. Of the 10 000 simulations 15 had $A_{15} < 17·5946$ and 69 had $A_{15} > 43·9864$. Accordingly we estimate the probability that the accumulation will be less than 60% of its expected value to be approximately 0·0015. (A 95% confidence interval for the probability is (0·0007, 0·0023).) Similarly, we estimate the probability that the accumulation will exceed its expected value by more than 50% to be 0·0069 (with a corresponding 95% confidence interval (0·0053, 0·0085)).

12.5 Dependent annual rates of return

So far we have discussed stochastic interest rate models in which the yields in distinct years are independent of one another. In section 12.4, using simulations, we gave illustrations of such models. Another area in which simulation techniques are particularly useful is models for which the yields in distinct years are *not* independent (see reference [57]). In such models the yield in any one year, although containing a 'random' component, depends in some clearly defined manner on the yields in previous years.

There exists a wide range of possible models which provide for some measure of interdependence of the annual yields in different years. A full discussion of such models is beyond the scope of this chapter, but, as the consequences of some form of dependence may be financially very significant, it is worth considering briefly an elementary example.

In the most simple form of dependent stochastic interest rate model, the yield in any one year is determined by (a) the yield in the preceding year, and (b) a random component. For example, the model might anticipate that, if the yield in a particular year is relatively high, then (because the relevant underlying economic and market factors may continue) the yield in the following year is likely to be higher than would otherwise be the case.

For the 'independent log-normal' model discussed in section 12.3 we assume that, for each value of t, $\log(1 + i_t)$ (where i_t is the yield in year t) has a normal distribution with specified mean μ and variance σ^2. One way of retaining a log-normal model while allowing also for some form of interdependence in the yields is provided by assuming that $\log(1 + i_t)$ has a normal distribution with constant variance σ^2 and mean

$$\mu_t = \mu + k[\log(1 + i_{t-1}) - \mu] \qquad (12.5.1)$$

where k is some positive constant.

By varying the value of k, we can allow for different levels of dependence between the yields in successive years. If $k = 0$, the model is simply the independent one. If $k = 1$, then $\mu_t = \log(1 + i_{t-1})$ and the normal distribution for $\log(1 + i_t)$ is centered on $\log(1 + i_{t-1})$. For $0 < k < 1$ the model reflects an intermediate position. The greater the value of k, the greater the influence of the yield in one year on the determination of the yield in the succeeding year.

In the 'independent uniform' model we assume that each year the distribution of the yield is uniform over a specified interval $[\mu - d, \mu + d]$ (see example 12.3.1). We may readily extend this model to allow for dependence in the annual yields by assuming that the distribution for i_t is uniform over the interval $[\mu_t - d, \mu_t + d]$, where

$$\mu_t = \mu + k(i_{t-1} - \mu) \qquad (12.5.2)$$

(In the above, μ, d, and k are specified constants and $0 \leqslant k \leqslant 1$.)

For this model the constant k is particularly simple to interpret. Each year the distribution for the yield is uniform over an interval of fixed length $2d$. The mean of the distribution for i_t is determined by linear interpolation (specified by the value of k) between i_{t-1} and μ. For example, if $k = 0.5$, the distribution for i_t is over an interval centred on a point midway between i_{t-1} and μ. When $k > 0$ the model implies that, if $i_{t-1} > \mu$, then i_t is more likely to exceed μ than to be less than μ and, if $i_{t-1} < \mu$, then i_t is more likely to be less than μ than to exceed μ.

As k increases from 0 to 1 in this 'dependent uniform' model there is greater dependence between the yields in successive years. This is reflected in an increasing spread in the distribution of both annual and single premium accumulations, and is illustrated by the following example.

Example 12.5.1

In order to assess the potential cost of certain minimum guarantees contained in its 'special growth fund' savings policies, a company wishes to consider the likely future accumulation of its investments on the assumption that the distribution of i_t, the yield on the fund in year t, will be uniform over the interval $[\mu_t - 0.03, \mu_t + 0.03]$, where

$$\mu_t = 0.06 + k(i_{t-1} - 0.06) \qquad (1)$$

and k is a specified constant. For the year just ended the yield on special growth fund was 6%.

Estimate by simulation, for $k = 0, 0.2, 0.4, 0.6, 0.8$, and 1, the mean value and the standard deviation of the accumulation in the special growth fund of 15 premiums of £1, payable annually in advance, the first premium being due now. Estimate also the upper and lower 5% and 10% points for the accumulation and find the yields corresponding to these extreme values. Ignore expenses.

Solution

Note that, since in the year just ended the yield on the special growth fund was 6% (i.e. $i_0 = 0.06$), it follows from equation 1 that for the first year of the accumulation period the distribution of the yield will be uniform over the interval [0·03, 0·09]. The second year's yield will depend on i_1 and k. (For example, if $k = 0.5$ and $i_1 = 0.04$, then the distribution of i_2 will be uniform over the interval [0·02, 0·08].)

Note also that, if $k = 0$, then each year the distribution of the yield is uniform over the interval [0·03, 0·09], and we may apply the results of section 12.3. The reader should verify that in this case the accumulation has mean value 24·673 and standard deviation 1·040.

For any given value of k, it is simple to simulate the accumulation by generating a sequence of 15 random numbers from the interval [0, 1], say $\{x_i : i = 1, 2, \ldots, 15\}$, and proceeding as indicated below.

Each year the distribution of the yield is uniform over an interval of length 0·06. The yield in the first year, i_1, is obtained as

$$i_1 = 0.03 + 0.06x_1$$

For $t \geqslant 2$, we determine i_t by (a) calculating μ_t from equation 1 and (b) defining

$$i_t = \mu_t - 0.03 + 0.06x_t$$

In this way the sequence of random numbers $\{x_i\}$ determines the yields for each year of the period. The final accumulation, A_{15}, is calculated from equation 12.3.3. For each of the given values of k we carried out 10 000 simulations. Our results are summarized in table 12.5.1, which gives the mean value and the standard deviation of the resulting accumulations and also the upper and lower 5% and 10% points for the resulting distributions.

Thus, for example when $k = 0.6$ the mean value of the accumulations was 24·774 and the standard deviation was 2·373. A total of 5% of the accumulations were less than 21·077 and 10% were less than 21·809; 10% of the accumulations exceeded 27·908 and 5% exceeded 28·901. Equivalently, we may say that the mean accumulation corresponded to an effective yield over the completed transaction of 6·05% per annum. Some 5% of the accumulations resulted in an effective yield over the completed transaction of less than 4·14% per annum, while 5% of the accumulations produced an overall effective yield of at least 7·83% per annum.

Table 12.5.1 *Dependent uniform model*; $\mu = 0.06$, $d = 0.03$, $i_0 = 0.06$; *term 15 years; annual premium £1; 10 000 simulations for each value of the dependency constant k*

Dependency constant	Mean accumu- lation	Standard deviation of accumu- lation	Lower tail for accumulation		Upper tail for accumulation	
			5%	10%	10%	5%
0·0	24·679	1·037	23·000	23·352	26·036	26·412
0·2	24·673	1·264	22·669	23·078	26·320	26·828
0·4	24·724	1·667	22·040	22·622	26·905	27·579
0·6	24·774	2·373	21·077	21·809	27·908	28·901
0·8	24·990	4·205	18·712	19·929	30·634	32·432
1·0	27·044	11·549	12·587	14·502	42·366	49·237

Table 12.5.2 *Overall effective annual yield per cent corresponding to values given in table 12.5.1*

Dependency constant	Yield for mean accumulation	Yield for lower tail		Yield for upper tail	
		5%	10%	10%	5%
0·0	6·00	5·18	5·36	6·63	6·79
0·2	6·00	5·01	5·22	6·75	6·97
0·4	6·02	4·67	4·98	7·01	7·29
0·6	6·05	4·14	4·55	7·43	7·83
0·8	6·15	2·72	3·47	8·50	9·15
1·0	7·07	− 2·23	− 0·42	12·16	13·83

For each value of k, table 12.5.2 shows the effective annual yield on the completed transaction for the mean accumulation and for each of the 'tail' values given in table 12.5.1. For values of k less than 0·6, the mean values of the accumulation are very similar. However, for larger values of k there is a marked increase in the mean value. In the extreme case, when $k = 1$, the mean accumulation is some 9·6% greater than in the independent case ($k = 0$). Note also that, as k increases, the range of values of the accumulation becomes wider. When $k = 0$, some 90% of the values lie between 23·000 and 26·412, a relative difference of 14·8%. When $k = 0·6$, the corresponding quantiles are 21·077 and 28·901, with a relative difference of 37·1%.

Suppose that the special growth fund policies were such that the policyholder received the accumulated amount of his premiums, with the proviso that there would be a minimum return of 4% per annum effective. The simulation results enable the costs of such guarantees to be assessed. (A separate 'guarantee fund' must be provided to meet these liabilities.) The simulation results show that, when $k = 0·6$, the expected amount to be paid from the guarantee fund is of the order of 1·5p per £1 annual premium. This figure is relatively small, since the probability that a policy will require supplementation from the guarantee fund is approximately 0·01. Of the 10 000 simulations, 114 produced accumulations less than $\bar{s}_{\overline{13}|}$ at 4%. For these 114 policies the average 'shortfall' in the accumulation was £1·36 per £1 annual premium.

12.6 An application of Brownian motion

One of the earliest applications of Brownian motion was to financial problems. In this section we indicate briefly how the theory of one-dimensional random walks with small steps may be used to help to describe the progress of a single premium investment. As in the case of some other topics, the results of the theory may not always be directly applicable to practical problems, but they give a starting point for further discussion.

Consider a random walk beginning at the origin and moving with independent steps of length δ at the rate of r steps per annum. At each of the times $0, 1/r, 2/r, \ldots$, (measured in years), the probability of an upward movement is p and that of a downward movement is $1 - p$.

Consider the distribution of the position $X(t)$ of the random walk at time t when δ becomes small in such a way that $(2p - 1)\delta r$ approaches a constant μ and $\delta^2 r$ approaches a constant σ^2. (Note that these conditions imply that p tends to $1/2$ and r tends to infinity.) $X(t)$ may then be considered to describe a *Brownian motion* with drift μ and variance σ^2. The distribution of $X(t)$ may be shown to be approximately normal with mean μt and variance $\sigma^2 t$. Further, the probability that, at *some* time in the first n years, $X(t)$ falls below the value $- \xi$ (where $\xi > 0$) may be shown to be approximately equal to

$$P_n = \frac{\xi}{\sigma\sqrt{2\pi}} \int_0^n \exp\left[-\frac{(\mu t + \xi)^2}{2\sigma^2 t} \right] t^{-3/2} dt$$

$$= \Phi\left(\frac{-\xi - \mu n}{\sigma\sqrt{n}} \right) + \exp\left(\frac{-2\mu\xi}{\sigma^2} \right) \Phi\left(\frac{-\xi + \mu n}{\sigma\sqrt{n}} \right) \qquad (12.6.1)$$

where $\Phi(x)$ denotes the standard normal distribution function. Hence the probability that $X(t)$ will *never* fall below $- \xi$ is

$$1 - P_\infty = \begin{cases} 0 & \text{if } \mu \leqslant 0 \\ 1 - \exp\left(\frac{-2\xi\mu}{\sigma^2} \right) & \text{if } \mu > 0 \end{cases} \qquad (12.6.2)$$

An application of these theoretical results to financial questions is as follows. Let $X(t)$ denote the logarithm of the proceeds at time t years of a single premium investment of 1 at time 0. Under idealized conditions $X(t)$ may be considered to follow a Brownian motion with drift μ and variance σ^2, where μ and σ^2 are the mean and variance respectively of the logarithm of the growth rate in any given year. In addition, the logarithms of the growth rates in years $1, 2, \ldots, n$ respectively are independent, identical, normally distributed variables, so that we obtain the mathematical model described in section 12.3. However, equation 12.6.1 allows us to compute the probability that, at some time during the first n years, the proceeds of the investment will fall below a specified level. This may be important in that guarantees may be given by the life office issuing the policy that the proceeds of the investment will not fall below a certain figure. Also, the

Brownian motion model may be used to help to determine the risk that certain share prices (which may sometimes be regarded as describing a Brownian motion) will, within a given time, fall below a fixed value (at which, perhaps, certain options may be exercised.)

Example 12.6.1

A single premium investment of £10 000 may be supposed to be such that the logarithm of the proceeds at time t per £1 initial investment follows a Brownian motion with $\mu = 0.045$ and $\sigma = 0.10$. Find the probability that, at some point in the next 15 years, the proceeds of this investment will be less than £8000.

Solution

Let

$$X(t) = \log \left[(\text{proceeds of investments at time } t)/10\,000 \right]$$

Note that the proceeds of the investment at time t are less than £8000 if and only if $X(t) < \log(0.8) = -0.223\,14$. We are required to find the probability that at some point in the next 15 years $X(t) < -0.223\,14$.

By equation 12.6.1, this probability is

$$P_{15} = \Phi \left[\frac{-0.223\,14 - (0.045 \times 15)}{0.1\sqrt{15}} \right] + 0.134\,22 \Phi \left[\frac{-0.22314 + (0.045 \times 15)}{0.1\sqrt{15}} \right]$$

$$= \Phi(-2.319) + 0.134\,22 \Phi(1.167)$$

$$= 0.0102 + 0.1179 = 0.1281$$

Note that $1 - P_\infty$, the probability that the proceeds of this investment will *never* fall below £8000 is, by equation 12.6.2,

$$1 - \exp \left[-\frac{2 \times 0.223\,14 \times 0.045}{(0.1)^2} \right] = 0.8658$$

A possible path taken by $X(t)$ is illustrated in Figure 12.6.1.

Similar calculations may be applied when the 'barrier', or minimum level of proceeds, is of the form $A(1+j)^t$ per unit invested, where j is a fixed escalation rate. In this case one may work in terms of 'escalating' units of currency, the rate of escalation being j per annum. In terms of these units of currency, $X(t)$ has drift $\mu - \log(1+j)$ and variance σ^2, and the probability that the proceeds fall below the 'barrier' level may be calculated as before. The technique is illustrated in the next example.

Example 12.6.2

Suppose that a single premium investment of £1 is such that the logarithm of the proceeds at time t describes a Brownian motion with $\mu = 0.07$ and $\sigma = 0.1395$. Find the probability that at some point within the next ten years the proceeds will fall below $(0.75)(1.04)^t$, where t is the time in years from the present.

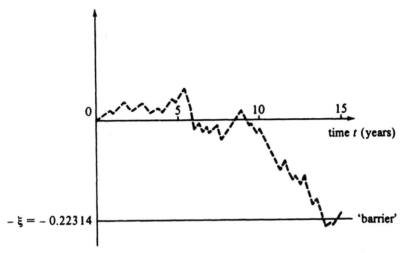

FIGURE 12.6.1 *Possible path of* X(t) *showing Brownian motion*

Solution

Work in monetary units which escalate at 4% per annum compound. Then $X(t)$, the logarithm of the proceeds at time t, describes a Brownian motion with drift $0{\cdot}07 - \log(1{\cdot}04) = 0{\cdot}030\,78$ and variance $(0{\cdot}1395)^2$. We require to find the probability that $X(t)$ falls below $\log(0{\cdot}75) = -0{\cdot}287\,68$ at some point in the next ten years. By equation 12.6.1, this is

$$P_{10} = \Phi\left(\frac{-0{\cdot}287\,68 - 0{\cdot}3078}{0{\cdot}1395\sqrt{10}}\right)$$

$$+ \exp\left[\frac{-2 \times 0{\cdot}030\,78 \times 0{\cdot}287\,68}{(0{\cdot}1395)^2}\right]\Phi\left(\frac{-0{\cdot}287\,68 + 0{\cdot}3078}{0{\cdot}1395\sqrt{10}}\right)$$

$$= \Phi(-1{\cdot}350) + 0{\cdot}4025\,\Phi(0{\cdot}0456) = 0{\cdot}2971$$

Exercises

Statistical results

The following elementary statistical definitions and results are relevant to certain of the problems below:

(a) The random variable X is said to have a *uniform distribution* over the interval $[a, b]$ if its probability density function is

$$f(x) = \begin{cases} \dfrac{1}{b - a} & \text{for } a \leqslant x \leqslant b \\ 0 & \text{otherwise} \end{cases}$$

(b) The random variable X is said to have a *triangular distribution* over the interval $[a, b]$ if its probability density function is

$$f(x) = \begin{cases} \dfrac{4(x - a)}{(b - a)^2} & \text{for } a \leqslant x < \dfrac{a + b}{2} \\[2ex] \dfrac{4(b - x)}{(b - a)^2} & \text{for } \dfrac{a + b}{2} \leqslant x \leqslant b \\[2ex] 0 & \text{otherwise} \end{cases}$$

(c) If the non-negative random variable X is such that $\log X$ has a normal distribution with mean μ and variance σ^2, then X is said to have a *log-normal distribution* with parameters μ and σ^2. For each positive integer n, we have

$$E[X^n] = \exp(n\mu + \tfrac{1}{2}n^2\sigma^2)$$

(d) If the random variable X has probability density function $f(x)$, and $g(X)$ is a function of X, the expected value of $g(X)$ may be calculated as

$$E[g(X)] = \int_{-\infty}^{\infty} g(x)f(x)\,\mathrm{d}x$$

The distribution function of $X, F(x)$, is defined as

$$F(x) = \int_{-\infty}^{x} f(t)\,\mathrm{d}t$$

12.1 (Statistical results required in subsequent problems)

(a) Show that if X has a uniform distribution over the interval $[a, b]$ then

$$E[X] = \frac{a + b}{2}$$

and, for each positive integer n,

$$E[(X - E[X])^n] = \begin{cases} \dfrac{(b - a)^n}{2^n(n + 1)} & \text{if } n \text{ is even} \\[2ex] 0 & \text{if } n \text{ is odd} \end{cases}$$

(b) Let X have a triangular distribution over the interval $[a, b]$.

(i) Show that

$$E[X] = \frac{a + b}{2}$$

and, for each positive integer n,

$$E[(X - E[X])^n] = \begin{cases} \dfrac{(b - a)^n}{2^{n-1}(n + 1)(n + 2)} & \text{if } n \text{ is even} \\[2ex] 0 & \text{if } n \text{ is odd} \end{cases}$$

(ii) Show further that if $F(x)$ is the distribution function for X and

$$F(x) = r \qquad 0 < r < 1$$

then

$$
x = \begin{cases} a + (b-a)\sqrt{\dfrac{r}{2}} & \text{if } r \leqslant \tfrac{1}{2} \\[3mm] b - (b-a)\sqrt{\left(\dfrac{1-r}{2}\right)} & \text{if } r > \tfrac{1}{2} \end{cases}
$$

12.2 A capital redemption policy has just been issued with premiums payable annually in advance throughout the duration of the policy. The term of the policy is 20 years and the sum assured is £10 000.

The premiums will be invested in a fund which earns interest at a constant effective annual rate *throughout the duration of the policy*. This rate will be 3%, 6% or 9%, each of these values being equally likely. There are no expenses.

(a) Find the expected value of the annual rate of interest. What would the annual premium be if it were calculated on the basis of this mean rate of interest?

(b) Let P be the annual premium.

   (i) Find, in terms of P, the expected value of the accumulated profit on the policy at the maturity date. For what value of P is this expected value zero? What is the value of this expected value if P is as calculated in (a)?

   (ii) Find, in terms of P, the expected value of the net present value of the policy immediately after it is effected. For what value of P is this expected value zero? What is the value of this expected value if P is as calculated in (a)?

12.3 The yields on a company's funds in different years are independently and identically distributed. Consider a single premium capital redemption policy with a term of ten years and sum assured £1000. Let P be the single premium for the policy and let i be the random variable denoting the yield in any given year. Expenses may be ignored.

(a) (i) Show that the expected value of the accumulated profit on the policy at the maturity date is

$$P(1 + E[i])^{10} - 1000 \tag{1}$$

and find an expression (in terms of P) for the standard deviation of the accumulated profit.

   (ii) Show that the expected value of the net present value of the policy immediately after it is effected is

$$P - 1000\left(E\left[\frac{1}{1+i}\right]\right)^{10} \tag{2}$$

Show further that the standard deviation of the net present value is

$$1000\left\{\left(E\left[\frac{1}{(1+i)^2}\right]\right)^{10} - \left(E\left[\frac{1}{1+i}\right]\right)^{20}\right\}^{1/2}$$

(b) For each of the three models for the distribution of i described below, find the value of P for which the expected value of the accumulated profit at the maturity date is zero and, using this value of P, calculate the standard deviation of the accumulated profit. For each model, find also the value of P for which the expected value of the net present value of the policy immediately after issue is zero, and calculate the standard deviation of the net present value.

Model I i takes each of the values 0·02, 0·04, and 0·06 with equal probability.

Model II i has a uniform distribution over the interval [0·02, 0·06].

Model III i has a triangular distribution over the interval [0·02, 0·06].

Compare your answers for the different models.

12.4   The yield i on a company's funds in any year is equally likely to be 3%, 6% or 9%. Yields in different years are independent.

(a) Show that $E[i] = 0·06$ and that $\text{var}[i] = 0·0006$.

(b) Hence, or otherwise, derive expressions for $E[S_n]$ and $\text{var}[S_n]$, the expected value and the variance of the accumulation of a single premium of £1 over a period of n years. For $n = 5, 10, 15,$ and 20 find the mean and standard deviation of S_n.

(c) Let A_n be the random variable representing the accumulation over n years of n annual premiums of £1. *Write down* an expression for $E[A_n]$. Using equation 12.3.13, or otherwise, find, for $n = 1, 2, \ldots, 15$, the standard deviation of A_n.

(d) For $n = 5, 10,$ and 15 find the value of (standard deviation $[A_n])/E[A_n]$ and comment on your answers.

12.5   Each year the yield on a company's funds is either 2%, 4% or 7% with corresponding probabilities 0·3, 0·5, and 0·2 respectively. Yields in different years are independent.

(a) Find the mean value and standard deviation of the accumulation of a single premium of £1000 over a period of 15 years.

Using simulation, estimate the probability that the accumulation will be (i) less than £1600, (ii) more than £2000. Give 95% confidence intervals for each probability.

(b) Find the mean value and standard deviation of the accumulation of 15 annual premiums, each of £100.

By simulation, or otherwise, estimate the probability that this accumulation will be (i) less than £1900, (ii) more than £2300. Give 95% confidence intervals for each probability.

12.6   (Comparison between uniform and triangular yield distribution)

Assume that, in any year, the yield per annum on an insurance company's funds is firstly *uniform* over the interval [0·03, 0·09], and secondly *triangular* over the interval [0·03, 0·09].

(a) For each distribution, find the mean and standard deviation of the yield each year. Again for each distribution, what is the probability that the yield in any given year will be between 5% and 7%?

(b) For both annual and single premium capital redemption policies, with terms 5, 10, and 15 years and premium £1, find the mean value and the standard deviation of the accumulated proceeds at the maturity date. Assume that yields in different years are independent.

Compare the results for the two distributions.

12.7 (Log-normal distribution)
The random variable $(1 + i)$ has a log-normal distribution with parameters μ and σ^2.

(a) Assuming standard results for the log-normal distribution (see the note at the beginning of these exercises), and using the facts that

$$E[1 + i] = 1 + E[i]$$

and

$$\text{var}[i] = \text{var}[1 + i] = E[(1 + i)^2] - (E[1 + i])^2$$

show that

$$E[i] = \exp(\mu + \tfrac{1}{2}\sigma^2) - 1$$

and

$$\text{var}[i] = \exp(2\mu + \sigma^2)[\exp(\sigma^2) - 1]$$

(b) Let $E[i] = j$ and $\text{var}[i] = s^2$. Show that

$$\sigma^2 = \log\left[1 + \left(\frac{s}{1+j}\right)^2\right]$$

and that

$$\mu = \log \frac{1+j}{\sqrt{\left[1 + \left(\frac{s}{1+j}\right)^2\right]}}$$

(c) (i) Given that $\mu = 0.08$ and $\sigma = 0.07$, find j and s.
    (ii) Given that $j = 0.08$ and $s = 0.07$, find μ and σ.

12.8 The yields on a company's fund in different years are independently and identically distributed. Each year the distribution of $(1 + i)$ is log-normal and i has mean value 0.06 and variance 0.0003.

(a) Find the parameters μ and σ^2 of the log-normal distribution for $(1 + i)$.
(b) Let S_{15} be the random variable denoting the accumulation of a single premium of £1 for a period of 15 years. Show that S_{15} has a log-normal distribution with parameters $0.872\,025$ and $0.063\,28^2$. Hence, using standard properties of the log-normal distribution, show that

$$E[S_{15}] = 2.3965$$

and

$$E[(S_{15})^2] = 5.7665$$

Deduce that

$$\text{var}[S_{15}] = 0.023$$

(c) Confirm the values of $E[S_{15}]$ and $\text{var}[S_{15}]$ by equations 12.3.5 and 12.3.7.

(d) Find the probability that a single premium of £1000 will accumulate over 15 years to (i) less than £2100, (ii) more than £2700. (Use tables of the normal distribution function.)

12.9 The yields on a company's fund in different years are independently and identically distributed. Each year the distribution of $(1 + i)$ is log-normal with parameters μ and σ^2.

Let V_n be the random variable denoting the present value of £1 due at the end of n years.

(a) Show that V_n has a log-normal distribution and find the parameters of the distribution.

(b) Assuming further that each year the yield has mean value 0·08 and standard deviation 0·05, find, for $n = 5, 10, 15$, and 20, the expected value and the standard deviation of the present value of £1000 due at the end of n years.

12.10 (A simulation exercise)

Let

$$X = \frac{X_1 + X_2 + X_3}{3}$$

where X_1, X_2, and X_3 are independent random variables, each having a uniform distribution over the interval $[0, 1]$.

(a) Find the mean and standard deviation of X.

(b) By repeatedly generating three random numbers on the unit interval, recording the average of these three numbers, and considering the resulting set of average values, estimate the probability that

  (i) $X \leqslant 1/4$;
  (ii) $X \geqslant 0\cdot8$;
  (iii) $1/3 \leqslant X \leqslant 2/3$;
  (iv) $0\cdot3 \leqslant X \leqslant 0\cdot4$.

(c) Again by simulation, estimate the value of t for which
  (i) The probability that $X \leqslant t$ is 0·05;
  (ii) The probability that $X \geqslant t$ is 0·01;
  (iii) The probability that $X \leqslant t$ is 0·25.

Note It is in fact quite simple to give exact answers to the above questions (see for example, reference [24]). (The answers are included in the solutions for this exercise.) The purpose of the exercise is to give the reader an elementary introduction to simulation. The reader should carry out the simulation exercise twice, firstly with a relatively small number of simulations (e.g. 100), and secondly with a considerably greater number. Note the difference between the estimated values and the true values in each case.

12.11 (Dependent annual rates of return)

In the year just ended (year 0) i_0, the yield on a company's funds, was 6%. For each integer t, i_t, the yield on the funds in year t, is equally likely to be

$$0.02 + k(i_{t-1} - 0.06), \quad 0.06 + k(i_{t-1} - 0.06), \quad \text{or} \quad 0.10 + k(i_{t-1} - 0.06)$$

where k is a known constant and i_{t-1} is the yield in the immediately preceding year.

(a) Find the expected values of i_1 and i_2.
(b) Consider a single premium of £1 invested now. Let S denote the accumulated amount of this investment after two years. Show that

(i) $$E[S] = 1.06^2 + \frac{0.0032}{3}k$$

(ii) $$\text{var}[S] = \frac{10^{-8}}{9}(2\,158\,336 + 2\,157\,312k + 1\,079\,168k^2)$$

12.12 (Dependent annual yields: 'uniform' model)
For each integer t, i_t, the yield on a company's funds in year t, has a uniform distribution over the interval

$$[0.05 + \tfrac{1}{2}(i_{t-1} - 0.05) - 0.03, \quad 0.05 + \tfrac{1}{2}(i_{t-1} - 0.05) + 0.03]$$

i.e. the interval

$$[\tfrac{1}{2}i_{t-1} - 0.005, \quad \tfrac{1}{2}i_{t-1} + 0.055]$$

In the year just ended, the yield on the company's funds was 5%.

(a) Using simulation, estimate the expected value and the standard deviation of the accumulated amount after 15 years of a single premium of £1000 invested now.

A single premium is invested now for a period of 15 years. At the end of this period the investor will receive the accumulated proceeds. Estimate the probability that over the completed transaction the effective annual yield will be (i) less than $3\frac{1}{2}\%$, (ii) greater than $6\frac{1}{2}\%$. Give a 95% confidence interval for the probability in each case.

Estimate also the effective annual yields corresponding to the mean accumulation and to both lower and upper 1% and 10% 'tails' of the distribution for the accumulation.

(b) Using simulation, estimate the expected value and the standard deviation of the accumulated amount after 15 years of 15 annual premiums of £100, the first premium being due now.

An investor will pay 15 level annual premiums for investment in the company's funds, the first premium being paid now. At the end of 15 years he will receive the accumulated proceeds. Estimate the probability that his effective annual yield over the completed transaction will be (i) less than $3\frac{1}{4}\%$, (ii) greater than $6\frac{1}{2}\%$. Give a 95% confidence interval for the probability in each case.

Estimate also the effective annual yields corresponding to the mean accumulation and to both lower and upper 1% and 10% 'tails' of the distribution of the accumulation.

12.13 (Dependent annual yields: 'triangular model')

Answer the previous question on the different assumption that i_t has a *triangular* distribution over the interval

$$[\tfrac{1}{2}i_{t-1} - 0\cdot005, \qquad \tfrac{1}{2}i_{t-1} + 0\cdot055]$$

Compare your new answers with those previously obtained. (The differences arise because of the difference between the uniform and triangular distribution each year.)

(See exercise 12.1 for one possible way of simulating a triangular distribution.)

12.14 (Comparison between dependent and independent annual yields: log-normal model)

For each integer t, i_t, the yield on a company's funds in year t, is such that $\log(1 + i_t)$ is normally distributed with standard deviation $0\cdot03$ and mean

$$0\cdot05 + k[\log(1 + i_{t-1}) - 0\cdot05]$$

where k is a known constant.

Let S be the random variable denoting the accumulation of a single premium of £1 over a 15-year period from the present time.

(a) Assume that $k = 0$. Find the mean and standard deviation of S. Find the probability that S will be (i) less than $1\cdot47$, (ii) greater than $3\cdot1$.
   Find the lower and upper 10% quantiles of the distribution for S. (Use tables of the normal distribution function.)

(b) Assume that $k = 0\cdot5$ and that i_0, the yield in the year just ended, is such that $\log(1 + i_0) = 0\cdot05$. Using simulation, estimate the mean and standard deviation of S. Estimate the probability that S will be (i) less than $1\cdot47$, (ii) greater than $3\cdot1$. Give 95% confidence intervals for each probability.
   Estimate the lower and upper 10% quantiles of the distribution for S.

Compare your answers for (a) and (b).

12.15 (Brownian motion)

An investor's portfolio of ordinary shares, currently worth £100 000, may be supposed to be such that

$$X(t) = \log(\text{market value at time } t \text{ of portfolio}/100\,000)$$

describes a Brownian motion with $\mu = 0\cdot05$ and $\sigma = 0\cdot15$. Calculate the probability that, at some point within the next ten years, the market value of his holding will fall below £90 000. Find also the probability that the market value will never fall below £90 000.

MISCELLANEOUS PROBLEMS

M.1 A certain security has just been issued bearing half-yearly interest at the rate of 6% per annum and redeemable in n years' time at $C\%$. The purchase price that would give a tax-free investor a yield of 4% per annum is 132·55% and the purchase price that would give an investor subject to income tax at 30% a net yield of 4% per annum is 107·85%. Find n and C. (n is an integer.)

M.2 The force of interest per annum at time t years, $\delta(t)$, will be a linear function of t for m years, and will thereafter be constant at the level then reached. Find an expression for the amount of 1 at the end of n years ($n > m$) and evaluate it when $m = 16$, $n = 39$, $\delta(0) = \log_e(1\cdot04)$ and $\delta(m) = \log_e(1\cdot03)$.

M.3 Holders of a certain loan stock are offered the following options:

(a) To take repayment of their loans now at par.
(b) To extend the loan for a further 40 years with interest, payable yearly in arrear, at
　　5% for the first five years
　　$4\frac{3}{4}\%$ for the next five years
　　$4\frac{1}{2}\%$ for the next ten years
　　4% thereafter.
The capital repayment will be at the rate of £105% at the end of the 40th year.
(c) To extend the loan for a further 40 years with interest, payable yearly in arrear, at
　　$4\frac{3}{4}\%$ for the first 20 years
　　$4\frac{1}{4}\%$ thereafter.
The capital repayment will be at the rate of £98% at the end of the 40th year.

If a holder (a financial institution) can invest now in other bonds to obtain interest during the next 40 years at $4\frac{1}{2}\%$ per annum, which of the above options should it select? Ignore taxation.

M.4 Mr Thomson moved into a rented flat on 1 January 1981. He planned to live there for two years. In order to pay his rent at the end of each month he put a sum of money, £X, into his bank deposit account on 1 January 1981 and instructed his bank manager to withdraw from this account each payment of rent when it was due. The rent was £360 per month and Mr Thomson calculated that, if his bank account earned interest at 6% p.a. effective, he would have just sufficient money in his account to pay for his rent for the next two years.

(a) Find X.

Throughout 1981 his rent remained constant at £360 per month and his bank account earned interest at 6% p.a. effective. On 1 January 1982 his bank manager informed him that throughout 1982 his bank account would earn interest at 4% p.a. convertible quarterly. On 1 April 1982 his landlord informed him that all future rent payments would be increased to £460 per month, starting with the payment due at the end of April.

(b) Suppose Mr Thomson does not put any extra money into his account. By calculating the amount in the account at the start of October 1982, or otherwise, show that the rent payment due at the end of October 1982 is the first payment for which there would not be sufficient money in his account.

(c) How much extra money should he have put into his bank account on 1 April 1982 in order that there should be just sufficient money to cover all his remaining payments of rent?

M.5   Ten thousand bonds each of £100 nominal bear interest at $4\frac{1}{2}\%$ per annum payable half-yearly, and are to be repaid according to the following schedule:

At the end of year	Number of bonds redeemed	Redemption price per bond
10	1000	80
13	1500	80
15	2000	100
18	2500	100
20	2000	120
23	1000	120

Find the net yield per annum obtained by a purchaser of the entire issue at a price of 85% if he is subject to income tax at 35% and to capital gains tax at 30%. (Capital losses may not be offset against capital gains for tax purposes.)

M.6   A certain building society pays interest at rate j per annum on the sum originally invested and at a higher rate k on accumulated interest.

(a) Show that the accumulation of 1 for n years with this society is $1 + js_{\overline{n}|k}$.

(b) Show mathematically that, as $n \to \infty$, the yield per annum obtainable from an investment for n years with this society tends to k.

M.7   Every year for well over 20 years an insurance company has issued 500 20-year capital redemption assurances and 500 ten-year capital redemption assurances each by annual premiums for a sum assured of £1000. The policies are issued uniformly over the year and the premiums are calculated at a rate of interest of 5% per annum net with an expense loading of 6% of each gross premium.

The ratio of the gross interest received by the company over a recent year to the average fund in that year was 8%, the tax suffered was at the rate of 20% of the gross interest income, and the expenses were 1·5 times those allowed for in the premium basis. If the fund at the beginning of the year was equal to the reserves, which were calculated by accumulating the premiums, less expenses allowed for in the premium basis, at 5% per annum net effective, find the fund at the end of the year. (Ignore the possibility of capital gains or losses.)

M.8 A loan of £20 000 nominal bears interest at $5\frac{1}{2}$% per annum payable quarterly in arrear and is redeemable in seven years' time at £110 per £100 nominal. The terms of the loan are to be altered so that repayment will be by a single sum of £29 700 at a certain future time. Calculate this time

(a) Using the approximate rule for the equated time (see example 3.2.3);
(b) Exactly, assuming a rate of interest of 6% per annum effective.

(Ignore taxation.)

M.9 X had a credit account with his tailor from 1 January 1978 to 1 January 1981. He paid a constant annual amount into the account on each 1 January. Purchases could be charged to the account only on each 1 January but might leave the account either in surplus or in deficit.

On 1 January 1981, after making his fourth deposit, X wished to close the account. He therefore made sufficient purchases to balance the account exactly. He later noticed that there were three distinct annual rates of interest i_1, i_2, and i_3 at which the value of his purchases was the same as the value of his deposits.

Derive an expression for the difference between the sum of his deposits and the sum of the costs of his purchases in terms of i_1, i_2, i_3 and the balance in the account at 1 January 1978 after the purchases and the deposit at that date had been made.

If X's annual payment was £327, the purchases at 1 January 1978 cost £427, the sum of the costs of all his purchases was £1307·94, i_1 was 10%, and i_2 was 12%, find i_3.

M.10 A certain irredeemable stock pays interest at $5\frac{1}{4}$% per annum, payable quarterly on 31 March, 30 June, 30 September, and 31 December. On 31 August 1982 this stock was quoted at a price of £49·50 per £100 nominal. Find (to the nearest 0·1%) the gross nominal yield per annum, convertible half-yearly.

An investor bought £20 000 nominal of this stock on 31 August 1982 at the price quoted above. Exactly one year later he sold his holding at a price such that the purchaser's gross nominal yield per annum, convertible half-yearly, was 10%. Find the net annual yield (to the nearest 0·25%) obtained by the investor on the entire transaction, given that he pays tax on income at 40% and on capital gains at 30%.

M.11 Y buys a house on 1 July 1982. He has been offered a £20 000 loan by a building society on either of the following two systems of repayment:

(a) Repayment by a level annuity payable half-yearly for 20 years, the first instalment falling due on 1 January 1983. The building society calculates the annuity payment using an interest rate of 12% per annum convertible half-yearly and without allowance for expenses; or
(b) Repayment of capital on 1 July 2002 by means of a capital redemption policy with premiums payable half-yearly in advance throughout the term of the policy. The premiums are calculated assuming an effective rate

of interest of 10% per annum and allowing for expenses of 10% of the first gross premium and 5% of each subsequent gross premium. Interest will be charged on the loan annually in arrear at the rate of $12\frac{1}{2}$% per annum.

Y will receive relief from income tax, which he will otherwise pay at the rate of 30%, on the interest content of his payments on either of these systems. Under system (b), Y will receive additional tax relief in the form of a 15% reduction on each annual premium.

As an alternative to a building society loan Y is prepared to pay the £20 000 in cash by withdrawing this sum from a bank account which will earn interest at 4% per half-year net effective over the coming 20 years.

Should Y pay cash, accept a building society loan under conditions (a), or a building society loan under conditions (b)?

M.12 A company owns a warehouse and a supply of barrels suitable for storing whisky. It will cost £10 000 to buy newly distilled whisky to stock the warehouse and a further £5000 to cover the cost of transport. After t years the company will be able to sell the whisky for a net price of £R_t, where

$$R_t = 10\,000(1 + t)^{1/2}$$

(a) For what period should the company hold a stock of whisky in order to maximize the present value of its profit? Assume a constant force of interest of 4·879% per annum and ignore tax, the costs of maintaining the warehouse, and depreciation.

(b) The company then realizes that when it sells the first consignment of whisky it will best employ the warehouse and barrels by restocking them. It therefore considers an infinite series of purchases of new whisky and sales of mature stock. On the same assumptions as for (a) and also assuming that the price of whisky and transport remains constant, for what period should the company hold each consignment of whisky in order to maximize the present value of its infinite stream of profits?

If your answer differs from that for (a) explain briefly what causes the difference.

M.13 A loan of £100 000 was issued on 1 January 1975 subject to repayment by an immediate annuity-certain payable yearly in arrear for 25 years and calculated at a rate of interest of 6% per annum effective. Income tax is payable on the interest content of each instalment.

What price should an investor, liable to income tax at 30%, offer on 1 January 1985 for the whole annuity if he wishes to realize 5% per annum net on his whole investment after replacing his capital by accumulating a fixed proportion of each annuity instalment in a bank account providing interest at 4% per annum net effective?

M.14 A capital redemption policy for £100 000 with term 50 years was issued by a financial institution on 1 January 1985 by annual premiums calculated at an interest rate of 6% per annum and ignoring expenses. The following table shows the rates of taxation which the institution expects to pay on interest income:

Period	Expected rate of tax
first 19 years	37·5%
thereafter	30%

The institution expects to be able to invest all the premiums to earn 9% gross interest per annum payable annually in arrear. Calculate the estimated profit or loss to the financial institution on the maturity of the policy (Ignore expenses.)

M.15 On January 1976 a depositor opened two savings accounts. On each 1 January from 1976 to 1981 inclusive he paid £100 into each account. He made no withdrawals. The terms of the two accounts were as follows:

Account 1   Interest was credited half-yearly in arrear on 30 June and 31 December. The rate of interest convertible half-yearly was 8% per annum for the first two years, 10% per annum for the next two years, and 12% per annum for the last two years.

Account 2   Each deposit and any interest falling due at the same date was invested at a rate of interest which was fixed for the period to 31 December 1981. Interest was credited annually in arrear. The rate of interest was 9% per annum for sums invested in 1976 and 1977, 11% per annum for sums invested in 1978 and 1979, and 13% per annum for sums invested in 1980 and 1981.

Which account gave the higher accumulation at 31 December 1981?

M.16 A loan of £500 000 nominal is issued in £100 nominal bonds bearing interest at 5% per annum payable quarterly and redeemable at 103% by drawings as follows:

    50 bonds at the end of each of the first ten years
    100 bonds at the end of each of the next ten years
    150 bonds at the end of each of the next ten years
    200 bonds at the end of each of the next ten years

At the price of issue the effective annual yield, after allowing for income tax at 25%, is $4\frac{1}{2}$% if the entire loan is purchased. Calculate the probability that the net yield on any given bond will be less than $4\frac{1}{2}$% per annum.

APPENDIXES

Appendix 1: Proof of theorem 2.4.1

Theorem

If $\delta(t)$ and $A(t_0, t)$ are continuous functions of t for $t \geqslant t_0$, and the principle of consistency holds, then, for $t_0 \leqslant t_1 \leqslant t_2$,

$$A(t_1, t_2) = \exp\left[\int_{t_1}^{t_2} \delta(t)dt \right]$$

Proof

Suppose that t_1 and t_2 are given with $t_0 \leqslant t_1 \leqslant t_2$. For $t \geqslant t_0$, let $f(t) = A(t_0, t)$. Note that f is continuous. For $t \geqslant t_0$ we have

$$\delta(t) = \lim_{h \to 0+} i_h(t)$$

$$= \lim_{h \to 0+} \frac{A(t, t + h) - 1}{h} \qquad \text{(by definition)}$$

$$= \lim_{h \to 0+} \frac{A(t_0, t)A(t, t + h) - A(t_0, t)}{h A(t_0, t)}$$

$$= \frac{1}{A(t_0, t)} \lim_{h \to 0+} \frac{A(t_0, t + h) - A(t_0, t)}{h} \qquad \text{(by consistency principle)}$$

$$= \frac{1}{f(t)} \lim_{h \to 0+} \frac{f(t + h) - f(t)}{h}$$

$$= \frac{1}{f(t)} [f'^+(t)]$$

where f'^+ denotes the right-sided derivative of f.

The last equation may be written as

$$f'^+(t) = f(t)\delta(t)$$

303

Since f and δ are both continuous functions, so too is f'^{+}. A continuous function which has a continuous right-sided derivative is in fact differentiable, so that we obtain

$$f'(t) = f(t)\delta(t).$$

By using the integrating factor $\exp\{-\int_{t_0}^{t}\delta(s)\mathrm{d}s\}$, we immediately obtain from this last equation

$$f(t) = c\exp\left\{\int_{t_0}^{t}\delta(s)\mathrm{d}s\right\} \tag{1}$$

where c is some constant.

Now the consistency principle implies that

$$
\begin{aligned}
A(t_1, t_2) &= A(t_0, t_2)/A(t_0, t_1) \\
&= f(t_2)/f(t_1) \qquad \text{(by definition of } f(t)\text{)} \\
&= \exp\left\{\int_{t_1}^{t_2}\delta(s)\mathrm{d}s\right\}
\end{aligned}
$$

(from 1 above), as required.

Appendix 2: The solution of non-linear equations

The problem of finding the solution (or solutions) of a non-linear equation occurs in many branches of mathematics. In compound interest such an equation is often in polynomial form, but this is not always the case. Simple examples of non-linear equations are

$$6x^4 - 3x - 1 = 0 \qquad (A.2.1)$$

$$x - \log x - 2 = 0 \qquad (A.2.2)$$

$$e^{-x} - \sin x = 0 \qquad (A.2.3)$$

Such an equation may be considered abstractly in its general form as

$$f(x) = 0$$

Although we may not be able to determine the *exact* value of a solution of the equation (i.e. the precise value of ξ such that $f(\xi) = 0$), in practice we may be able to estimate the value of ξ to any desired degree of accuracy. This means that, given any very small positive number ε (such as 10^{-8}), we can find a number η such that $|\xi - \eta| < \varepsilon$. For practical purposes η is then regarded as a solution of the equation.

Special techniques exist for solving polynomial equations (see reference [9], pp. 110–27). However, the methods which are most suitable in the more general case can be applied easily to polynomial equations. Accordingly we restrict our attention in this appendix to three of the principal methods which may be applied to solve the general non-linear equation. Students who wish to study this topic further should refer to chapter 3 of reference [9].

Before trying to find the root (or roots) of any equation, it is important to confirm by as elementary methods as possible that the equation does indeed have at least one root. (Much time-wasting may thereby be avoided!) In addition, one should try to determine the number of roots and find some information about their location.

In almost all practical problems the function f will be continuous, and henceforth we shall assume that this is so. (If f is not continuous, the equation $f(x) = 0$ may require more careful analysis.) If $a < b$ and $f(a)$ is of opposite sign to $f(b)$, there exists ξ with $a < \xi < b$ such that $f(\xi) = 0$. This well-known property can be used in most cases to find an interval containing a solution of the equation and leads naturally to one method for finding it.

Bisection method

Suppose that $a_0 < b_0$ and $f(a_0)f(b_0) < 0$. Put $m = (a_0 + b_0)/2$. Then either

(a) $f(a_0)f(m) \leqslant 0$, in which case there is a solution in the interval $[a_0, m]$; or
(b) $f(a_0)f(m) > 0$, in which case $f(m)f(b_0) < 0$ so that there is a solution in the interval $[m, b_0]$.

We have thus determined an interval, $[a_1, b_1]$ say, contained in the interval $[a_0, b_0]$ and one half its length, in which the solution lies. Repeating this procedure we may construct a series of nested intervals $[a_i, b_i]$, each containing the root, such that $(b_i - a_i) = (b_0 - a_0)/2^i$. By making i sufficiently large we can thus find the solution to any desired degree of accuracy.

The principal advantage of the bisection method lies in the fact that, once an interval has been found which contains a solution, then the method guarantees that, provided we are willing to carry out a sufficient number of iterations, the value of the solution may be found to any specified degree of accuracy. The major drawback is that, if very high accuracy is required, the necessary number of iterations may be large. With a modern computer this may not be a serious objection, but it is natural to seek faster methods.

We describe below two further methods, both of which are generally satisfactory in nearly all practical situations. The last method, although usually the most rapid, requires the derivative of f to be found.

Secant method

(This is also known as the 'modified regula falsi' method.)
A sequence $\{x_0, x_1, x_2, \ldots\}$ is constructed in such a way that as n increases x_n tends to the desired solution.

The values of x_0 and x_1 are chosen arbitrarily, but in the neighbourhood of the solution. For $n \geqslant 1$, x_{n+1} is defined to be the x-coordinate of the point of intersection of the x-axis and the line joining the points $[x_{n-1}, f(x_{n-1})]$ and $[x_n, f(x_n)]$ (see Figure A.2.1).

It is easily verified that

$$x_{n+1} = x_n - f(x_n)\frac{x_n - x_{n-1}}{f(x_n) - f(x_{n-1})}$$

$$= \frac{f(x_n)x_{n-1} - f(x_{n-1})x_n}{f(x_n) - f(x_{n-1})}$$

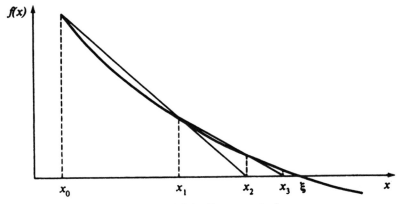

FIGURE A.2.1 *Secant method*

Given $\varepsilon > 0$, we continue the iteration until

$$f(x_n - \varepsilon)f(x_n + \varepsilon) \leqslant 0$$

at which stage we know that x_n is within ε of the solution.

Newton–Raphson method

Finally, we consider a technique which gives theoretically faster convergence to a solution than either of the two preceding methods, but which requires that we are able to evaluate the derivative of f.

We try to construct a sequence $\{x_n\}$ which converges rapidly to the required solution. Intuitively, the rationale for the method may be described as follows.

Suppose that x_n is close to the solution, so that $|f(x_n)|$ is small. We wish $(x_n + h)$ to be a better approximation. If h is small,

$$f(x_n + h) \simeq f(x_n) + hf'(x_n)$$

Hence, if

$$h = -f(x_n)/f'(x_n)$$

it follows that $f(x_n + h) \simeq 0$.

Accordingly, we let

$$x_{n+1} = x_n - f(x_n)/f'(x_n)$$

and hope that the sequence thus defined (from a given starting point x_0) converges. If it does, the limit point ξ is a solution of the equation. The method is illustrated in Figure A.2.2.

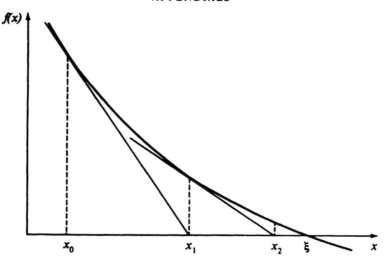

FIGURE A.2.2   *Newton–Raphson method*

The principal advantage of this method lies in its speed of convergence (see reference [9]), but the choice of the initial point x_0 may be of critical importance in the search for a particular solution. A 'bad' starting value may lead to a solution other than that sought or even to a non-convergent sequence. Also, the method does not produce a small interval containing the solution. Usually the iterative procedure is continued until $|x_{n+1} - x_n|$ is 'sufficiently small'. In practice this is generally an acceptable criterion, but is not entirely foolproof and, theoretically, an additional test (such as that described above for the secant method) should be applied. Although the calculation of the derivative may be somewhat laborious and extra care may be needed initially, the method may produce significant savings in time when many similar calculations are required.

Example

Find the positive solution of equation A.2.1.

Solution

Here we have

$$f(x) = 6x^4 - 3x - 1$$

and

$$f'(x) = 24[x^3 - \tfrac{1}{8}]$$

Hence f is a decreasing function for $x < 1/2$ and an increasing function for $x > 1/2$. Also $f(1/2) = -17/8$. Since $f(x)$ is positive for large positive and large negative

Table A.2.1 *Solution to $6x^4 - 3x - 1 = 0$*

| Method of solution | Accuracy* | | | |
| | $\varepsilon = 10^{-4}$ | | $\varepsilon = 10^{-8}$ | |
	Solution	Number of iterations	Solution	Number of iterations
Bisection	0·8831	12	0·883 111 66	25
Secant	0·8831	5	0·883 111 66	7
Newton–Raphson	0·8831	3	0·883 111 67	4

*The true root lies within ε of the value given.

values of x, it follows that the equation has two real solutions. By our remarks above we see that one solution lies in the interval $(-1, 0)$ and the other in the interval $(1/2, 1)$.

The number of iterations required by each method depends on the accuracy to which the solution is required. For the bisection method, we put $a_0 = 0\cdot5$, $b_0 = 1$; for the secant method, we let $x_0 = 0\cdot5$ and $x_1 = 1$; for the Newton–Raphson method, we let $x_0 = 0\cdot75$. Our results are summarized in table A.2.1. The speed of convergence of the three methods should be noted.

Exercises

A.1 Find the negative solution of equation A.2.1
A.2 Find all the real solutions of equation A.2.2
A.3 How many real solutions does equation A.2.3 have?
 Find the two smallest solutions.

Solutions

(To 5 significant figures.)
A.1 $-0\cdot31391$.
A.2 $0\cdot15859$, $3\cdot1462$.
A.3 An infinite number. $0\cdot58853$, $3\cdot0964$.

REFERENCES

1 AITCHISON, J. and BROWN, J. A. C. *The Log-normal Distribution* (Cambridge University Press, 1963).

2 BOYLE, P. P. 'Rates of return as random variables', *Journal of Risk and Insurance*, 1976, 53(4), pp. 693–713.

3 BRIGHAM, E. F. *Fundamentals of Financial Management* 3rd edn (Holt Saunders, 1983).

4 BUTCHER, M. V. and NESBITT, C. J. *Mathematics of Compound Interest* (Ulrich's Books: Ann Arbor, 1971).

5 CARSBERG, B. *Analysis for Investment Decisions* (Accountancy Age Books, 1974).

6 CLARKSON, R. S. 'A mathematical model for the gilt-edged market', *Transactions of the Faculty of Actuaries*, 1978, 36(2), pp. 85–138; or *Journal of the Institute of Actuaries*, 1979, 106(2), pp. 85–132.

7 CONARD, J. W. *The Behaviour of Interest Rates* (National Bureau of Economic Research: New York, 1966).

8 *Consumer Credit Tables* (HMSO: London, 1977) 15 parts.

9 CONTE, S. D. and DE BOOR, C. *Elementary Numerical Analysis – an Algorithmic Approach* 3rd edn (McGraw-Hill: Kogakusha, 1980).

10 COPELAND, T. E. and WESTON, J. F. *Financial Theory and Corporate Policy* (Addison-Wesley, 1979).

11 DAY, J. G. and JAMIESON, A. T. *Institutional Investment* 2nd edn (Institute of Actuaries and Faculty of Actuaries, 1980).

12 DAYKIN, C. F. 'Consumer credit calculations', *Journal of the Institute of Actuaries*, 1980, 107(3), pp. 327–44.

13 DOBBIE, G. M. and WILKIE, A. D. 'The *FT*-Actuaries fixed interest indices' *Journal of the Institute of Actuaries*, 1978, 105(1), pp. 15–26: and *Transactions of the Faculty of Actuaries*, 1979, 36(3), pp. 203–13.

14 DODDS, J. C. and FORD, J. L. *The Term Structure of Interest Rates* (Martin Robertson, 1974).

15 DONALD, D. W. A. *Compound Interest and Annuities-certain* 2nd edn (Heinemann, 1975)

16 DONALD, D. W. A. 'Variation in the price of redeemable securities by yield and term' *Transactions of the Faculty of Actuaries*, 1963, 28(2), pp. 99–101.

17 FISHER, I. *The Theory of Interest* (Macmillan, 1930: reprinted by Augustus M. Kelly, 1970).

18 GUTTENTAGE, J. M. and CAGAN, P. (eds) *Essays on Interest Rates* (National Bureau of Economic Research: New York, 1969).

19 HAHN, F. H. and BRECHLING, F. P. R. (eds) *The Theory of Interest Rates* (Macmillan, 1965: reprinted by Augustus M. Kelly, 1970).

20 HIRSHLEIFER, J, *Investment, Interest and Capital* (Prentice-Hall, Englewood Cliffs, 1970).

21 HOMER, S. *A History of Interest Rates* (Rutgers University Press, 1963).

22 HOMER, S. *et al. The Five Year Outlook for Interest Rates* (Rand McNally, 1963).

23 KELLISON, S. *Theory of Interest* (Irwin, 1971).

24 KENDALL, M. G. and STUART, A. *The Advanced Theory of Statistics* vol. 1, 4th edn (Griffin, 1983).

25 KNOX, D. M., ZIMA, P. and BROWN, R. L. *Mathematics of Finance* (McGraw-Hill, Sydney, 1984).

26 KNUTH, D. E. *The Art of Computer Programming* vol. 2 (Addison-Wesley, 1969).

27 LEVY, H. and SARNAT, M. *Capital Investment and Financial Decisions* (Prentice-Hall International, 1978).

28 LUMBY, S. *Investment Appraisal* (Nelson, 1981).

29 MALKIEL, B. G. *The Term Structure of Interest Rates* (Princeton University Press, 1966).

30 MAO, J. C. T. *Quantitative Analysis of Financial Decisions* (Macmillan, 1969).

31 McCUTCHEON, J. J. 'Some remarks on Stoodley's formula' *Transactions of the Faculty of Actuaries*, 1982, 38(2), pp. 182–91.

32 MERRETT, A. J. and SYKES, A. *Finance and Analysis of Capital Projects* (Longman, 1963).

33 MICHAELSEN, J. B. *The Term Structure of Interest Rates* (Intext Educational Publishers, 1973).

34 NELSON, C. R. *The Term Structure of Interest Rates* (Basic Books, New York, 1972).

35 PANJER, H. H. and BELLHOUSE, D. R. 'Theory of stochastic mortality and interest rates', Actuarial Research Clearing House, Society of Actuaries, 1978, 2, pp. 123–53.

36 POLLARD, J. H. 'On fluctuating interest rates', *Bulletin de l'Association Royale des Actuaires Belges*, 1971, 66, pp. 68–97.

37 POLLARD, J. H. 'Premium loadings for non-participating business', *Journal of the Institute of Actuaries*, 1976, 103(2), pp. 205–12.

38 REDINGTON, F. M. 'Review of the principles of life office valuations', *Journal of the Institute of Actuaries*, 1952, 78(3), pp. 286–315.

39 RENWICK, F. B. *Introduction to Investments and Finance* (Macmillan, 1971).

40 ROLL, R. *The Behaviour of Interest Rates* (Basic Books, New York, 1970).

41 RUBINSTEIN, R. V. *Simulation and the Monte-Carlo Method* (John Wiley and Sons, 1981).

42 RUSSELL, A. M. and RICKARD, J. A. 'Uniqueness of non-negative internal rate of return', *Journal of the Institute of Actuaries*, 1982, 109(3), pp. 435–45.

43 SCOTT, W. F. 'A note on varying rates of interest', *Transactions of the Faculty of Actuaries*, 1976, 34(4), pp. 443–9.

44 SCOTT, W. F. 'A reserve basis for maturity guarantees in unit-linked life assurance', *Transactions of the Faculty of Actuaries*, 1977, 35(4), pp. 365–91.

45 SHEDDEN, A. D. 'A practical approach to applying immunisation theory', *Transactions of the Faculty of Actuaries*, 1977, 35(4), pp. 313–36.

46 SHEDDEN, A. D. Addendum to reference [45], *Transactions of the Faculty of Actuaries*, 1978, 36(1), pp. 20–6.

47 STOODLEY, C. L. 'The effect of a falling interest rate on the values of certain actuarial functions', *Transactions of the Faculty of Actuaries*, 1934, 14, pp. 137–75.

48 TAYLOR, G. C. 'Determination of internal rates of return in respect of

an arbitrary cash flow', *Journal of the Institute of Actuaries*, 1980, 107(4), pp. 487–97.

49 TODHUNTER, R. *Textbook on Compound Interest and Annuities-certain* 3rd edn (Cambridge University Press, 1931).

50 VAN HORNE, J. C. *Function and Analysis of Capital Markets* (Prentice-Hall, 1970).

51 VAN HORNE, J. C. *Financial Market Rates and Flows* (Prentice-Hall, Englewood Cliffs, 1978).

52 WATERS, H. R. 'The moments and distributions of actuarial functions', *Journal of the Institute of Actuaries*, 1978, 105(1), pp. 61–75.

53 WESTCOTT, D. A. 'Moments of compound interest functions under fluctuating interest rates', *Scandinavian Actuarial Journal*, 1981, 4, pp. 237–44.

54 WILKIE, A. D. 'On the calculation of "real" investment returns', *Journal of the Institute of Actuaries*, 1984, 111(1), pp. 149–72.

55 WILKIE, A. D. 'The rate of interest as a stochastic process – theory and applications', Proceedings 20th ICA, Tokyo, 1976, vol. 1, pp. 325–38.

56 WILKIE, A. D. 'Using a stochastic model to estimate the distribution of real rates of return on ordinary shares', Proceedings 22nd ICA, Sydney, 1984, vol. 5, pp. 1–10.

57 WILKIE, A. D. 'A stochastic investment model for actuarial use', *Transactions of the Faculty of Actuaries*, 1986, 39(3), pp. 341–77.

58 WILKIE, A. D. '*FT*-actuaries British Government securities indices 1976', *Journal of the Institute of Actuaries*, 1978, 105(1), pp. 27–33; and *Transactions of the Faculty of Actuaries*, 1979, 36(3), pp. 214–31.

SOLUTIONS TO EXERCISES

Chapter 2

2.1 Let t be a given positive number. For $i \geqslant 0$, define

$$f(i) = (1 + i)^t - (1 + it)$$

Note that $f(0) = 0$ and that

$$f'(i) = t[(1 + i)^{t-1} - 1]$$

(a) Suppose that $t < 1$. Then, for $i > 0$, $f'(i) < 0$, which means that, for positive values of i, $f(i)$ is a decreasing function. Thus, if $i > 0$, $f(i) < f(0) = 0$, so that $(1 + i)^t < (1 + it)$, as required.

(b) Suppose that $t > 1$. Then, for $i > 0$, $f'(i) > 0$, which means that, for positive values of i, $f(i)$ is an increasing function. Thus if $i > 0$, $f(i) > 0$, which establishes the desired result.

2.2 Let time be measured in years from the beginning of the year. By formula 2.4.4,

$$i_h(t_0) = \text{the nominal rate of interest per annum at time } t_0 \text{ on transactions of term } h$$

$$= \frac{\exp\left[\int_{t_0}^{t_0+h} \delta(t)dt\right] - 1}{h}$$

Now

$$\delta(t) = 0 \cdot 15 - 0 \cdot 03t \qquad \text{for } 0 \leqslant t \leqslant 1$$

so

$$\int_{t_0}^{t_0+h} \delta(t)dt = h\delta(t_0 + \tfrac{1}{2}h) = h[0 \cdot 15 - 0 \cdot 03(t_0 + \tfrac{1}{2}h)]$$

When $t_0 = 0$ we obtain the following answers:

$$\frac{\exp[h(0 \cdot 15 - 0 \cdot 015h)] - 1}{h} \qquad \text{for } h = \tfrac{1}{4}, \tfrac{1}{12}, \tfrac{1}{365}$$

which gives (a) 0·148 957, (b) 0·149 676, and (c) 0·149 990.
When $t_0 = 1/2$, we obtain the following answers:

$$\frac{\exp[h(0·135 - 0·015h)] - 1}{h} \qquad \text{for } h = \tfrac{1}{4}, \tfrac{1}{12}, \tfrac{1}{365}$$

which gives (a) 0·133 427, (b) 0·134 498, and (c) 0·134 984.

Note $\delta(1/2) = 0·135$.

2.3   For $0 \leqslant t \leqslant 1$ let $F(t)$ denote the accumulation at time t of an investment of 1 at time 0. Note that $F(1/2) = 20\,596·21/20\,000$ and $F(1) = 21\,183·70/20\,000$. Hence $\log F(1/2) = 0·029\,375$ and $\log F(1) = 0·057\,500$.

By hypothesis, for $0 \leqslant t \leqslant 1$, $\delta(t) = a + bt$. Hence

$$\int_0^t \delta(s)\mathrm{d}s = at + \tfrac{1}{2}bt^2$$

and so, by formula 2.4.10,

$$\log F(t) = at + \tfrac{1}{2}bt^2$$

Letting $t = \tfrac{1}{2}$ in the last equation, we obtain

$$0·029\,375 = \tfrac{1}{2}a + \tfrac{1}{8}b$$

and, letting $t = 1$, we have

$$0·057\,500 = a + \tfrac{1}{2}b$$

These last two equations imply that $a = 0·06$ and $b = -0·005$. Thus

$$\delta(t) = 0·06 - 0·005t$$

Also

$$\log F(\tfrac{3}{4}) = \tfrac{3}{4}a + \tfrac{9}{32}b = 0·043\,594$$

from which it follows that $F(3/4) = 1·044\,558$. The accumulated amount of the account at time 3/4 is 20 000 $F(3/4)$, which equals £20 891·16.

2.4   Measure time in years from the present. By formula 2.5.5,

$$v(t) = v^t = (1·08)^{-t} \qquad \text{for all } t \geqslant 0$$

(a) Let X be the single payment at time 5. We must have

$$6280v^4 + 8460v^7 + 7350v^{13} = Xv^5$$

from which we obtain $X = £18\,006$.

(b) Let t be the appropriate future time. We find t from the equation

$$6280v^4 + 8460v^7 + 7350v^{13} = 22\,090v^t$$

from which we obtain $t = 7\cdot66$ years.

2.5 Measure time in years from the start of the given year.

(a) Note that $\delta(0) = 0\cdot15$ and let

$$\delta(t) = 0\cdot15 + bt + ct^2$$

Putting $t = 1/2$ in this equation, we obtain

$$0\cdot10 = 0\cdot15 + \tfrac{1}{2}b + \tfrac{1}{4}c$$

and, letting $t = 1$, we have

$$0\cdot08 = 0\cdot15 + b + c$$

The last two equations imply that $b = -0\cdot13$ and $c = 0\cdot06$, so that

$$\delta(t) = 0\cdot15 - 0\cdot13t + 0\cdot06t^2$$

This implies that $\int_0^1 \delta(t)dt = 0\cdot105$. By formula 2.4.9 the accumulated amount at the end of the year is $5000\exp(0\cdot105)$, which equals £5553·55.

(b) In this case $\delta(t)$ is linear over $0 \leqslant t \leqslant 1/2$ and also over $1/2 \leqslant t \leqslant 1$, so

$$\int_0^1 \delta(t)dt = \int_0^{1/2} \delta(t)dt + \int_{1/2}^1 \delta(t)dt$$

$$= \frac{1}{2}\left[\frac{\delta(0) + \delta(\tfrac{1}{2})}{2}\right] + \frac{1}{2}\left[\frac{\delta(\tfrac{1}{2}) + \delta(1)}{2}\right]$$

$$= \tfrac{1}{4}(0\cdot15 + 0\cdot10) + \tfrac{1}{4}(0\cdot10 + 0\cdot08) = 0\cdot1075$$

Hence the accumulated amount is

$$5000\exp(0\cdot1075) = £5567\cdot45$$

2.6 (a) By equation 2.6.3,

$$v(t) = \tfrac{3}{4}v_1^t + \tfrac{1}{4}v_2^t$$

where

$$v_1 = \exp[-(p + s)] = \exp(-0\cdot095\,31) = (1\cdot1)^{-1}$$

$$v_2 = \exp(-p) = \exp(-0\cdot058\,269) = (1\cdot06)^{-1}$$

Hence

$$v(t) = \tfrac{1}{4}(1\cdot06)^{-t} + \tfrac{3}{4}(1\cdot1)^{-t}$$

(b) (i) Let S be the single payment offered to the investor. The equation of value at the present time is

$$600[v(0) + v(1) + \cdots + v(11)] = Sv(12)$$

i.e.

$$S = \frac{600[1 + \tfrac{1}{4}(a_{\overline{11}} \text{ at } 6\%) + \tfrac{3}{4}(a_{\overline{11}} \text{ at } 10\%)]}{[\tfrac{1}{4}(1\cdot06)^{-12} + \tfrac{3}{4}(1\cdot1)^{-12}]}$$

$$= £12\,956$$

(ii) Let X be the annual payment offered to the investor. The equation of value at the present time is

$$600[v(0) + v(1) + \cdots + v(11)] = X[v(12) + v(13) + \cdots + v(23)]$$

i.e.

$$X = \frac{600[1 + \tfrac{1}{4}(a_{\overline{11}} \text{ at } 6\%) + \tfrac{3}{4}(a_{\overline{11}} \text{ at } 10\%)]}{[\tfrac{1}{4}(a_{\overline{23}} - a_{\overline{11}} \text{ at } 6\%) + \tfrac{3}{4}(a_{\overline{23}} - a_{\overline{11}} \text{ at } 10\%)]}$$

$$= £1625$$

2.7 (a) Let $\delta(0) = \delta_0$, $\delta(m) = \delta_m$. By formula 2.4.3, the accumulation is

$$\exp\left[\int_0^n \delta(t)dt\right]$$

Now, for $0 \leqslant t \leqslant m$,

$$\delta(t) = \delta_0 + \frac{t}{m}(\delta_m - \delta_0)$$

so that, if $n \leqslant m$, the accumulation is (by the trapezoidal rule)

$$\exp\left(\frac{n}{2}\left\{\delta_0 + \left[\delta_0 + \frac{n}{m}(\delta_m - \delta_0)\right]\right\}\right)$$

$$= \exp\left(n\left\{\delta_0 + \frac{n}{2m}(\delta_m - \delta_0)\right\}\right)$$

When $n \geqslant m$, we have the accumulation

$$\exp\left[\int_0^n \delta(t)dt\right] = \exp\left[\int_0^m \delta(t)dt + \int_m^n \delta(t)dt\right]$$

$$= \exp\left[\frac{m}{2}(\delta_0 + \delta_m) + (n - m)\delta_m\right]$$

(b) (i) The accumulation for 15 years is

$$\exp\{15[0\!\cdot\!08 + \tfrac{15}{32}(0\!\cdot\!048 - 0\!\cdot\!08)]\} = 2\!\cdot\!6512$$

(ii) The accumulation for 40 years is

$$\exp[\tfrac{16}{2}(0\!\cdot\!08 + 0\!\cdot\!048) + (24 \times 0\!\cdot\!048)] = 8\!\cdot\!8110$$

(c) (i) We find δ such that $e^{15\delta} = 2\!\cdot\!6512$. i.e. $\delta = 0\!\cdot\!065$.
(ii) We find δ such that $e^{40\delta} = 8\!\cdot\!8110$, i.e. $\delta = 0\!\cdot\!0544$.

2.8 (a) By formula 2.5.3,

$$v(t) = \exp\left[-\int_0^t \delta(s)ds \right]$$

Now

$$\int_0^t \delta(s)ds = \begin{cases} 0\!\cdot\!08t & \text{for} \quad 0 \leqslant t \leqslant 5 \\ 0\!\cdot\!1 + 0\!\cdot\!06t & \text{for} \quad 5 \leqslant t \leqslant 10 \\ 0\!\cdot\!3 + 0\!\cdot\!04t & \text{for} \quad t \geqslant 10 \end{cases}$$

Hence

$$v(t) = \begin{cases} \exp(-0\!\cdot\!08t) & \text{for} \quad 0 \leqslant t \leqslant 5 \\ \exp(-0\!\cdot\!1 - 0\!\cdot\!06t) & \text{for} \quad 5 \leqslant t \leqslant 10 \\ \exp(-0\!\cdot\!3 - 0\!\cdot\!04t) & \text{for} \quad t \geqslant 10 \end{cases}$$

(b) (i) Let the single payment be denoted by S. The equation of value, at the present time, is

$$600[v(0) + v(1) + \cdots + v(14)] = Sv(15)$$

Hence we obtain

$$S = \frac{600(1 + e^{-0\!\cdot\!08} + \cdots + e^{-0\!\cdot\!86})}{e^{-0\!\cdot\!9}} \qquad \text{(using (a))}$$

$$S = \pounds 14\,119.$$

(ii) Let A be the annual payment. The equation of value, at the present time, is

$$600[v(0) + v(1) + \cdots + v(14)] = A[v(15) + v(16) + \cdots + v(22)]$$

Hence

$$A = \frac{600(1 + e^{-0\!\cdot\!08} + \cdots + e^{-0\!\cdot\!86})}{(e^{-0\!\cdot\!9} + e^{-0\!\cdot\!94} + \cdots + e^{-1\!\cdot\!18})} \qquad \text{(using (a))}$$

$$= \pounds 2022$$

2.9 (a)   (i) By formula 2.5.3

$$\int_0^t \delta(s)ds = -\log v(t)$$

So

$$\delta(t) = -\frac{v'(t)}{v(t)}$$

$$= \frac{2t + 2\alpha + 1}{(t + \alpha)(t + \alpha + 1)}$$

(ii) By formula 2.4.4, the effective rate of interest for the period from time n to time $n + 1$ is

$$\exp\left[\int_n^{n+1} \delta(t)dt\right] - 1 = \exp\left[\int_0^{n+1} \delta(t)dt - \int_0^n \delta(t)dt\right] - 1$$

$$= \frac{v(n)}{v(n + 1)} - 1$$

$$= \frac{(n + 1 + \alpha)(n + 2 + \alpha)}{(n + \alpha)(n + 1 + \alpha)} - 1$$

$$= \frac{2}{n + \alpha}$$

(iii) Note that

$$v(t) = \alpha(\alpha + 1)\left[\frac{1}{(t + \alpha)} - \frac{1}{(t + \alpha + 1)}\right]$$

So

$$a(n) = \alpha(\alpha + 1)\left\{\left[\frac{1}{(1 + \alpha)} - \frac{1}{(2 + \alpha)}\right] + \left[\frac{1}{(2 + \alpha)} - \frac{1}{(3 + \alpha)}\right]\right.$$

$$\left. + \cdots + \left[\frac{1}{(n + \alpha)} - \frac{1}{(n + \alpha + 1)}\right]\right\}$$

$$= \alpha(\alpha + 1)\left[\frac{1}{(1 + \alpha)} - \frac{1}{(n + \alpha + 1)}\right]$$

$$= \frac{n\alpha}{(n + \alpha + 1)}$$

(b) Let P be the level annual premium. The equation of value, at the present time, is

$$P[1 + a(11)] = 1800[a(21) - a(11)]$$

So

$$P = \frac{1800\left(\dfrac{21 \times 15}{22 + 15} - \dfrac{11 \times 15}{12 + 15}\right)}{1 + \dfrac{11 \times 15}{12 + 15}} = £608 \cdot 11$$

Let the value at time 12 of the series of annual payments be X. The equation of value at the present time is

$$X v(12) = 1800[a(21) - a(11)]$$

Hence

$$X = \frac{1800\left(\dfrac{21 \times 15}{22 + 15} - \dfrac{11 \times 15}{12 + 15}\right)}{\dfrac{15 \times 16}{(15 + 12)(16 + 12)}} = £13\,622$$

2.10 (a) Let the rate of payment at time t be $£\rho(t)$ per annum. We are given that $\rho(t) = p + qt$, where p and q are certain constants. The cash payable in the rth year is

$$I_r = \int_{r-1}^{r} (p + qt)\mathrm{d}t = p + \left(\frac{2r - 1}{2}\right)q \qquad r = 1, 2, \ldots, n$$

Hence

$$I_1 = p + \tfrac{1}{2}q, \qquad I_2 = p + \tfrac{3}{2}q$$

which gives

$$p = \tfrac{3}{2}I_1 - \tfrac{1}{2}I_2, \qquad q = -I_1 + I_2$$

Hence

$$\rho(t) = (\tfrac{3}{2}I_1 - \tfrac{1}{2}I_2) + (I_2 - I_1)t \qquad 0 \leqslant t \leqslant n$$

Now let $0 < t \leqslant n$. The cash payable up to time t is

$$\int_0^t \rho(s)\mathrm{d}s = pt + \tfrac{1}{2}qt^2$$

$$= t[\tfrac{3}{2}I_1 - \tfrac{1}{2}I_2 + \tfrac{1}{2}t(I_2 - I_1)]$$

(b) (i) By formula 2.7.5, the present value is

$$\int_0^n (p + qt)\mathrm{e}^{-\delta t}\mathrm{d}t = p \int_0^n \mathrm{e}^{-\delta t}\mathrm{d}t + q \int_0^n t\mathrm{e}^{-\delta t}\mathrm{d}t$$

Now,

$$\int_0^n e^{-\delta t}dt = \frac{1-e^{-\delta n}}{\delta}$$

and, by integration by parts,

$$\int_0^n te^{-\delta t}dt = \frac{1-e^{-n\delta}-n\delta e^{-n\delta}}{\delta^2}$$

Hence the present value is

$$\left(\frac{3}{2}I_1 - \frac{1}{2}I_2\right)\left(\frac{1-e^{-\delta n}}{\delta}\right) + (I_2 - I_1)\left(\frac{1-e^{-n\delta}-n\delta e^{-n\delta}}{\delta^2}\right)$$

(ii) We equate the last expression to 9047 and solve for I_1. This gives

$$[\tfrac{3}{2}I_1 - (\tfrac{1}{2} \times 1 \cdot 07 I_1)](11 \cdot 810\,68) + (0 \cdot 07 I_1)(95 \cdot 669\,94) = 9047$$

whence $I_1 = £500$.

The cash payable each year increases in arithmetical progression, and $I_2 = I_1 + (0 \cdot 07 \times 500) = I_1 + 35$. The cash payable in the final year is thus

$$I_1 + [(n-1) \times 35] = 500 + (19 \times 35) = £1165$$

2.11 (a) By formula 2.5.3,

$$v(t) = \exp\left[-\int_0^t \delta(s)ds \right]$$

$$= \exp\left[-\int_0^t ae^{-bs}ds \right]$$

$$= \exp\left[\frac{a}{b}(e^{-bt} - 1) \right]$$

(b) (i) Since $\delta(0) = 0 \cdot 1$, $a = 0 \cdot 1$, and $\delta(10)/\delta(0) = e^{-10b} = 0 \cdot 5$, we find that $b = 0 \cdot 069\,315$. Hence the present value of the payments is

$$1000[v(1) + v(2) + v(3) + v(4)] = £3205 \cdot 43$$

(ii) We find δ by solving the equation

$$1000(e^{-\delta} + e^{-2\delta} + e^{-3\delta} + e^{-4\delta}) = 3205 \cdot 43$$

that is,

$$e^{-\delta} + e^{-2\delta} + e^{-3\delta} + e^{-4\delta} = 3 \cdot 205\,43$$

By numerical methods (see appendix 2), we obtain $\delta = 0.090\,63$.

2.12 (a) By formula 2.5.3,

$$v(t) = \exp\left[- \int_0^t \delta(y)dy \right]$$

$$= \exp\left[- \int_0^t (r + se^{-ry})dy \right]$$

$$= \exp\left[- rt + \frac{s}{r}(e^{-rt} - 1) \right]$$

$$= \exp\left(-\frac{s}{r} \right)\exp(-rt)\exp\left(\frac{s}{r}e^{-rt} \right)$$

(b) (i) The present value of the payment stream is, by formula 2.7.5,

$$1000 \int_0^n v(t)dt = 1000\exp\left(-\frac{s}{r} \right)\int_0^n \exp(-rt)\exp\left(\frac{s}{r}e^{-rt} \right)dt$$

$$= 1000\exp\left(-\frac{s}{r} \right)\left[-\frac{1}{s}\exp\left(\frac{s}{r}e^{-rt} \right) \right]_{t=0}^{t=n}$$

$$= 1000\exp\left(-\frac{s}{r} \right)\left(-\frac{1}{s} \right)\left[\exp\left(\frac{s}{r}e^{-rn} \right) \right.$$

$$\left. - \exp\left(\frac{s}{r} \right) \right]$$

$$= \frac{1000}{s}\left\{ 1 - \exp\left[\frac{s}{r}(e^{-rn} - 1) \right] \right\}$$

(ii) On substituting $n = 50$, $r = \log 1.01$, and $s = 0.03$, the present value is found to be £23 109.

Chapter 3

3.1 (a) The time, n years, for money to double itself is the solution of the equation

$$\left(1 + \frac{K}{100} \right)^n = 2$$

whence

$$n = \frac{\log 2}{\log(1 + K/100)}$$

If x is small, $\log(1 + x) \simeq x$ (on ignoring the remainder after the term in x). Hence

$$n \simeq \frac{100 \log 2}{K} \simeq \frac{70}{K}$$

The corresponding rule-of-thumb for the time taken for money to treble itself is

$$n \simeq \frac{100 \log 3}{K} \simeq \frac{110}{K}$$

(b) Let n_1 and n_2 denote respectively the times taken for money to double and treble itself. The following table gives these values, calculated exactly, and the corresponding rule-of-thumb values $70/K$ and $110/K$:

	Rate of interest per cent, K					
	5		10		20	
	n_1	n_2	n_1	n_2	n_1	n_2
Exact	14·207	22·517	7·273	11·527	3·802	6·026
Approximate	14	22	7	11	3·5	5·5

(c) The approximation is clearly less accurate at high rates of interest than at low rates.

3.2 (a) By formula 3.1.9, $i = e^{\delta} - 1$ and $\delta = \log(1 + i)$. By equation 3.1.5, $v = e^{-\delta}$ and $\delta = -\log v$. Also, $v = (e^{\delta})^{-1} = (1 + i)^{-1}$, and $i = (1 + i) - 1 = v^{-1} - 1$. By equation 3.1.6, $d = 1 - e^{-\delta}$, and $e^{-\delta} = 1 - d$ shows that $\delta = -\log(1 - d)$. Also, equations 3.1.6 and 3.1.9 show that $d = 1 - (1 + i)^{-1} = i(1 + i)^{-1}$ and $i = (1 + i) - 1 = (1 - d)^{-1} - 1$. Equation 3.1.6 shows that $1 - d = e^{-\delta} = v$ (by equation 3.1.5), and hence $1 - v = d$.

(b) The answers are found by substituting the appropriate values in the formulae of table 3.1.1.

(i) When $\delta = 0.08$, we obtain $i = 0.083\,287$, $d = 0.076\,884$, and $v = 0.923\,116$.

(ii) When $d = 0.08$, we obtain $v = 0.92$, $i = 0.086\,957$, and $\delta = 0.083\,382$.

(iii) When $i = 0.08$, we obtain $v = 0.925\,926$, $d = 0.074\,074$, and $\delta = 0.076\,961$.

(iv) When $v = 0.95$, we obtain $d = 0.05$, $i = 0.052\,632$, and $\delta = 0.051\,293$.

3.3 (a) The equation of value for each of the three savings plans, and the corresponding yield per annum (i.e. solution of the equation of value), are as follows:

   (i) $-1330 + 1000(1 + i)^3 = 0,$ $i \simeq 0.0997$ or 9.97%
   (ii) $-1550 + 1000(1 + i)^5 = 0,$ $i \simeq 0.0916$ or 9.16%
   (iii) $-1000(1 + i)^4 + 425a_{\overline{3}|} = 0,$ $i \simeq 0.0858$ or 8.58%

   (b) Let this rate of interest per annum be i. We must have

$$1330(1 + i)^2 \geqslant 1550$$

Hence $i \geqslant 0.0795$, i.e. at least 7.95%.

   (c) Let the rate of interest per annum used to calculate the amount of the annuity-certain be i. Then i must be such that

$$1550 \geqslant 425\ddot{a}_{\overline{3}|i}$$

i.e.

$$\ddot{a}_{\overline{4}|i} \leqslant 3.647\,06$$

i.e.

$$a_{\overline{3}|i} \leqslant 2.647\,06$$

By interpolation between 6% and 7%, we find that $a_{\overline{3}|i} = 2.647\,06$ when $i \simeq 6.53\%$, so we require i to be at least 6.53%.

3.4 (a) The equation of value for plan (i) is

$$100\ddot{s}_{\overline{10}|i} = 1700$$

where i is the yield per annum. We thus have

$$\ddot{s}_{\overline{10}|i} = 17$$

i.e.

$$s_{\overline{11}|i} = 18$$

By linear interpolation between 9% and 10%, we obtain $i \simeq 9.45\%$.
    The equation of value for plan (ii) is

$$100\ddot{s}_{\overline{15}|i} = 3200$$

where i is yield per annum. We thus have

$$\ddot{s}_{\overline{15}|} = 32$$

i.e.

$$s_{\overline{16}|} = 33$$

By linear interpolation between 8% and 9%, we obtain $i \simeq 9.00\%$.

(b) Let i be the minimum fixed rate of interest which will give the desired proceeds. The equation of value, 15 years after the date on which the first premium was paid, is

$$1700(1 + i)^5 + 100\ddot{s}_{\overline{5}|i} - 3200 = 0.$$

By linear interpolation between 8% and 9%, we obtain $i \simeq 8 \cdot 50\%$. The fixed rate of interest must therefore be at least $8 \cdot 50\%$.

3.5 (a) (i) The equation of value one year before the date of payment of the first premium is

$$- X a_{\overline{n}|} + Y_n | a_{\overline{m}|} = 0,$$

i.e.

$$- X a_{\overline{n}|} + Y(a_{\overline{m+n}|} - a_{\overline{n}|}) = 0$$

i.e.

$$Y a_{\overline{m+n}|} - (X + Y) a_{\overline{n}|} = 0$$

(ii) The equation of value at the date of the last payment is

$$Y s_{\overline{m}|} - X s_{\overline{n}|}(1 + i)^m = 0$$

i.e.

$$Y s_{\overline{m}|} - X(s_{\overline{m+n}|} - s_{\overline{m}|}) = 0$$

i.e.

$$(X + Y) s_{\overline{m}|} - X s_{\overline{m+n}|} = 0$$

(b) (i) The equation of value is

$$2000 a_{\overline{20}|} - 3000 a_{\overline{10}|} = 0$$

i.e.

$$2 a_{\overline{20}|} - 3 a_{\overline{10}|} = 0$$

By linear interpolation between 7% and 8%, the yield per annum is approximately $7 \cdot 19\%$.

(ii) The equation of value is now

$$2000 a_{\overline{10+m}|} - 3000 a_{\overline{10}|} = 0$$

i.e.

$$a_{\overline{10+m}|} - 1 \cdot 5 a_{\overline{10}|} = 0.$$

It is clear by general reasoning that the yield increases as m increases. By trials we see that the yield first equals or exceeds 8% when $m = 12$, and first exceeds 10% when $m = 17$. The required range of values of m is thus 12 to 16 years (inclusive).

(c) The equation of value is

$$3000s_{\overline{20}|} - 1000s_{\overline{n+20}|} = 0$$

i.e.

$$3s_{\overline{20}|} - s_{\overline{n+20}|} = 0$$

It is clear by general reasoning that the yield decreases as n increases. By trials we see that the yield first equals or falls below 10% when $n = 11$, and first falls below 8% when $n = 13$. The required range of values of n is thus 11 to 12 years.

3.6 The equations of value are

(a) $-10\,000(1 + i)^2 + 21\,500(1 + i) - 11\,550 = 0$
(b) $-10\,000(1 + i)^2 + 20\,400(1 + i) - 10\,395 = 0$

Theorem 3.2.1 is not relevant, since the net cash flow changes sign twice.

Theorem 3.2.2 gives no information in respect of project (a), but shows that (b) has exactly one positive solution to the equation of value. This follows by a consideration of the accumulated net cash flows, which are (a) $\{-10\,000,\ 11\,500,\ -50\}$ and (b) $\{-10\,000,\ 10\,400,\ 5\}$.

The equation of value for project (a) may be written as

$$-10\,000i^2 + 1500i - 50 = 0$$

which gives $i = 0.05$ or 0.10.

The equation of value for project (b) may be written as

$$-10\,000i^2 + 400i + 5 = 0$$

which gives $i = 0.05$ or -0.01.

3.7 (a) Consider the purchase of an article with retail price £100. The cash price is £70, or one may pay £75 in 6 months' time. The effective rate of discount per annum d is found by solving the equation

$$70 = 75(1 - d)^{1/2}$$

whence $d = 0.1289$ or 12.89%.
The effective annual rate of interest is

$$i = \frac{d}{1 - d} = 0.1480 \qquad \text{or} \qquad 14.80\%$$

(b) One may pay £72.50 in three months' time instead of £70 now, so the effective annual rate of discount d is found from the equation

$$70 = 72 \cdot 50(1 - d)^{1/4}$$

which gives $d = 0 \cdot 1310$ or $13 \cdot 10\%$.

The new arrangement therefore offers a greater effective annual rate of discount to cash purchasers.

3.8   (a)

$$_5|a_{\overline{32}|} = a_{\overline{37}|} - a_{\overline{5}|} = 14 \cdot 6908$$

$$\ddot{a}_{\overline{62}|} = \frac{1 - v^{62}}{d} = 23 \cdot 7149$$

$$\bar{a}_{\overline{62}|} = \frac{1 - v^{62}}{\delta} = \frac{d}{\delta}\ddot{a}_{\overline{62}|} = 23 \cdot 2560$$

$$_{12}|\ddot{a}_{\overline{50}|} = v^{12}\ddot{a}_{\overline{50}|} = v^{12}(1 + a_{\overline{49}|}) = 13 \cdot 9544$$

$$s_{\overline{62}|} = \frac{(1 + i)^{62} - 1}{i} = 259 \cdot 451$$

$$\ddot{s}_{\overline{61}|} = s_{\overline{62}|} - 1 = 258 \cdot 451$$

$$(I\ddot{a})_{\overline{62}|} = \frac{\ddot{a}_{\overline{62}|} - 62v^{62}}{d} = 474 \cdot 905$$

$$_5|(Ia)_{\overline{20}|} = v^5(Ia)_{\overline{20}|} = 102 \cdot 867$$

$$(I\bar{a})_{\overline{23}|} = \frac{\ddot{a}_{\overline{23}|} - 25v^{25}}{\delta} = 175 \cdot 136$$

$$(\bar{I}\bar{a})_{\overline{23}|} = \frac{\bar{a}_{\overline{23}|} - 25v^{25}}{\delta} = 167 \cdot 117.$$

(b) Since $\ddot{a}_{\overline{2n}|} = \ddot{a}_{\overline{n}|} + v^n\ddot{a}_{\overline{n}|} = \ddot{a}_{\overline{n}|}(1 + v^n)$

we have

$$1 + v^n = \frac{10 \cdot 934\,563}{7 \cdot 029\,584}$$

So

$$v^n = 0 \cdot 555\,506$$

Hence

$$\ddot{a}_{\overline{n}|} = \frac{1 - v^n}{d} = \frac{1 - 0 \cdot 555\,506}{d} = 7 \cdot 029\,584$$

Hence $d = 0 \cdot 063\,232$ and $i = d/(1 - d) = 0 \cdot 0675$ or $6\frac{3}{4}\%$.

Therefore

$$n = \frac{\log v^n}{\log v} = \frac{\log(0.555\,506)}{\log(1 - 0.063\,232)} = 9.$$

3.9 (a) It is clear at once that the value of the annuity is

$$5\ddot{a}_{\overline{5}|} + 7_6|\ddot{a}_{\overline{9}|} + 10_{15}|\ddot{a}_{\overline{5}|} \qquad \text{(i)}$$

$$= 5\ddot{a}_{\overline{5}|} + 7(\ddot{a}_{\overline{13}|} - \ddot{a}_{\overline{6}|}) + 10(\ddot{a}_{\overline{20}|} - \ddot{a}_{\overline{13}|})$$

$$= 10\ddot{a}_{\overline{20}|} - 3\ddot{a}_{\overline{13}|} - 2\ddot{a}_{\overline{6}|} \qquad \text{(ii)}$$

$$= 10(1 + a_{\overline{19}|}) - 3(1 + a_{\overline{14}|}) - 2(1 + a_{\overline{5}|})$$

$$= 5 + 10a_{\overline{19}|} - 3a_{\overline{14}|} - 2a_{\overline{5}|} \qquad \text{(iii)}$$

(b) The accumulated amount of the annuity payments is

$$5(1 + i)^{14}s_{\overline{6}|} + 7(1 + i)^5 s_{\overline{9}|} + 10s_{\overline{5}|} \qquad \text{(i)}$$

$$= 5(s_{\overline{20}|} - s_{\overline{14}|}) + 7(s_{\overline{14}|} - s_{\overline{5}|}) + 10s_{\overline{5}|}$$

$$= 5s_{\overline{20}|} + 2s_{\overline{14}|} + 3s_{\overline{5}|} \qquad \text{(ii)}$$

3.10 (a) Let $\ddot{a}^*_{\overline{n}|i}$ denote the value of this annuity at an annual rate of interest i. We have

$$\ddot{a}^*_{\overline{n}|i} = 1 + (1 + r)(1 + i)^{-1} + (1 + r)^2(1 + i)^{-2} + \cdots$$
$$+ (1 + r)^{n-1}(1 + i)^{-(n-1)}$$

$$= 1 + (1 + j)^{-1} + (1 + j)^{-2} + \cdots + (1 + j)^{-(n-1)}$$

since

$$1 + j = 1 + \left(\frac{i - r}{1 + r}\right) = \frac{1 + i}{1 + r}$$

Hence

$$\ddot{a}^*_{\overline{n}|i} = \ddot{a}_{\overline{n}|j}$$

(b) Let $a^*_{\overline{n}|i}$ denote the value of this annuity at an annual rate of interest i. We have

$$a^*_{\overline{n}|i} = (1 + i)^{-1} + (1 + r)(1 + i)^{-2} + (1 + r)^2(1 + i)^{-3} + \cdots$$
$$+ (1 + r)^{n-1}(1 + i)^{-n}$$

$$= (1 + r)^{-1}[(1 + j)^{-1} + (1 + j)^{-2} + \cdots + (1 + j)^{-n}]$$

$$= (1 + r)^{-1}a_{\overline{n}|j}$$

Hence the present value of this annuity is not equal to $a_{\overline{n}|j}$ (assuming, of course, that $r \neq 0$).

(c) Let the first annuity payment be X. The equation of value is

$$10\,000 = X(1 \cdot 05)^{-1} a_{\overline{20}|j}$$

where

$$1 + j = \frac{1 \cdot 09}{1 \cdot 05}$$

i.e.

$$j = 0 \cdot 038\,10$$

Now $a_{\overline{20}|j} = 13 \cdot 822\,455$, so that

$$10\,000 = X(1 \cdot 05)^{-1}(13 \cdot 822\,455)$$

which gives $X = £759 \cdot 63$.

3.11 The accumulation of the premiums until time 5 years, just before the payment of the premium then due, is

$$100\,\ddot{s}_{\overline{5}|0 \cdot 08}$$

Thus the accumulation of the premiums until just before the payment of the premium due at time 12 is

$$100\,\ddot{s}_{\overline{5}|0 \cdot 08}(1 \cdot 06)^{7} + 100\,\ddot{s}_{\overline{7}|0 \cdot 06}$$

The accumulation until time 20 is therefore

$$[100\,\ddot{s}_{\overline{5}|0 \cdot 08}(1 \cdot 06)^{7} + 100\,\ddot{s}_{\overline{7}|0 \cdot 06}](1 \cdot 05)^{8} + 100\,\ddot{s}_{\overline{8}|0 \cdot 05}$$

$$= 100[\ddot{s}_{\overline{5}|0 \cdot 08}(1 \cdot 06)^{7}(1 \cdot 05)^{8} + \ddot{s}_{\overline{7}|0 \cdot 06}(1 \cdot 05)^{8} + \ddot{s}_{\overline{8}|0 \cdot 05}]$$

$$= £3724 \cdot 77$$

The yield per annum i obtained by the investor on the completed transaction is found by solving the equation of value

$$100\,\ddot{s}_{\overline{20}|} = 3724 \cdot 77$$

i.e.

$$s_{\overline{21}|} = 38 \cdot 2477$$

By interpolation between 5% and 6%, we obtain $i = 5 \cdot 59\%$.

3.12 (a) Since

$$L = (X - jL)s_{\overline{n}|i}$$

we have

$$X = L\left(j + \frac{1}{s_{\overline{n}|i}}\right)$$

$$= L\left(j - i + \frac{1}{a_{\overline{n}|i}}\right)$$

(using equation 3.3.10)

Let the effective rate of interest per annum paid by the borrower be j'. We have the equation of value

$$L = X a_{\overline{n}|j'}$$

from which we obtain

$$L = L\left(j - i + \frac{1}{a_{\overline{n}|i}}\right) a_{\overline{n}|j'} \qquad \text{(by above)}$$

Therefore

$$\frac{1}{a_{\overline{n}|j'}} = j - i + \frac{1}{a_{\overline{n}|i}}$$

$$= j + \frac{1}{|s_{\overline{n}|i}}$$

$$> j + \frac{1}{s_{\overline{n}|j}} \qquad \text{(since } j > i)$$

$$= \frac{1}{a_{\overline{n}|j}}$$

which implies that $j' > j$. The annual yield to the lender is, of course, j.

(b) We must solve the equation

$$a_{\overline{10}|j'} = \frac{1}{j + \dfrac{1}{s_{\overline{10}|0\cdot04}}} = \frac{1}{j + 0\cdot083\,292}$$

(i) When $j = 0\cdot06$, we have

$$a_{\overline{10}|j'} = \frac{1}{0\cdot06 + 0\cdot083\,292} = 6\cdot9788.$$

By linear interpolation between 7% and 8%, $j' = 7\cdot14\%$.

(ii) When $j = 0.07$, we have

$$a_{\overline{10}|j'} = \frac{1}{0.07 + 0.083\,292} = 6.5235$$

By linear interpolation between 8% and 9%, $j' = 8.64\%$.

3.13 (a) (i) Let the annual repayment be X. We have

$$X a_{\overline{25}|} = 3000 \qquad \text{at } 12\%$$

So

$$X = \frac{3000}{a_{\overline{25}|}} = £382.50$$

(ii) (1) Using equation 3.8.1, the loan outstanding just after the payment at time 9 years is

$$382.50 a_{\overline{16}|} = 2667.56$$

and just after the payment at time 10 years it is

$$382.50 a_{\overline{15}|} = 2605.17$$

Hence the capital repaid at the end of the tenth year is

$$2667.56 - 2605.17 = £62.39$$

and the interest paid at this time is

$$382.50 - 62.39 = £320.11$$

(2) The capital outstanding just after the payment at time 24 years is

$$382.50 a_{\overline{1}|} = £341.52$$

The capital repaid at the end of the 25th year is thus £341.52, and the interest paid at this time is

$$382.50 - 341.52 = £40.98$$

Note   These answers may also be obtained from the loan schedule (see section 3.8). In this case the amount of the loan is £3000, so the values in the loan schedule should be multiplied by $3000/a_{\overline{25}|}$; for example, the interest content of the 10th repayment is

$$\frac{3000}{a_{\overline{25}|}}(1 - v^{16}) = £320.11 \qquad \text{as above}$$

(iii)   The loan outstanding after the tth repayment is $382 \cdot 50 a_{\overline{25-t}|}$. This first falls below 1800 when $a_{\overline{25-t}|}$ (at 12%) first falls below 4·7059. By the compound interest tables, the smallest value of t for which $a_{\overline{25-t}|} < 4 \cdot 7059$ is 18, so the answer is the 18th repayment.

(iv)   Using the loan schedule (see section 3.8), the capital content exceeds the interest content of the tth instalment when

$$1 - v^{26-t} < v^{26-t}$$

i.e. when

$$v^{26-t} > 0 \cdot 5$$

This first occurs when $t = 20$, i.e. for the 20th payment.

(b) The loan outstanding just after the 15th annual payment has been made is

$$382 \cdot 50 a_{\overline{10}|} \qquad \text{at } 12\% \qquad = 2161 \cdot 20$$

Let Y be the revised annual payment. The equation of value is

$$Y a_{\overline{16}|} = 2161 \cdot 20 \qquad \text{at } 12\%$$

which gives $Y = £309 \cdot 89$.

3.14 (a) Let X be the initial annual repayment. The equation of value is

$$16\,000 = X a_{\overline{10}|} \qquad \text{at } 8\%$$

from which we obtain $X = £2384 \cdot 46$.

(b) The loan outstanding just after the fourth payment is made is

$$2384 \cdot 46 a_{\overline{6}|} \qquad \text{at } 8\% \qquad = 11\,023 \cdot 14$$

Let Y be the revised annual instalment. The equation of value is

$$Y a_{\overline{6}|} = 11\,023 \cdot 14 \qquad \text{at } 10\%$$

which gives $Y = £2530 \cdot 99$.

(c) The loan outstanding just after the seventh payment is made is

$$2530 \cdot 99 a_{\overline{3}|} \qquad \text{at } 10\% \qquad = 6294 \cdot 32$$

Let Z be the revised annual instalment; the equation of value is

$$Z a_{\overline{3}|} = 6294 \cdot 32 \qquad \text{at } 9\%$$

whence $Z = £2486 \cdot 60$.

The equation of value for the entire transaction is

$$16\,000 = 2384 \cdot 46 a_{\overline{7}|} + 2530 \cdot 99 \,_4|a_{\overline{7}|} + 2486 \cdot 60 \,_7|a_{\overline{7}|}$$

i.e.

$$16\,000 + 146 \cdot 53 a_{\overline{7}|} - 44 \cdot 39 a_{\overline{7}|} - 2486 \cdot 60 a_{\overline{10}|} = 0$$

By interpolation between 8% and 9%, the yield, or effective rate of interest, per annum is 8·60%.

3.15 (a) (i) Let X be the annual repayment. The amount lent is the present value, on the stated interest basis, of the repayments. Thus

$$X[a_{\overline{6}|0 \cdot 10} + v^6_{0 \cdot 10} a_{\overline{120}|0 \cdot 09}] = 2000$$

from which we obtain $X = £238 \cdot 17$.

(ii) The loan outstanding just after payment of the third payment is (see equation 3.7.4)

$$2000(1 \cdot 1)^3 - 238 \cdot 17 s_{\overline{3}|0 \cdot 10} = 1873 \cdot 65$$

The loan outstanding just after payment of the fourth payment is

$$2000(1 \cdot 1)^4 - 238 \cdot 17 s_{\overline{4}|0 \cdot 10} = 1822 \cdot 84$$

The capital repaid at time 4 is therefore

$$1873 \cdot 65 - 1822 \cdot 84 = £50 \cdot 81 \tag{1}$$

The loan outstanding just after payment of the eleventh payment is (see equation 3.7.5)

$$238 \cdot 17 a_{\overline{7}|} \qquad \text{at } 9\% \qquad = 1198 \cdot 71$$

The loan outstanding just after payment of the twelfth payment is

$$238 \cdot 17 a_{\overline{6}|} \qquad \text{at } 9\% \qquad = 1068 \cdot 42$$

Hence the capital repaid at time 12 years is

$$1198 \cdot 71 - 1068 \cdot 42 = £130 \cdot 29 \tag{2}$$

(b) After the special payment, the capital outstanding is £968·42. The revised annual repayment, Y, is found from the equation of value

$$Ya_{\overline{6}|0 \cdot 09} = 968 \cdot 42$$

which gives

$$Y = £215 \cdot 88$$

3.16 Let X be the initial annuity payment, which is found from the equation of value

$$1000 = 2Xa_{\overline{10}|} - Xa_{\overline{5}|} \qquad \text{at } 10\%$$
$$= X(2a_{\overline{10}|} - a_{\overline{5}|})$$

whence $X = £117\cdot67$.

The loan outstanding just *after* the tth repayment is the value at time t of the future instalments, i.e.

$$L_t = \begin{cases} X(2a_{\overline{10-t}|} - a_{\overline{5-t}|}) & \text{for } 0 \leqslant t \leqslant 5 \\ X(2a_{\overline{10-t}|}) & \text{for } 5 \leqslant t \leqslant 10 \end{cases}$$

where we define $a_{\overline{0}|} = 0$ (see section 3.3). The capital repaid at time t years is $L_{t-1} - L_t$; when $1 \leqslant t \leqslant 5$, this is

$$X[(2a_{\overline{11-t}|} - a_{\overline{6-t}|}) - (2a_{\overline{10-t}|} - a_{\overline{5-t}|})]$$

$$= X\left[\frac{2(v^{10-t} - v^{11-t})}{0\cdot10} - \frac{(v^{5-t} - v^{6-t})}{0\cdot10}\right]$$

$$= 10X[2(v^{10} - v^{11}) - (v^5 - v^6)](1\cdot10)^t$$

$$= 17\cdot67(1\cdot1)^{t-1}$$

and when $6 \leqslant t \leqslant 10$, this is

$$2X(a_{\overline{11-t}|} - a_{\overline{10-t}|}) = 2X\left(\frac{v^{10-t} - v^{11-t}}{0\cdot10}\right)$$

$$= 90\cdot73(1\cdot1)^{t-1}$$

Check

$$17\cdot67(1 + 1\cdot1 + 1\cdot1^2 + \cdots + 1\cdot1^4) + 90\cdot73(1\cdot1^5 + 1\cdot1^6 + \cdots + 1\cdot1^9)$$

$$= 90\cdot73s_{\overline{10}|} - 73\cdot06s_{\overline{5}|} \qquad \text{at } 10\%$$

$$= 999\cdot96, \quad \text{i.e.} \quad 1000 \qquad \text{(ignoring round-off error)}$$

3.17 The total cash payable is

$$\int_0^n t\,dt = \tfrac{1}{2}n^2$$

We find n from the equation

$$(I\bar{a})_{\overline{n}|} = \tfrac{1}{4}n^2 \qquad \text{at } 5\%$$

i.e.

$$\frac{\bar{a}_{\overline{n}|} - nv^n}{\delta} = \tfrac{1}{4}n^2$$

i.e.

$$\frac{1-v^n}{\delta} - nv^n - \tfrac{1}{4}\delta n^2 = 0 \qquad \text{at } 5\%$$

By trials and interpolation, we find that $n = 22{\cdot}37$.

Chapter 4

4.1   (a) By formula 4.1.3,

$$\left[1 + \frac{i^{(4)}}{4}\right]^4 = 1 + i$$

whence $i^{(4)} = 0{\cdot}061\,086$.
    By formula 3.1.9,

$$e^\delta = 1 + i$$

whence $\delta = \log(1 + i) = 0{\cdot}060\,625$
By formula 4.1.6,

$$\left[1 - \frac{d^{(2)}}{2}\right]^2 = 1 - d = v = (1 + i)^{-1}$$

whence $d^{(2)} = 0{\cdot}059\,715$.
  (b) Since

$$\left[1 + \frac{i^{(12)}}{12}\right]^{12} = \left[1 + \frac{i^{(2)}}{2}\right]^2 = 1 + i$$

we obtain $i^{(12)} = 0{\cdot}061\,701$
    By formulae 3.1.9 and 4.1.3,

$$e^\delta = 1 + i = \left[1 + \frac{i^{(2)}}{2}\right]^2$$

from which we obtain $\delta = 0{\cdot}061\,543$.
    We have

$$\left[1 - \frac{d^{(4)}}{4}\right]^4 = 1 - d = (1 + i)^{-1} = \left[1 + \frac{i^{(2)}}{2}\right]^{-2}$$

which gives $d^{(4)} = 0{\cdot}061\,072$.
  (c) Since

$$\left[1 + \frac{i^{(2)}}{2}\right]^2 = e^\delta = (1 - d)^{-1} = \left[1 - \frac{d^{(12)}}{12}\right]^{-12}$$

we obtain $i^{(2)} = 0{\cdot}063\,655$, $\delta = 0{\cdot}062\,663$, and $d = 0{\cdot}060\,740$.

(d) Since

$$\left[1+\frac{i^{(4)}}{4}\right]^4 = \left[1-\frac{d^{(2)}}{2}\right]^{-2} = e^{\delta}$$

we obtain $i^{(4)} = 0.062\,991$ and $d^{(2)} = 0.061\,534$

4.2
$$a_{\overline{67}|}^{(4)} = \frac{1-v^{67}}{i^{(4)}} = 23.5391$$

$$\ddot{s}_{\overline{18}|}^{(12)} = \frac{i}{d^{(12)}} s_{\overline{18}|} = 26.1977$$

$$_{14|}\ddot{a}_{\overline{10}|}^{(2)} = \ddot{a}_{\overline{24}|}^{(2)} - \ddot{a}_{\overline{14}|}^{(2)} = \frac{i}{d^{(2)}}(a_{\overline{24}|} - a_{\overline{14}|}) = 4.8239$$

$$\bar{s}_{\overline{56}|} = \frac{(1+i)^{56}-1}{\delta} = 203.7755$$

$$a_{\overline{16.5}|}^{(4)} = \frac{1-v^{16.5}}{i^{(4)}} = 12.0887$$

$$\ddot{s}_{\overline{15.25}|}^{(12)} = \frac{(1+i)^{15.25}-1}{d^{(12)}} = 20.9079$$

$$_{4.25|}a_{\overline{3.75}|}^{(4)} = v^{4.25}\left(\frac{1-v^{3.75}}{i^{(4)}}\right) = 2.9374$$

$$\bar{a}_{\overline{26/3}|} = \frac{1-v^{26/3}}{\delta} = 7.3473$$

4.3 (a) (i) We work in time units of one year. The payments are illustrated in the following diagram:

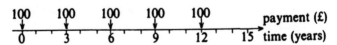

As in section 4.3, the accumulation at time 15, just before the payment then due is made, is

$$\frac{100}{a_{\overline{3}|}} s_{\overline{15}|} \quad \text{at } 10\% \quad = £1277.59, \quad \text{say} \quad £1278$$

(ii) We now work in units of a half-year. The position is illustrated in the following diagram:

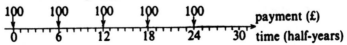

The accumulation at time 30, just before the payment then due is made, is (see section 4.3)

$$\frac{100}{a_{\overline{5}|}} s_{\overline{30}|} \quad \text{at } 5\% \quad = £1308 \cdot 96, \quad \text{say} \quad £1309$$

(b) Let us proceed from first principles. Let time be measured in years, and let $i = 0 \cdot 08$. The present value is

$$240(v + v^4 + v^7 + \cdots + v^{46})$$
$$= 240v[1 + (v^3) + (v^3)^2 + \cdots + (v^3)^{15}]$$
$$= 240v\frac{[1 - (v^3)^{16}]}{1 - v^3}$$
$$= 240v\frac{a_{\overline{48}|}}{a_{\overline{3}|}}$$
$$= £1051 \cdot 07, \quad \text{say} \quad £1051$$

Note The value of the payments two years before the present time is easily seen to be $(240/s_{\overline{3}|}) a_{\overline{48}|}$ at 8%. Multiplying this value by $1 \cdot 08^2$, we obtain the present value.

4.4 (a) $$f'(x) = e^{\delta/x}(1 - \delta/x) - 1 \qquad \text{for all } x > 0$$

Now it follows from elementary calculus that

$$e^y(1 - y) - 1 < 0 \qquad \text{for all } y \neq 0 \qquad (1)$$

so $f'(x) < 0$ for all $x > 0$. Hence $f(x)$ is a decreasing function on $(0, \infty)$ and, for each $m = 1, 2, \ldots$,

$$f(m + 1) < f(m)$$

Hence $i^{(m+1)} < i^{(m)}$ for each $m = 1, 2, \ldots$, (since $f(m) = i^{(m)}$ for each $m = 1, 2, \ldots$).

(b) We have

$$g'(x) = 1 - e^{-\delta/x}(1 + \delta/x)$$
$$= -[e^y(1 - y) - 1]$$

where $y = -\delta/x$. It follows from 1 that $g'(x) > 0$ for all $x > 0$ and hence that $g(x)$ is an increasing function on $(0, \infty)$. Since $g(m) = d^{(m)}$ for each $m = 1, 2, \ldots$, we have the result that $d^{(m+1)} > d^{(m)}$ for each $m = 1, 2, \ldots$.

(c) We have

$$\frac{1}{d^{(m)}} - \frac{1}{i^{(m)}} = \frac{1}{m[1 - (1 + i)^{-1/m}]} - \frac{1}{m[(1 + i)^{1/m} - 1]}$$

$$= \frac{(1+i)^{1/m} - 1}{m[(1+i)^{1/m} - 1]} = \frac{1}{m}$$

4.5 (a) Work in time units of one year. The present values are:

    (i) $600a_{\overline{20}|}$ at 12%   = £4482

    (ii) $600a_{\overline{20}|}^{(4)}$ at 12% $= 600[i/i^{(4)}]a_{\overline{20}|}$ at 12%

                                          = £4679

    (iii) $600a_{\overline{20}|}^{(12)}$ at 12% $= 600[i/i^{(12)}]a_{\overline{20}|}$ at 12%

                                          = £4723

    (iv) $600\bar{a}_{\overline{20}|}$ at 12% $= 600(i/\delta)a_{\overline{20}|}$ at 12%

                                          = £4745

(b) We now work in time units of a quarter-year with an interest rate of 3%. The present values are:

    (i) $(600/s_{\overline{4}|})\, a_{\overline{80}|}$ at 3%   = £4331   (see section 4.3)

    (ii) $150a_{\overline{80}|}$ at 3%   = £4530

    (iii) $150a_{\overline{80}|}^{(3)}$ at 3% $= 150[i/i^{(3)}]a_{\overline{80}|} = £4575$

    (iv) $150\bar{a}_{\overline{80}|}$ at 3%   = £4598

4.6 (a) Work in time units of one year. The present values are

    (i) $a_{\overline{\infty}|}^{(2)}$ at 12% $= 1/i^{(2)}$ at 12%   = 8·5763

    (ii) $v^{1/4}\, \ddot{a}_{\overline{\infty}|}^{(2)}$ at 12% $= v^{1/4}/d^{(2)}$ at 12%   = 8·8227

        (or, using (i), $v^{1/4}[(1/2) + a_{\overline{\infty}|}^{(2)}]$.)

(b) Work in time units of a half-year. The present values are

    (i) $(1/2)a_{\overline{\infty}|}$ at 6% $= 1/2i$ at 6%   = 8·3333

    (ii) $(1/2)v^{1/2}\, \ddot{a}_{\overline{\infty}|}$ at 6% $= v^{1/2}/2d$ at 6%   = 8·5797

(c) Work in time units of a quarter-year. The present values are

    (i) $(1/2)(1/s_{\overline{2}|})a_{\overline{\infty}|}$ at 3% $= 1/(2is_{\overline{2}|})$ at 3%

                                          = 8·2102

    (ii) $(1+i)\left(\dfrac{1}{2}\dfrac{a_{\overline{\infty}|}}{s_{\overline{2}|}}\right)$ at 3% = 8·4565

4.7 Let X be the initial annual amount of the annuity. We illustrate the payments and the rate of interest on the following diagram:

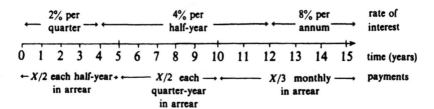

The present value of the annuity is found by summing the present values of the payments in each of the periods 1–4 years, year 5, 6–10 years, 11–12 years and 13–15 years. This gives the equation of value:

$$2049 = \frac{X}{2}\left(\frac{a_{\overline{4}|0\cdot02}}{s_{\overline{2}|0\cdot02}}\right) + v^{16}_{0\cdot02}\left(\frac{X}{2}\right)a_{\overline{2}|0\cdot04}$$

$$+ Xv^{16}_{0\cdot02}v^{2}_{0\cdot04}a^{(2)}_{\overline{10}|0\cdot04} + 2Xv^{16}_{0\cdot02}v^{12}_{0\cdot04}a^{(6)}_{\overline{2}|0\cdot04}$$

$$+ 4Xv^{16}_{0\cdot02}v^{16}_{0\cdot04}a^{(12)}_{\overline{3}|0\cdot08} = 17\cdot0763X$$

Hence $X = £119\cdot99$, say £120.

Note At 4%, $i^{(6)} = 0\cdot039\,349$.

4.8 (a) Let the present value of this annuity be A. We have

$$A = 1 + 4v + 9v^2 + \cdots + 361v^{18} + 400v^{19}$$

So

$$vA = v + 4v^2 + 9v^3 + \cdots + 361v^{19} + 400v^{20}$$

and

$$A - vA = 1 + 3v + 5v^2 + \cdots + 39v^{19} - 400v^{20}$$
$$= 1 + v + v^2 + \cdots + v^{19} + 2(v + 2v^2 + 3v^3 + \cdots + 19v^{19})$$
$$- 400v^{20}$$
$$= 1 + a_{\overline{19}|} + 2(Ia)_{\overline{19}|} - 400v^{20}$$

Hence

$$A = \frac{1 + a_{\overline{19}|} + 2(Ia)_{\overline{19}|} - 400v^{20}}{1 - v} \qquad \text{at 5\%} \qquad = £1452\cdot26$$

(b) The present value of the annuity is $[d/d^{(4)}]\,A$ at 5%, i.e. £1426·06.

(c) The present value of the annuity is $[d/i^{(2)}]\,A$ at 5%, i.e. £1400·18.

(d) The present value of the annuity is $(d/\delta)\,A$ at 5%, i.e. £1417·40.

4.9 (a) Work in time units of one year. The accumulation is

$$200\,s^{(4)}_{\overline{13\cdot5}|} \qquad \text{at 12\%} \qquad = 200\left[\frac{(1 + i)^{13\cdot5} - 1}{d^{(4)}}\right] \qquad \text{at 12\%}$$

$$= £6475\cdot64$$

(b) Work in time units of a half-year. The accumulation is

$$100\,s^{(2)}_{\overline{27}|} \qquad \text{at 6\%} \qquad = £6655\cdot86$$

(c) Work in time units of a quarter-year. The accumulation is

$$50\,s_{\overline{54}|} \qquad \text{at 3\%} \qquad = £6753\cdot58$$

(d) Work in time units of one month. The accumulation is (see section 4.3)

$$\frac{50s_{\overline{16}|}}{a_{\overline{5}|}} \quad \text{at } 1\% \quad = £6821 \cdot 85$$

4.10 Let X be the amount of the revised annuity, and consider the position at the date of the request. The equation of value is

$$X(1+i)^{1/4}a^{(2)}_{\overline{21 \cdot 5}|} = 200v^{1/4}\ddot{a}_{\overline{22}|} + 320(1+i)^{1/12}a^{(4)}_{\overline{16 \cdot 25}|}$$

$$+ 180a^{(12)}_{\overline{18 \cdot 75}|} \quad \text{at } 8\%$$

from which we obtain

$$X = \frac{6899 \cdot 89}{10 \cdot 5091} = £656 \cdot 56$$

4.11 (a) (i) Let X be the monthly annuity payment. Working in time units of one year, we have the equation of value

$$400 s^{(2)}_{\overline{20}|} = 12 X \ddot{a}^{(12)}_{\overline{15}|} \quad \text{at } 12\%$$

whence $X = £361 \cdot 01$, say £361.

(ii) Let Y be the monthly annuity payment. Working in time units of a half-year, we obtain the equation

$$200 \ddot{s}_{\overline{40}|} = 6 Y \ddot{a}^{(6)}_{\overline{30}|} \quad \text{at } 6\%$$

whence (on calculating $i^{(6)} = 6[(1+i)^{1/6} - 1]$ at $6\% = 0 \cdot 058\,553$) $Y = £383 \cdot 93$, say £384.

(iii) Let Z be the monthly annuity payment. Working in time units of one month, we have

$$200\left(\frac{s_{\overline{240}|}}{a_{\overline{6}|}}\right) = Z\ddot{a}_{\overline{180}|} \quad \text{at } 1\%$$

whence $Z = £405 \cdot 67$, say £406.

(b) The monthly annuity payments are now:

(i) $Xv^{5/12}$ at $12\% = £344 \cdot 36$, say £344

(ii) $Yv^{5/6}$ at $6\% = £365 \cdot 73$, say £366

(iii) Zv^5 at $1\% = £385 \cdot 98$, say £386

4.12 (a) Let $X/12$ be the monthly repayment. We solve the equation of value

$$X a^{(12)}_{\overline{25}|} = 9880 \quad \text{at } 7\%$$

to obtain $X = 821 \cdot 76$. Hence the monthly repayment is £68·48.

(b) The loan outstanding is $X a^{(12)}_{\overline{34/3}|}$ at $7\% = £6486$.

(c) The loan outstanding just after the repayment on 10 September 1989 is made is $Xa^{(12)}_{83/6}$ at 7%.

The loan outstanding just after the repayment on 10 October 1989 is made is $Xa^{(12)}_{13.75}$ at 7%.

The capital repaid on 10 October 1989 is thus

$$X(a^{(12)}_{83/6} - a^{(12)}_{13.75})   \quad \text{at } 7\%$$

$$= X\left(\frac{v^{13.75} - v^{83/6}}{i^{(12)}}\right) \quad \text{at } 7\% \quad = £26.86$$

(d) (i) The capital repayment contained in these 12 instalments is

$$X(a^{(12)}_{22/31} - a^{(12)}_{19/31}) \quad \text{at } 7\% \quad = £516.20$$

(ii) The total interest in these 12 instalments is (by (i))

$$X - 516.20 = £305.56$$

(e) The capital and interest contents of each instalment may be found by means of the loan schedule (see section 3.8), using the rate of interest $j = i^{(12)}_{0.07}/12$ per month. The capital and interest contents of the tth repayment are thus

$$\frac{9880}{a_{\overline{300}|}} v^{301-t} \quad \text{at rate } j$$

and

$$\frac{9880}{a_{\overline{300}|}} (1 - v^{301-t}) \quad \text{at rate } j$$

respectively. Note that $j = 0.005\,654$. We therefore require the smallest t such that

$$v^{301-t} > \tfrac{1}{2}(1 - v^{301-t}) \quad \text{at rate } j$$

i.e.

$$(1+j)^t > \frac{(1+j)^{301}}{3} \quad \text{at rate } j$$

i.e.

$$t > \frac{-\log 3 + 301 \log(1+j)}{\log(1+j)} = 106.15$$

The required value of t is thus 107. The capital repaid first exceeds one-half of the interest content at the 107th instalment, payable on 10 June 1987. (The capital content of this instalment is £22.94 and the interest content is £45.54.)

4.13 (a) Let $X/12$ be the original monthly repayment, and let $Y/12$ be the monthly payment payable if the borrower had elected not to extend the term of the loan. We have

$$X a_{\overline{20}|}^{(12)} = 19\,750 \qquad \text{at } 9\%$$

whence $X = 2079 \cdot 12$ and $X/12 = £173 \cdot 26$.

Also, the loan outstanding just after the 87th monthly payment had been made is

$$X a_{\overline{12 \cdot 75}|}^{(12)} \qquad \text{at } 9\% \qquad = 16\,027 \cdot 52$$

and hence

$$Y a_{\overline{12 \cdot 75}|}^{(12)} = 16\,027 \cdot 52 \text{ at } 10\%$$

This gives $Y = 2180 \cdot 52$ and $Y/12 = £181 \cdot 71$.

The increase in the monthly instalment would thus have been $(Y - X)/12 = £8 \cdot 45$.

(b) Let the revised outstanding term of the loan be n months. We find n from the inequalities

$$2079 \cdot 12 a_{\overline{(n-1)/12}|}^{(12)} < 16\,027 \cdot 52 \leqslant 2079 \cdot 12 a_{\overline{n/12}|}^{(12)} \qquad \text{at } 10\%$$

(Notice that n must be an integer, and the final repayment may be of reduced amount.) By trials, we obtain $n = 169$, so that the revised outstanding term is 14 years and 1 month. The loan outstanding just after the penultimate instalment is paid is

$$[16\,027 \cdot 52 - 2079 \cdot 12 a_{\overline{14}|}^{(12)}](1 + i)^{14} \qquad \text{at } 10\% \qquad = 81 \cdot 13.$$

The final instalment must comprise this sum plus interest on it for one month; the final payment is therefore

$$81 \cdot 13(1 + i)^{1/12} \qquad \text{at } 10\% \qquad = £81 \cdot 78$$

Note: an alternative method of finding this reduced final payment is from the equation of value

$$2079 \cdot 12 a_{\overline{14}|}^{(12)} + \text{(final payment)} \, v^{169/12} = 16\,027 \cdot 52 \qquad \text{at } 10\%$$

4.14 (a) Let X be the initial quarterly payment. We have the equation of value

$$4X a_{\overline{13}|}^{(4)} + 160 \,_5| a_{\overline{10}|}^{(4)} + 160 \,_{10}| a_{\overline{3}|}^{(4)} = 11\,820 \qquad \text{at } 12\%$$

This gives $X = 389 \cdot 96$.

(b) (i) The initial annual amount of the annuity is $4X$, i.e. £1559·84. The loan outstanding just after payment of the eighth instalment (at the end of the second year) is

$$11\,820(1 + i)^2 - 1559\cdot84s^{(4)}_{\overline{2}|} \qquad \text{at } 12\% \qquad = 11\,374\cdot85$$

and, just after the twelfth instalment is paid, the loan outstanding is

$$11\,820(1 + i)^3 - 1559\cdot84s^{(4)}_{\overline{3}|} \qquad \text{at } 12\% \qquad = 11\,111\cdot45$$

The capital repaid during the third year of the loan is therefore

$$11\,374\cdot85 - 11\,111\cdot45 = \pounds263\cdot40$$

(ii) In the final five years, it is easier to calculate the outstanding loan by valuing the *future* annuity instalments than by accumulating the original loan less past instalments of the annuity. Thus we obtain the expressions

$$1879\cdot84a^{(4)}_{\overline{3}|} \qquad \text{at } 12\%$$

$$1879\cdot84a^{(4)}_{\overline{2}|} \qquad \text{at } 12\%$$

for the loan outstanding at the end of the twelfth and thirteenth years respectively (after payment of the annuity instalments due at these times). The capital repaid in the thirteenth year is thus

$$1879\cdot84(a^{(4)}_{\overline{3}|} - a^{(4)}_{\overline{2}|}) \qquad \text{at } 12\% = \pounds1396\cdot82$$

(c) The loan outstanding just after the 33rd instalment is paid is (on valuing future payments)

$$1879\cdot84a^{(4)}_{\overline{6\cdot75}|} - 160a^{(4)}_{\overline{1\cdot75}|} \qquad \text{at } 12\% \qquad = 8493\cdot09$$

Let the revised quarterly annuity payment be Y. We have the equation of value

$$4Ya^{(4)}_{\overline{6\cdot75}|} = 8493\cdot09 \qquad \text{at } 12\%$$

which gives $Y = 456\cdot50$

4.15 Let X_t denote the gross annuity payment at time t; let c_t denote the capital repayment at time t; and let L_t denote the loan outstanding just after the payment at time t is made. We have the relations

$$X_t = 0\cdot1L_{t-1} + c_t \qquad 1 \leqslant t \leqslant 10 \qquad (1)$$

(where L_0 denotes the amount of the original loan), and

$$(0\cdot7 \times 0\cdot1L_{t-1}) + c_t = 5000 \qquad 1 \leqslant t \leqslant 10 \qquad (2)$$

We also have, for $t \geqslant 1$,

$$
\begin{aligned}
L_t &= L_{t-1} - c_t \\
&= L_{t-1} - 5000 + 0\cdot07 L_{t-1} \qquad \text{(by equation 2)} \\
&= 1\cdot07 L_{t-1} - 5000 \qquad\qquad\qquad\qquad (3)
\end{aligned}
$$

Now this shows that

$$
\begin{aligned}
L_1 &= 1\cdot07 L_0 - 5000 \\
L_2 &= 1\cdot07^2 L_0 - 5000(1 + 1\cdot07) \\
&= 1\cdot07^2 L_0 - 5000 s_{\overline{2}|0\cdot07}
\end{aligned}
$$

and so on, giving

$$
L_t = 1\cdot07^t L_0 - 5000 s_{\overline{t}|0\cdot07} \qquad 1 \leqslant t \leqslant 10 \qquad (4)
$$

(This result may be rigorously established from equation 3 by finite induction, or by solving a first-order difference equation.) But $L_{10} = 0$, so

$$
1\cdot07^{10} L_0 - 5000 s_{\overline{10}|0\cdot07} = 0
$$

whence

$$
L_0 = 5000 a_{\overline{10}|0\cdot07} = 35\,117\cdot91
$$

Equations 1, 2 and 4 now show that for $1 \leqslant t \leqslant 10$,

$$
\begin{aligned}
X_t &= 0\cdot1 L_{t-1} + (5000 - 0\cdot07 L_{t-1}) \\
&= 5000 + 0\cdot03 L_{t-1} \\
&= 5000 + 0\cdot03 \left\{ (1\cdot07)^{t-1} L_0 - 5000 \left[\frac{(1\cdot07)^{t-1} - 1}{0\cdot07} \right] \right\} \\
&= 7142\cdot86 - 1089\cdot32(1\cdot07)^{t-1}
\end{aligned}
$$

Hence the price that the investor should pay to obtain a yield of 8% effective per annum is

$$
\sum_{t=1}^{10} X_t v^t \qquad \text{at } 8\%
$$

$$
= 7142\cdot86 a_{\overline{10}|0\cdot08} - (1089\cdot32/1\cdot07) \sum_{t=1}^{10} \left(\frac{1\cdot07}{1\cdot08} \right)^t
$$

$$
= 7142\cdot86 a_{\overline{10}|0\cdot08} - 1089\cdot32 \left[\frac{1 - \left(\dfrac{1\cdot07}{1\cdot08} \right)^{10}}{1\cdot08 - 1\cdot07} \right]
$$

$$
= £38\,253
$$

4.16 (a) The annuity payments are illustrated in the following diagram, in which time is measured in years from 1 May 1986:

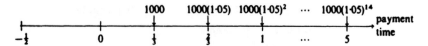

The value, at time 0, of the annuity payments is

$$1000(v^{1/3} + 1{\cdot}05v^{2/3} + 1{\cdot}05^2 v^{3/3} + \cdots + 1{\cdot}05^{14} v^{15/3})$$

$$= 1000v^{1/3}[1 + 1{\cdot}05v^{1/3} + (1{\cdot}05v^{1/3})^2 + \cdots$$
$$+ (1{\cdot}05v^{1/3})^{14}] \qquad \text{at } 6\%$$

$$= 1000v^{1/3}\left[\frac{1 - (1{\cdot}05v^{1/3})^{15}}{1 - (1{\cdot}05v^{1/3})}\right] \qquad \text{at } 6\%$$

$$= 1000\frac{1{\cdot}05^{15}1{\cdot}06^{-5} - 1}{1{\cdot}05 - 1{\cdot}06^{1/3}}$$

$$= 18\,214{\cdot}79$$

Hence the purchase price is

$$18\,214{\cdot}79v^{1/2} \qquad \text{at } 6\% \qquad = £17\,691{\cdot}77, \qquad \text{say} \qquad £17\,692$$

(b) (i) The loan outstanding just after payment of the sixth instalment is

$$17\,691{\cdot}77(1{\cdot}06)^{5/2} - 1000(1{\cdot}06)^{5/3}[1 + 1{\cdot}05v^{1/3} + \cdots$$
$$+ (1{\cdot}05v^{1/3})^5] \qquad \text{at } 6\%$$

$$= 17\,691{\cdot}77(1{\cdot}06)^{5/2} - 1000\frac{1{\cdot}05^6 - 1{\cdot}06^2}{1{\cdot}05 - 1{\cdot}06^{1/3}}$$

$$= 13\,341{\cdot}57$$

The interest content of the seventh instalment is thus

$$13\,341{\cdot}57[(1{\cdot}06)^{1/3} - 1] = £261{\cdot}67$$

(ii) We note that the fifteenth instalment equals

$$1000(1{\cdot}05)^{14} = 1979{\cdot}93$$

and this must provide the capital outstanding just after the fourteenth payment is made, L, and interest on this amount. That is,

$$L + L[(1{\cdot}06)^{1/3} - 1] = 1979{\cdot}93$$

whence

$$L = \frac{1979 \cdot 93}{(1 \cdot 06)^{1/3}} = 1941 \cdot 85$$

and the interest content of the fifteenth instalment is

$$L[(1 \cdot 06)^{1/3} - 1] = £38 \cdot 08$$

Chapter 5

5.1 (a) *Project A* The internal rate of return i is the solution of the equation

$$-1\,000\,000 + 270\,000 a_{\overline{5}|} = 0 \qquad \text{at rate } i$$

i.e.

$$a_{\overline{5}|} \qquad \text{at rate } i \qquad = 3 \cdot 7037$$

which gives $i = 0 \cdot 2120$ or $21 \cdot 20\%$.

Project B The internal rate of return i is the solution of the equation

$$-1\,200\,000 - 20\,000 a_{\overline{5}|} + 1\,350\,000\,_5|a_{\overline{5}|} = 0 \qquad \text{at rate } i,$$

which gives $i = 0 \cdot 1854$ or $18 \cdot 54\%$.

(b) We have (in the notation of section 5.3)

$$NPV_A(0 \cdot 15) = 1000[-1000 + 270 a_{\overline{5}|}] \qquad \text{at } 15\%$$
$$= £211\,577$$

$$NPV_B(0 \cdot 15) = 1000[-1200 - 20 a_{\overline{5}|} + 1350\,_5|a_{\overline{5}|}] \qquad \text{at } 15\%$$
$$= £286\,814$$

Comment Project A is viable when money can be borrowed at up to $21 \cdot 2\%$ per annum interest, but project B is viable only for interest rates up to $18 \cdot 5\%$ per annum. At 15% per annum interest, however, project B has the greater net present value: the graphs of the functions $NPV_A(i)$ and $NPV_B(i)$ must be as shown in Figure S.5.1.

5.2 (a) *Project A* The internal rate of return i is the solution of the equation

$$-160 - 80v + 60(a_{\overline{11}|} - a_{\overline{2}|}) = 0 \qquad \text{at rate } i$$

(where we have worked in money units of £1000). This gives $i = 0 \cdot 0772$, or $7 \cdot 72\%$.

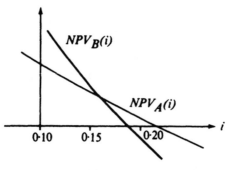

FIGURE S.5.1

Project B The internal rate of return i is the solution of the equation

$$-193 - 80v + 70(a_{\overline{11}|} - a_{\overline{4}|}) = 0 \qquad \text{at rate } i$$

(where we again work in money units of £1000). This gives $i = 0.0804$, or 8.04%.

(b) The purchase money of £33000 must buy at least £10000 per annum for seven years, beginning at time 5 years, for the combined transaction to be more favourable to the businessman than project B. The minimum rate of interest per annum i must therefore satisfy

$$-33\,000 + 10\,000(a_{\overline{11}|} - a_{\overline{4}|}) = 0 \qquad \text{at rate } i$$

i.e.

$$a_{\overline{11}|} - a_{\overline{4}|} = 3.3 \qquad \text{at rate } i$$

This gives $i = 0.1011$, or 10.11%.

5.3 (a) Working in money units of £1000, we have

$$-10 - 3v + (\tfrac{1}{2}v^2 + \tfrac{2}{2}v^3 + \tfrac{3}{2}v^4 + \cdots + \tfrac{8}{2}v^9) = 0 \qquad \text{at rate } i$$

where i is the internal rate of return. That is,

$$-10 - 3v + \tfrac{1}{2}v(Ia)_{\overline{8}|} = 0 \qquad \text{at rate } i$$

Hence $i = 0.0527$, or 5.27%.

(b) The investor should proceed with the venture, since its internal rate of return exceeds the rate of interest he must pay on borrowed money. The profit in nine years' time is found by noting that the investor's bank account, just *before* receiving the final income (but *after* paying interest on his borrowing), is

$$-10\,000(1{\cdot}05)^9 - 3000(1{\cdot}05)^8 + 500(1{\cdot}05)^8(Ia)_{\overline{10}|0{\cdot}05} = -3680$$

so the final income payment (of £4000) extinguishes the loan, and leaves a profit of £320.

5.4 (a) *Sheep rearing* The internal rate of return i is the solution of the equation of value

$$20\,000 = 1100a_{\overline{20}|} + 20\,000v^{20}$$

which gives $i = 5{\cdot}5\%$ exactly. This is also clear from the fact that income is sufficient to pay interest at $5{\cdot}5\%$ p.a. on the initial cost, and the capital is repaid at the end of 20 years.

Goat breeding The internal rate of return i is found by solving

$$20\,000 = 900a_{\overline{20}|} + 25\,000v^{20} \qquad \text{at rate } i$$

which gives $i = 5{\cdot}24\%$.

Forestry The internal rate of return i is found by solving

$$20\,000(1 + i)^{20} = 57\,300$$

which gives $i = 5{\cdot}40\%$.

(b) We calculate the profit after 20 years from each of the three ventures:

Sheep rearing The 'surplus' income is £100 p.a., so the profit is $100s_{\overline{20}|}$ at $4\% = $£2978.

Goat breeding A further £100 must be borrowed annually in arrear for 20 years, and these loans grow at 5% p.a. compound. The total indebtedness at time 20 years is thus

$$20\,000 + 100s_{\overline{20}|} \qquad \text{at } 5\% \qquad = 23\,307$$

Hence the profit to the investor at the end of the project is $25\,000 - 23\,307 = $£1693.

Forestry The original debt of £20000 will have grown to $20\,000(1{\cdot}05)^{20} = 53\,066$ after 20 years, so the profit is $57\,300 - 53\,066 = $£4234.

Hence the largest profit is obtained from forestry.

5.5 Let us work in money units of £1 million, and take i as the rate of interest:

(a) The present value of the outlay is

$$2{\cdot}7 + 0{\cdot}2a_{\overline{3}|} \qquad \text{at rate } i$$

The expected value of the present value of the income is

$$0{\cdot}1v^3 + \bar{a}_{\overline{10}|}(0{\cdot}25v + 0{\cdot}2v^2 + 0{\cdot}1v^3) \qquad \text{at rate } i$$

where we have multiplied the values of the income from a strike in years 1, 2, and 3 respectively by the probability of striking oil in these years. The rate of interest per annum which will make the present value of the company's outlays equal the expected value of the present value of its income is thus the solution of the equation

$$2{\cdot}7 + 0{\cdot}2a_{\overline{3}|} = 0{\cdot}1v^3 + \bar{a}_{\overline{10}|}(0{\cdot}25v + 0{\cdot}2v^2 + 0{\cdot}1v^3) \qquad \text{at rate } i$$

This gives $i = 9{\cdot}33\%$.

(b) On these assumptions, the present value of the company's income is

$$0{\cdot}1v^3 + 0{\cdot}85v^3\bar{a}_{\overline{10}|}$$

So the effective annual rate of return i is found by solving

$$2{\cdot}7 + 0{\cdot}2a_{\overline{3}|} = 0{\cdot}1v^3 + 0{\cdot}85v^3\bar{a}_{\overline{10}|} \qquad \text{at rate } i$$

which gives $i = 14{\cdot}55\%$.

5.6 Let us work in money units of £1000.

(a) We must find any values of i, the rate of interest per annum, satisfying

$$1 - 2v + 2v^2 = 0 \qquad \text{for project } A$$

and

$$-4 + 7v - 1{\cdot}5v^2 = 0 \qquad \text{for project } B$$

Now the equation $1 - 2v + 2v^2 = 0$ has no real solution, so there is no value of i satisfying the given condition for project A. But

$$-4 + 7v - 1{\cdot}5v^2 = (-1{\cdot}5v + 1)(v - 4)$$

which equals 0 for $v = 2/3$ or 4. Since $i = v^{-1} - 1$, the required values of i for project B are $0{\cdot}5$ and $-0{\cdot}75$, i.e. 50% and -75%.

(b) In the notation of section 5.3,

$$NPV_A(i) - NPV_B(i) = (1 - 2v + 2v^2) - (-4 + 7v - 1{\cdot}5v^2)$$
$$= 5 - 9v + 3{\cdot}5v^2$$
$$= v^2(5x^2 - 9x + 3{\cdot}5) \qquad \text{(where } x = 1 + i\text{).}$$

By the properties of quadratics,

$$5x^2 - 9x + 3{\cdot}5 > 0 \qquad \text{for } x > 1{\cdot}231\,66 \text{ and } x < 0{\cdot}568\,33$$

So

$$NPV_A(i) - NPV_B(i) > 0 \quad \text{for } i > 23 \cdot 166\% \text{ and } i < -43 \cdot 167\%$$

The range of positive interest rates for which $NPV_A(i) > NPV_B(i)$ is therefóre $i > 23 \cdot 166\%$.

(c) It follows from (b) that project B is the more profitable in case (i), and project A is the more profitable in case (ii). The accumulated profits, at an interest rate of i per annum, of projects A and B are (in units of £1000)

$$(1 + i)^2 - 2(1 + i) + 2$$

and

$$-4(1 + i)^2 + 7(1 + i) - 1 \cdot 5$$

respectively. This gives the accumulations
(i) £1040 for A, £1140 for B
(ii) £1062·5 for A, £1000 for B.

5.7 (a) The internal rate of return i is the solution of the equation of value

$$1000[-10 + 6(1 + i)^{-1} + 6 \cdot 6(1 + i)^{-2}] = 0$$

which gives $i = 0 \cdot 1660$, say 16·6%.

(b) (i) Since the internal rate of return exceeds 16%, the person should proceed with the investment. The balance at the end of one year (just after receipt of the payment then due) is

$$-10\,000(1 \cdot 16) + 6000 = -5600$$

So the balance at the end of two years (just after receipt of the payment then due) is

$$-5600(1 \cdot 16) + 6600 = £104$$

(ii) Under the conditions of the loan, the borrower must pay £1600 in interest at time 1 year, and £11 600 in interest and capital repayment at time 2 years. He will therefore have £4400 to invest at time 1 year, which gives £4400(1·13) = £4972 at time 2 years. Together with the second payment of £6600, he will thus have £11 572 available at time 2 years. But this is less than £11 600, so the investor should *not* proceed.

5.8 (a) Let us work in money units of £1000. The net cash flows associated with the project are shown in the following diagram:

```
-500    200        200        200       -250      -250
        200  200   200   200  200  200        -250        net cash flow
  |----+----+----+----+----+----+----+----+----+----+---->
  0    1    2    3    4    5    6    7    8    9    10     time (years)
```

Until the initial loan is repaid, the chemical company's balance just after the transaction at time t years is

$$-500(1+i)^t + 200s_{\overline{t}|} \qquad \text{at } 15\%$$

This is negative if $a_{\overline{t}|} < 2\cdot5$ at 15%, i.e. if $t \leqslant 3$. The bank loan is therefore paid off after four years.

(b) The balance at time 4 years, just after repaying the bank loan, is

$$-500(1+i)^4 + 200s_{\overline{4}|} \qquad \text{at } 15\% \qquad = 124\cdot172$$

The balance at time 10 years, when the project ends, is thus

$$124\cdot172(1+i)^6 + 200(s_{\overline{6}|} - s_{\overline{3}|}) - 250s_{\overline{3}|} \qquad \text{at } 12\%$$

$$= 349\cdot651 \qquad \text{or} \qquad \pounds349\,651$$

5.9 (a) (i) Let us work in money units of £1000. By formula 5.4.1, the DPP is the value of t such that

$$-80(1+i)^t - 5(1+i)^{t-1} + 10\bar{s}_{\overline{t-2}|} \qquad \text{at } 7\% \qquad = 0$$

By interpolation, we obtain $t = 17\cdot767$, say 17·77 years.

(ii) The accumulated profit after 22 years is found by accumulating the income, after the DPP has elapsed, at 6% per annum. This gives

$$10\bar{s}_{\overline{4\cdot233}|} \qquad \text{at } 6\% \qquad = 48\cdot00$$

The accumulated amount is thus £48 000.

(b) (i) The businessman may accumulate his rental income for each year at 6%, giving $10\bar{s}_{\overline{1}|}$ at 6% = 10·2971 at the end of the year. The DPP is thus the smallest integer t such that

$$-80(1+i)^t - 5(1+i)^{t-1} + 10\cdot2971s_{\overline{t-2}|} \qquad \text{at } 7\% \quad \geqslant 0$$

By trials, this gives $t = 18$ years.

(ii) The balance in the businessman's account just after the transaction at time 18 years is

$$-80(1+i)^{18} - 5(1+i)^{17} + 10\cdot2971s_{\overline{16}|} \qquad \text{at } 7\% \qquad = 0\cdot977\,14.$$

Hence the accumulated amount in his account after 22 years is

(see equation 5.4.5)

$0.977\,14(1+i)^4 + 10\bar{s}_{\overline{7}|}$ at 6%
$= 46.279,$ i.e.   £46 279.

5.10 (a) $NPV(i) = -1000 + C(1+i)^{-1} - 600(1+i)^{-4}$
(b) Let $f(i) = NPV(i)$. We have

$$f'(i) = C(1+i)^{-5}\left[\frac{2400}{C} - (1+i)^3\right]$$

$$\left.\begin{array}{ll} >0 & \text{for } i < i_0 \\ <0 & \text{for } i > i_0 \\ =0 & \text{for } i = i_0 \end{array}\right\} \text{ where } i_0 = \left(\frac{2400}{C}\right)^{1/3} - 1$$

Also, $f(0) = C - 1600$ and $f(i) \to -1000$ as $i \to \infty$. We also have

$$f(i_0) = -1000 + C\left(\frac{C}{2400}\right)^{1/3} - 600\left(\frac{C}{2400}\right)^{4/3}$$

$$= -1000 + \frac{3C}{4}\left(\frac{C}{2400}\right)^{1/3}$$

(i) When $C > 1600$, $f(0) > 0$ and the graph of $f(i)$ must have the appearance of Figure S.5.2(a). Note: i_0 may be positive, zero or negative.

When $C = 1600$, $f(0) = 0$ and the graph of $f(i)$ has the appearance of Figure S.5.2(b).

In both cases, there is exactly one positive solution of the equation $f(i) = 0$.

(ii) When $\alpha < C < 1600$, we have $f(0) < 0$, $i_0 > 0$ and

$$f(i_0) > -1000 + \frac{3\alpha}{4}\left(\frac{\alpha}{2400}\right)^{1/3}$$

$$= 0$$

Hence the graph of $f(i)$ has the appearance of Figure S.5.2(c). It is clear that there are precisely two positive solutions of the equation $f(i) = 0$.

(iii) When $C = \alpha$, we have $f(0) < 0$, $i_0 > 0$ and $f(i_0) = 0$. It is clear that the curve $f(i)$ touches the i-axis at i_0, and is negative elsewhere. Hence there is exactly one positive solution of the equation (namely i_0).

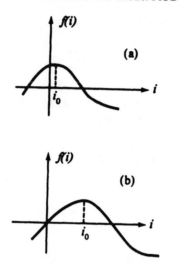

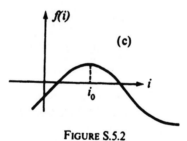

FIGURE S.5.2

   (iv) When $C < \alpha$, $f(i_0) < 0$, and hence the equation $f(i) = 0$ has no
        solutions (and therefore no positive solutions).
(c) (i) When $C = 1600$, there is exactly one positive solution,
        namely 0·3631, or 36·31%.
   (ii) When $C = 1550$, there are precisely two positive solutions,
        namely 0·1036 and 0·2177, or 10·36% and 21·77%.
  (iii) When $C = \alpha$, we have precisely one positive solution, namely
        $i_0 \simeq 0·1588$, or 15·88%.
5.11 (a) By equation 5.7.5, the time-weighted rates of return are

   $\frac{164}{124} - 1 = 0·3226$,   or   32·26%   for the property fund

   and

   $\frac{155}{121} - 1 = 0·2810$,   or   28·10%   for the equity fund.

(b) We first consider the property fund.

(i) Assuming that the investor buys 100 units per quarter, we obtain the equation of value

$$124(1 + i) + 131(1 + i)^{3/4} + 148(1 + i)^{1/2} + 158(1 + i)^{1/4} = 4 \times 164$$

where i is the yield per annum. This gives $i = 0.2932$, or 29.32%.

(ii) Now assume that the investor purchases £1 worth of units each quarter. The equation of value is

$$4\ddot{s}^{(4)}_{\overline{1}|} = 1.64\left(\frac{1}{1.24} + \frac{1}{1.31} + \frac{1}{1.48} + \frac{1}{1.58}\right)$$

i.e. $\ddot{s}^{(4)}_{\overline{1}|} = 1.1801$. This gives $i = 0.2979$, or 29.79%.

(c) We now consider the equity fund.

(i) The equation of value is

$$121(1 + i) + 92(1 + i)^{3/4} + 103(1 + i)^{1/2} + 131(1 + i)^{1/4} = 4 \times 155,$$

which gives $i = 0.6731$, or 67.31%.

(ii) The equation of value is

$$4\ddot{s}^{(4)}_{\overline{1}|} = 1.55\left(\frac{1}{1.21} + \frac{1}{0.92} + \frac{1}{1.03} + \frac{1}{1.31}\right)$$

i.e. $\ddot{s}^{(4)}_{\overline{1}|} = 1.4135$. This yield is thus 0.7088, or 70.88%.

Comment For each fund, the yield to the investor depends only slightly on whether a fixed sum was invested, or a fixed number of units was purchased each quarter. The equity fund has the lower time-weighted rate of return, but much higher yields: this is because the price of the equity fund fluctuates more than that of the property fund, and the equity fund price at the end of the year was relatively high.

5.12 (a) (i) By equation 5.7.5, the time-weighted rate of return is

$$\left(\frac{352}{186}\right)^{1/6} - 1 = 0.1122, \qquad \text{or} \qquad 11.22\%$$

(ii) As in example 5.7.2, let U_r be the unit price on 1 April 1979 + r. The yield per annum i is found from the equation

$$200 \sum_{r=0}^{5} U_r(1 + i)^{6-r} = 1200 U_6$$

i.e.
$$186(1 + i)^6 + 211(1 + i)^5 + 255(1 + i)^4 + 249(1 + i)^3$$
$$+ 288(1 + i)^2 + 318(1 + i) = 2112$$

which gives $i = 0.1060$, or 10.60%.

(iii) The yield per annum i is found from the equation

$$500\ddot{s}_{\overline{6}|} = 500\left(\sum_{r=0}^{5}\frac{1}{U_r}\right)U_6 \qquad \text{at rate } i$$

i.e. $\ddot{s}_{\overline{6}|i} = 8\cdot6839$. Hence $i = 0\cdot1067$, or $10\cdot67\%$.

(b) The revised answer to (a) (ii) is

$$200\sum_{r=0}^{5}(1\cdot02U_r)(1 + i)^{6-r} = 1200 \times 0\cdot98U_6$$

which gives

$$186(1 + i)^6 + 211(1 + i)^5 + 255(1 + i)^4 + 249(1 + i)^3 + 288(1 + i)^2$$
$$+ 318(1 + i) = 2029\cdot18$$

Hence $i = 0\cdot0933$, or $9\cdot33\%$.

The revised answer to (a) (iii) is

$$500\ddot{s}_{\overline{6}|} = 500\left[\sum_{r=0}^{5}\left(\frac{1}{1\cdot02U_r}\right)\right](0\cdot98\,U_6)$$

i.e. $\ddot{s}_{\overline{6}|i} = 8\cdot3434$, which gives $i = 0\cdot0950$, or $9\cdot50\%$.

Chapter 6

6.1 (a) The annual premium is

$$P = \frac{10\,000}{\ddot{s}_{\overline{20}|}} \qquad \text{at } 6\% \qquad = \pounds256\cdot46$$

(b) The policy value is

$$V = 10\,000\,{}_6V_{\overline{20}|} = P\ddot{s}_{\overline{6}|} \qquad \text{at } 6\% \qquad = \pounds1896\cdot21$$

(c) The paid-up sum assured is

$$10\,000\,{}_6W_{\overline{20}|} = V(1 + i)^{14} \qquad \text{at } 6\% \qquad = \pounds4287\cdot15$$

6.2 The equation of value, at the maturity date, is

$$0\cdot96P'(\ddot{s}_{\overline{40}|} - \tfrac{1}{2}\ddot{s}_{\overline{30}|} - \tfrac{1}{4}\ddot{s}_{\overline{20}|}) - (0\cdot06P' + 50)(1 + i)^{40} = 10\,000 \qquad \text{at } 5\%$$

Hence

$$P' = \frac{10\,000 + 50(1 + i)^{40}}{0\cdot96(\ddot{s}_{\overline{40}|} - \tfrac{1}{2}\ddot{s}_{\overline{30}|} - \tfrac{1}{4}\ddot{s}_{\overline{20}|}) - 0\cdot06(1 + i)^{40}} \qquad \text{at } 5\%$$
$$= \pounds130\cdot17$$

6.3 (a) Let P' be the level annual premium. As in equation 6.5.1, we have

$$(0.98P' - 15)[\ddot{s}_{\overline{5}|0.08}(1.07)^5(1.06)^{10} + \ddot{s}_{\overline{5}|0.07}(1.06)^{10} + \ddot{s}_{\overline{10}|0.06}]$$
$$+ (0.02P' - 115)(1.08)^5(1.07)^5(1.06)^{10} = 26\,000$$

Hence

$$P' = \frac{27\,038}{40.161\,26} = £673.24$$

(b) The yield per annum obtained by the policyholder is found by the equation

$$673.24\ddot{s}_{\overline{20}|} = 26\,000 \qquad \text{at rate } i$$

This gives $i = 0.0592$, or 5.92%.

6.4 Let P'' be the (gross) annual premium. As in equation 6.5.1, we obtain the equation of value

$$0.925P''[\ddot{s}_{\overline{10}|0.04}(1.03)^{20} + \ddot{s}_{\overline{20}|0.03}]$$
$$- (250 - 0.075P'')(1.04)^{10}(1.03)^{20} = 20\,000$$

where expenses are regarded as consisting of 7.5% of *all* premiums plus 'additional' initial expenses of $250 - 0.075P''$. We thus obtain

$$46.4612P'' + 0.2005P'' = 20\,668.37$$

and hence $P'' = £442.94$.

6.5 (a) Let us consider an annual premium policy with sum assured 1. The annual premium is $P_{\overline{n}|} = 1/\ddot{s}_{\overline{n}|}$, and we have

$$_1W_{\overline{n}|} = P_{\overline{n}|}(1 + i)^n = \frac{(1 + i)^n}{\ddot{s}_{\overline{n}|}} = \frac{1}{\ddot{a}_{\overline{n}|}}$$

Note: Since $\ddot{a}_{\overline{n}|}$ decreases with i, $_1W_{\overline{n}|}$ increases with i.

(b) Let $1 \leqslant t \leqslant n - 1$. By equation 6.4.3,

$$_tW_{\overline{n}|} = \frac{\ddot{s}_{\overline{t}|}(1 + i)^{n-t}}{\ddot{s}_{\overline{n}|}}$$

so

$$1 - _tW_{\overline{n}|} = \frac{\ddot{s}_{\overline{n}|} - \ddot{s}_{\overline{t}|}(1 + i)^{n-t}}{\ddot{s}_{\overline{n}|}} = \frac{\ddot{s}_{\overline{n-t}|}}{\ddot{s}_{\overline{n}|}} = \frac{s_{\overline{n-t}|}}{s_{\overline{n}|}}$$

We now have

$$1 - {}_iW_{\overline{n}|} = \frac{s_{\overline{n-i}|}}{s_{\overline{n}|}}$$

$$= \left(\frac{s_{\overline{n-i}|}}{s_{\overline{n-i+1}|}}\right)\left(\frac{s_{\overline{n-i+1}|}}{s_{\overline{n-i+2}|}}\right)\cdots\left(\frac{s_{\overline{n-1}|}}{s_{\overline{n}|}}\right)$$

$$= (1 - {}_1W_{\overline{n-i+1}|})(1 - {}_1W_{\overline{n-i+2}|})\cdots(1 - {}_1W_{\overline{n}|})$$

(c) Since, as established in (a), each of the terms ${}_1W_{\overline{n-i+1}|}, \ldots, {}_1W_{\overline{n}|}$ increases with i, each of the terms in brackets on the right-hand side of the last equation decreases with i. Thus $1 - {}_iW_{\overline{n}|}$ decreases with i, so that ${}_iW_{\overline{n}|}$ increases with i.

Since, when $i = 0$, ${}_iW_{\overline{n}|} = t/n$, it follows that ${}_iW_{\overline{n}|} > t/n$ for each $i > 0$.

6.6 (a) Let the office annual premium be P'. We have the equation of value

$$0.98\,P'[\ddot{s}_{\overline{5}|0.08}(1.07)^{10}(1.06)^{15} + \ddot{s}_{\overline{10}|0.07}(1.06)^{15} + \ddot{s}_{\overline{15}|0.06}]$$
$$- 0.48P'(1.08)^5(1.07)^{10}(1.06)^{15} = 100\,000$$

Hence $P' = £1178.58$.

(b) The paid-up sum assured is the greater of
   (i) $\frac{12}{30} \times 100\,000 = 40\,000$
   (ii) $(1.06)^{18}[0.5P'(1.06)^{12} + 0.95P'\ddot{s}_{\overline{10}|0.06}] = 54\,102.64$
   The paid-up sum assured is thus £54 103.

   The office's accumulation on 1 January 2000 is

$$P'\{0.98[\ddot{s}_{\overline{5}|0.08}(1.07)^7 + \ddot{s}_{\overline{7}|0.07}]$$
$$- 0.48(1.08)^5(1.07)^7\}(1.07)^3(1.06)^{15} = 61\,981$$

The office will therefore make a profit of $61\,981 - 54\,103 = £7878$

6.7 We first calculate the half-yearly premium P' by the equation of value

$$0.97P'\frac{s_{\overline{96}|}}{a_{\overline{2}|}} + 0.9P'[(1 + i)^{100} + (1 + i)^{98}] = 36\,000 \qquad \text{at } 1\tfrac{1}{2}\%$$

(where we are working in time units of a quarter-year). This gives

$$P' = \frac{36\,000}{\left(0.97 \times \dfrac{211.720\,23}{1.955\,88}\right) + 0.9(4.432\,05 + 4.302\,02)}$$

$$= £318.98$$

The surrender value is thus

$$0.95[P'(1+i)^{13.5} + 2P's^{(2)}_{\overline{13}|}] \quad \text{at } 5\%$$
$$= 0.95(2P's^{(2)}_{\overline{13.5}|}) \quad \text{at } 5\%$$
$$= £11\ 722$$

Let the rate of interest paid on the deposit account be i per annum effective. We must have

$$11\ 722(1+i)^{10.5} = P'(2s^{(2)}_{\overline{14}|0.03886} + 1)(1.038\ 86)^{11}$$
$$= 18\ 608$$

which gives $i = 0.0450$ or $4\frac{1}{2}\%$.

6.8 Let S be the sum assured (which equals the purchase price of the ground rent), and let P' be the annual premium of the capital redemption policy. We illustrate A's net cash flow in the following diagram:

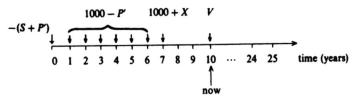

The selling price, at time 7 years, of the remaining instalments of the ground rent is

$$X = 1000a_{\overline{18}|} \quad \text{at } 6\% \quad = £10\ 827.6.$$

We also have the relationship

$$(0.9P' - 15) + (0.98P' - 2)a_{\overline{19}|} = Sv^{25} \quad \text{at } 3\%$$

i.e.

$$P' = 0.031\ 97S + 2.922\ 05 \tag{1}$$

The paid-up sum assured is

$$W = (0.9P' - 15)(1.03)^{25} + (0.98P' - 2)(s_{\overline{25}|0.03} - s_{\overline{19}|0.03})$$
$$= 12.999\ 95P' - 54.091\ 47$$

Hence the surrender value is

$$V = Wv^{15} \quad \text{at } 4\%$$

i.e.

$$V = 0.230\,77S - 8.9432 \tag{2}$$

Now A obtains a yield of 4% per annum effective over the whole transaction, so we must have (on considering the net cash flow)

$$-(S + P') + (1000 - P')a_{\overline{5}|} + 11\,827.6v^7 + Vv^{10} = 0 \qquad \text{at } 4\%$$

Using equations 1 and 2, we obtain an equation for S, which gives $S = £13\,612$.

6.9 (a) Let S be the sum assured, and let P_1', P_2' be the annual premiums (payable half-yearly) quoted by office A and office B respectively. We have the equation

$$0.95P_1'[s^{(2)}_{\overline{10}|0.07}(1.06)^{10} + s^{(2)}_{\overline{10}|0.06}] = S + 20(1.06)^{10}(1.07)^{10}$$

which gives

$$P_1' = 0.026\,444S + 1.863\,18 \tag{1}$$

We also have

$$P_2'\left[\frac{i}{d^{(2)}}\right]\{0.96s_{\overline{20}|} - 0.005[(1 + i)^{19} + 2(1 + i)^{18} + \cdots$$
$$+ 20(1 + i)^0]\}$$
$$= S + 30(1 + i)^{20} \qquad \text{at } 7\%$$

i.e.

$$P_2'\left[\frac{i}{d^{(2)}}\right][0.96s_{\overline{20}|} - 0.005(1 + i)^{20}(Ia)_{\overline{20}|}]$$
$$= S + 30(1 + i)^{20} \qquad \text{at } 7\%$$

whence

$$P_2' = 0.025\,242S + 2.930\,55 \tag{2}$$

Now we are told that

$$P_1' - P_2' = 2 \times 0.67$$

hence (using equations 1 and 2) $S = 2003$, or £2000 (since it is a multiple of £100). The annual premium is (using equation 2)

$$P_2' = (0.025\,242 \times 2000) + 2.930\,55 = £53.41$$

(b) The surrender value is

$$\left(\frac{k}{100}\right)P_2's^{(2)}_{\overline{8.5}|} \qquad \text{at } 4\% \qquad = 0.101\,88kP_2'$$

Working in time units of a half-year, we have the equation of value

$$\tfrac{1}{2}P_2'\,\bar{s}_{\overline{17}|0\cdot015} = 0\cdot101\,88\,kP_2'$$

whence

$$k = \frac{0\cdot5\bar{s}_{\overline{17}|0\cdot015}}{0\cdot101\,88} = 95\cdot65$$

6.10 (a) Let us work in time units of a half-year, and let P' be the quarterly gross premium. We have the equation of value

$$0\cdot965(2P')\ddot{a}^{(2)}_{\overline{20}|} - 0\cdot02(2P')\ddot{a}^{(2)}_{\overline{2}|} = 25\,000v^{20} \qquad \text{at } 4\%.$$

Hence we obtain $P' = £423\cdot59$.

(b) The surrender value V is the greater of

  (i) $0\cdot94(4P')\ddot{s}^{(4)}_{\overline{10}|0\cdot06} = 7226\cdot78$

  (ii) $\tfrac{16}{40}25\,000v^6_{0\cdot1} = 5644\cdot70$

That is, $V = £7226\cdot78$. The effective annual yield i obtained by the policyholder is found from the equation

$$4P'\,\ddot{s}^{(4)}_{\overline{4}|} = 7226\cdot78 \qquad \text{at rate } i$$

which gives $i = 0\cdot0304$ or $3\cdot04\%$.

6.11 (a) Let P'' be the office annual premium. We have

$$0\cdot97P''\bar{s}_{\overline{15}|} - 100(1+i)^{15} = 10\,000 \qquad \text{at } 4\%$$

which gives $P'' = £503\cdot97$.

(b) By formula 6.2.12, the Zillmerized reserve just *before* payment of the fifth premium is

$$10\,000(1\cdot01\,_4V_{\overline{15}|} - 0\cdot01) \qquad \text{at } 4\% \qquad = £2041\cdot94$$

The Zillmerized reserve just *after* payment of the fifth premium is

$$£2041\cdot94 + (£503\cdot97 \times 0\cdot97) = £2530\cdot79$$

(Note: $0\cdot03P''$ is absorbed in expenses.)

6.12 (a) Let P'' be the office annual premium (which is payable monthly in advance). By considering values at the maturity date, we have

$$P''[\bar{s}^{(12)}_{\overline{30}|0\cdot10}(1\cdot08)^5(1\cdot05)^2 + \bar{s}^{(12)}_{\overline{30}|0\cdot08}(1\cdot05)^2 + \bar{s}^{(12)}_{\overline{20}|0\cdot05}]$$
$$- 80(1\cdot1)^3(1\cdot08)^5(1\cdot05)^2$$
$$- (0\cdot05P'' + 5)[\bar{s}_{\overline{31}|0\cdot10}(1\cdot08)^5(1\cdot05)^2 + \bar{s}_{\overline{31}|0\cdot08}(1\cdot05)^2 + \bar{s}_{\overline{31}|0\cdot05}]$$
$$= 4040\,\ddot{a}^{(2)}_{\overline{50}|0\cdot05}$$

This gives $P'' = 1306\cdot15$, and hence the monthly premium is $P''/12 = £108\cdot85$.

(b) Let S be the revised sum assured. The reserve credited by the office in respect of the old policy is

$$V = P''[\ddot{s}^{(12)}_{\overline{30}|0\cdot10}(1\cdot08)^3 + \ddot{s}^{(12)}_{\overline{30}|0\cdot08}] - 80(1\cdot1)^3(1\cdot08)^3$$
$$- (0\cdot05P'' + 5)[\ddot{s}_{\overline{3}|0\cdot10}(1\cdot08)^3 + \ddot{s}_{\overline{3}|0\cdot08}]$$

and we must have

$$V - 50 = (S + 30)v^2_{0\cdot08}v^2_{0\cdot05}$$

This gives $S = £12\,065$.

6.13 Let P' be the office premium without allowance for lapses. We have

$$0\cdot975P'\ddot{a}_{\overline{20}|} = 1000v^{20} + 20 \qquad \text{at } 5\%$$

Therefore

$$P' = \frac{376\cdot89 + 20}{0\cdot975 \times 13\cdot0853} = £31\cdot11$$

Now let P'' be the office premium allowing for lapses, and let $D = 0\cdot018\,69$. We have

$$0\cdot975P''[1 + (1 - D)v + (1 - D)^2v^2 + \cdots + (1 - D)^{19}v^{19}]$$
$$= 20 + 1000(1 - D)^{19}v^{20} + \sum_{t=1}^{19} (tP'')(1 - D)^{t-1}Dv^t \qquad \text{at } 5\%.$$

Note that

$$(1 - D)v_{0\cdot05} = v_{0\cdot07}$$

Hence

$$0\cdot975\,P''\ddot{a}_{\overline{20}|0\cdot07} = 20 + \frac{1000}{(1 - D)}v^{20}_{0\cdot07} + P''\left(\frac{D}{1 - D}\right)(Ia)_{\overline{19}|0\cdot07}$$

Thus

$$11\cdot0522P'' = 20 + 263\cdot34 + 1\cdot5796P''$$

and so $P'' = £29\cdot91$.

6.14 Let $0 \leqslant t < n - 1$. An investment of 1 at time t will produce income of i_t at each of the times $t + 1, t + 2, \ldots, n$, plus a return of capital of 1 at time n. The interest payments at times $t + 1, t + 2, \ldots, n - 1$ will, however, accumulate to $i_t(S_{t+1} + S_{t+2} + \cdots + S_{n-1})$ at time n, so the total proceeds at time n are

$$1 + i_t + i_t(S_{t+1} + S_{t+2} + \cdots + S_{n-1}) \tag{1}$$

By definition this also equals S_t, so we have proved that the given equation holds.

To establish equation 6.7.5 for $0 \leqslant t < n - 1$, we first note that (see equation 6.7.4)

$$S_{n-1} = 1 + i_{n-1}$$

Hence the relationship

$$S_t + S_{t+1} + \cdots + S_{n-1} = (1 + i_t) + (1 + i_t)(1 + i_{t+1}) + \cdots$$
$$+ (1 + i_t)\cdots(1 + i_{n-1}) \tag{2}$$

holds for $t = n - 1$. Now suppose that it is true for $t = m + 1$, where $0 \leqslant m \leqslant n - 2$. We have

$$S_m + S_{m+1} + \cdots + S_{n-1} = S_m + (S_{m+1} + \cdots + S_{n-1})$$
$$= [1 + i_m + i_m(S_{m+1} + \cdots + S_{n-1})]$$
$$+ (S_{m+1} + \cdots + S_{n-1})$$
$$\text{(by equation 1)}$$
$$= (1 + i_m) + (1 + i_m)[S_{m+1} + \cdots + S_{n-1}]$$
$$= (1 + i_m) + (1 + i_m)(1 + i_{m+1}) + \cdots$$
$$+ (1 + i_m)\cdots(1 + i_{n-1})$$
$$\text{(by inductive hypothesis)}$$

Hence equation 2 also holds for $t = m$.

By finite induction, we have shown that equation 2 holds for $t = 0, 1, \ldots, n - 1$. Hence we have, for $0 \leqslant t < n - 1$,

$$S_t = 1 + i_t(1 + S_{t+1} + S_{t+2} + \cdots + S_{n-1}) \quad \text{(by equation 1)}$$
$$= 1 + i_t[1 + (1 + i_{t+1}) + (1 + i_{t+1})(1 + i_{t+2}) + \cdots$$
$$+ (1 + i_{t+1})\cdots(1 + i_{n-1})]$$

by equation 2, as required.

6.15 Let X be the purchase price of the ground rent, and let P' be the gross annual premium for the capital redemption policy. Working in time units of one year, we see that the accumulation at time 15 years of the first premium, less initial expenses, is

$$(0 \cdot 88P' - 100)(1 + 0 \cdot 1s_{\overline{15}| 0 \cdot 08})$$

Hence we obtain the equation of value

$$(0{\cdot}88P' - 100)(1 + 0{\cdot}1s_{\overline{30}|0{\cdot}08}) + 0{\cdot}92P's_{\overline{20}|0{\cdot}08} = X$$

Therefore

$$27{\cdot}3293P' = X + 371{\cdot}52$$

or

$$P' = \frac{X + 371{\cdot}52}{27{\cdot}3293} \tag{1}$$

Let us now work in time units of a half-year. The equation of value at the date on which the policy is made paid-up is

$$P'\frac{\ddot{s}_{\overline{20}|}}{\ddot{a}_{\overline{2}|}} + X(1 + i)^{20} = 1000[s^{(2)}_{\overline{20}|} + a^{(2)}_{\overline{10}|}] + \frac{2X}{3}v^{10} \qquad \text{at } 5\%$$

Substituting the expression for P' given by equation 1 leads to $X = £14181$.

6.16 Let the premium payable at time t years be $P' + 50t$. The accumulation, at time 20 years, of the premiums is (see section 6.7)

$$1{\cdot}1P'[s_{\overline{10}|0{\cdot}08} + (1{\cdot}08)^{10}s_{\overline{10}|0{\cdot}06}]$$
$$+ 50(1{\cdot}08)\{[s_{\overline{9}|0{\cdot}08} + (1{\cdot}08)^9 s_{\overline{10}|0{\cdot}06}]$$
$$+ [s_{\overline{8}|0{\cdot}08} + (1{\cdot}08)^8 s_{\overline{10}|0{\cdot}06}]$$
$$+ \quad \vdots \qquad\qquad \vdots$$
$$+ [s_{\overline{1}|0{\cdot}08} + (1{\cdot}08)s_{\overline{10}|0{\cdot}06}] + [s_{\overline{10}|0{\cdot}06}]\}$$
$$+ 50(1{\cdot}06)(s_{\overline{9}|0{\cdot}06} + s_{\overline{8}|0{\cdot}06} + \cdots + s_{\overline{1}|0{\cdot}06})$$
$$= 1{\cdot}1P'(s_{\overline{10}|0{\cdot}08} + (1{\cdot}08)^{10}s_{\overline{10}|0{\cdot}06})$$
$$+ 50(1{\cdot}08)[(Is)_{\overline{9}|0{\cdot}08} + (s_{\overline{10}|0{\cdot}08})(s_{\overline{10}|0{\cdot}06})]$$
$$+ 50(1{\cdot}06)(Is)_{\overline{9}|0{\cdot}06}$$
$$= 47{\cdot}237\,18P' + 16\,149$$

Equating this to $63\,386$ gives $P' = £1000$. That is, the first annual premium is £1000.

6.17 (a) *Former basis* The single premium A' and the annual premium P' for an n-year policy are given by the equations

$$0{\cdot}925A' - 80 = 10\,000v^n \qquad \text{at } 10\%$$

and

$$0.95P' + 0.975P'a_{\overline{n-1}|} - 80 = 10\,000v^n \qquad \text{at } 10\%$$

whence

$$A' = \frac{10\,000v^n + 80}{0.925} \qquad \text{at } 10\%$$

and

$$P' = \frac{10\,000v^n + 80}{0.95 + 0.975a_{\overline{n-1}|}} \qquad \text{at } 10\%$$

(b) *Current basis* The single premium A' and the annual premium P' for an n-year policy are given by the equations

$$(0.925A' - 80)(1 + 0.13s_{\overline{n}|0.10}) = 10\,000$$

and

$$(0.95P' - 80)(1 + 0.13s_{\overline{n}|0.10}) + 0.975P'\ddot{s}_{\overline{n-1}|0.10} = 10\,000$$

whence

$$A' = \frac{10\,000 + 80(1 + 0.13s_{\overline{n}|0.10})}{0.925(1 + 0.13s_{\overline{n}|0.10})}$$

and

$$P' = \frac{10\,000 + 80(1 + 0.13s_{\overline{n}|0.10})}{(1.0985s_{\overline{n}|0.10} - 0.025)}$$

We now calculate A' and P' on both bases for terms $n = 5, 15$, and 25 years.

	Annual premium		Single premium	
Term	Former	Current	Former	Current
5	1556·48	1518·16	6799·15	6113·71
15	304·20	298·49	2674·49	2193·68
25	103·29	102·79	1084·32	870·72

Comment As might be expected, at each term the proportionate reduction in premium (from the former basis to the current basis) is greater for single premium policies than for annual premium policies. For single premium policies the proportionate reduction in premium increases with the term. The opposite is true for annual premiums.

6.18 *Note* All functions are at 10% interest.

(a)
$$_{15}V_{\overline{20}|} = 1 - \frac{\ddot{a}_{\overline{5}|}}{\ddot{a}_{\overline{20}|}} = 0{\cdot}554\,73$$

$$_{16}V_{\overline{20}|} = 1 - \frac{\ddot{a}_{\overline{4}|}}{\ddot{a}_{\overline{20}|}} = 0{\cdot}627\,66$$

$$_{15\cdot5}V_{\overline{20}|} = (_{15}V_{\overline{20}|} + P_{\overline{20}|})(1 + i)^{1/2} = 0{\cdot}598\,46$$

Linear interpolation between $_{15}V_{\overline{20}|} + P_{\overline{20}|}$ and $_{16}V_{\overline{20}|}$ gives 0·599 13.

(b)
$$_{15\cdot5}V^{(4)}_{\overline{20}|} = P^{(4)}_{\overline{20}|}\ddot{s}^{(4)}_{\overline{15\cdot5}|} = 0{\cdot}590\,33$$

$$_{15\cdot75}V^{(4)}_{\overline{20}|} = P^{(4)}_{\overline{20}|}\ddot{s}^{(4)}_{\overline{15\cdot75}|} = 0{\cdot}608\,77$$

$$_{15\cdot65}V^{(4)}_{\overline{20}|} = (_{15\cdot5}V^{(4)}_{\overline{20}|} + \tfrac{1}{4}P^{(4)}_{\overline{20}|})(1 + i)^{0\cdot15} = 0{\cdot}603\,00$$

Linear interpolation between $_{15\cdot5}V^{(4)}_{\overline{20}|} + (1/4)P^{(4)}_{\overline{20}|}$ and $_{15\cdot75}V^{(4)}_{\overline{20}|}$ gives 0·603 01.

Comments The interpolated answer is reasonably accurate in (a), and very accurate in (b). (This is not unexpected, since the interval of interpolation is smaller.)

Chapter 7

7.1 Let us work in time units of a half-year. We note that the period from 14 August 1984 to 1 December 1984 is 109 days, and the period from 1 June 1984 to 1 December 1984 is 183 days. Define $t = 109/183$. The yield per half-year, i, is the solution of the equation of value

$$35{\cdot}125 = 1{\cdot}75(1 + i)^{-t}\left(\frac{1+i}{i}\right)$$

which gives $i = 0{\cdot}050\,83$. The yield per annum is thus $(1 + i)^2 - 1 = 0{\cdot}1042$, or 10·42%.

Expressed as a nominal rate of interest convertible half-yearly, the yield per annum is $2(0{\cdot}050\,83)$ i.e. 0·101 66 or 10·17%.

7.2 Let us work in time units of one year, measured from 31 July 1986, and in money units of £500. The following diagram illustrates the position:

capital repayments

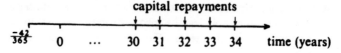

In the notation of this chapter, we have $C = 525$, $p = 2$, and $g = 0.1/1.05$.

(a) We first find the value of the loan just after payment of the interest on° 31 July 1986. We have for K, the value of the capital repayments,

$$K = 105(a_{\overline{34}|} - a_{\overline{29}|}) \qquad \text{at } 9\%$$

$$= 33.553$$

and, by Makeham's formula, the value of the loan on 31 July 1986 is

$$K + \frac{0.1\ (1 - 0.4)}{1.05\ \ 0.09^{(2)}}(525 - K) = 352.45$$

Hence the 'ex dividend' value on 19 June 1986 is

$$352.45v^{42/365} \qquad \text{at } 9\% \qquad = 348.975, \qquad \text{or} \qquad £69.79\%$$

(b) The 'cum dividend' value on 19 June 1986 is

$$[352.452 + (0.6 \times 25)]v^{42/365} = 363.826, \qquad \text{or} \qquad £72.77\%$$

7.3 Let us work in time units of one year and in money units of £1000. Assume first that redemption is always at par: an adjustment to allow for the actual redemption prices will be added later. In the notation of this chapter, we have $C = 300$ and

$$K = 10a_{\overline{30}|} \qquad \text{at } 7\% \qquad = 124.0904$$

By Makeham's formula, the value of the loan is

$$K + \frac{0.08(0.6)}{0.07^{(4)}}(300 - K) = 247.836$$

We now add the value of the premiums on redemption, namely

$$2(a_{\overline{30}|} - a_{\overline{15}|}) \qquad \text{at } 7\% \qquad = 6.602$$

which gives a total value of 254.438, or £84.81%.

7.4 (a) Let us work in time units of one year. In the notation of chapter 7, we have $R = 1.1$, $C = 1\,100\,000$, $g = 0.08/1.1$, and $p = 4$. Since £250 000 nominal is redeemed at time 25, we have

$$K = 1.1[75\,000(a_{\overline{24}|} - a_{\overline{14}|}) + 250\,000v^{25}] \qquad \text{at} \qquad 7\%$$

$$= 275\,387.5$$

Hence, by Makeham's formula, the price is

$$A = 275\,387{\cdot}5 + \frac{0{\cdot}08(0{\cdot}7)}{1{\cdot}1 \times 0{\cdot}07^{(4)}}(1\,100\,000 - 275\,387{\cdot}5)$$

$$= £890\,626$$

(b) Let the price payable be A^*. We have

$A^* = A$ − present value of 'lost' premiums on redemption

$$= A - (75\,000 \times 0{\cdot}01)(v^{16} + 2v^{17} + \cdots + 9v^{24})$$
$$- (250\,000 \times 0{\cdot}1)v^{25} \qquad \text{at } 7\%$$

$$= A - 750v^{15}(Ia)_{\overline{9}|} - 25\,000v^{25} \qquad \text{at } 7\%$$

$$= A - 12\,668$$

$$= £877\,958$$

7.5 (a) Let us work in terms of £100 nominal. We have $R = 1$, $C = 100$, $p = 2$, $D = 0{\cdot}11$, $g = 0{\cdot}11$, and $t_1 = 0{\cdot}5$. Since $g(1 - t_1) = 0{\cdot}055 > i^{(2)}$ at 4%, we value the loan assuming that redemption will take place as soon as possible. We thus have

$$K = 100v^{31{\cdot}5} \qquad \text{at } 4\% \qquad = 29{\cdot}0703$$

and the price is (by Makeham's formula)

$$K + \frac{0{\cdot}055}{0{\cdot}04^{(2)}}(C - K) = £127{\cdot}564$$

The greatest net yield will be obtained if the stock is *never* redeemed, i.e. if it is taken to be a perpetuity. The net yield per annum is, in this case, the solution of the equation

$$127{\cdot}564 = 5{\cdot}5a_{\overline{\infty}|}^{(2)} \qquad \text{at rate } i$$

which gives $i^{(2)} = 0{\cdot}043\,116$, and hence $i = 4{\cdot}358\%$.

(b) We have $g(1 - t_1) < i^{(2)}$ at 7%, so we value the loan on the assumption that it will be redeemed as late as possible, i.e. that it may be treated as a perpetuity. The price is thus $5{\cdot}5a_{\overline{\infty}|}^{(2)}$ at 7% = £79·923.

The greatest net yield will now be obtained if the stock is redeemed as early as possible, that is, on 15 May 2018. The net yield i is the solution of the equation

$$79{\cdot}923 = 5{\cdot}5a_{\overline{31{\cdot}5}|}^{(2)} + 100v^{31{\cdot}5} \qquad \text{at rate } i$$

By trials, we obtain $i = 7{\cdot}230\%$.

7.6 We first value the loan on the assumption that redemption is at par, then add the value of the premiums. Assuming redemption is at par, we have

$$K = 1000a_{\overline{10}|} \quad \text{at } 7\% \quad = 7023 \cdot 58$$

and the value of the loan is (by Makeham)

$$K + \frac{0 \cdot 045}{0 \cdot 07^{(4)}}(10\,000 - K) = 8986 \cdot 51$$

The value of the premiums on redemption is

$$X = v + 2^2 v^2 + 3^2 v^3 + \cdots + 10^2 v^{10} \quad \text{at } 7\%$$

Now

$$vX = v^2 + 2^2 v^3 + 3^2 v^4 + \cdots + 10^2 v^{11}$$

So

$$(1 - v)X = v + 3v^2 + 5v^3 + \cdots + 19v^{10} - 100v^{11}$$
$$= 2(v + 2v^2 + 3v^3 + \cdots + 10v^{10})$$
$$\quad - (v + v^2 + \cdots + v^{10}) - 100v^{11}$$
$$= 2(Ia)_{\overline{10}|} - a_{\overline{10}|} - 100v^{11}$$

whence

$$X = \frac{2(Ia)_{\overline{10}|} - a_{\overline{10}|} - 100v^{11}}{1 - v} \quad \text{at } 7\%$$

$$= 228 \cdot 44$$

Hence the price to be paid for the entire loan is $8986 \cdot 51 + 228 \cdot 44 = \pounds 9214 \cdot 95$.

7.7 (a) Let us work in money units of £1 million. The loan outstanding on 17 April 1915 is 130 nominal, and, on that date, the value of the capital repayments is

$$K = 2a_{\overline{10}|} + 3v^{10}a_{\overline{13}|} + 4v^{23}a_{\overline{7}|} + 5v^{30}a_{\overline{5}|} + 6v^{35}a_{\overline{5}|} \quad \text{at } 3\tfrac{1}{2}\%$$

$$= 62 \cdot 717$$

By Makeham's formula, the value of the outstanding loan is

$$K + \frac{0 \cdot 03}{0 \cdot 035^{(4)}}(130 - K) = 121 \cdot 14$$

Hence the price per £100 nominal on 16 April 1915 is

$$\frac{121 \cdot 14}{130} \times 100 = £93 \cdot 18$$

(b) The yield to the investor if redemption occurs in n years' time is the solution of the equation of value

$$93 \cdot 18 = 3a^{(4)}_{\overline{n}|} + 100v^n$$

Since the stock is below par, the yield decreases with the term to redemption, so we find n such that the yield is more than 5% for term n years but less than 5% for term $n + 1$ years. When $i = 0 \cdot 05$, the equation of value may be written as

$$93 \cdot 18 = \frac{3(1 - v^n)}{0 \cdot 05^{(4)}} + 100v^n$$

or

$$v^n_{0 \cdot 05} = \frac{[93 \cdot 18 \times 0 \cdot 05^{(4)}] - 3}{[100 \times 0 \cdot 05^{(4)}] - 3} = 0 \cdot 8246$$

Since (at 5%) $v^3 > 0 \cdot 8246 > v^4$, the yield will be at least 5% only for bonds redeemed within three years (i.e. in the years 1916 to 1918 inclusive). The required probability is thus

$$\frac{2 + 2 + 2}{130} = \frac{3}{65} = 0 \cdot 0462$$

7.8 (a) Let us work in time units of one year, and let the number of drawings be m. We have

$$\sum_{t=1}^{m} (900 + 100t) = 16\,500$$

i.e.

$$900m + 50m(m + 1) = 16\,500$$

which gives $m = 11$. Hence the drawings are made at times $5, 6, 7, \ldots, 15$.

We have $R = 1 \cdot 1$, $g = 0 \cdot 055/1 \cdot 1 = 0 \cdot 05$, $p = 2$, $C = 1\,815\,000$, and $i = 0 \cdot 04$. The value of the capital repayments is

$$K = \sum_{t=1}^{11} (900 + 100t)110v^{4+t} \qquad \text{at } 4\%$$

$$= (900 \times 110)(a_{\overline{15}|} - a_{\overline{4}|}) + 11\,000v^4(Ia)_{\overline{11}|} \qquad \text{at } 4\%$$

$$= 1\,203\,394$$

Hence, by Makeham's formula, the syndicate should pay

$$K + \frac{0.05(0.75)}{0.04^{(2)}}(1\,815\,000 - K) = £1\,782\,452 \qquad \text{or} \qquad £108.03\%$$

(b) Since the price paid is less than £110, the yield decreases as the term to redemption, n years, increases. The net yield per annum if redemption occurs at time n years is the solution of the equation

$$107 = (0.7 \times 5.5)a_{\overline{n}|}^{(2)} + 110v^n \qquad \text{at rate } i$$

When $i = 0.04$, this equation may be written as

$$107 = (0.7 \times 5.5)\frac{(1 - v_{0.04}^n)}{0.04^{(2)}} + 110v_{0.04}^n$$

which gives $v_{0.04}^n = 0.7656$. Since (at 4%) $v^6 > 0.7656 > v^7$, the yield is at least 4% for bonds drawn up to and including time 6 (i.e. the first two redemption dates). The required probability is thus

$$\frac{1000 + 1100}{16\,500} = \frac{7}{55} = 0.1273$$

7.9 (a) Let $I_0 = 187.52$ (the index value for August 1985) and let $I_1 = 192.10$ (the index value for February 1986). Let r denote the assumed rate of increase per annum in the index. Thus the index value increases by the factor $(1 + r)^{1/2}$ every six months. Measure time in half-years from the issue date. Let Q_0 be the index value at the issue date. (Note that Q_0 is the value at the *precise* date of issue and *not* the index value 'for April 1986'.)

Consider a stock of term n. (We have $n = 40$ or 60.) Let the issue price per cent nominal be A. Consider a purchase of £100 nominal of stock. The purchaser has an outlay of A at time 0 (when the index value is Q_0).

On 7 October 1986 (i.e. at time 1, when the index value will be $Q_0(1 + r)^{1/2}$) the purchaser will receive an indexed interest payment of amount

$$\frac{3}{2}\frac{(\text{index value for February 1986})}{(\text{index value for August 1985})} = \frac{3}{2}\frac{I_1}{I_0}$$

On 7 April 1987 (i.e. at time 2, when the index value will be $Q_0(1 + r)$) the purchaser will receive interest of

$$\frac{3}{2}\frac{(\text{index value for August 1986})}{(\text{index value for August 1985})}$$

$$= \frac{3}{2}\frac{(1 + r)^{1/2}(\text{index value for February 1986})}{(\text{index value for August 1985})} = \frac{3}{2}(1 + r)^{1/2}\frac{I_1}{I_0}$$

On 7 October 1987 (i.e. at time 3, when the index value will be $Q_0(1 + r)^{3/2}$) the purchaser will receive interest of

$$\frac{3(\text{index value for February 1987})}{2(\text{index value for August 1985})}$$

$$= \frac{3(1 + r)(\text{index value for February 1986})}{2(\text{index value for August 1985})}$$

$$= \tfrac{3}{2}(1 + r)\frac{I_1}{I_0}$$

More generally, for $1 \leqslant t \leqslant n$, at time t (when the index value will be $Q_0(1 + r)^{t/2}$) the purchaser will receive an indexed interest payment of amount

$$\tfrac{3}{2}(1 + r)^{(t-1)/2}\frac{I_1}{I_0}$$

The indexed amount of the capital repayment at time n is

$$100(1 + r)^{(n-1)/2}\frac{I_1}{I_0}$$

Let i be the effective real yield per half-year. By combining the above remarks, we see that i is obtained from the equation

$$-\frac{A}{Q_0} + \sum_{t=1}^{n}\left[\tfrac{3}{2}(1 + r)^{(t-1)/2}\frac{I_1}{I_0}\right]\left[\frac{1}{Q_0(1 + r)^{t/2}}\right](1 + i)^{-t}$$

$$+ \left[100(1 + r)^{(n-1)/2}\frac{I_1}{I_0}\right]\left[\frac{1}{Q_0(1 + r)^{n/2}}\right](1 + i)^{-n} = 0$$

which gives

$$A = \frac{I_1}{(1 + r)^{1/2}I_0}(\tfrac{3}{2}a_{\overline{n}|} + 100v^n) \qquad \text{at rate } i \qquad (1)$$

For a real yield of 3% per annum convertible half-yearly $i = 0\cdot015$, and it follows from equation 1 that

$$A = \frac{I_1}{(1 + r)^{1/2}I_0}100$$

for *any* value of n. Since, on the assumption that $r = 0\cdot06$ there is an effective real yield of $1\tfrac{1}{2}\%$ per half-year, by letting $r = 0\cdot06$ in the last expression we find the issue price of either stock to be

$$\frac{(192\cdot10)(100)}{1\cdot06^{1/2}(187\cdot52)} = 99\cdot50\%$$

(b) It follows from equation 1 that, if the index increases at the rate r per annum, the real yield for a stock with term n and issue price 99·50% is i per half-year, where

$$\tfrac{5}{2}a_{\overline{n}|} + 100v^n = \frac{99·50(1 + r)^{1/2}\,187·52}{192·10} \qquad \text{at rate } i$$

When $r = 0·04$ the right-hand side of this last equation equals 99·05, which is *less* than 100. Hence (see page 169) the shorter term gives the higher yield.

When $r = 0·08$ the right-hand side of the last equation equals 100·94, which is *greater* than 100. In this case, therefore, the shorter term gives the lower yield.

(When $r = 0·04$ the values of i for $n = 40$ and for $n = 60$ are 1·5319% and 1·5242% respectively. The corresponding values when $r = 0·08$ are 1·4688% and 1·4763%.)

7.10 (a) The interest payment in October 1984 per £100 nominal of stock is

$$\frac{5/2}{2}\frac{(\text{RPI for February 1984})}{(\text{RPI for February 1983})} = 1·25\frac{344·0}{327·3}$$

$$= £1·3137$$

(Note this value is calculated to four decimal places 'rounded down'.)

(b) We refer to calendar months as follows:

June 1984 is 'month 0'
July 1984 is 'month 1'
August 1984 is 'month 2', etc.

Let r be the assumed rate of growth in the RPI per half-year. (If it is assumed that the RPI will increase by 10% per annum, then $(1 + r)^2 = 1·1$.) Measure time in half-years from 14 August 1984. Note that the period from 14 August 1984 to 16 October 1984 is a fraction f of the half-year from 16 April 1984 to 16 October 1984, where $f = 63/183$.

The first interest payment is at time f. (Note: we could use the value $63/182·5$ for f. For practical purposes either value may be adopted.) Let Q_0 be the RPI value at time 0. (Note: this is the value at the *precise* date of purchase and *not* the value 'for August 1984'.)

The purchaser of the stock will receive 72 interest payments (if he holds the stock until it is redeemed). For $1 \leqslant j \leqslant 72$ the jth payment of interest will be at time $(j - 1 + f)$, when the index value will be $Q_0(1 + r)^{j-1+f}$. The first interest payment will be of amount £1·3137 (see (a)). For $2 \leqslant j \leqslant 72$ the jth interest payment will be in month $(6j - 2)$ and will be indexed by the RPI value for month $(6j - 10)$. Let $I_0 = 351 \cdot 9$, so that I_0 is the RPI value for month 0. Since, by hypothesis, the RPI increases continuously at rate r per half-year, it is reasonable to assume that (for $j \geqslant 0$) the RPI value for month j will be $I_0(1 + r)^{j/6}$.

Let $I_b = 327 \cdot 3$, i.e. the RPI value for the base month. For $j \geqslant 2$ the indexed amount of the jth interest payment will be

$$\frac{5/2}{2}\frac{\text{(index value for month } (6j - 10))}{\text{(index value for base month)}} = 1 \cdot 25\,\frac{I_0(1 + r)^{(6j - 10)/6}}{I_b}$$

Allowing for the fact that the redemption money is also index-linked, the purchaser calculated his effective real yield as i per half-year from the equation

$$0 = -\frac{85 \cdot 625}{Q_0} + \frac{1 \cdot 3137}{Q_0(1 + r)^f}(1 + i)^{-f}$$

$$+ \sum_{j=2}^{72}\left[1 \cdot 25\,\frac{I_0(1 + r)^{j - (10/6)}}{I_b}\right]$$

$$\times\left[\frac{1}{Q_0(1 + r)^{j - 1 + f}}\right](1 + i)^{-(j - 1 + f)}$$

$$+ \left[100\,\frac{I_0(1 + r)^{72 - (10/6)}}{I_b}\right]\left[\frac{1}{Q_0(1 + r)^{71 + f}}\right](1 + i)^{-(71 + f)}$$

which gives

$$85 \cdot 625 = (1 + i)^{-f}\left[1 \cdot 3137(1 + r)^{-f}\right.$$

$$\left. + \frac{I_0}{I_b}(1 + r)^{-f - (2/3)}(1 \cdot 25a_{\overline{71}|i} + 100v_i^{71})\right]$$

Putting $(1 + r) = \sqrt{1 \cdot 1}$ and using the known values for f, I_0 and I_b we find from the last equation that $i = 0 \cdot 016\,651\,96$. Then $(1 + i)^2 = 1 \cdot 033\,581$, so that the effective real yield is $3 \cdot 36\%$ per annum.

7.11 (a) Since the price paid per bond exceeds £100 and is less than £150, the yield on redemption after n years, i_n, increases for $1 \leqslant n \leqslant 20$ and decreases for $21 \leqslant n \leqslant 40$. The yield i_n is found by solving the equations of value

$$125 = 100v^n + 6a_{\overline{n}|}\qquad\text{at rate } i\qquad 1 \leqslant n \leqslant 20\qquad(1)$$

or

$$125 = 150v^n + 6a_{\overline{n}|}\qquad\text{at rate } i\qquad 21 \leqslant n \leqslant 40\qquad(2)$$

(i) By trials, we find that $i_9 < 0.03$ and $i_{10} > 0.03$; hence $i_n > 3\%$ for $10 \leqslant n \leqslant 20$, and $i_n < 3\%$ for $1 \leqslant n \leqslant 9$. It is also clear from equation 1 that $i_n < 5\%$ for $1 \leqslant n \leqslant 20$.

By further trials, we find that $i_{36} > 0.05$ and $i_{37} < 0.05$. We thus have $i_n < 5\%$ for $37 \leqslant n \leqslant 40$, and $i_n > 5\%$ for $21 \leqslant n \leqslant 36$. It is also clear, by equation 2, that $i_n > 3\%$ for $21 \leqslant n \leqslant 40$. Hence the yield lies between 3% and 5% if and only if $10 \leqslant n \leqslant 20$ or $37 \leqslant n \leqslant 40$. The probability that this will occur is

$$\frac{(10 \times 11) + (4 \times 15)}{(10 \times 20) + (20 \times 15)} = 0.34$$

(ii) Since $i_n > 3\%$ for $21 \leqslant n \leqslant 40$, we need consider $1 \leqslant n \leqslant 20$ only. It is clear that $i_4 < 0$ and $i_5 > 0$. Hence the yield is negative if and only if $n \leqslant 4$. The probability of redemption in the first four years is

$$\frac{10 \times 4}{500} = 0.08$$

(b) We solve equation 1 with $n = 4$, i.e.

$$125 = 100v^4 + 6a_{\overline{4}|} \qquad \text{at rate } i$$

By trials, we obtain $i = -0.0022$ or -0.22%.

7.12 Let us work in time units of a quarter-year and let L be the original loan. The capital repaid in 1985 is, according to the original schedule,

$$\frac{L}{a_{\overline{48}|}}(a_{\overline{36}|} - a_{\overline{32}|}) \qquad \text{at } 3\%$$

and hence the interest paid in 1985 is

$$\frac{4L}{a_{\overline{48}|}} - \frac{L}{a_{\overline{48}|}}(a_{\overline{36}|} - a_{\overline{32}|}) \qquad \text{at } 3\%, \qquad \text{i.e.} \qquad 0.101\,181L$$

Equating this expression to 6374.41 gives $L = £63\,000$. Hence the loan outstanding at the date of purchase was, according to the original schedule,

$$\left[\frac{L}{a_{\overline{48}|}}\right]a_{\overline{32}|} \qquad \text{at } 3\%, \qquad \text{i.e.} \qquad £50\,837.35$$

and each quarterly instalment is of amount

$$\frac{L}{a_{\overline{48}|}} \qquad \text{at } 3\% = 2493.40$$

(a) The net yield per quarter is 2%. Accordingly we value the outstanding instalments, less tax, at this rate of interest. If there were no income tax, we would have the value of outstanding instalments as

$$A = 2493 \cdot 40 a_{\overline{32}|} \qquad \text{at } 2\% \qquad = 58\,515 \cdot 95$$

and hence we may find K, the value of the capital repayments, by Makeham's formula; that is,

$$58\,515 \cdot 95 = K + \frac{0 \cdot 03}{0 \cdot 02}(50\,837 \cdot 35 - K)$$

(Note that £50 837·35 is the outstanding loan at the date of purchase.) Hence $K = 35\,480 \cdot 16$, and the purchase price paid by the investor is (again by Makeham's formula)

$$K + \frac{0 \cdot 03(0 \cdot 6)}{0 \cdot 02}(50\,837 \cdot 35 - K) = £49\,302$$

(b) The net yield per quarter is now $0 \cdot 08^{(4)}/4$. Ignoring income tax, the value of the outstanding instalments is

$$2493 \cdot 40 a_{\overline{32}|} \qquad \text{at rate } 0 \cdot 08^{(4)}/4$$
$$= (4 \times 2493 \cdot 40) a_{\overline{8}|}^{(4)} \qquad \text{at rate } 0 \cdot 08$$
$$= 59\,006 \cdot 55$$

We may find K (the value of the capital repayments) by Makeham's formula, i.e.

$$59\,006 \cdot 55 = K + \frac{0 \cdot 03}{0 \cdot 08^{(4)}/4}(50\,837 \cdot 35 - K)$$

This gives $K = 35\,828 \cdot 12$, and the price paid by the investor is (again using Makeham)

$$K + \frac{0 \cdot 03 \times 0 \cdot 6}{0 \cdot 08^{(4)}/4}(50\,837 \cdot 35 - K) = £49\,735$$

Note The value may also be obtained as $K + 0 \cdot 6(59\,006 \cdot 55 - K)$.

7.13 Let us work in time units of four years. Let X be the amount of each repayment, which is made every four years. We have the equation of value

$$X a_{\overline{10}|} = 100\,000 \qquad \text{at rate } J$$

where J is the effective rate of interest per time unit. Since $1 + J = (1 \cdot 04)^4$, we have

$$X = \frac{100\,000[(1 \cdot 04)^4 - 1]}{1 - (1 \cdot 04)^{-40}}$$

$$= \frac{100\,000 s_{\overline{4}|}}{a_{\overline{40}|}} \qquad \text{at } 4\% \qquad = 21\,454 \cdot 6$$

Using the theory of the loan schedule (see section 3.8), we see that the capital repayments are $X v_J^{10}, X v_J^9, \ldots, X v_J$ at times $1, 2, \ldots, 10$ respectively. On changing to time units of one year, we note that these payments are at times $4, 8, \ldots, 40$ respectively and their value, at 4% per annum interest, is

$$K = X(v_J^{10} v_{0 \cdot 04}^4 + v_J^9 v_{0 \cdot 04}^8 + \cdots + v_J v_{0 \cdot 04}^{40})$$
$$= X(v_{0 \cdot 04}^{40} v_{0 \cdot 04}^4 + v_{0 \cdot 04}^{36} v_{0 \cdot 04}^8 + \cdots + v_{0 \cdot 04}^4 v_{0 \cdot 04}^{40})$$
$$= 10 X v_{0 \cdot 04}^{44}$$
$$= 38\,199 \cdot 1$$

Assuming that income tax is at 25% throughout, the value to the investor of the annuity payments is (by Makeham)

$$K + 0 \cdot 75(100\,000 - K) = 84\,549 \cdot 8$$

We now find the value of the additional income tax payments in the first 16 years. According to the loan schedule, the interest contents of the first four payments are

$$X(1 - v_J^{10}), \quad X(1 - v_J^9), \quad X(1 - v_J^8), \quad X(1 - v_J^7)$$

respectively: the value of the extra income tax payments is therefore

$$0 \cdot 25 X[(1 - v_J^{10}) v_{0 \cdot 04}^4 + (1 - v_J^9) v_{0 \cdot 04}^8 + (1 - v_J^8) v_{0 \cdot 04}^{12} + (1 - v_J^7) v_{0 \cdot 04}^{16}]$$
$$= 0 \cdot 25 X[(v_{0 \cdot 04}^4 + v_{0 \cdot 04}^8 + v_{0 \cdot 04}^{12} + v_{0 \cdot 04}^{16}) - 4 v_{0 \cdot 04}^{44}]$$

$$= 0 \cdot 25 X \left[\frac{a_{\overline{16}|}}{s_{\overline{4}|}} - 4 v^{44} \right] \qquad \text{at } 4\%, \qquad \text{i.e.} \qquad 10\,897 \cdot 9$$

Hence the price paid for the loan is $84\,549 \cdot 8 - 10\,897 \cdot 9 = \pounds 73\,652$.

7.14 Let i be the net annual yield. The value of the capital repayments is

$$K = 10\,000(v^8 + v^{16} + v^{24}) \qquad \text{at rate } i$$

The value of the net interest payments is found, by first principles, to be

$$I = 30\,000 \times 0.06 \times 0.7 a_{\overline{5}|}$$
$$+ 20\,000 \times 0.04 \times 0.7(a_{\overline{12}|} - a_{\overline{5}|})$$
$$+ 20\,000 \times 0.04 \times 0.5(a_{\overline{16}|} - a_{\overline{12}|})$$
$$+ 10\,000 \times 0.02 \times 0.5(a_{\overline{24}|} - a_{\overline{16}|}) \qquad \text{at rate } i$$
$$= 700 a_{\overline{5}|} + 160 a_{\overline{12}|} + 300 a_{\overline{16}|} + 100 a_{\overline{24}|} \qquad \text{at rate } i.$$

We find i by solving the equation of value

$$26\,000 = K + I$$

A rough solution gives $i \simeq 4.5\%$, and by successive trials and interpolation we find that $i \simeq 4.63\%$.

7.15 (a) Let the amount of the monthly instalment be X. Working in time units of a quarter-year, we have $3 X a_{\overline{60}|}^{(3)} = 100\,000$ at 4%. This gives $X = £1454.17$.

(b) Work in time units of one month. Let $j = 0.04^{(3)}/3 = 0.013\,159$. Note that the interest payable at the end of each month is j times the loan outstanding at the start of the month.

The net yield is somewhat less than 5% p.a. Let us value the loan to obtain a net yield of $4\frac{1}{2}\%$ p.a. On this basis, if income tax is ignored, the value of the annuity payments is

$$X a_{\overline{180}|} \text{ at rate } \frac{0.045^{(12)}}{12} = 12 X a_{\overline{15}|}^{(12)} \text{ at } 4\frac{1}{2}\% = 191\,240$$

By Makeham's formula, the value of the capital repayments K is such that

$$191\,240 = K + \frac{0.013\,159}{0.045^{(12)}/12}(100\,000 - K)$$

from which we obtain $K = 64\,645$. Hence the value of the annuity payments, net of tax, is (again using Makeham)

$$K + \frac{0.2 \times 0.013\,159}{0.045^{(12)}/12}(100\,000 - K) = £89\,964$$

Note This value may also be obtained as $K + 0.2(191\,240 - K)$.

The corresponding value if the net annual yield were 5% is given as £86 467. By interpolation between $4\frac{1}{2}\%$ and 5%, we find that the net annual yield i is approximately equal to 4.78%.

7.16 Let the redemption prices per cent be $100 + \lambda$, $100 + 2\lambda$, $100 + 3\lambda$, and $100 + 4\lambda$ at times 2, 4, 6, and 8 years respectively. We shall first value the loan on the assumptions that redemption is at par and that income tax is at 50%, and later apply 'corrections' to allow for the premiums on redemption and the 'excess' tax. On these assumptions, we have

$$K = 2000(v^2 + v^4 + v^6 + v^8) \qquad \text{at } 7\%$$

$$= \frac{2000a_{\overline{8}|}}{s_{\overline{2}|}} \qquad \text{at } 7\% \qquad = 5769 \cdot 37$$

By Makeham's formula, the net value of the loan is

$$K + \frac{0 \cdot 1 \times 0 \cdot 5}{0 \cdot 07^{(4)}}(8000 - K) = 7403 \cdot 91$$

The value of the 'excess' income tax in the first five years is (by first principles)

$$20[a^{(4)}_{\overline{2}|} + a^{(4)}_{\overline{4}|} + 2a^{(4)}_{\overline{5}|}] \qquad \text{at } 7\% = 274 \cdot 85$$

and the premiums on redemption have value

$$20\lambda(v^2 + 2v^4 + 3v^6 + 4v^8) \qquad \text{at } 7\% \qquad = 134 \cdot 526\lambda$$

Hence we have the equation

$$7403 \cdot 91 + 274 \cdot 85 + 134 \cdot 526\lambda = 7880 \cdot 55$$

whence $\lambda = 1 \cdot 5$. The redemption prices are therefore 101.5%, 103%, 104.5%, and 106% respectively.

7.17 (a) We see at once that the amount of each annuity payment is

$$\frac{100\,000}{a_{\overline{13}|}} \qquad \text{at } 8\% \qquad = 11\,682 \cdot 95$$

Ignoring income tax, the value of the loan repayments is

$$11\,682 \cdot 95a_{\overline{13}|} \qquad \text{at } 7\% \qquad = 106\,407$$

By Makeham's formula, if the value of the capital payments is K, then

$$K + \frac{0 \cdot 08}{0 \cdot 07}(100\,000 - K) = 106\,407$$

Hence we may find K, which is 55 149. The price to be paid may be found, by Makeham's formula, to be

$$K + \frac{0 \cdot 08 \times 0 \cdot 6}{0 \cdot 07}(100\,000 - K) = £85\,904$$

Note This value may also be found as $K + 0 \cdot 6(106\,407 - K)$.

(b) Since $g = i$, we cannot use Makeham's formula to find K, the value of the capital. We must find K by the direct method, which gives

$$K = 11\,682 \cdot 95(v_{0 \cdot 08}^{15} v_{0 \cdot 08} + v_{0 \cdot 08}^{14} v_{0 \cdot 08}^{2} + \cdots + v_{0 \cdot 08} v_{0 \cdot 08}^{15})$$
$$= 11\,682 \cdot 95 v_{0 \cdot 08}^{16} \times 15$$
$$= 51\,152 \cdot 08$$

Hence, by Makeham's formula, the price to be paid for the loan is

$$K + \frac{0 \cdot 6(0 \cdot 08)}{0 \cdot 08}(100\,000 - K) = £80\,461$$

7.18 Let P_n denote the net value of £100 nominal of this stock, at 6% per annum interest, assuming that redemption takes place at time $n\,(10 \leqslant n \leqslant 15)$. We have

$$P_n = [85 + \tfrac{1}{2}(n - 10)(n - 9)]v^n + (0 \cdot 3 \times 8 \cdot 5 a_{\overline{n}|}) \qquad \text{at } 6\%.$$

The values of P_n are as follows:

n	10	11	12	13	14	15
P_n	66·23	65·42	65·11	65·24	65·72	66·49

We now consider the value, per £600 nominal, on the assumption that the borrower exercises his option at time n years ($n = 10, 11, 12, 13, 14,$ or 15).

Time at which option is exercised, n years	Value per £600 nominal at 6% interest	
10	$6P_{10}$	$= 397 \cdot 38$
11	$P_{10} + 5P_{11}$	$= 393 \cdot 33$
12	$P_{10} + P_{11} + 4P_{12}$	$= 392 \cdot 09$
13	$P_{10} + P_{11} + P_{12} + 3P_{13}$	$= 392 \cdot 48$
14	$P_{10} + P_{11} + P_{12} + P_{13} + 2P_{14}$	$= 393 \cdot 44$
15	$P_{10} + P_{11} + P_{12} + P_{13} + P_{14} + P_{15}$	$= 394 \cdot 21$

From this table, we see that the smallest value (392·09) occurs if the borrower exercises his option at time 12 years. The investor should therefore calculate his offer price on this assumption. The maximum price is thus £392·09 per £600 nominal, or £65·35%.

7.19 Measure time in years from the date of issue. The investor purchases the loan at time 1/2. If the loan is repaid at time $n (1 \leqslant n \leqslant 5)$, the yield per annum is that value of i for which

$$102 = (1 + i)^{1/2}(100v^n + 10a_{\overline{n}|}) \qquad \text{at rate } i$$

Using numerical methods, for $n = 1, 2, \ldots, 5$ we find the corresponding value of i as given in the following table:

n	1	2	3	4	5
i	0·1630	0·1220	0·1138	0·1103	0·1084

The least yield is 10·84% p.a., which will be achieved if redemption occurs five years after the date of issue.

Although the redemption price is less than the purchase price, in this situation early redemption will lead to a higher yield for the purchaser. This is because of the relatively short time to the first interest payment.

Chapter 8

8.1 (a) Let us work in terms of £100 nominal of loan. We have $C = 110$, and the value of capital repayments (ignoring tax) is

$$K = 110v^{15} \qquad \text{at } 8\% \qquad = 34\cdot6766$$

By Makeham's formula, with allowance for capital gains tax (see formula 8.2.7), the price A must satisfy the relationship

$$A = K + \frac{0\cdot09(0\cdot55)}{1\cdot1 \times 0\cdot08^{(2)}}(110 - K) - 0\cdot3\left(\frac{110 - A}{110}\right)K$$

which gives $A = £74\cdot52$ per £100 nominal.

(b) We note that the capital gain, per £100 nominal, is £30, of which £9 must be paid in tax. The net proceeds on redemption are therefore £101, so that the net yield per annum is the solution of the equation

$$80 = 0\cdot55 \times 9a_{\overline{15}|}^{(2)} + 101v^{15} \qquad \text{at rate } i.$$

This gives $i = 7\cdot32\%$.

8.2 Let us work in terms of £100 nominal. In the notation of chapter 7,

$C = 120$ and

$$K = \frac{120}{15}(a_{\overline{15}|} - a_{\overline{7}|}) \quad \text{at } 7\% \quad = 55\cdot5871$$

By a straightforward modification of Makeham's formula, to allow for the fact that interest is payable in advance rather than in arrear, the value of the loan, ignoring capital gains tax, is

$$K + \frac{0\cdot7(0\cdot07)}{d^{(2)}_{0\cdot07}}(C - K)$$

Let the price per cent, allowing for capital gains tax, be A. We have

$$A = K + \frac{0\cdot7(0\cdot07)}{d^{(2)}_{0\cdot07}}(C - K) - 0\cdot25\left(\frac{C - A}{C}\right)K$$

from which we obtain $A = £100\cdot81\%$.

8.3   We first calculate A_0, the price paid by investor A for the bond. This is given by the equation of value

$$A_0 = 60a^{(2)}_{\overline{10}|} + 1000v^{10} - 0\cdot4(1000 - A_0)v^{10} \qquad \text{at } 10\%$$

whence $A_0 = 720\cdot04$.

(a) Let A_1 be the price paid by investor B. We have

$$A_1 = 60a^{(2)}_{\overline{5}|} + 1000v^5 - 0\cdot4(1000 - A_1)v^5 \qquad \text{at } 10\%$$

which gives $A_1 = 805\cdot65$. Hence investor A makes a capital gain of $805\cdot65 - 720\cdot04 = 85\cdot61$ on selling the bond, and thus pays capital gains tax of $0\cdot4(85\cdot61) = 34\cdot24$ at the date of sale. His net proceeds on the sale of the bond are therefore $805\cdot65 - 34\cdot24 = 771\cdot41$, and his net annual yield is the solution of the equation

$$720\cdot04 = 60a^{(2)}_{\overline{5}|} + 771\cdot41v^5 \qquad \text{at rate } i.$$

By trials and interpolation, $i = 9\cdot71\%$.

(b) Let X be the *net* proceeds to investor A on the sale of the bond. We find X by means of the equation of value

$$720\cdot04 = 60a^{(2)}_{\overline{5}|} + Xv^5 \qquad \text{at } 10\%$$

which gives $X = 784\cdot39$. The price paid by B, A_2, must therefore satisfy

$$A_2 - 0\cdot4(A_2 - 720\cdot04) = 784\cdot39$$

which gives $A_2 = 827\cdot29$.

Hence the net proceeds to investor B on redemption of the bond are $1000 - 0.4(1000 - 827.29) = 930.92$, so that his net annual yield is the solution of

$$827.29 = 60a_{\overline{5}|}^{(2)} + 930.92v^5 \qquad \text{at rate } i$$

By trials and interpolation, $i = 9.50\%$.

8.4 (a) Let P_1 be the purchase price per unit redemption money. We have

$$P_1(1 + i)^n = 1 - t(1 - P_1)$$

from which we obtain

$$P_1 = \frac{1 - t}{(1 + i)^n - t}$$

(b) Let P_2 be the sale price per unit redemption money. We have the equation of value

$$P_1(1 + i)^m = P_2 - t(P_2 - P_1)$$

whence

$$P_2 = P_1\left[\frac{(1 + i)^m - t}{1 - t}\right]$$

$$= \frac{(1 + i)^m - t}{(1 + i)^n - t} \qquad \text{(using (a))}$$

(c) Let P_3 be the price paid by B. Using (a) (with an outstanding term of $(n - m)$ years), we find that

$$P_3 = \frac{1 - t}{(1 + i)^{n-m} - t}$$

(Note that $P_3 > P_1$.)

We now obtain the equation of value

$$P_1(1 + j)^m = P_3 - t(P_3 - P_1)$$

from which we have

$$(1 + j)^m = \frac{P_3(1 - t) + tP_1}{P_1}$$

$$= t + \frac{(1 - t)[(1 + i)^n - t]}{[(1 + i)^{n-m} - t]}$$

i.e.

$$j = \left\{ t + \frac{(1-t)[(1+i)^n - t]}{[(1+i)^{n-m} - t]} \right\}^{1/m} - 1$$

Substituting $n = 10$, $m = 5$, $t = 0\cdot4$, and $i = 0\cdot1$ in this last equation, we obtain $j = 8\cdot26\%$.

8.5   We first find the number of capital repayments n by solving the equation

$$115\,000 = 1000\{15 + (15+2) + (15+6) + \cdots + [15 + n(n-1)]\}$$

By trials, $n = 5$. Hence there are five repayments of capital, the cash amounts being $15\,000$, $17\,000$, $21\,000$, $27\,000$, and $35\,000$ respectively. The value of the capital repayments is

$$K = 15\,000v^{10} + 17\,000v^{13} + 21\,000v^{16} + 27\,000v^{19}$$
$$+ 35\,000v^{22} \qquad \text{at } 7\%$$
$$= 37\,158\cdot5$$

Let the price paid be P per unit nominal. We have $g = 0\cdot06/1\cdot15$, and the equation of value is

$$100\,000P = K + 0\cdot7\frac{(0\cdot06/1\cdot15)}{0\cdot07^{(2)}}(115\,000 - K)$$
$$- 0\cdot4\left(\frac{1\cdot15 - P}{1\cdot15}\right)K$$

which gives $P = 0\cdot730\,49$. Hence the value of the whole loan is £73 049.

8.6   The price paid by the investor is

$$100v^{15\cdot5} + 3a^{(2)}_{\overline{15\cdot5}|} + a^{(2)}_{\overline{5\cdot5}|} \qquad \text{at } 6\% \qquad = 75\cdot34$$

The net yield per annum obtained by the investor is the solution of the equation

$$75\cdot34 = 0\cdot6[3a^{(2)}_{\overline{15\cdot5}|} + a^{(2)}_{\overline{5\cdot5}|}] + [100 - 0\cdot2(100 - 75\cdot34)]v^{15\cdot5} \quad \text{at rate } i$$

By trials, $i = 4\cdot00\%$.

8.7   (a) Divide the loan into two sections:

     Section 1   First seven years:

$$C_1 = \text{cash repaid} = 400 \times 7 \times 100 = 280\,000$$

K_1 = value of capital repayments = $400 \times 100a_{\overline{7}|0 \cdot 06} = 223\,296$

$g_1 = 0 \cdot 04$

Section 2 Next twelve years:

$$C_2 = 600 \times 12 \times 115 = 828\,000$$

$$K_2 = 600 \times 115(a_{\overline{19}|0 \cdot 06} - a_{\overline{7}|0 \cdot 06}) = 384\,723$$

$$g_2 = 4/115 = 0 \cdot 034\,78$$

Let P be the price per bond. The equation of value is

$$10\,000P = K_1 + \frac{0 \cdot 65g_1}{i^{(2)}}(C_1 - K_1) - 0 \cdot 3\frac{(C_1 - 2800P)}{C_1}K_1$$

$$+ K_2 + \frac{0 \cdot 65g_2}{i^{(2)}}(C_2 - K_2) - 0 \cdot 3\frac{(C_2 - 7200P)}{C_2}K_2$$

i.e.

$$10\,000P = 223\,296 + 24\,934 - 66\,989(1 - 0 \cdot 01P)$$

$$+ 384\,723 + 169\,488 - 115\,417(1 - 0 \cdot 008\,695\,6P)$$

i.e.

$$(10\,000 - 669 \cdot 89 - 1003.62)P = 620\,035$$

which gives $P = £74 \cdot 47$.

(b) During years 1–7 and also years 8–19 yields decrease (because one must wait longer for the capital gain), so the graph of yield against time to redemption is roughly as in Figure S.8.1. The yield on a

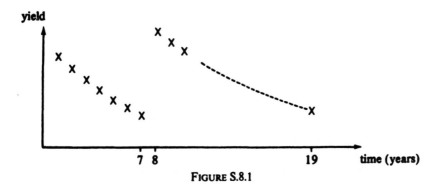

FIGURE S.8.1

bond redeemed at time n years is found by solving the equation

$$74 \cdot 47 = 100v^n + [0 \cdot 65 \times 4a_{\overline{n}|}^{(2)}] - \frac{0 \cdot 3(100 - 74 \cdot 47)}{100}100v^n$$

$$= 92 \cdot 34v^n + 2 \cdot 6a_{\overline{n}|}^{(2)} \qquad \text{for } 1 \leqslant n \leqslant 7$$

or

$$74 \cdot 47 = 115v^n + [0 \cdot 65 \times 4a_{\overline{n}|}^{(2)}] - 0 \cdot 3\left(\frac{115 - 74 \cdot 47}{115}\right)115v^n$$

$$= 102 \cdot 84v^n + 2 \cdot 6a_{\overline{n}|}^{(2)} \qquad \text{for } 8 \leqslant n \leqslant 19$$

For $n = 7$, the RHS at 6% is 76·14, so yield $> 6\%$ for $1 \leqslant n \leqslant 7$.

For $n = 11$, the RHS at 6% is 74·98.

For $n = 12$, the RHS at 6% is 73·23.

Hence, if $n \leqslant 11$ the yield is greater than 6%, and if $n \geqslant 12$ the yield is less than 6%. The number of bonds yielding less than 6% is thus $8 \times 600 = 4800$.

8.8 Assuming that redemption is always at par, the value of the capital repayments is

$$25\,000a_{\overline{20}|} \text{ at } 8\% \qquad = 245\,454$$

Hence the value of the net interest payments is, by Makeham,

$$(1 - 0 \cdot 4)\frac{0 \cdot 12}{0 \cdot 08^{(12)}}(500\,000 - 245\,454) = 237\,375$$

The value of the premiums on redemption is

$$25\,000 \times 0 \cdot 1(a_{\overline{20}|} - a_{\overline{10}|}) \qquad \text{at } 8\% \qquad = 7770$$

so the value of the capital and net interest repayments, ignoring capital gains tax, is

$$245\,454 + 7770 + 237\,375 = 490\,599$$

Let A be the price paid for the loan. We have

$$A = 490\,599 - \text{value of capital gains tax payments}$$

It is clear from this equation that the price must be below par, so capital gains tax will be paid on redemption of all the stock, not merely on the part redeemed at 110%. The value of the capital gains tax payments is

$$0{\cdot}4\left[\left(25\,000-\frac{A}{20}\right)v+\left(25\,000-\frac{A}{20}\right)v^2+\cdots+\left(25\,000-\frac{A}{20}\right)v^5\right]$$

$$+0{\cdot}3\left[\left(25\,000-\frac{A}{20}\right)v^6+\left(25\,000-\frac{A}{20}\right)v^7+\cdots\right.$$

$$\left.+\left(25\,000-\frac{A}{20}\right)v^{10}\right]$$

$$+0{\cdot}3\left\{\left[(25\,000\times1{\cdot}1)-\frac{A}{20}\right]v^{11}+\left[(25\,000\times1{\cdot}1)-\frac{A}{20}\right]v^{12}+\cdots\right.$$

$$\left.+\left[(25\,000\times1{\cdot}1)-\frac{A}{20}\right]v^{20}\right\} \qquad \text{at } 8\%$$

$$=10\,000a_{\overline{5}|}+7500(a_{\overline{10}|}-a_{\overline{5}|})+8250(a_{\overline{20}|}-a_{\overline{10}|})$$

$$-\frac{A}{20}[0{\cdot}4a_{\overline{5}|}+0{\cdot}3(a_{\overline{10}|}-a_{\overline{5}|})+0{\cdot}3(a_{\overline{20}|}-a_{\overline{10}|})] \qquad \text{at } 8\%$$

$$=2500a_{\overline{5}|}-750a_{\overline{10}|}+8250a_{\overline{20}|}-\frac{A}{20}(0{\cdot}1a_{\overline{5}|}+0{\cdot}3a_{\overline{20}|}) \qquad \text{at } 8\%$$

$$=85\,949-0{\cdot}167\,236A$$

Hence we have

$$A=490\,599-(85\,949-0{\cdot}167\,236A)$$

from which we obtain $A=£485\,912$.

8.9 Let us work in terms of £100 nominal of loan. The value of the capital repayments, ignoring capital gains tax, is

$$K=\frac{5}{s_{\overline{2}|}}(a_{\overline{48}|}-a_{\overline{8}|}) \qquad \text{at } 8\% \qquad =15{\cdot}486\,77$$

Hence, if interest were at 6% throughout, the value of the gross interest payments would be

$$\frac{0{\cdot}06}{0{\cdot}08}(100-K)=63{\cdot}384\,92$$

We now find the value of the 'extra' gross interest in the final 18 years.

The value, at time 30 years, of the outstanding capital repayments is

$$K' = \frac{5}{s_{\overline{7}|}} a_{\overline{18}|} \qquad \text{at } 8\% \qquad = 22.528\,54$$

Note that the loan outstanding at time 30 years is £45 nominal. The value at time 30 of the 'extra' gross interest is thus

$$\frac{0.01}{0.08}(45 - K') = 2.808\,93$$

Hence the value at time 0 of the 'extra' gross interest is $2.808\,93v^{30}$ at $8\% = 0.279\,85$. The price per cent A therefore satisfies the equation

$$A = 15.486\,77 + 0.65(63.384\,92 + 0.279\,85) - 0.3(100 - A)\frac{15.486\,77}{100}$$

This gives $A = £54.77$.

8.10 We first find the term of the loan, n years. The capital repaid at time $t + 4$ is $44\,000 + 6000t$ $(t = 1, 2, \ldots, n)$. Hence the capital repaid by time n is

$$44\,000n + 6000 \sum_{t=1}^{n} t = 44\,000n + 3000n(n + 1)$$

$$= 880\,000 \qquad \text{for} \qquad n = 11.$$

Hence the loan has term 15 years, and the value of capital is

$$K = 50\,000v^5 + 56\,000v^6 + \cdots + 110\,000v^{15} \qquad \text{at rate } i$$
(where i is the net annual yield)

$$= 44\,000(a_{\overline{15}|} - a_{\overline{7}|}) + 6000v^4(Ia)_{\overline{11}|}$$

We now solve the equation of value

$$831\,600 = 0.945 \times 880\,000$$

$$= K + \frac{0.625 \times 0.055}{i^{(2)}}(880\,000 - K)$$

$$- 0.3 \times \frac{880\,000 - (0.945 \times 880\,000)}{880\,000}K \qquad \text{at rate } i \qquad (1)$$

A rough solution is

$$i \simeq (0.055 \times 0.625) + \frac{(100 - 94.5) \times 0.7}{94.5 \times 11} = 3.8\%$$

When $i = 4\%$, $K = 581\,513$ and the right-hand side of equation 1 is $830\,969$. The right-hand side of equation 1 at $3\frac{1}{2}\%$ is £867 387, and hence $i = 4\cdot0\%$ (to the nearest $0\cdot1\%$).

8.11 Let us work in terms of £300 nominal, so £10 nominal is redeemed each year, and let us consider the loan to be in three sections, referring to years 1–10, 11–15, and 16–30 respectively. Assuming that redemption is at par throughout, we have

$$K_1 = 10a_{\overline{10}|} \qquad \text{at } 7\% = 70\cdot235\,82$$
$$K_2 = 10(a_{\overline{15}|} - a_{\overline{10}|}) \qquad \text{at } 7\% = 20\cdot843\,32$$
$$K_3 = 10(a_{\overline{30}|} - a_{\overline{15}|}) \qquad \text{at } 7\% = 33\cdot011\,27$$

Let

$$K = K_1 + K_2 + K_3 = 10a_{\overline{30}|} \text{ at } 7\% = 124\cdot090\,41$$

Then the value of the net interest is (by Makeham)

$$\frac{0\cdot08}{0\cdot07^{(4)}}(0\cdot6)(300 - 124\cdot090\,41) = 123\cdot7455$$

The value of the loan, ignoring capital gains tax (CGT), is therefore

$$K_1 + K_2 + 1\cdot2K_3 + 123\cdot7455 = 254\cdot438\,16$$

Let A be the value of the loan allowing for CGT. We have

$$A = 254\cdot438\,16 - \text{value of CGT payments.}$$

This shows that $A \leqslant 254\cdot438\,16$, so CGT is payable on all redemptions. The value of the CGT payments is

$$0\cdot4 \sum_{t=1}^{10} \left(10 - \frac{A}{30}\right)v^t + 0\cdot25 \sum_{t=11}^{15} \left(10 - \frac{A}{30}\right)v^t$$
$$+ 0\cdot25 \sum_{t=16}^{30} \left(12 - \frac{A}{30}\right)v^t \qquad \text{at } 7\%$$
$$= 0\cdot4\frac{300 - A}{300}K_1 + 0\cdot25\frac{300 - A}{300}K_2$$
$$+ 0\cdot25\frac{360 - A}{360}1\cdot2K_3$$

Hence we have

$$A = 254\cdot438\,16 - 0\cdot4\frac{300 - A}{300}K_1 - 0\cdot25\frac{300 - A}{300}K_2$$
$$- 0\cdot25\frac{360 - A}{360}1\cdot2K_3$$

which gives $A = 245 \cdot 196$. The price paid by the investor is thus £81·73 per bond.

8.12 (a) We first note that the total nominal amount of the loan is

$$1000(15 + 20 + 25 + \cdots + 85) = 750\,000$$

We have $C = 750\,000$, and the value of capital repayments is

$$K = 1000[10(a_{\overline{20|}} - a_{\overline{3|}}) + 5v^5(Ia)_{\overline{13|}}] \quad \text{at } 10\% \qquad = 195\,498$$

Hence the price paid by B is, by Makeham,

$$K + \frac{0 \cdot 07(1 - 0 \cdot 33)}{i^{(2)}}(C - K) \qquad \text{at } 10\% \qquad = £461\,906$$

(b) Let $k = \lambda/100$. We find k by means of the relationship

value at 31 December 1974 of the premiums on redemption, net of capital gains tax at the rate paid by B

= value at 31 December 1974 of net interest given up by B

Let N_t denote the nominal amount redeemed at time t years (measured from 31 December 1974). Note that the purchase price per unit nominal is $461\,906/750\,000 = 0 \cdot 615\,875$. Hence, for $t > 10$, the *net* premium on redemption at time t is

$$N_t(1 + k) - 0 \cdot 1[N_t(1 + k) - 0 \cdot 615\,875\,N_t] - N_t$$
$$= N_t(0 \cdot 9k - 0 \cdot 0384\,13)$$

Hence the value at time 0 of the net premiums on redemption after time 10 is

$$\sum_{t=11}^{20} v^t N_t(0 \cdot 9k - 0 \cdot 0384\,13)$$
$$= 1000(0 \cdot 9k - 0 \cdot 0384\,13)v^{10}(40v + 45v^2 + \cdots + 85v^{10})$$
$$= (0 \cdot 9k - 0 \cdot 0384\,13)v^{10}K'$$

where

$$K' = 1000(40v + 45v^2 + \cdots + 85v^{10}) \qquad \text{at } 10\%$$
$$= 1000[35a_{\overline{10|}} + 5(Ia)_{\overline{10|}}]$$
$$= 360\,239$$

Now we also note that the nominal amount of loan outstanding on 1 January 1985 is

$$1000(40 + 45 + \cdots + 85) = 625\,000$$

and K' is the value at that date of the outstanding capital repayments, assuming that these are at par. Hence the value *at 1 January 1985* of the net income given up by B is (by Makeham)

$$(625\,000 - K')\left[\frac{0.07(1 - 0.33) - 0.05(1 - 0.5)}{0.1^{(2)}}\right] = 59\,398$$

Hence we may find k by the equation

$$(0.9k - 0.038\,413)v^{10}K' = v^{10}(59\,398) \qquad \text{at } 10\%$$

i.e.

$$0.9k - 0.038\,413 = \frac{59\,398}{360\,239}$$

which gives $k = 0.2259$, and hence $\lambda = 100k = 22.59$.

8.13 We first assume redemption at par and ignore capital gains tax. The value of the issue is

$$K + \frac{0.65 \times 0.04}{i^{(2)}}(1\,000\,000 - K) \qquad \text{at rate } i \text{ (the yield)}$$

where

$$K = 100\,000(v^{10} + 1.5v^{13} + 2v^{15} + 2.5v^{18} + 2v^{20} + v^{23}) \qquad \text{at rate } i$$

But we must allow for (a) actual redemption money, i.e. *add*

$$-20\,000v^{10} - 30\,000v^{13} + 40\,000v^{20} + 20\,000v^{23}$$

and for (b) capital gains tax (*on bonds redeemable above 85*), i.e. *subtract*

$$0.3[(100 - 85)2000v^{15} + (100 - 85)2500v^{18}$$
$$+ (120 - 85)2000v^{20} + (120 - 85)1000v^{23}]$$

Hence we finally obtain the equation of value

$$850\,000 = K + \frac{0.65 \times 0.04}{i^{(2)}}(1\,000\,000 - K)$$
$$- 20\,000v^{10} - 30\,000v^{13} + 40\,000v^{20} + 20\,000v^{23}$$
$$- 9000v^{15} - 11\,250v^{18} - 21\,000v^{20} - 10\,500v^{23}$$

i.e.

$$850\,000 = K + \frac{0.026}{i^{(2)}}(1\,000\,000 - K)$$

$$-20\,000v^{10} - 30\,000v^{13} - 9000v^{15} - 11\,250v^{18}$$

$$+ 19\,000v^{20} + 9500v^{23} \tag{1}$$

A rough solution (assume redemption at par in about 16 years' time) gives

$$\text{yield} = \frac{2.6 + [(100 - 85)/16]0.7}{85} = 3.8\%$$

When $i = 4\%$, the right-hand side of equation 1 is £806 894. When $i = 3.5\%$, the right-hand side of equation 1 is £860 416. By interpolation, $i = 3.59\%$.

Chapter 9

9.1 (a) Let us work in terms of £100 nominal of loan. We have $R = 1.05$, $g = 0.04/1.05$, $z = 0.03$, and $C = 105$. By formula 9.3.9, the annual service is

$$S = 105\left(\frac{0.04}{1.05} + 0.03\right) = 7.15$$

In terms of the total loan of £1 000 000, the annual service is thus £71 500. The term of the loan, n years, is found by solving

$$7.15a_{\overline{n-1}} < 105 \leqslant 7.15a_{\overline{n}} \qquad \text{at rate } g$$

from which we obtain $n = 22$ years.

(b) By formula 9.3.14, the total service required in year 22 is

$$(1 + g)(105 - 3.15s_{\overline{21}|g}) = 6.6208.$$

Hence, in terms of £1 000 000 nominal, the final year's service is £66 208.

(c) By formulae 9.3.2 and 9.3.3, the *total cash* indebtedness repaid after m years is $3.15\,s_{\overline{m}|}$ at rate g. One half of the original cash indebtedness is 52.5, and (since the redemption price is constant) the required number of years is m, where

$$3.15s_{\overline{m-1}|} \leqslant 52.5 < 3.15s_{\overline{m}|} \qquad \text{at rate } g.$$

This gives $m = 14$; that is, more than 50% of the nominal amount of the loan will be repaid after 14 years.

9.2 (a) The value of the loan is simply

$$71\,500a_{\overline{31}} + 66\,208v^{22} \qquad \text{at } 5\% \qquad = £939\,346.$$

(b) We first find K, the value of the capital repayments, by the indirect method (see example 7.6.5). We have

$$939\,346 = K + \frac{g}{0 \cdot 05}(1\,050\,000 - K)$$

where $g = 0 \cdot 04/1 \cdot 05$. This gives $K = 585\,252$, and the value of the loan is (by Makeham's formula)

$$K + \frac{g}{0 \cdot 05^{(2)}}(1\,050\,000 - K) = £943\,718$$

(c) Again using Makeham's formula, we obtain the value

$$K + \frac{g(0 \cdot 625)}{0 \cdot 05^{(2)}}(1\,050\,000 - K) = £809\,293$$

9.3 (a) (i) The term, n years, is found by solving

$$9 \cdot 9a_{\overline{n-1}} < 110 \leqslant 9 \cdot 9a_{\overline{n}} \qquad \text{at rate } g$$

where $g = 5 \cdot 5/110 = 0 \cdot 05$. This gives $n = 17$ years. By formula 9.3.14, the final year's service is

$$1 \cdot 05(110 - 4 \cdot 4s_{\overline{16}|0 \cdot 05}) = 6 \cdot 202\,35$$

The value of the loan, ignoring tax and assuming that interest is payable annually, is

$$A = 9 \cdot 9a_{\overline{16}} + 6 \cdot 202\,35v^{17} \qquad \text{at } 6\% \qquad = 102 \cdot 352$$

Hence we may find K, the value of the capital repayments, by solving

$$A = K + \frac{0 \cdot 05}{0 \cdot 06}(110 - K)$$

which leads to $K = £64 \cdot 11$.

(ii) By Makeham's formula, the value of the interest payments is

$$\frac{0 \cdot 05}{0 \cdot 06^{(2)}}(110 - K) = £38 \cdot 81$$

This value may also be obtained as

$$[0 \cdot 06/0 \cdot 06^{(2)}](102 \cdot 352 - 64 \cdot 11).$$

(b) Let A' be the price payable by the investor. We have the equation

$$A' = 64{\cdot}11 + (0{\cdot}65 \times 38{\cdot}81) - 0{\cdot}15\frac{110 - A'}{110}64{\cdot}11$$

whence $A' = £87{\cdot}36$.

9.4 The definition of n and f implies that

$$1 = (g + z)a_{\overline{n+f}|g}$$
$$= (g + z)(a_{\overline{n}|g} + fv_g^{n+1}) \qquad \text{(by definition of } a_{\overline{n+f}|g})$$

Hence

$$f(g + z)v_g^{n+1} = 1 - ga_{\overline{n}|g} - za_{\overline{n}|g}$$
$$= v_g^n - za_{\overline{n}|g}$$

so that

$$f(g + z) = (1 + g)(1 - zs_{\overline{n}|g}) \qquad (1)$$

Note that the amount of the service payment at time $(n + 1)$, comprising both capital and interest, is

$$(1 + g)(C - zCs_{\overline{n}|g})$$

The value (at rate of interest i per annum) of the loan is simply the value of the service payments, i.e.

$$(g + z)Ca_{\overline{n}|i} + (1 + g)[C - zCs_{\overline{n}|g}]v_i^{n+1}$$
$$= (g + z)Ca_{\overline{n}|i} + f(g + z)Cv_i^{n+1} \qquad \text{(by equation 1)}$$
$$= (g + z)C[a_{\overline{n}|i} + fv_i^{n+1}]$$
$$= (g + z)C[a_{\overline{n}|i} + f(a_{\overline{n+1}|i} - a_{\overline{n}|i})]$$
$$= (g + z)Ca_{\overline{n+f}|i} \qquad \text{(by definition)}$$

as required.

9.5 (a) We have $g = 0{\cdot}05$ and $C = 1\,050\,000$. The term, n years, is found by solving

$$74\,500a_{\overline{n-1}} < 1\,050\,000 \leqslant 74\,500a_{\overline{n}} \qquad \text{at } 5\%$$

This gives a term of 25 years exactly (i.e. there is no reduction in the final year's service). Ignoring taxation and assuming that interest is paid annually, we value the loan as

$$A = 74\,500a_{\overline{25}} \qquad \text{at } 6\% \qquad = 952\,360.$$

Hence we find K by means of the relationship

$$952\,360 = K + \frac{0\cdot05}{0\cdot06}(1\,050\,000 - K)$$

which gives $K = 464\,160$. The value of the loan A' to give the investor 6% net per annum is the solution of

$$A' = K + \frac{0\cdot05 \times 0\cdot6}{0\cdot06^{(2)}}(1\,050\,000 - K)$$

$$- 0\cdot3\frac{1\,050\,000 - A'}{1\,050\,000}K$$

from which we obtain $A' = £717\,286$.

(b) The capital repaid after one year is £22 000 (in cash terms). Hence the total capital repaid after 15 years is (see equation 9.3.2)

$$22\,000s_{\overline{15}|} \quad \text{at } 5\% \quad = £474\,728$$

The total *nominal* capital repaid after 15 years is thus

$$\frac{474\,728}{1\cdot05} = £452\,122$$

9.6 Let us work in terms of £100 nominal of loan. We have $g = 0\cdot02$, $C = 125$, and the annual service is

$$S = (z + g)C = (0\cdot04 + 0\cdot02)125 = 7\cdot5$$

The term, n years, is found from

$$7\cdot5a_{\overline{n-1}|} < 125 \leqslant 7\cdot5a_{\overline{n}|} \quad \text{at } 2\%$$

from which we obtain $n = 21$ years. The redemption money in the final year is (by equation 9.3.13)

$$125 - 5s_{\overline{20}|0\cdot02} = 3\cdot5132$$

Hence the value of the capital repayments, ignoring capital gains tax, is

$$K = 5v + 5v^2(1\cdot02) + \cdots + 5v^{20}(1\cdot02)^{19} + 3\cdot5132v^{21} \quad \text{at } 2\%$$
$$= (5v \times 20) + 3\cdot5132v^{21}$$

$$= 100\cdot3571$$

(Note that, because $g = i$, the indirect method of valuing the capital does not work.) The issue price A satisfies the relationship

$$A = K + 0.625\frac{0.02}{0.02^{(2)}}(125 - K) - 0.2\frac{125 - A}{125}K$$

from which we find that $A = £114.0825$, or £1 140 825 for the entire loan.

9.7 (a) We have

$$R_t = \begin{cases} 1.1 & \text{for } 1 \leqslant t \leqslant 10 \\ 1.2222 & \text{for } 11 \leqslant t \leqslant 20 \end{cases}$$

Equation 9.2.3 shows that the cash amount of the rth capital repayment is

$$X\left(1 + \frac{0.11}{1.1}\right)^{r-1} \qquad \text{for } r = 1, 2, \ldots, 11$$

$$X\left(1 + \frac{0.11}{1.1}\right)^{10}\left(1 + \frac{0.11}{1.2222}\right)^{r-11} \qquad \text{for } 12 \leqslant r \leqslant 20$$

This proves the desired result.

(b) The annual annuity payment is $X + 27\,500$. We may find X by observing that the loan outstanding (in cash terms) at the beginning of the 20th year is the capital payment at time 20, i.e.

$$X(1.1)^{10}(1.09)^9$$

Interest in the final year is thus $X(1.1)^{10}(1.09)^9(0.09)$, so that the total payment in the final year is

$$X(1.1)^{10}(1.09)^{10}$$

But this must equal $X + 27\,500$, giving $X = 5350$. The annual annuity payment is therefore £32 850.

(c) The value of the capital repayments, ignoring capital gains tax, is

$$K = K_1 + K_2$$

$$= 5350(v + 1.1v^2 + \cdots + 1.1^9v^{10})$$

$$\quad + 5350(1.1)^{10}v^{10}(v + 1.09v^2 + \cdots + 1.09^9v^{10}) \qquad \text{at } 8\%$$

$$= 53\,871 + 62\,050$$

$$= 115\,921$$

The value of the entire loan, ignoring taxation, is

$$32\,850a_{\overline{20|}} \qquad \text{at } 8\% \qquad = 322\,525$$

Hence the value of the gross interest payments is

$$322\,525 - 115\,921 = 206\,604$$

Let the price per £1 nominal of loan be P'. The value of the capital gains tax payments is

$$0{\cdot}2\frac{1{\cdot}1 - P'}{1{\cdot}1}K_1 + 0{\cdot}2\frac{1{\cdot}2222 - P'}{1{\cdot}2222}K_2$$

$$= 0{\cdot}2(K_1 + K_2) - P'\left(\frac{0{\cdot}2K_1}{1{\cdot}1} + \frac{0{\cdot}2K_2}{1{\cdot}2222}\right)$$

$$= 23\,184 - 19\,948{\cdot}5P'$$

Hence we obtain the equation of value

$$250\,000P' = 115\,921 + (0{\cdot}7 \times 206\,604) - (23\,184 - 19\,948{\cdot}5P')$$

This gives $P' = 1{\cdot}031\,77$; the total price is $250\,000P' = £257\,942$.

9.8 (a) Consider the loan to have originally had nominal amount £100, so that £80 nominal remains outstanding. The cash indebtedness outstanding is thus $80 \times 1{\cdot}125 = 90$. The annual service is now £10 and the cash available for repaying part of the loan in one year's time is $10 - (80 \times 0{\cdot}09) = 2{\cdot}8$, which increases at rate $g = 0{\cdot}09/1{\cdot}125 = 0{\cdot}08$ per annum compound.

The outstanding term is thus found by solving the inequality

$$2{\cdot}8s_{\overline{n-1}|} < 90 \leqslant 2{\cdot}8s_{\overline{n}|} \qquad \text{at } 8\%$$

This gives an outstanding term of 17 years, the capital payment at time 17 being $90 - 2{\cdot}8s_{\overline{16}|0{\cdot}08} = 5{\cdot}092$.

The value (at rate i) of the outstanding capital payments is

$$K = 2{\cdot}8(v + 1{\cdot}08v^2 + \cdots + 1{\cdot}08^{15}v^{16}) + 5{\cdot}092v^{17}$$

$$= 2{\cdot}8\frac{1 - \left(\dfrac{1{\cdot}08}{1+i}\right)^{16}}{i - 0{\cdot}08} + 5{\cdot}092(1+i)^{-17}$$

The value of the outstanding loan is, by Makeham's formula,

$$A = K + \frac{0{\cdot}08}{i^{(2)}}(90 - K)$$

This is equated to the purchase price, $80 \times 1{\cdot}02 = 81{\cdot}6$, and solved for i. By trial and interpolation we obtain $i \simeq 0{\cdot}0975$ or $9\frac{3}{4}\%$.

(b) For the next five years the company pays only interest of 7.2 per annum, payable half-yearly. We now have $g = 0.09/1.5 = 0.06$ and the new outstanding cash indebtedness is $80 \times 1.5 = 120$.

Under the suggested arrangement the new outstanding term is $5 + n$ years, where

$$2.8 s_{\overline{n-1}} < 120 \leqslant 2.8 s_{\overline{n}} \qquad \text{at } 6\%$$

This gives $n = 22$, so that the capital payment at time 27 is $120 - 2.8 s_{\overline{21}|0.06} = 8.020$.

Ignoring the fact that interest is payable half-yearly we may value the loan to yield $9\frac{3}{4}\%$ per annum as

$$7.2 a_{\overline{5}|} + 10(a_{\overline{26}|} - a_{\overline{5}|}) + (1.06 \times 8.020 v^{27}) = 83.44$$

which is greater than 81.6. As the value allowing for half-yearly interest is greater still, the revised scheme offers the better yield.

9.9   Consider the original loan. $C = 1\,000\,000$, $g = 0.05$, $zs_{\overline{30}|0.05} = 1$, so $z = 0.015\,051\,4$. Annual service $= (g + z)C = 65\,051$, and K, the value of capital repayments, is given by

$$K = 15\,051[v + (1.05)v^2 + \cdots + (1.05)^{29}v^{30}] \quad \text{at rate } i \text{ (net yield p.a.)}$$

$$= 15\,051v\frac{1 - (1.05v)^{30}}{1 - (1.05)v} = 15\,051\frac{1 - \left(\dfrac{1.05}{1+i}\right)^{30}}{i - 0.05}$$

The yield i is found by solving the equation of value

$$950\,890 = K + \frac{0.625 \times 0.05}{i}(1\,000\,000 - K) \qquad \text{at rate } i$$

and it is found that $i = 3.5\%$.

Now consider the new loan, of nominal amount N, set up on 1 January 1970; $z's_{\overline{15}|0.05} = 1$, so $z' = 0.046\,3422$ and the new annual service is $(g + z')N = 0.096\,342\,2N$. We now value the net payments under the altered arrangement at $3\frac{1}{2}\%$ p.a., and equate this to 950 890. The capital outstanding on 1 January 1965 is

$$1\,000\,000 - 15\,051 s_{\overline{10}|0.05} = 810\,690$$

so the interest in each of the years from 1965 to 1970 is $0.05 \times 810\,690 = 40\,534$.

The value of the capital repayments is

$$K' = 15\,051v\frac{1 - (1.05v)^{10}}{1 - (1.05v)}$$

$$+ v^{15}(0.046\,342\,2N)v\frac{1-(1.05v)^{15}}{1-(1.05v)} \qquad \text{at } 3\tfrac{1}{2}\%$$

$$= 155\,276 + 0.444\,22N$$

The value of all the service payments, ignoring tax, is

$$S = 65\,051a_{\overline{10}|} + 40\,534v^{10}a_{\overline{5}|} + 0.096\,342\,2Nv^{15}a_{\overline{13}|} \qquad \text{at } 3\tfrac{1}{2}\%$$

Hence the value of the net payments is

$$K' + 0.625(S - K') = 950\,890$$

Solving for N gives $N = £815\,540$.

9.10 Consider the original loan at the date of issue. We have $C = 1\,000\,000$, $g = 0.07$, and $zs_{\overline{20}|} = 1$ at 7%, so $z = 0.024\,393$ and the service is

$$S = (g + z)C = 94\,393 \text{ per annum}$$

The nominal amount repaid by 1 January 1985 is

$$1\,000\,000\,zs_{\overline{10}|} \qquad \text{at } 7\% \qquad = 337\,025$$

so the nominal amount outstanding on 1 January 1985 is £662 975. We now value the future net payments, at 5% interest p.a., on both the original terms and the new terms.

Original terms The first year's capital repayment is

$$94\,393 - (0.07 \times 662\,975) = 47\,985$$

Hence K, the value of the capital repayments (ignoring capital gains tax), is given by

$$K = 47\,985(v + 1.07v^2 + \cdots + 1.07^9v^{10}) \qquad \text{at } 5\%$$

$$= 47\,985v\frac{1-(1.07v)^{10}}{1-1.07v} \qquad \text{at } 5\%$$

$$= 498\,229$$

Hence the value of the outstanding loan is

$$K + \frac{0.625(0.07)}{i^{(2)}}(662\,975 - K) - 0.3\left(\frac{1-0.9}{1}\right)K \qquad \text{at } 5\%$$

$$= £629\,215$$

New terms The first year's capital repayment is now $60\,000 - (0.07 \times 662\,975) = 13\,592$. Hence the new outstanding term, m years, is given by the relationship

$$13\,592s_{\overline{m-1}|} < 662\,975 \leqslant 13\,592s_{\overline{m}|} \qquad \text{at } 7\%$$

This gives $m = 22$, and the capital repayment in the final year is

$$662\,975 - 13\,592s_{\overline{21}|} \qquad \text{at } 7\% \qquad = 53\,167$$

Hence

$$K = 13\,592v[1 + 1{\cdot}07v + \cdots + (1{\cdot}07v)^{20}]$$
$$+ 53\,167v^{22} \qquad \text{at } 5\%$$

$$= 13\,592v\frac{1 - (1{\cdot}07v)^{21}}{1 - 1{\cdot}07v}, + 53\,167v^{22} \qquad \text{at } 5\%$$

$$= 348\,612$$

The value of the outstanding loan is therefore

$$K + \frac{0{\cdot}07 \times 0{\cdot}625}{i^{(2)}}(662\,975 - K)$$

$$- 0{\cdot}3\left(\frac{1 - 0{\cdot}9}{1}\right)K \text{ at } 5\% \qquad = £616\,618$$

The reduction in the value of the loan is thus £12 597.

9.11 (a) Because of the varying redemption price, we work in terms of *nominal* amounts of loan. The redemption money available in the first year is 25 000, so the nominal amount redeemed in the first year is $25\,000/1{\cdot}15 = 21\,739$.

Let $g_1 = 0{\cdot}04/1{\cdot}15$; the total nominal capital redeemed in the first ten years is

$$21\,739[1 + (1 + g_1) + (1 + g_1)^2 + \cdots + (1 + g_1)^9]$$
$$= 21\,739s_{\overline{10}|} \qquad \text{at rate } g_1$$
$$= 254\,773$$

which leaves $500\,000 - 254\,773 = 245\,227$ nominal outstanding. The redemption money available in the 11th year is thus

$$45\,000 - (0{\cdot}04 \times 245\,227) = 35\,191$$

and this is also equal to the nominal amount redeemed at this time. Let $g_2 = 0{\cdot}04$. At time 10 the outstanding term, n years, is such that

$$35\,191s_{\overline{n-1}|} < 245\,227 \leqslant 35\,191s_{\overline{n}|} \qquad \text{at rate } g_2 = 0{\cdot}04$$

This gives $n = 7$, so the total term is 17 years. The nominal amount outstanding after 16 years is $245\,227 - 35\,191\,s_{\overline{8}|}$ at 4%, i.e. 11 805, and hence the service in the 17th year is $11\,805 \times 1{\cdot}04 = 12\,278$.

On the issue date the value of the capital repayments is (see equation 9.2.3)

$$K = 25\,000[v + (1 + g_1)v^2 + \cdots + (1 + g_1)^9 v^{10}]$$
$$\quad + 35\,191\,v^{10}[v + (1 + g_2)v^2 + \cdots + (1 + g_2)^5 v^6]$$
$$\quad + 11\,805\,v^{17} \qquad \text{at rate } i \text{ (yield p.a.)}$$

$$= 25\,000\frac{1 - [v(1 + g_1)]^{10}}{i - g_1} + 35\,191\,v^{10}\frac{1 - [v(1 + g_2)]^6}{i - g_2}$$
$$\quad + 11\,805\,v^{17}$$

Now the value of the entire loan, assuming interest payments to be made yearly, is

$$A = 45\,000a_{\overline{16}|} + 12\,278\,v^{17} \qquad \text{at rate } i$$

so the value of the actual loan is

$$A' = K + \frac{i}{i^{(2)}}(A - K) \qquad \text{at rate } i$$

We now solve $A' = 461\,000$ by trial and error. A rough solution suggests $i \simeq 6\%$. By trial and interpolation we obtain $i = 0{\cdot}0600$, so that the yield is indeed 6%.

(b) We consider the value on 1 February 1981 of the payments actually received by the pension fund. The value of the capital repayments is

$$K' = 25\,000[v + (1 + g_1)v^2 + \cdots + (1 + g_1)^3 v^4]$$
$$\qquad \qquad \qquad \qquad \qquad \text{at rate } i \text{ (yield p.a.)}$$

$$= 25\,000\frac{1 - [(1 + g_1)v]^4}{i - g_1} \qquad \text{at rate } i$$

The value of the net interest payments is thus

$$\frac{i}{i^{(2)}}(45\,000a_{\overline{7}|} - K') \qquad \text{at rate } i$$

and the equation of value for i is

$$461\,000 = K' + \frac{i}{i^{(2)}}(45\,000a_{\overline{7}|} - K') + 403\,000v^4 \qquad \text{at rate } i$$

By trials, $i = 7{\cdot}00\%$.

9.12 Work with £100 nominal of loan, so that $C = 110$. Note that $g = 0.055/1.1 = 0.05$. Let the annual service (as originally arranged) be $5.5 + X$, so that X is the cash applied to redemption at the first drawing. Since the term was to be 20 years, we have

$$X s_{\overline{20}|0.05} = 110$$

from which we obtain $X = 3.32669$.

After five years the outstanding indebtedness was $110 - X s_{\overline{5}|0.05} = 91.618$. In years 6, 7 and 8 only interest was paid. Let Y be the cash applied to redemption in year 9. Since the loan is to be repaid by the originally agreed date, it follows that

$$Y s_{\overline{12}|0.05} = 91.618$$

so that $Y = 5.75594$.

The value of the actual capital repayments for the complete loan is thus

$$K = X(v + 1.05v^2 + \cdots + 1.05^4 v^5) + Y v^8(v + 1.05v^2 + \cdots + 1.05^{11}v^{12})$$
$$= 3.32669\frac{1 - (1.05v)^5}{i - 0.05} + 5.75594v^8\frac{1 - (1.05v)^{12}}{i - 0.05}$$

where i is the net yield per annum.

By Makeham's formula, the net value of the loan to the investor is

$$K + \frac{0.05 \times 0.6}{i}(110 - K)$$

By equating this last expression to 100, we obtain i. Clearly the net annual yield exceeds $0.6 \times 5.5 = 3.3\%$. The capital gain of 10% is spread over an average term of about 13 years, so that the net yield is approximately $3.3 + 0.8 = 4.1\%$.

By trials and interpolation we obtain $i = 0.0394$ or 3.94%.

Chapter 10

10.1 (a) Measure time in years throughout. By formula 10.3.2, the DMT is

$$\frac{(100 \times 0v^0) + (230 \times 5v^5) + (600 \times 13v^{13})}{100v^0 + 230v^5 + 600v^{13}}$$

which equals (i) 8.42 years when $i = 5\%$, (ii) 5.90 years when $i = 15\%$.

(b) (i) The DMT is

$$\frac{1000(v + 2v^2 + 3v^3 + \cdots + 20v^{20})}{1000(v + v^2 + \cdots + v^{20})} \quad \text{at } 8\%$$

$$= \frac{(Ia)_{\overline{20}|}}{a_{\overline{20}|}} \quad \text{at } 8\% \quad = 8 \cdot 04 \text{ years}$$

(ii) The DMT is

$$\frac{\begin{array}{c}1000[v + (1\cdot 1 \times 2v^2) + (1\cdot 2 \times 3v^3) + \cdots \\ + (2\cdot 9 \times 20v^{20})]\end{array}}{\begin{array}{c}1000(v + 1\cdot 1v^2 + 1\cdot 2v^3 + \cdots \\ + 2\cdot 9v^{20})\end{array}} \quad \text{at } 8\%$$

$$= \frac{0\cdot 9(Ia)_{\overline{20}|} + 0\cdot 1 \sum_{t=1}^{20} t^2 v^t}{0\cdot 9a_{\overline{20}|} + 0\cdot 1(Ia)_{\overline{20}|}} \quad \text{at } 8\%$$

Let

$$X = \sum_{t=1}^{n} t^2 v^t = v + 4v^2 + 9v^3 + \cdots$$
$$+ (n^2 - 2n + 1)v^{n-1} + n^2 v^n$$

Note that

$$vX = v^2 + 4v^3 + \cdots + (n^2 - 2n + 1) v^n + n^2 v^{n+1}$$

whence

$$(1 - v)X = v + 3v^2 + 5v^3 + \cdots + (2n - 1)v^n - n^2 v^{n+1}$$
$$= 2(Ia)_{\overline{n}|} - a_{\overline{n}|} - n^2 v^{n+1}$$

It follows that

$$X = \sum_{t=1}^{n} t^2 v^t = \frac{2(Ia)_{\overline{n}|} - a_{\overline{n}|} - n^2 v^{n+1}}{d}$$

hence we may calculate the DMT, which is found to be 9·78 years.

(iii) The DMT is

$$\frac{\begin{array}{c}1000[(v + (1\cdot 08 \times 2v^2) + (1\cdot 08^2 \times 3v^3) + \cdots \\ + (1\cdot 08^{19} \times 20v^{20})]\end{array}}{\begin{array}{c}1000(v + 1\cdot 08v^2 + 1\cdot 08^2 v^3 + \cdots \\ + 1\cdot 08^{19}v^{20})\end{array}} \quad \text{at } 8\%$$

$$= \frac{v(1 + 2 + 3 + \cdots + 20)}{20v} \quad \text{at } 8\% \quad = \frac{20 \times 21}{2 \times 20}$$

$$= 10 \cdot 5 \text{ years}$$

(iv) The DMT is

$$\frac{\sum\limits_{t=1}^{20} 1000t(1\cdot1)^{t-1}v^t}{\sum\limits_{t=1}^{20} 1000(1\cdot1)^{t-1}v^t} \qquad \text{at } 8\%$$

$$= \left(\sum_{t=1}^{20} t\theta^{t-1}\right)\Bigg/\left(\sum_{t=1}^{20}\theta^{t-1}\right) \qquad \text{(where } \theta = 1\cdot1/1\cdot08)$$

$$= \frac{(Ia)_{\overline{20}|}}{a_{\overline{20}|}} \qquad \text{at rate } j \text{ (where } \theta = 1/1+j)$$

$$= \frac{\ddot{a}_{\overline{20}|} - 20v^{20}}{1 - v^{20}} \qquad \text{at rate } j$$

$$= \frac{\left(\dfrac{1-\theta^{20}}{1-\theta}\right) - 20\theta^{20}}{1 - \theta^{20}} \qquad \text{(since } \theta = v \text{ at rate } j)$$

$$= 11\cdot11 \text{ years}$$

10.2 (a) Let the cash flows occur at times r_1, $r_2, \ldots, r_k$ where $0 \leqslant r_1 \leqslant r_2 \leqslant \cdots \leqslant r_k$. Assume that for the first series of payments the cash flow at time r_i is x_i and that the cash flow for the second series at this time is y_i. (Note that some of the $\{x_i\}$ and $\{y_i\}$ may be zero.) The discounted mean term for the first series is

$$t_1 = \frac{\sum x_i r_i v^{r_i}}{\sum x_i v^{r_i}} = \frac{\sum x_i r_i v^{r_i}}{V_1} \qquad \text{(say, defining } V_1)$$

The discounted mean term for the second series is

$$t_2 = \frac{\sum y_i r_i v^{r_i}}{\sum y_i v^{r_i}} = \frac{\sum y_i r_i v^{r_i}}{V_2} \qquad \text{(say, defining } V_2)$$

These equations imply that

$$V_1 t_1 + V_2 t_2 = \sum (x_i + y_i) r_i v^{r_i}$$

so that, if $V_1 + V_2$ is non-zero,

$$\frac{V_1 t_1 + V_2 t_2}{V_1 + V_2} = \frac{\sum (x_i + y_i) r_i v^{r_i}}{\sum (x_i + y_i) v^{r_i}}$$

which is the discounted mean term for the combined series.

For the more general result, relating to n series of payments, let V_j and t_j denote the present value and discounted mean term of the jth series. Let t be the discounted mean term of the cash flow obtained by combining all n series. Then, if $\sum_{j=1}^{n} V_j \neq 0$,

$$t = \sum_{j=1}^{n} V_j t_j \bigg/ \sum_{j=1}^{n} V_j$$

(b) (i) (1) Receipts by investor: $V_1 = 10000 a_{\overline{10}|} = 61\,445 \cdot 67$ (at 10%) and $t_1 = (Ia)_{\overline{10}|}/a_{\overline{10}|} = 4 \cdot 725$, say $4 \cdot 73$.

      (2) Payments by investor: $V_2 = -30000(v^5 + v^{15}) = -25\,809 \cdot 40$ and $t_2 = (5v^5 + 15v^{15})/(v^5 + v^{15}) = 7 \cdot 783$, say $7 \cdot 78$.

  (ii)   By above, $t = (V_1 t_1 + V_2 t_2)/(V_1 + V_2) = 2 \cdot 511$, say $2 \cdot 51$. (Note that, since V_2 relates to payments *by* the investor, V_2 is negative.)

10.3 (a) The value of the annuity is

$$\sum_{t=1}^{n} [(n+1)t - t^2]v^t = (n+1)(Ia)_{\overline{n}|} - X$$

where $X = \sum_{t=1}^{n} t^2 v^t$. By a result given in the solution of exercise 10.1,

$$X = \frac{2(Ia)_{\overline{n}|} - a_{\overline{n}|} - n^2 v^{n+1}}{1 - v}$$

from which the required result follows immediately. When $n = 9$ and $i = 0 \cdot 09$, the value of this special annuity is $109 \cdot 162$.

(b) (i) The total of the sums assured for the policies issued on 1 January 1985 is $20\,000\,\ddot{s}_{\overline{10}|}$ at 9%, i.e. £331 206. Just *after* payment of the premiums due on 1 January 1985, the discounted mean term of the *future* net cash flow for the block of business is

$$\frac{\sum_{t=1}^{9} 20\,000 t v^t - (331\,206 \times 10 v^{10})}{\sum_{t=1}^{9} 20\,000 v^t - 331\,206 v^{10}}$$

$$= \frac{20\,000(Ia)_{\overline{9}|} - 3\,312\,060 v^{10}}{20\,000 a_{\overline{9}|} - 331\,206 v^{10}}$$

$$= 43 \cdot 39 \qquad \text{(at 9\%)}$$

(ii) For the business transacted in the years 1976 to 1985 (inclusive), the corresponding discounted mean term of the future net cash flow on 1 January 1985, is

$$\frac{20\,000 \sum_{t=1}^{9} (10-t)tv^t - 331\,206 \sum_{t=1}^{10} tv^t}{20\,000 \sum_{t=1}^{9} (10-t)v^t - 331\,206 \sum_{t=1}^{10} v^t}$$

$$= \frac{(20\,000 \times 109 \cdot 162) - 331\,206(Ia)_{\overline{10|}}}{20\,000\{10a_{\overline{9|}} - (Ia)_{\overline{9|}}\} - 331\,206a_{\overline{10|}}}$$

$$= 5 \cdot 50$$

Note The value 109·162 is obtained from (a).

10.4 (a) Consider a nominal amount of £1 of stock. Income is 0·05 per annum, payable continuously until the stock is redeemed at time n. By equation 10.5.5 we calculate the volatility as

$$\frac{0 \cdot 05(\bar{I}\bar{a})_{\overline{n|}} + nv^n}{0 \cdot 05\bar{a}_{\overline{n|}} + v^n} \qquad \text{at } force \text{ of interest } 0.07$$

Noting that, at force of interest δ, $(\bar{I}\bar{a})_{\overline{n|}} = (\bar{a}_{\overline{n|}} - nv^n)/\delta$, we find that the volatility is 11·592 when $n = 20$ and 14·345 when $n = 60$.

(b) By equation 10.5.9, the maximum volatility (at force of interest 0·07) occurs when, at this force of interest,

$$0 \cdot 05(n - \bar{a}_{\overline{n|}}) + 0 \cdot 07\bar{a}_{\overline{n|}} = \frac{0 \cdot 07}{0 \cdot 07 - 0 \cdot 05}$$

i.e. when

$$0 \cdot 05n + 0 \cdot 02\frac{1 - e^{-0 \cdot 07n}}{0 \cdot 07} = 3 \cdot 5$$

By numerical methods, we obtain the solution as $n = 64 \cdot 3486$.
With this value of n, the volatility at force of interest 7% is 14·349 (see equation 10.5.5).

10.5 (a) (i) Present value is

$$V = \sum_{t=1}^{4} (1000 + 100t)v^{5t} = 2751 \cdot 54 \qquad \text{(at 5\%)}$$

(ii) Discounted mean term is

$$\frac{\sum_{t=1}^{4} 5t(1000 + 100t)v^{5t}}{V} \qquad \text{(at 5\%)} \qquad = \frac{31\,609}{2751 \cdot 54} = 11 \cdot 4877 \text{ years}$$

(b) Suppose that the amounts invested in the ten-year stock and in the 30-year stock are A and B respectively. Then

$$A + B = 2751 \cdot 54 \qquad \text{(from (a))}$$

The cash flows arising from the ten-year stock have present value

$$V_1 = A(0 \cdot 05 a_{\overline{10}|} + v^{10})$$

and discounted mean term

$$t_1 = \frac{0 \cdot 05 (Ia)_{\overline{10}|} + 10 v^{10}}{0 \cdot 05 a_{\overline{10}|} + v^{10}}$$

Note that, at an interest rate of 5%, these values are simply

$$V_1 = A \qquad \text{and} \qquad t_1 = \ddot{a}_{\overline{10}|0 \cdot 05} = 8 \cdot 107\,82$$

Similarly, at an interest rate of 5%, the cash flows arising from the 30-year stock have present value $V_2 = B$ and discounted mean term $t_2 = \ddot{a}_{\overline{30}|0 \cdot 05} = 16 \cdot 141\,07$.

Using the result of exercise 10.2, we calculate the discounted mean term of the total asset-proceeds as

$$t = \frac{(A \times 8 \cdot 107\,82) + (B \times 16 \cdot 141\,07)}{A + B}$$

$$= 16 \cdot 141\,07 - 0 \cdot 002\,919\,55 A \qquad \text{(since } A + B = 2751 \cdot 54)$$

Equating this expression to $11 \cdot 4877$, we obtain $A = £1593 \cdot 87$ and hence $B = £1157 \cdot 67$.

10.6 (a) Let P' be the office annual premium. Then

$$0 \cdot 9375 P' \ddot{s}_{\overline{10}|0 \cdot 05} = 100\,000$$

from which we obtain $P' = £8076 \cdot 66$

(b) (i) Let the term of the stock purchased be n years. The discounted mean term of the asset proceeds, at 5% per annum interest, is

$$\frac{(1 \times 3 \cdot 125 v) + (2 \times 3 \cdot 125 v^2) + \cdots + (n \times 3 \cdot 125 v^n) + (n \times 100 v^n)}{3 \cdot 125 v + 3 \cdot 125 v^2 + \cdots + 3 \cdot 125 v^n + 100 v^n}$$
$$\text{at } 5\%$$

$$= \frac{3 \cdot 125 (Ia)_{\overline{n}|} + 100 n v^n}{3 \cdot 125 a_{\overline{n}|} + 100 v^n} = \frac{(Ia)_{\overline{n}|} + 32 n v^n}{a_{\overline{n}|} + 32 v^n}$$

We must find the value of n which makes the last expression closest to 10. This happens for $n = 12$, when the discounted mean term of the asset proceeds is 9·98. The stock purchased had, therefore, an outstanding term of 12 years.

(ii) The amount invested was $(1 - 0.0625)8076.66 = £7571.86$. The market price (per cent nominal) of the stock purchased was

$$3.125a_{\overline{12}|} + 100v^{12} \qquad \text{at } 5\% \qquad = 83.381$$

The nominal amount purchased was therefore

$$\frac{7571.86}{83.381} 100 = £9081$$

10.7 (a) Consider the purchase of £100 nominal of stock with an outstanding term of n years. Interest, payable continuously, is £k per annum. The discounted mean term of the asset proceeds is (see equation 10.3.4)

$$\frac{\displaystyle\int_0^n kte^{-\delta t}dt + 100ne^{-\delta n}}{\displaystyle\int_0^n ke^{-\delta t}dt + 100e^{-\delta n}} \qquad \text{(where } \delta = 0.05\text{)}$$

$$= \left[k\frac{\left(\dfrac{1-v^n}{0.05} - nv^n\right)}{0.05} + 100nv^n \right] \Big/ \left[k\left(\frac{1-v^n}{0.05}\right) + 100v^n \right]$$

$$\text{(where } v = e^{-0.05}\text{)}$$

$$= \frac{20k + [(5-k)n - 20k]v^n}{k + (5-k)v^n}$$

Equating the last expression to 5, we obtain

$$[25 + 15k + n(k - 5)]v^n = 15k$$

from which the required equation follows immediately.

(i) If $k = 5$, then $e^{0.05n} = 4/3$, from which we obtain $n = 5.753\,64$.
(ii) If $k = 10$, then $e^{0.05n} = (35 + n)/30$, from which we obtain $n = 6.480\,54$.

(b) The discounted mean terms of the total assets and liabilities are equal. Since the 'spread' of the liabilities about the DMT is zero, while that of the assets is positive, it follows from section 10.7 that the man is immunized against small changes in the force of interest. The price (per cent) of the 10% coupon stock with term 6·480 54 years is

$$10\bar{a}_{\overline{6\cdot48054}} + 100v^{6\cdot48054} \qquad \text{at force of interest 0·05}$$

$$= 127\cdot68$$

The amount of money invested is $100\,000e^{-0\cdot25} = 77\,880$. Hence, at any given force of interest,

$$V_A - V_L = \frac{77\,880}{127\cdot68}[10\bar{a}_{\overline{6\cdot48054}} + 100v^{6\cdot48054}] - 100\,000v^5$$

(i) When $\delta = 0\cdot055$, this equals £2·26.
(ii) When $\delta = 0\cdot045$, this equals £2·35.

10.8 Suppose that the purchaser buys nominal amounts £A and £B of the five-year and fifteen-year stock respectively. For 'full' immunization we require that equations 10.8.1 and 10.8.2 be satisfied. (In the notation of section 10.8, $t_1 = 10, a = b = 5$, and $S = 1\,000\,000$.) Hence

$$Ae^{5\delta_0} + Be^{-5\delta_0} = 1\,000\,000$$

and

$$5Ae^{5\delta_0} = 5Be^{-5\delta_0}$$

These equations imply that $A = 500\,000e^{-5\delta_0}$ and $B = 500\,000e^{5\delta_0}$.
Assume now that $\delta_0 = 0\cdot05$. Then $A = 389\,400$ and $B = 642\,013$. Thus the investor should buy a nominal amount £389 400 of the five-year stock and a nominal amount £642 013 of the fifteen-year stock (in each case at a cost of £303 265). The total amount invested, £606 530, is the present value (at a force of interest of 5% p.a.) of the liability.

The difference between the value, at force of interest δ, of the investor's assets and liabilities is

$$V_A - V_L = 389\,400e^{-5\delta} + 642\,013e^{-15\delta} - 1\,000\,000e^{-10\delta}$$

(a) When $\delta = 0\cdot07$, this profit is £2485.
(b) When $\delta = 0\cdot03$, this profit is £3707.

10.9 (a) The cash available for investment now is $100\,000v^8$ at a force of interest of 5%, i.e. £67 032. Note that the price per unit nominal of the 20-year stock is v^{20} (at $\delta = 0\cdot05$), i.e. 0·367 879.

Suppose that the company buys a nominal amount $£X$ of the 20-year stock (at a cost of $0.367879X$). This means that an amount $(67032 - 0.367879X)$ is held in cash, which has a discounted mean term of zero. Using exercise 10.2, we calculate the discounted mean term of the company's assets as

$$\frac{(0.367879X \times 20) + (67032 - 0.367879X) \times 0}{0.367897X + (67032 - 0.367879X)} = \frac{7.357580X}{67032}$$

For this to equal eight (the term of the liability) we need $X = 72885$. The company should buy £72885 nominal of 20-year zero coupon bonds at a cost of £26813 and hold £40219 in cash.

(b) At force of interest δ p.a. the difference between the value of the company's assets and liabilities is

$$V_A - V_L = 72885e^{-20\delta} + 40219 - 100000e^{-8\delta}$$

(1) When $\delta = 0.03$, this is £1556
(2) When $\delta = 0.07$, this is £1071.

10.10 (a) The single premium is $10000v^{15}$ at 8%, i.e. £3152.42. Let $£X$ be the nominal amount of 20-year stock purchased. The purchase price is Xv^{20} at 8%, i.e. $0.214548X$. For the discounted mean term of the assets (at 8%) to equal that of the liability we need (see exercise 10.2)

$$\frac{(0.214548X)20 + (3152.42 - 0.214548X)0}{3152.42} = 15$$

from which we obtain $X = 11019.98$, say 11020. The company buys a nominal amount £11020 of 20-year stock at a cost of £2364.32 and retains cash of £788.10.

(b) (i) $V_A - V_L = 788.10 + 11020v^{20} - 10000v^{15}$ at   5%
$$= £131.25$$

say £131 (profit).
(ii) As in (i), but at 10%, i.e. £32.24, say £32 (profit).
(iii) We assume that, for $0 \leqslant t \leqslant 20$,

$$\delta(t) = \log(1.05) + \frac{t}{20}[\log(1.1) - \log(1.05)]$$

This implies that, if

$$v(t) = \exp\left(-\int_0^t \delta(s)ds\right)$$

then $v(15) = 0{\cdot}370\,268$ and $v(20) = 0{\cdot}236\,690$.
Hence

$$V_A - V_L = 788{\cdot}10 + 11\,020v(20) - 10\,000v(15)$$

$$= -306{\cdot}26$$

that is, a *loss* of £306.

10.11 Note that (at force of interest δ)

$$\bar{a}_{\overline{n}|} = \int_0^n e^{-\delta t}dt$$

$$(\overline{I}\bar{a})_{\overline{n}|} = \int_0^n te^{-\delta t}dt$$

so that

$$\left.\begin{array}{l} \dfrac{\partial}{\partial n}\bar{a}_{\overline{n}|} = e^{-n\delta} \\[2mm] \dfrac{\partial}{\partial n}(\overline{I}\bar{a})_{\overline{n}|} = ne^{-n\delta} \end{array}\right\} \qquad (1)$$

Also, from equation 10.5.5,

$$T(\delta) = \frac{g(\overline{I}\bar{a})_{\overline{n}|} + ne^{-n\delta}}{g\bar{a}_{\overline{n}|} + e^{-n\delta}}$$

Hence, differentiating and using the results in equations 1, we have

$$\frac{\partial T(\delta)}{\partial n} = X[g\bar{a}_{\overline{n}|} + e^{-n\delta}]^{-2}$$

where

$$X = (g\bar{a}_{\overline{n}|} + e^{-n\delta})(gne^{-n\delta} + e^{-n\delta} - n\delta e^{-n\delta})$$

$$- [g(\overline{I}\bar{a})_{\overline{n}|} + ne^{-n\delta}](ge^{-n\delta} - \delta e^{-n\delta})$$

$$= \frac{e^{-n\delta}}{\delta}\{(g\delta\bar{a}_{\overline{n}|} + \delta e^{-n\delta})(gn + 1 - n\delta)$$

$$- [g\delta(\overline{I}\bar{a}_{\overline{n}|}) + n\delta e^{-n\delta}](g - \delta)\} \qquad (2)$$

Note now that

$$e^{-n\delta} = 1 - \delta\bar{a}_{\overline{n}|}$$

and that

$$\delta(\bar{I}\bar{a})_{\overline{n}|} = \bar{a}_{\overline{n}|} - ne^{-n\delta} = \bar{a}_{\overline{n}|}(1 + n\delta) - n$$

Substitution of the last two expressions for $e^{-n\delta}$ and $\delta(\bar{I}\bar{a})_{\overline{n}|}$ in the right-hand side of equation 2 gives (on simplification)

$$X = \frac{e^{-n\delta}}{\delta}[\delta + ng(g - \delta) - (g - \delta)^2\bar{a}_{\overline{n}|}]$$

as required.

10.12 (a)   For brevity we write δ rather than δ_0.

(i) $Aae^{\delta a} = bBe^{-\delta b} = b(S - Ae^{\delta a}) \Rightarrow A = \dfrac{bS}{(a + b)e^{\delta a}}$

$Bbe^{-\delta b} = aAe^{\delta a} = a(S - Be^{-\delta b}) \Rightarrow B = \dfrac{aS}{(a + b)e^{-\delta b}}$

(ii) $Bbe^{-\delta b} = aAe^{\delta a} = a(S - Be^{-\delta b}) \Rightarrow a = \dfrac{Bbe^{-\delta b}}{S - Be^{-\delta b}}$

Then, using this value of a, calculate A as

$$A = [S - Be^{-\delta b}]e^{-\delta a}$$

(iii) $Aae^{\delta a} = bBe^{-\delta b} = b(S - Ae^{\delta a}) \Rightarrow b = \dfrac{Aae^{\delta a}}{S - Ae^{\delta a}}$

Then, using this value of b, calculate B as

$$B = [S - Ae^{\delta a}]e^{\delta b}$$

(iv) $S - Ae^{\delta a} = Be^{-\delta b} = \dfrac{Aae^{\delta a}}{b} \Rightarrow (a + b)e^{\delta a} = \dfrac{bS}{A}$

i.e.

$$f(a) = \frac{bS}{A}$$

where

$$f(x) = (x + b)e^{\delta x}$$

Note that, for $x \geqslant 0$, $f(x)$ is an *increasing* function which tends to infinity as x tends to infinity. Also, $f(0) = b < bS/A$ (since, by hypothesis, $A < S$). Hence there is a unique positive value of x such that $f(x) = bS/A$.

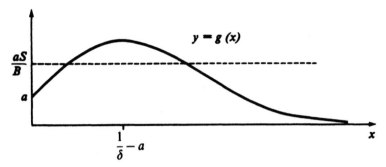

FIGURE S.10.1

Having found a, calculate B as

$$B = \frac{Aae^{\delta a}}{be^{-\delta b}}$$

(b) $\qquad S - Be^{-\delta b} = Ae^{\delta a} = \frac{Bbe^{-\delta b}}{a} \Rightarrow (a + b)e^{-\delta b} = \frac{aS}{B}$

i.e. $g(b) = aS/B$, where

$$g(x) = (a + x)e^{-\delta x}$$

If $\delta a < 1$, the graph of $g(x)$ is as in Figure S.10.1. In this case the maximum value of $g(x)$ occurs when $x = (1/\delta) - a$ and is $[\delta e^{1-\delta a}]^{-1}$. If $\delta a < 1$ and

$$a < \frac{aS}{B} < \frac{1}{\delta e^{1-\delta a}}$$

i.e. if $\delta a < 1$ and

$$a\delta e^{1-\delta a} < \frac{B}{S} < 1$$

there are then *two* positive values of x for which $g(x) = aS/B$.

When $\delta = 0.05, S = 1, a = 15$, and $B = 0.98$, the above conditions are satisfied. The solutions are
(i) $b = 1.489\,885$ and $A = 0.042\,679$
(ii) $b = 8.976\,051$ and $A = 0.176\,843$
(These values were obtained using the secant method: see appendix 2.)

Chapter 11

11.1   The APR is obtained by solving the equation

$$200 = 208a_{\overline{1}|}^{(12)}$$

i.e. $a_{\overline{1}|}^{(12)} = 0.961\,54$. The values of $a_{\overline{1}|}^{(12)}$ at 8·05% and 8·1% interest are 0·961 55 and 0·961 33 respectively. The value of i is thus between 8.05% and 8.1%. It follows that the APR is 8·1%.

11.2   Work in time units of three months and let the effective rate of interest per quarter be j. Then

$$-100(1+j)^3 + 1760(1+j)^2 - 605(1+j) - 1331 = 0$$

i.e.

$$[10(1+j) - 11][10(1+j)^2 - 165(1+j) - 121] = 0$$

so that $j = 0.1$, 16·203, or -1.703.

   The APR is defined to be the *smallest* positive root of the yield equation (with a time unit of one year). It is therefore given by the value of

$$i = (1+j)^4 - 1$$

when $j = 0.1$. Thus $i = 0.4641$ and the APR is 46·4%.

11.3   (a) Let P be the appropriate monthly payment. Measure time in months from the start of the year and assume that the nth payment is the first to reduce the outstanding loan to less than £1 000.

   Immediately after the nth monthly instalment is paid the outstanding loan is L_n, where (for $0 \leqslant t \leqslant n$)

$$L_t = 2000(1.015)^t - Ps_{\overline{t}|0.015} \qquad (1)$$

Subsequent interest is charged at the rate of 1% per month. Since the final monthly instalment must repay the entire outstanding loan, it follows that

$$L_n(1.01)^{12-n} - Ps_{\overline{12-n}|0.01} = 0$$

i.e.

$$[2000(1.015)^n - Ps_{\overline{n}|0.015}](1.01)^{12-n} - Ps_{\overline{12-n}|0.01} = 0 \qquad (2)$$

The last equation must be solved by trial and error, as the value of n is unknown. We require that $L_{n-1} \geqslant 1000$ and $L_n < 1000$, where L_{n-1} and L_n are given by equation 1.

Putting $n = 7$, we obtain $P = 182 \cdot 29$ (from equation 2). With this value of P we obtain, from equation 1, $L_6 = 1051 \cdot 30$ and $L_7 = 884 \cdot 78$, which is consistent with our assumptions. The monthly payment is thus £182·29.

(b) The APR is given by the equation

$$2000 = (182 \cdot 29)(12)a_{\overline{1}|i}^{(12)}$$

or $a_{\overline{1}|i}^{(12)} = 0 \cdot 914\,29$. The values of $a_{\overline{1}|i}^{(12)}$ at 18·2% and 18·25% are 0·914 47 and 0·914 26 respectively, which shows that i lies between these two rates of interest. The APR is thus 18·2%.

11.4 The annual repayment is £335·04 (i.e. $12 \times 27 \cdot 92$), so that the APR is obtained from the equation

$$1000 = 335 \cdot 04 a_{\overline{4}|i}^{(12)}$$

i.e. $a_{\overline{4}|i}^{(12)} = 2 \cdot 984\,72$. The values of $a_{\overline{4}|i}^{(12)}$ at 25·0% and 25·05% are 2·985 05 and 2·982 53 respectively. The APR is thus 25·0%.

11.5 The charge for credit is $(36 \times 48) - 1000$, i.e. £728. Since the rebate is determined on the basis of a two-month deferment of the settlement date, the amount of the rebate is

$$\frac{(6-2)(7-2)}{(36)(37)}728, \qquad \text{i.e.} \qquad £10 \cdot 93$$

The amount payable to settle the loan is the total of the outstanding repayments less the rebate, namely

$$(6 \times 48) - 10 \cdot 93, \qquad \text{i.e.} \qquad £277 \cdot 07$$

11.6 (a) The charge for credit is $2 \times 0 \cdot 1 \times 5000 = 1000$, so that the total of the repayments is $(5000 + 1000)$, i.e. £6000. Each monthly repayment is therefore of amount £250.

The APR is obtained from the equation

$$5000 = 3000 a_{\overline{2}|i}^{(12)}$$

i.e. $a_{\overline{2}|i}^{(12)} = 1 \cdot 6667$. The values of $a_{\overline{2}|i}^{(12)}$ at 19·7% and 19·75% interest are 1·6673 and 1·6666 respectively, so that the APR is 19·7%.

(b) Note that there are 12 instalments outstanding at the time of settlement.

(i) $(\alpha = 0)$ The amount of the rebate is

$$\frac{12 \times 13}{24 \times 25}1000 = 260$$

so that the sum payable to settle the loan is $(12 \times 250) - 260$, i.e. £2740. The yield is obtained from the equation

$$5000 - 3000a_{\overline{1}|}^{(2)} - 2740v = 0 \qquad \text{at rate } i$$

This gives $i = 0.201\,60$ or 20.16% (the APR is 20.1%).

(ii) $(\alpha = 2)$ The rebate is

$$\frac{10 \times 11}{24 \times 25}1000 = 183.33$$

so that the sum payable to settle the loan is £2816·67. The yield is obtained from the equation

$$5000 - 3000a_{\overline{1}|}^{(2)} - 2816.67v = 0 \qquad \text{at rate } i$$

This gives $i = 0.222\,24$ or 22.22% (the APR is 22.2%).

11.7 (a) The charge for credit is $6 \times 0.02 \times 1200 = £144$, so that the monthly payment is $(1200 + 144)/6 = £224$. The effective rate of interest per month is i, where $1200 = 224a_{\overline{6}|i}$, from which we obtain $i = 0.033\,373\,32$ or 3.337%.

(b) (i) *Compound interest schedule* (see section 3.8)

Month	Loan o/s at start	Interest	Capital repaid	Loan o/s at end
1	1200·00	40·05	183·95	1016·05
2	1016·05	33·91	190·09	825·96
3	825·96	27·56	196·44	629·52
4	629·52	21·01	202·99	426·53
5	426·53	14·23	209·77	216·76
6	216·76	7·24	216·76	0·00

(ii) *Rule of 78 schedule* (see section 11.3) Since there are six monthly repayments, the total charge for credit is divided into $6 \times 7/2 = 21$ 'units' of interest. One unit of interest is thus of amount $144/21 = £6.857\,143$.

Month	Loan o/s at start	No. of units of interest	Interest	Capital repaid	Loan o/s at end
1	1200·00	6	41·14	182·86	1017·14
2	1017·14	5	34·29	189·71	827·43
3	827·43	4	27·43	196·57	630·86
4	630·86	3	20·57	203·43	427·43
5	427·43	2	13·71	210·29	217·14
6	217·14	1	6·86	217·14	0·00

11.8 Under a compound interest schedule the interest content of the repayment decreases each month. The tax relief therefore decreases. This means that for a compound interest schedule the *net* monthly payments (after tax relief) form an *increasing* sequence.

Under a flat interest schedule the net monthly payments form a *level* series, as the interest content of each repayment is constant in this case.

The totals of the net repayments for the two schedules are equal. (Both totals equal $P + [n(1 - t)PF/12]$.) In such a situation the higher yield is provided by the level sequence, i.e. by the flat interest schedule.

Consider the given example. The charge for credit is $0.13 \times 6000 = £780$, so the gross monthly repayment is $6780/12 = £565$.

The gross effective monthly rate of interest is i, where $6000 = 565a_{\overline{12}|i}$, from which it follows that $i = 0.019\,323$ or 1.9323%

With tax relief at 30%, the net interest payable by the borrower at the end of each month is simply $0.7 \times 1.9323\%$ (i.e. 1.3526%) of the loan outstanding at the start of the month. The net effective monthly rate of interest is thus 1.3526%.

Under a flat interest schedule the net monthly repayment is $500 + (0.7 \times 65) = £545.5$. The net effective monthly rate of interest in this case is given by the equation $6000 = 545.5a_{\overline{12}|i}$, from which we obtain $i = 0.013\,660$ or 1.3660%. Note that this rate is greater than the corresponding net rate for the compound interest schedule.

11.9 The term of the loan is n/m years. Thus, if the flat rate of interest for the loan is F per annum, the charge for credit is

$$D = PF\frac{n}{m}$$

Hence

$$F = \frac{mD}{nP}$$

Substituting this value of F in equation 11.2.7, we obtain

$$i \simeq \frac{2\dfrac{mD}{nP}}{\dfrac{n+1}{n} + \dfrac{mD}{nP}\left(\dfrac{n-3m+2}{3m}\right)}$$

$$= \frac{2mD}{(n+1)P + \dfrac{D}{3}(n-3m+2)}$$

as required.

11.10 The original term of the loan was 365 days, the repayment date being 15 April 1986. The charge for credit was £200.

(a) *No deferment*   The settlement date was taken as the actual date of early repayment (17 July 1985), at which time the outstanding period of the loan was 272 days. The rebate allowed was thus $(272/365)200 = 149{\cdot}04$, so that the sum paid to settle the loan was $(2200 - 149{\cdot}04) = £2050{\cdot}96$.

Since the loan was repaid after 93 days, the APR is obtained from the equation

$$2000(1 + i)^{93/365} = 2050{\cdot}96$$

from which $i = 0{\cdot}103\,79$. The APR is thus $10{\cdot}4\%$.

(b) *Two-month deferment*   In calculating the rebate, the lender took the settlement date as 17 September 1985 (two months later than the actual date of repayment). From this date until the original repayment date (15 April 1986) is a period of 210 days, so that the rebate allowed was $(210/365)200 = 115{\cdot}07$. The sum paid to settle the loan was thus $(2200 - 115{\cdot}07) = £2084{\cdot}93$.

The APR is obtained from the equation

$$2000(1 + i)^{93/365} = 2084{\cdot}93$$

which gives $i = 0{\cdot}177\,30$. The APR is thus $17{\cdot}7\%$.
(Note the increase in the value of the APR in (b).)

11.11 The original term of the loan was 273 days. The charge for credit was £25.

(a) The APR for the original contract is given by equation

$$160(1 + i)^{273/365} = 185,$$

from which $i = 0.214\,23$. The APR for the original contract was thus 21·4%.

(b) The deferred settlement date was taken as 14 July 1986, which is 81 days before the originally agreed repayment date. The rebate allowed was therefore $(81/273)25 = £7.42$, so the sum paid by the borrower to settle the loan was £177·58.

The loan was repaid on 14 May 1986, i.e. 131 days after it was granted. Since the equation

$$160(1 + i)^{131/365} = 177.58$$

implies that $i = 0.337\,04$, the APR on the completed transaction was 33·7%.

Chapter 12

12.1 (a)

$$E[X] = \int_{-\infty}^{\infty} x f(x) dx$$

where $f(x)$ is the probability density function of X. Using the appropriate definition of $f(x)$ (see statistical result (a) preceding the exercises) we obtain

$$E[X] = \int_a^b \frac{x}{b - a} dx = \frac{a + b}{2}$$

as required. Then

$$E[(X - E[X])^n] = \int_a^b \left(x - \frac{a + b}{2}\right)^n \frac{1}{b - a} dx$$

Letting $\mu = (a + b)/2$, $d = (b - a)/2$, and $y = x - \mu$, we evaluate this integral as

$$\frac{1}{2d} \int_{-d}^{d} y^n dy = \frac{d^{n + 1} - (-d)^{n + 1}}{2(n + 1)d}$$

from which the required result follows immediately.

(b) (i) Let μ and d be as defined above. Then, from the definition of $f(x)$ in statistical result (b), we have

$$E[X] = \int_a^{(a + b)/2} \frac{x4(x - a)}{(b - a)^2} dx + \int_{(a + b)/2}^b \frac{x4(b - x)}{(b - a)^2} dx$$

$$= \frac{a + b}{2} \qquad \text{(on simplification)}.$$

Also

$$E[(X - E[X])^n] = \int_a^b (x - \mu)^n f(x) dx$$

where $f(x)$ is defined in statistical result (b). Since $b - a = 2d$, the last integral is equal to

$$\int_a^\mu \frac{(x - \mu)^n(x - a)}{d^2} dx + \int_\mu^b \frac{(x - \mu)^n(b - x)}{d^2} dx$$

This expression is easily evaluated as follows. Let $x - \mu = y$. Note that $x - a = y + d$ and that $b - x = d - y$. Making the appropriate substitution in each integral, we obtain

$$\int_{-d}^0 \frac{y^n(y + d)}{d^2} dy + \int_0^d \frac{y^n(d - y)}{d^2} dy$$

which is readily found to equal

$$d^n \left\{ \frac{1}{n + 1} [1 - (-1)^{n+1}] - \frac{1}{n + 2} [1 + (-1)^{n+2}] \right\}$$

Since $d = (b - a)/2$, the required result follows immediately from this expression.

(ii)
$$F(x) = \int_{-\infty}^x f(t) dt$$

where $f(t)$ is defined by statistical result (b). It is readily verified that, on carrying out the integration, we obtain

$$F(x) = \begin{cases} \dfrac{2(x - a)^2}{(b - a)^2} & \text{if } a \leqslant x \leqslant \dfrac{a + b}{2} \\[3mm] 1 - \dfrac{2(b - x)^2}{(b - a)^2} & \text{if } \dfrac{a + b}{2} < x \leqslant b \end{cases}$$

Note that $F(x)$ is an increasing function and that $F[(a + b)/2] = 1/2$. Let $F(x) = r$. Then, if $0 < r \leqslant 1/2$, x lies between a and $(a + b)/2$, and thus

$$\frac{2(x - a)^2}{(b - a)^2} = r$$

from which we obtain, as required,

$$x = a + (b - a) \sqrt{\frac{r}{2}}$$

If $1/2 < r < 1$, then x lies between $(a + b)/2$ and b and thus

$$1 - \frac{2(b-x)^2}{(b-a)^2} = r$$

from which we obtain, as required,

$$x = b - (b-a)\sqrt{\left(\frac{1-r}{2}\right)}$$

12.2 (a) $E[i] = \frac{1}{3}(0.03) + \frac{1}{3}(0.06) + \frac{1}{3}(0.09) = 0.06$ or 6%

If the annual premium were calculated at this mean rate of interest, it would be $10\,000/\ddot{s}_{\overline{20}|0.06} = £256.46$.

(b) (i) The expected value of the accumulated profit is

$$\frac{1}{3}(P\ddot{s}_{\overline{20}|0.03} - 10\,000) + \frac{1}{3}(P\ddot{s}_{\overline{20}|0.06} - 10\,000)$$
$$+ \frac{1}{3}(P\ddot{s}_{\overline{20}|0.09} - 10\,000)$$

which equals $40.811\,25P - 10\,000$. This will be zero if $P = 245.03$.

If $P = 256.46$ (as in (a)), the expected value of the accumulated profit is £466.45.

(ii) The expected value of the net present value of the policy immediately before the first premium is paid is

$$\frac{1}{3}(P\ddot{a}_{\overline{20}|0.03} - 10\,000v_{0.03}^{20}) + \frac{1}{3}(P\ddot{a}_{\overline{20}|0.06} - 10\,000v_{0.06}^{20})$$
$$+ \frac{1}{3}(P\ddot{a}_{\overline{20}|0.09} - 10\,000v_{0.09}^{20})$$

which equals $12.477\,34P - 3479.70$. This will be zero if $P = £278.88$.

If $P = 256.46$ (as in (a)), the expected value of the net present value of the policy is $-£279.76$.

12.3 (a) (i) Using the notation of section 12.3, we note that, if θ denotes the accumulated profit at the maturity date, then

$$\theta = PS_{10} - 1000$$

Hence

$$E[\theta] = PE[S_{10}] - 1000$$
$$= P(1 + E[i])^{10} - 1000 \qquad \text{(by equation 12.3.5)}$$

Let $E[i] = j$ and $\text{var}[i] = s^2$. Then

standard deviation of $\theta = P$(standard deviation of S_{10})

$$= P\{[(1+j)^2 + s^2]^{10}$$
$$- (1+j)^{20}\}^{1/2} \quad \text{(by 12.3.7)}$$

(ii) Let ϕ be the net present value of the policy immediately after it is effected. Then

$$\phi = P - 1000(1 + i_1)^{-1}(1 + i_2)^{-1} \cdots (1 + i_{10})^{-1}$$

where i_t denotes the yield in the tth year. Thus

$$E[\phi] = P - 1000E[(1 + i_1)^{-1} \cdots (1 + i_{10})^{-1}]$$

Since (by assumption) $i_1, i_2, \ldots, i_{10}$ are independently and identically distributed, the last equation implies that

$$E[\phi] = P - 1000\left(E\left[\frac{1}{1+i}\right]\right)^{10}$$

Also, since P is a constant,

standard deviation of ϕ
$= 1000 \times$ standard deviation of $[(P - \phi)/1000]$
$= 1000 \times$ standard deviation of $[(1 + i_1)^{-1} \cdots (1 + i_{10})^{-1}]$

Using the fact that, for any random variable X, var $[X] = E[X^2] - (E[X])^2$, we immediately obtain from the last equation:

standard deviation of ϕ

$$= 1000\left\{\left(E\left[\frac{1}{(1+i)^2}\right]\right)^{10} - \left(E\left[\frac{1}{1+i}\right]\right)^{20}\right\}^{1/2}$$

(b) The reader should verify the following results for each of the given models:

Model I $j = E[i] = 0\cdot04$

$$s^2 = \text{var}[i] = \tfrac{1}{3}(-0\cdot02)^2 + \tfrac{1}{3}(0)^2 + \tfrac{1}{3}(0\cdot02)^2 = \tfrac{8}{3}10^{-4}$$

$$E\left[\frac{1}{1+i}\right] = \frac{1}{3}\left(\frac{1}{1\cdot02}\right) + \frac{1}{3}\left(\frac{1}{1\cdot04}\right) + \frac{1}{3}\left(\frac{1}{1\cdot06}\right) = 0\cdot961\,775\,61$$

thus

$$\left(E\left[\frac{1}{1+i}\right]\right)^{10} = 0\cdot677\,232\,23$$

$$E\left[\frac{1}{(1+i)^2}\right] = \frac{1}{3}\left(\frac{1}{1\cdot02}\right)^2 + \frac{1}{3}\left(\frac{1}{1\cdot04}\right)^2 + \frac{1}{3}\left(\frac{1}{1\cdot06}\right)^2$$

$$= 0\cdot92\,524\,048$$

and thus

$$\left(E\left[\frac{1}{(1+i)^2}\right]\right)^{10} = 0\cdot459\,775\,94$$

Model II $j = E[i] = 0\cdot04$

$$s^2 = \text{var}[i] = \tfrac{4}{3}10^{-4} \qquad \text{(see exercise 12.1 (a))}$$

$$E\left[\frac{1}{1+i}\right] = \int_{0\cdot02}^{0\cdot06} \frac{1}{1+x}25\,dx = 0\cdot961\,657\,01$$

and thus

$$\left(E\left[\frac{1}{1+i}\right]\right)^{10} = 0\cdot676\,397\,61$$

$$E\left[\frac{1}{(1+i)^2}\right] = \int_{0\cdot02}^{0\cdot06} \frac{1}{(1+x)^2}25\,dx = 0\cdot924\,898\,26$$

and thus

$$\left(E\left[\frac{1}{(1+i)^2}\right]\right)^{10} = 0\cdot458\,078\,21$$

Model III $j = E[i] = 0\cdot04$

$$s^2 = \text{var}[i] = \tfrac{2}{3}10^{-4} \qquad \text{(see exercise 12.1(b))}$$

Note that the probability density function of i is (see statistical result (b))

$$f(x) = \begin{cases} 2500(x - 0\cdot02) & \text{if } 0\cdot02 \leqslant x < 0\cdot04 \\ 2500(0\cdot06 - x) & \text{if } 0\cdot04 \leqslant x \leqslant 0\cdot06 \end{cases}$$

$$E\left[\frac{1}{1+i}\right] = \int_{0\cdot02}^{0\cdot06} \frac{1}{1+x}f(x)\,dx = 0\cdot961\,597\,48$$

and thus

$$\left(E\left[\frac{1}{1+i}\right]\right)^{10} = 0\cdot675\,978\,90$$

$$E\left[\frac{1}{(1+i)^2}\right] = \int_{0\cdot02}^{0\cdot06} \frac{1}{(1+x)^2}f(x)\,dx = 0\cdot924\,728\,15$$

and thus

$$\left(E\left[\frac{1}{(1+i)^2}\right]\right)^{10} = 0{\cdot}457\,236\,39$$

Combining these results with those of (a), we obtain the values in the following table:

	Value of P for which the expected value of the accumulated profit is zero (£)	Standard deviation of accumulated profit (P as in previous column) (£)	Value of P for which the expected value of the net present value at outset is zero (£)	Standard deviation of net present value (£)
Model I	675·56	49·68	677·23	33·65
Model II	675·56	35·12	676·40	23·76
Model III	675·56	24·83	675·98	17·00

12.4 (a) $\quad j = E[i] = \frac{1}{3}(0{\cdot}03) + \frac{1}{3}(0{\cdot}06) + \frac{1}{3}(0{\cdot}09) = 0{\cdot}06$

$\qquad s^2 = \text{var}[i] = \frac{1}{3}(-0{\cdot}03)^2 + \frac{1}{3}(0)^2 + \frac{1}{3}(0{\cdot}03)^2 = 0{\cdot}0006$

(b) Hence, using equations 12.3.5 and 12.3.7, we obtain

$$E[S_n] = 1{\cdot}06^n$$

$$\text{var}[S_n] = 1{\cdot}1242^n - 1{\cdot}06^{2n}$$

This gives the following values:

n	5	10	15	20
$E[S_n]$	1·338 23	1·790 85	2·396 56	3·207 14
$\sqrt{(\text{var}[S_n])}$	0·069 19	0·131 02	0·214 89	0·332 28

(c) From equation 12.3.12 it follows that $E[A_n] = \ddot{s}_{\overline{n}|0{\cdot}06}$. Using equations 12.3.13 and 12.3.14, we calculate the values in the following table:

n	$\sqrt{(\text{var}[A_n])}$	n	$\sqrt{(\text{var}[A_n])}$	n	$\sqrt{(\text{var}[A_n])}$
1	0·024 49	6	0·280 93	11	0·802 84
2	0·056 75	7	0·361 94	12	0·946 24
3	0·098 50	8	0·453 92	13	1·104 76
4	0·149 63	9	0·557 55	14	1·279 48
5	0·210 32	10	0·673 58	15	1·471 55

(d)

n	5	10	15
$\sqrt{(\mathrm{var}[A_n])}/E[A_n]$	0·0352	0·0482	0·0596

Note that, as n increases, the value of the standard deviation increases relative to the value of the mean.

12.5   (a) $j = E[i] = 0.3(0.02) + 0.5(0.04) + 0.2(0.07) = 0.04$

$s^2 = \mathrm{var}[i] = 0.3(-0.02)^2 + 0.5(0)^2 + 0.2(0.03)^2 = 0.0003$

Using the notation of section 12.3, we have

$$1000\,E[S_{15}] = 1000(1.04)^{15} = 1800.94$$

$$1000\sqrt{(\mathrm{var}[S_{15}])} = 1000(1.0819^{15} - 1.04^{30})^{1/2} = 116.28$$

We carried out 10 000 simulations. Of these 254 gave an accumulation of less than £1600 and 468 an accumulation greater than £2000. Our estimated probabilities and corresponding 95% confidence intervals are
(i) 0·0254; (0·0223, 0·0285)
(ii) 0·0468; (0·0427, 0·0509)

(b) $100\,E[A_{15}] = 100\ddot{s}_{\overline{15}|0.04} = 2082.45$

$$100\sqrt{(\mathrm{var}[A_{15}])} = 86.87$$

Note $\mathrm{var}[A_{15}]$ was found from equations 12.3.13 and 12.3.14

Of the 10 000 simulations, 60 gave an accumulation of less than £1900 and 90 an accumulation greater than £2300. The estimated probabilities and 95% confidence intervals are therefore
(i) 0·006; (0·0045, 0·0075)
(ii) 0·009; (0·0071, 0·0109)

12.6   (a) *Uniform model*

$$j = E[i] = 0.06$$

$$s^2 = \mathrm{var}[i] = 0.0003 \text{ (see exercise 12.1)}$$

$$s = 0.017\,321$$

$$\mathrm{prob}\{0.05 < i \leqslant 0.07\} = \frac{0.07 - 0.05}{0.09 - 0.03} = \frac{1}{3}$$

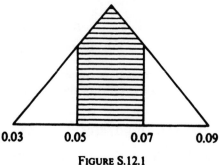

Triangular model

$$j = E[i] = 0·06$$

$$s^2 = \text{var}[i] = 0·000\,15 \quad (\text{see exercise 12.1})$$

$$s = 0·012\,25$$

Consider a graph of the probability density function (Figure S.12.1). The shaded area is easily seen to be five-ninths of the area of the triangle, which is 1. Hence the required probability is 5/9.

(b) By using the above results we may easily calculate the values given in the table below:

	$E[S_n]$	$\sqrt{(\text{var}[S_n])}$		$E[A_n]$	$\sqrt{(\text{var}[A_n])}$	
n	Both models	Uniform	Triangular	Both models	Uniform	Triangular
5	1·338 23	0·048 91	0·034 58	5·975 32	0·148 70	0·105 14
10	1·790 85	0·092 59	0·065 45	13·971 64	0·476 15	0·336 63
15	2·396 56	0·151 81	0·107 29	24·672 53	1·040 00	0·735 20

Note that for the triangular model the standard deviations are consistently some 30% less than those for the uniform model. This is a measure of the reduced variability in interest rates under the triangular model.

12.7 (a) We are given that $(1 + i)$ has a log-normal distribution with parameters μ and σ^2. Hence (see statistical result (c))

$$E[(1 + i)^n] = \exp(n\mu + \tfrac{1}{2}n^2\sigma^2) \tag{1}$$

Thus

$$1 + E[i] = E[1 + i] = \exp(\mu + \tfrac{1}{2}\sigma^2)$$

so that

$$E[i] = \exp(\mu + \tfrac{1}{2}\sigma^2) - 1$$

Also

$$
\begin{aligned}
\mathrm{var}[i] &= E[(1 + i)^2] - (E[1 + i])^2 \\
&= \exp(2\mu + 2\sigma^2) - [\exp(\mu + \tfrac{1}{2}\sigma^2)]^2 \qquad \text{(by equation 1)} \\
&= \exp(2\mu + \sigma^2)[\exp(\sigma^2) - 1]
\end{aligned}
$$

(b) We have

$$1 + j = \exp(\mu + \tfrac{1}{2}\sigma^2)$$

and

$$s^2 = \exp(2\mu + \sigma^2)[\exp(\sigma^2) - 1]$$

Therefore

$$\frac{s^2}{(1 + j)^2} = \exp(\sigma^2) - 1$$

so that

$$\sigma^2 = \log\left[1 + \left(\frac{s}{1 + j}\right)^2\right]$$

Also

$$
\begin{aligned}
\log(1 + j) &= \mu + \tfrac{1}{2}\sigma^2 \\
&= \mu + \tfrac{1}{2}\log\left[1 + \left(\frac{s}{1 + j}\right)^2\right]
\end{aligned}
$$

so that, as required,

$$\mu = \log \frac{1 + j}{\sqrt{\left[1 + \left(\frac{s}{1 + j}\right)^2\right]}}$$

(c) (i) $j = 0.085\,944$; $s = 0.076\,109$
(ii) $\mu = 0.074\,865$; $\sigma = 0.064\,747$

12.8 (a) In our earlier notation, $j = 0.06$ and $s^2 = 0.0003$. Hence (see exercise 12.7) $\mu = 0.058\,135$ and $\sigma^2 = 0.000\,267$.

(b) $$S_{15} = (1 + i_1)(1 + i_2)\cdots(1 + i_{15})$$

where i_t denotes the yield in year t. Hence

$$\log S_{15} = \log(1 + i_1) + \log(1 + i_2) + \cdots + \log(1 + i_{15})$$

The terms on the right-hand side of this equation are independent normally distributed random variables, each with mean μ and variance σ^2 (as found in (a)). This implies that $\log S_{15}$ is normally distributed with mean $\mu' = 15\mu = 0.872\,025$ and variance $\sigma'^2 = 15\sigma^2 = 0.004\,005 = (0.063\,28)^2$.

Hence (see exercise 12.7)

$$E[S_{15}] = \exp(\mu' + \tfrac{1}{2}\sigma'^2) = 2.396\,544$$

$$E[(S_{15})^2] = \exp(2\mu' + 2\sigma'^2) = 5.766\,469$$

Hence

$$\text{var}[S_{15}] = 5.766\,469 - (2.396\,544)^2 = 0.023$$

(c) By equation 12.3.5, $E[S_{15}] = 1.06^{15} = 2.3965$.
By equation 12.3.7, $\text{var}[S_{15}] = (1.06^2 + 0.0003)^{15} - 1.06^{30} = 0.023$.

(d) (i) $S_{15} < 2.1 \Leftrightarrow \log S_{15} < \log 2.1 = 0.741\,937$.

Since $\log S_{15}$ has a normal distribution with mean $0.872\,025$ and standard deviation $0.063\,28$ (as found above), the required probability is

$$\Phi\left(\frac{0.741\,937 - 0.872\,025}{0.063\,28}\right) = \Phi(-2.0557)$$

(Here $\Phi(x)$ is used to denote the standard normal distribution function.) From standard tables, by interpolation, we evaluate this probability as 0.0199.

(ii) $S_{15} > 2.7 \Leftrightarrow \log S_{15} > \log 2.7 = 0.993\,252$. The required probability is thus

$$1 - \Phi\left(\frac{0.993\,252 - 0.872\,025}{0.063\,28}\right)$$

$$= 1 - \Phi(1.9157) = 0.0277$$

12.9 (a) $V_n = (1 + i_1)^{-1}(1 + i_2)^{-1}\ldots(1 + i_n)^{-1}$
Hence

$$\log V_n = -\log(1 + i_1) - \log(1 + i_2) - \cdots - \log(1 + i_n)$$

Since, for each value of t, $\log(1 + i_t)$ is normally distributed with mean μ and variance σ^2, each term of the right-hand side of the last equation is normally distributed with mean $-\mu$ and variance σ^2. Also the terms are independently distributed. This means that $\log V_n$ is normally distributed with mean $-n\mu$ and variance $n\sigma^2$, i.e. V_n has log-normal distribution with parameters $-n\mu$ and $n\sigma^2$.

(b) We are given that $E[i] = 0.08$ and $\text{var}[i] = (0.05)^2 = 0.0025$. This implies (see exercise 12.7) that $\mu = 0.075\,890\,52$ and $\sigma^2 = 0.002\,141\,05$.

Let $\mu' = -n\mu$ and $\sigma'^2 = n\sigma^2$. Then (see statistical result (c))

$$E[V_n] = \exp(\mu' + \tfrac{1}{2}\sigma'^2) = \exp(-n\mu + \tfrac{1}{2}n\sigma^2)$$
$$= [\exp(-\mu + \tfrac{1}{2}\sigma^2)]^n$$

and

$$E[(V_n)^2] = \exp(2\mu' + 2\sigma'^2) = \exp(-2n\mu + 2n\sigma^2)$$
$$= [\exp(-\mu + \sigma^2)]^{2n}$$

The standard deviation of V_n is $\{E[(V_n)^2] - (E[V_n])^2\}^{1/2}$.

Using the last two equations (with μ and σ^2 as calculated above) we may readily calculate the values in the following table:

n	$1000 E[V_n]$	$1000\sqrt{(\text{var}[V_n])}$
5	687·91	71·37
10	473·22	69·62
15	325·53	58·81
20	223·93	46·84

12.10 (a) For $i = 1, 2, 3$ we have (see exercise 12.1)

$$E[X_i] = \tfrac{1}{2} \quad \text{and} \quad \text{var}[X_i] = \tfrac{1}{12}$$

Thus

$$E[X] = \frac{1}{3}\sum_{i=1}^{3} E[X_i] = \tfrac{1}{2} \text{ and var}[X] = \left(\frac{1}{3}\right)^2 \sum_{i=1}^{3} \text{var}[X_i] = \tfrac{1}{36}$$

Hence the standard deviation of X is $1/6$.

(b) *Exact* values for the probabilities are
  (i) $9/128 \simeq 0.070\,313$
  (ii) 0.036
  (iii) $\tfrac{2}{3} \simeq 0.666\,667$
  (iv) 0.1625
  The reader should compare his estimated values with the above true values.

(c) *Exact* values of t are
  (i) $[\sqrt[3]{90}]^{-1} \simeq 0.223\,144$

(ii) $1 - \sqrt[3]{(2/900)} \simeq 0.869\,504$

(iii) The unique real root of the equation

$$36t^3 - 54t^2 + 18t - 1 = 0.$$

This root is approximately $0.382\,380$.

Again the reader should compare these values with his estimates.

12.11 (a) We are given that $i_0 = 0.06$. Hence i_1 is equally likely to be 0.02, 0.06, or 0.10. Thus

$$E[i_1] = \tfrac{1}{3}(0.02) + \tfrac{1}{3}(0.06) + \tfrac{1}{3}(0.10) = 0.06$$

In order to find the expected value of i_2 we may rely on first principles, noting that i_2 is equally likely to be one of nine values $\{0.02 - 0.04k, 0.06 - 0.04k, \ldots, 0.10 + 0.04k\}$. This immediately gives $E[i_2] = 0.06$.

Alternatively we may work with conditional expectations. Let X be the set consisting of the three numbers 0.02, 0.06, and 0.10. For $\alpha \in X$,

$$E[i_2 | i_1 = \alpha] = \tfrac{1}{3}[0.02 + k(\alpha - 0.06)] + \tfrac{1}{3}[0.06 + k(\alpha - 0.06)]$$
$$+ \tfrac{1}{3}[0.10 + k(\alpha - 0.06)]$$
$$= 0.06 + k(\alpha - 0.06)$$

Then

$$E[i_2] = \sum_{\alpha \in X} \mathrm{prob}(i_1 = \alpha)E[i_2 | i_1 = \alpha]$$
$$= \tfrac{1}{3}[0.06 - 0.04k] + \tfrac{1}{3}[0.06] + \tfrac{1}{3}[0.06 + 0.04k]$$
$$= 0.06$$

as before.

(b) (i) Note that $S = (1 + i_1)(1 + i_2)$. Hence

$$E[S | i_1 = \alpha] = (1 + \alpha)\{1 + E[i_2 | i_1 = \alpha]\}$$
$$= (1 + \alpha)\{1.06 + k(\alpha - 0.06)\}$$

by the result in (a) for the conditional expectation of i_2. Then

$$E[S] = \sum_{\alpha \in X} \mathrm{prob}(i_1 = \alpha)E[S | i_1 = \alpha]$$
$$= \tfrac{1}{3}[1.02(1.06 - 0.04k)] + \tfrac{1}{3}[1.06^2]$$
$$+ \tfrac{1}{3}[1.1(1.06 + 0.04k)]$$
$$= 1.06^2 + \frac{0.0032}{3}k$$

(ii) $$S^2 = (1 + i_1)^2 (1 + i_2)^2$$

Hence

$$E[S^2 | i_1 = \alpha] = (1 + \alpha)^2 E[(1 + i_2)^2 | i_1 = \alpha]$$

$$= (1 + \alpha)^2 \{ [1 \cdot 02 + k(\alpha - 0 \cdot 06)]^2$$
$$+ [1 \cdot 06 + k(\alpha - 0 \cdot 06)]^2$$
$$+ [1 \cdot 1 + k(\alpha - 0 \cdot 06)]^2 \} / 3$$

$$= (1 + \alpha)^2 \left[\frac{3 \cdot 374}{3} + 2 \cdot 12 k(\alpha - 0 \cdot 06) + k^2(\alpha - 0 \cdot 06)^2 \right]$$

on simplification. We now calculate $E[S^2]$ as

$$E[S^2] = \sum_{\alpha \in X} \text{prob}(i_1 = \alpha) E[S^2 | i_1 = \alpha]$$

$$= \tfrac{1}{3} \sum_{\alpha \in X} E[S^2 | i_1 = \alpha]$$

$$= \frac{1}{3} \left[(1 \cdot 02)^2 \left(\frac{3 \cdot 374}{3} - 2 \cdot 12 k 0 \cdot 04 + k^2 0 \cdot 04^2 \right) \right.$$

$$+ (1 \cdot 06)^2 \left(\frac{3 \cdot 374}{3} \right)$$

$$\left. + (1 \cdot 1)^2 \left(\frac{3 \cdot 374}{3} + 2 \cdot 12 k 0 \cdot 04 + k^2 0 \cdot 04^2 \right) \right]$$

$$= \tfrac{1}{9}(11 \cdot 383\,876 + 0 \cdot 043\,146\,24 k + 0 \cdot 010\,801\,92 k^2)$$

on simplification
   Having calculated $E[S]$ and $E[S^2]$, we evaluate var $[S]$ as $E[S^2] - (E[S])^2$. This leads immediately to the required result.

12.12 *Note*   The estimates below are based on 10 000 simulations.

(a) *Single premium £1000*
   expected value of accumulation $\simeq £2086$
   standard deviation of accumulation $\simeq £252$

   prob(yield $< 3\tfrac{1}{2}\%$) $\simeq 0 \cdot 0428$; 95% confidence interval
   $(0 \cdot 0388, 0 \cdot 0468)$
   prob(yield $> 6\tfrac{1}{2}\%$) $\simeq 0 \cdot 0318$; 95% confidence interval
   $(0 \cdot 0284, 0 \cdot 0352)$

Distribution of percentage annual yield as in table below:

Mean value	Lower tail		Upper tail	
	1%	*10%*	*10%*	*1%*
5·02	3·03	3·87	6·06	6·90

(b) *Annual premium £100*
   expected value of accumulation $\simeq £2269$
   standard deviation of accumulation $\simeq £179$

   prob(yield $< 3\frac{1}{4}\%$) $\simeq 0.0329$; 95% confidence interval
                   (0·0294, 0·0364)
   prob(yield $> 6\frac{1}{2}\%$) $\simeq 0.0479$; 95% confidence interval
                   (0·0437, 0·0521)

Distribution of percentage annual yield as in table below:

Mean value	Lower tail		Upper tail	
	1%	*10%*	*10%*	*1%*
5·02	2·83	3·77	6·17	7·13

Note that although the mean yield is the same for both single and annual premium cases, the spread of the distribution is greater for the annual premium case.

12.13 *Note* The estimates below are based on 10 000 simulations

(a) *Single premium £1000*
   expected value of accumulation $\simeq £2084$
   standard deviation of accumulation $\simeq £177$

   prob(yield $< 3\frac{1}{2}\%$) $\simeq 0.0067$; 95% confidence interval
                   (0·0051, 0·0083)
   prob(yield $> 6\frac{1}{2}\%$) $\simeq 0.0056$; 95% confidence interval
                   (0·0041, 0·0071)

Distribution of percentage annual yield as in table below:

Mean value ·	Lower tail		Upper tail	
	1%	*10%*	*10%*	*1%*
5·02	3·60	4·23	5·76	6·39

Note the lower spread in comparison with exercise 12.12.

(b) *Annual premium £100*
   expected value of accumulation $\simeq £2269$

standard deviation of accumulation $\simeq £127$

prob(yield $< 3\frac{1}{4}\%) \simeq 0.0036$; 95% confidence interval
$$(0.0024, 0.0048)$$
prob(yield $> 6\frac{1}{2}\%) \simeq 0.0098$; 95% confidence interval
$$(0.0079, 0.0117)$$

Distribution of percentage annual yield as in table below:

Mean	Lower tail		Upper tail	
value	1%	10%	10%	1%
5·02	3·47	4·16	5·85	6·50

12.14 (a)
$$\log S = \sum_{t=1}^{15} \log(1 + i_t)$$

Since $k = 0$, the terms in the summation are independently and normally distributed, each with mean 0·05 and variance 0·03². This means that $\log S$ is normally distributed with mean $\mu = 15(0.05) = 0.75$ and variance $\sigma^2 = 15(0.03)^2 = (0.11619)^2$

Hence (see statistical result (c))

$$E[S] = \exp(\mu + \tfrac{1}{2}\sigma^2) = 2.1313$$

$$E[S^2] = \exp(2\mu + 2\sigma^2) = 4.6043$$

from which we obtain the standard deviation of S as 0·2485.

Let $\Phi(x)$ denote the standard normal distribution function.

$$S < 1.47 \Leftrightarrow \log S < 0.385\,262$$

Thus

$$\text{prob}(S < 1.47) = \Phi\left(\frac{0.385\,262 - 0.75}{0.116\,19}\right) = \Phi(-3.139)$$

$$= 0.0008$$

Also,

$$S > 3.1 \Leftrightarrow \log S > 1.131\,402$$

Thus

$$\text{prob}(S > 3.1) = 1 - \Phi\left(\frac{1.131\,402 - 0.75}{0.116\,19}\right) = 1 - \Phi(3.283)$$

$$= 0.0006$$

Note that $\Phi(1.2816) = 0.9$. Also

$$0.75 + 1.2816(0.116\,19) = 0.898\,908$$

$$0.75 - 1.2816(0.116\,19) = 0.601\,092$$

The lower 10% tail value is $\exp(0.601\,092) = 1.8241$.
The upper 10% tail value is $\exp(0.898\,908) = 2.4569$.
(b) We now have $k = 0.5$ and $\log(1 + i_0) = 0.05$. The estimates below are based on 10 000 simulations:

expected value of $S \simeq 2.1660$
standard deviation of $S \simeq 0.4835$: very much greater than in (a)

$\text{prob}(S < 1.47) \simeq 0.0034$; 95% confidence interval $(0.0023, 0.0045)$
$\text{prob}(S > 3.1) \simeq 0.0382$; 95% confidence interval $(0.0344, 0.0420)$

Lower 10% tail $\simeq 1.600$; upper 10% tail $\simeq 2.799$.

Note that the dependent case has greater spread and longer tails.

12.15 Let $M(t)$ denote the market value of the portfolio at time t, that is

$$X(t) = \log\left[\frac{M(t)}{100\,000}\right]$$

Note that $M(t) < 90\,000$ if and only if $X(t) < -0.105\,36$.
 Let $\zeta = -\log 0.9 = 0.105\,36$. Then, by equation 12.6.1, the probability that the market value will fall below £90 000 within ten years is

$$P_{10} = \Phi\left[\frac{-0.105\,36 - (0.05 \times 10)}{0.15\sqrt{10}}\right]$$
$$+ \exp\left[\frac{-2 \times 0.05 \times 0.105\,36}{(0.15)^2}\right]\Phi\left[\frac{-0.105\,36 + (0.05 \times 10)}{0.15\sqrt{10}}\right]$$
$$= \Phi(-1.276) + 0.6261\Phi(0.8320)$$
$$= 0.1010 + (0.6261 \times 0.7973) = 0.600$$

By equation 12.6.2, the probability that the market value will *never* fall below £90 000 is

$$1 - P_\infty = 1 - \exp\left[\frac{-2 \times 0.05 \times 0.105\,36}{(0.15)^2}\right]$$
$$= 1 - 0.6261 = 0.3739$$

Miscellaneous problems

M.1 Consider a nominal amount £100 of the stock. The value of income tax payments (at 4% interest) is

$$6 \times 0.3 \times a_{\overline{n}|}^{(2)} = 132.55 - 107.85 = 24.70$$

Hence

$$a_{\overline{n}|} = \frac{i^{(2)}}{i} \frac{24\cdot70}{1\cdot8} = 13\cdot59$$

Therefore $n = 20$ from the compound interest tables. Therefore the value of the interest payments to the tax-free investor is

$$6a_{\overline{20}|}^{(2)} = 6 \times 13\cdot725 = 82\cdot35$$

The value of the capital repayment is $132\cdot55 - 82\cdot35 = 50\cdot20$. Therefore $Cv^{20} = 50\cdot20$, and hence $C = 110$.

M.2 Let $\delta(0) = \delta_0, \delta(m) = \delta_m$. We have

$$\delta(t) = \begin{cases} \delta_0 + \dfrac{t}{m}(\delta_m - \delta_0) & \text{for } 0 \leqslant t \leqslant m \\ \\ \delta_m & \text{for } \quad t > m \end{cases}$$

So the accumulation after n years is

$$\exp\left(\int_0^n \delta(t)dt\right) = \exp\left\{\int_0^m \left[\delta_0 + \frac{t}{m}(\delta_m - \delta_0)\right]dt + \delta_m(n-m)\right\}$$

$$= \exp\left\{\left[\delta_0 t + \frac{1}{2}\frac{t^2}{m}(\delta_m - \delta_0)\right]_{t=0}^{t=m} + \delta_m(n-m)\right\}$$

$$= \exp[\delta_0 m + \tfrac{1}{2}m(\delta_m - \delta_0) + \delta_m(n-m)]$$

$$= \exp[\tfrac{1}{2}m(\delta_0 - \delta_m) + \delta_m n]$$

$$= [\exp(\delta_0 - \delta_m)]^{m/2}[\exp(\delta_m)]^n$$

$$= \left(\frac{1\cdot04}{1\cdot03}\right)^8 (1\cdot03)^{39} = (1\cdot04)^8(1\cdot03)^{31}$$

$$= 3\cdot4215$$

M.3 We value the payments under each scheme at $4\frac{1}{2}\%$ p.a. interest. (Alternatively, we may accumulate the cash flow from each scheme at $4\frac{1}{2}\%$ p.a. interest until time 40: this will give the present values $\times$ $(1\cdot045)^{40}$.) Consider £100 nominal of loan.

(a) Present value = £100
(b) Present value = $4a_{\overline{40}|} + 0\cdot5a_{\overline{30}|} + 0\cdot25a_{\overline{10}|} + 0\cdot25a_{\overline{31}|} + 105v^{40}$
    = £101·24.

(c) Present value $= 4{\cdot}25a_{\overline{40}|} + 0{\cdot}5a_{\overline{50}|} + 98v^{40}$ at $4\tfrac{1}{2}\%$
$\qquad\qquad\quad = 101{\cdot}56$

Therefore choose option (c).

M.4 (a) $\qquad\qquad X = 12 \times 360 \times a^{(12)}_{\overline{20}|0\cdot06} = \pounds8135{\cdot}8$

(b) Balance at 1 January 1982 is

$$Y = 8135{\cdot}8(1{\cdot}06) - [12 \times 360s^{(12)}_{\overline{1}|0\cdot06}] = 4186{\cdot}4$$

Now consider quarterly periods, commencing at 1 January 1982, with interest at 1% per quarter. The balance at 1 April 1982 is
$$Z = Y(1{\cdot}01) - (3 \times 360s^{(3)}_{\overline{1}|0\cdot01})$$

$$= 3144{\cdot}7 \qquad (i^{(3)} \text{ at } 1\% = 0{\cdot}009\,966\,9)$$

The balance at 1 October 1982 is thus

$$Z(1{\cdot}01)^2 - [3 \times 460s^{(3)}_{\overline{2}|0\cdot01}] = 424{\cdot}9$$

With interest for a month, this is still less than £460, so there will be insufficient money for rent at the end of October 1982.

(c) Let balance at 1 April 1982 be increased to Z', sufficient to provide rent until the end of 1982. Therefore

$$Z'(1{\cdot}01)^3 - [3 \times 460s^{(3)}_{\overline{3}|0\cdot01}] = 0$$

whence $Z' = 4071{\cdot}9$.

Extra money needed at 1 April 1982 is $Z' - Z = \pounds927{\cdot}2$.

M.5 We first assume redemption at par and ignore capital gains tax. Value is

$$A = K + \frac{0{\cdot}65 \times 0{\cdot}045}{i^{(2)}}(1\,000\,000 - K) \qquad \text{at rate } i \text{ (the yield)}$$

where

$$K = 100\,000(v^{10} + 1{\cdot}5v^{13} + 2v^{15} + 2{\cdot}5v^{18} + 2v^{20} + v^{23})$$

But we must allow for (a) actual redemption money, i.e. *add*

$$-20\,000v^{10} - 30\,000v^{13} + 40\,000v^{20} + 20\,000v^{23}$$

and for (b) capital gains tax (*on bonds redeemable above 85*): i.e. *subtract*

$$0{\cdot}3[(100 - 85)2000v^{15} + (100 - 85)2500v^{18}$$
$$+ (120 - 85)2000v^{20} + (120 - 85)1000v^{23}]$$

Hence we finally obtain the equation of value

$$850\,000 = K + \frac{0.65 \times 0.045}{i^{(2)}}(1\,000\,000 - K)$$

$$- 20\,000v^{10} - 30\,000v^{13} + 40\,000v^{20} + 20\,000v^{23}$$

$$- 9000v^{15} - 11\,250v^{18} - 21\,000v^{20} - 10\,500v^{23}$$

i.e.

$$850\,000 = K + \frac{0.029\,25}{i^{(2)}}(1\,000\,000 - K) - 20\,000v^{10}$$

$$- 30\,000v^{13} - 11\,250v^{18} + 19\,000v^{20}$$

$$+ 9500v^{23} - 9000v^{15} \tag{1}$$

A rough solution (assume redemption at par in about 16 years' time) gives

$$\text{yield} \simeq \frac{2.9 + [(100 - 85)/16]0.7}{85} \simeq 4.2\%$$

Try $i = 4\%$ right-hand side of equation 1 is £845 955.
When $i = 3.5\%$ right-hand side of equation 1 is £901 008.
By interpolation, $i \simeq 3.96\%$.

M.6 (a) This follows easily from the fact that the initial investment of 1 produces interest of j at times at $1, 2, \ldots$ and n, and these sums are accumulated at rate k.

(b) We require to show

$$\lim_{n \to \infty} (1 + js_{\overline{n}|k})^{1/n} = 1 + k$$

or equivalently,

$$\lim_{n \to \infty} \left[\frac{\log(1 + js_{\overline{n}|k})}{n} \right] = \log(1 + k)$$

But by l'Hospital's rule (∞/∞ form) we have

$$\lim_{n \to \infty} \left\{ \frac{\log(1 + js_{\overline{n}|k})}{n} \right\} = \lim_{n \to \infty} \frac{1}{(1 + js_{\overline{n}|k})} \left[\frac{j}{k}(1 + k)^n \log(1 + k) \right]$$

(differentiating above and below the line)

$$= \lim_{n \to \infty} \frac{j}{k} \left[\frac{(1+k)^n}{1 + \frac{j}{k}[(1+k)^n - 1]} \right] \log(1+k)$$

$$= \frac{j}{k} \left(\frac{k}{j} \right) \log(1+k) = \log(1+k)$$

M.7 The net and gross premiums at 5% are:

10-year term:

$$\text{net premium} = \frac{1000}{\ddot{s}_{\overline{10|}}} = 75 \cdot 72$$

$$\text{gross premium} = \frac{1000}{0 \cdot 94 \ddot{s}_{\overline{10|}}} = 80 \cdot 55$$

20-year term:

$$\text{net premium} = \frac{1000}{\ddot{s}_{\overline{20|}}} = 28 \cdot 80$$

$$\text{gross premium} = \frac{1000}{0 \cdot 94 \ddot{s}_{\overline{20|}}} = 30 \cdot 64$$

Let A, B be the funds at the beginning and end of the year respectively. A is the accumulation of the net premiums, i.e. (since policies are sold uniformly over the year)

$$A = 500 \times 28 \cdot 80(\bar{s}_{\overline{20|}} + \bar{s}_{\overline{19|}} + \cdots + \bar{s}_{\overline{1|}})$$
$$+ 500 \times 75 \cdot 72(\bar{s}_{\overline{10|}} + \bar{s}_{\overline{9|}} + \cdots + \bar{s}_{\overline{1|}}) \qquad \text{at } 5\%$$

$$= \left(28 \cdot 80 \times 500 \frac{\bar{s}_{\overline{20|}} - 20}{\delta} \right) + \left(75 \cdot 72 \times 500 \frac{\bar{s}_{\overline{10|}} - 10}{\delta} \right) \qquad \text{at } 5\%$$

$$= 6\,833\,674$$

using $\bar{s}_{\overline{n|}} = [(1+i)^n - 1]/\delta$.

Alternatively, using stationary fund properties,

$$A = \frac{1}{\delta}[\underbrace{1\,000\,000}_{\substack{\text{rate of} \\ \text{claims} \\ \text{payments}}} - \underbrace{(5000 \times 75 \cdot 72) - (10\,000 \times 28 \cdot 80)]}_{\text{rate of net premium income}} = 6\,833\,367$$

(The final figures are unreliable due to round-off error.)

Note that the gross premiums received in the year are

$$(500 \times 10 \times 80 \cdot 55) + (500 \times 20 \times 30 \cdot 64) = 709\,150$$

and that the expenses for the year are 9% of 709 150, i.e. £63 824. Let I denote the gross interest income in the year. Then, since the claims amount to £1 000 000, it follows that

$$6\,833\,367 + 0 \cdot 8I + 709\,150 = 1\,000\,000 + 63\,824 + B$$

Also

$$\frac{2I}{6\,833\,367 + B} = 0 \cdot 08$$

Solving these last two equations for I and B, we obtain $B =$ £6 918 761, or to the nearest £1000, £6 919 000.

M.8 (a)

$$n = \frac{\dfrac{1100}{4}\left(\dfrac{1}{4} + \dfrac{2}{4} + \cdots + \dfrac{28}{4}\right) + (22\,000 \times 7)}{29\,700}$$

$$= \frac{\dfrac{275}{4}\left(\dfrac{28 \times 29}{2}\right) + (22\,000 \times 7)}{29\,700} = 6 \cdot 125$$

(b) $29\,700v^n = 1100a_{\overline{n}|}^{(4)} + 22\,000v^7$ at 6%

$$v^n = \frac{(1100 \times 1 \cdot 022\,227 \times 5 \cdot 5824) + (22\,000 \times 0 \cdot 665\,09)}{29\,700}$$

$$1 \cdot 06^{-n} = 0 \cdot 703\,978$$

Therefore $n = 6 \cdot 024$.

M.9 Let the annual deposit be P and the purchases at time t (measured in years from 1 January 1978) be B_t. Then the rates of interest $i_1, i_2,$ and i_3 satisfy

$$(P - B_0)(1 + i)^3 + (P - B_1)(1 + i)^2 + (P - B_2)(1 + i) + P - B_3 = 0$$

(1)

Putting $(1 + i) = X$, this cubic must factorize; hence

$$(P - B_0)[X - (1 + i_1)][X - (1 + i_2)][X - (1 + i_3)] = 0 \quad (2)$$

We are to find $P - B_0 + P - B_1 + P - B_2 + P - B_3$, i.e. $f(0)$, where $f(i) =$ LHS of equation 1. From equation 2,

$$f(0) = -(P - B_0)i_1 i_2 i_3$$

Since $-(P - B_0) = 100, 4P = 1308, 4P - \Sigma B = 0.06$, we obtain

$$0.06 = 100 i_1 i_2 i_3 = 100(0.1)(0.12)i_3$$

Hence $i_3 = 0.05$.

M.10 Work in time units of a half-year. The gross yield per period on 31 August 1982 is the solution of

$$49.50 = 2.75 v^{1/6} \ddot{a}^{(2)}_{\overline{\infty}|}$$

$$= 2.75 v^{1/6} / d^{(2)}$$

$$= \frac{2.75 v^{1/6}}{2(1 - v^{1/2})}$$

By trial and interpolation, the yield per half-year is approximately 5.75%, so the gross annual yield, convertible half-yearly, is 11.5% (to the nearest 0.1%).

The selling price one year later was

$$\frac{2.75 v^{1/6}}{2(1 - v^{1/2})} \qquad \text{at rate } 5.0\% \text{ (gross yield per half-year)} = £56.59$$

The equation of value, working in years from 31 August 1982, to find the investor's net yield is

$$49.5 = [0.6 \times 5.5 v^{1/12} \ddot{a}^{(4)}_{\overline{1}|}] + [56.59 - 0.3(56.59 - 49.5)]v$$

$$= 3.3 v^{1/12}\left[\frac{d}{d^{(4)}}\right] + 54.46 v$$

By trials, the solution (to the nearest 0.25%) is $i = 17.25\%$.

M.11 *Loan* (a)   Half-yearly instalment is $20\,000/a_{\overline{40}|0.06} = 1329.23$.
Value at 4% per half-year ignoring tax relief is $1329.23\, a_{\overline{40}|0.04} = 26\,309.18$.
If K is value of capital,

$$K + \frac{0.06}{0.04}(20\,000 - K) = 26\,309.18$$

Thus

$$\tfrac{1}{2}K = 3690.82$$

$$K = 7381.64$$

Or use

$$K = \sum_{t=1}^{40} 1329 \cdot 23 \, v_{0 \cdot 04}^{t} v_{0 \cdot 06}^{41-t}$$

Value of tax on interest payments is $0 \cdot 3(0 \cdot 06/0 \cdot 04)(12\,618 \cdot 4) = 5678 \cdot 28$.

Net cost of the loan is £20 631.

Loan (b)   Value of net interest payments is $0 \cdot 7 \times 2500 \times (a_{\overline{20}}/s_{\overline{1}})$ at $4\% = 16\,979 \cdot 12$.

Capital redemption policy premium P per half-year:

$$0 \cdot 95 \times 2P \, s_{\overline{20}}^{(2)} - 0 \cdot 05 P(1 + i)^{20} = 20\,000 \qquad \text{at } 10\%$$

$$P(116 \cdot 919 - 0 \cdot 336) = 20\,000$$

Thus $P = 171 \cdot 552$.

Net value of premiums is $0 \cdot 85 \times 171 \cdot 552 \times \ddot{a}_{\overline{40}|0 \cdot 04} = 3001 \cdot 69$.

Net cost of loan is £19 980·81.

He should accept loan under conditions (b).

M.12 Let $c = 15\,000$ and $k = \log 1 \cdot 05 = 0 \cdot 048\,79$

(a) Present value of profit, if whisky is sold at time t, is

$$R_t e^{-kt} - c$$

By differentiation, we note that this has its maximum value when

$$\frac{dR_t}{dt} e^{-kt} - R_t k e^{-kt} = 0$$

i.e. when

$$\frac{1}{R_t} \frac{dR_t}{dt} = k$$

$$\Leftrightarrow \frac{1}{2(1 + t)} = k$$

$$\Rightarrow t = \frac{1}{2k} - 1 = 9 \cdot 25, \qquad \text{say} \qquad 9 \cdot 25 \text{ years.}$$

(b) Present value of profit (under new arrangement), if each consignment is held for t years, is

$$-c + (R_t - c)e^{-kt} + (R_t - c)e^{-2kt} + (R_t - c)e^{-3kt} + \cdots$$

$$= -c + \frac{R_t - c}{e^{kt} - 1}$$

By differentiation, we find that, for a maximum value,

$$\frac{dR_t}{dt} = \frac{k(R_t - c)}{1 - e^{-kt}}$$

or

$$\frac{5000}{(1 + t)^{1/2}} - \frac{0.048\,79[10\,000(1 + t)^{1/2} - 15\,000]}{1 - \exp(-0.048\,79t)} = 0$$

When $t = 4$ the left-hand side of the last equation equals 210.50, and when $t = 5$ it equals -98.78. By interpolation we estimate the solution as 4.7 years.

The difference is caused by the greater frequency of operation (larger profits on more operations).

M.13 Let A be the gross annuity payment per annum. Then

$$100\,000 = Aa_{\overline{25}|} \qquad \text{at } 6\%$$

$$\text{So } A = 7822.6.$$

Ignoring income tax (and replacement of capital), value at 1 January 1985 per £1 p.a. of gross annuity is $a_{\overline{13}|}$ at 5%.

If K is the capital value per £1 p.a. of gross annuity, then

$$K + \frac{0.06}{0.05}(a_{\overline{30}|0.06} - K) = a_{\overline{30}|0.05}$$

Hence

$$K + \frac{0.06}{0.05}(9.7122 - K) = 10.3797$$

$$0.2K = 1.274\,94$$

$$K = 6.3747$$

Present value of net payments per £1 p.a. of gross annuity is (by Makeham)

$$K + \frac{0.06 \times 0.7}{0.05}(a_{\overline{30}|0.06} - K) = 9.1782$$

If he pays X per £1 annuity,

$$X = 9.1782 - \left(\frac{X}{s_{\overline{30}|0.04}}\right)a_{\overline{30}|0.05} + Xv_{0.05}^{15}$$

i.e.

$$X(1 + 0.5184 - 0.481\,02) = 9.1782$$

i.e.

$$X = 8.8475$$

Therefore

$$\text{total price} = 8 \cdot 8475 \times 7822 \cdot 6 = \pounds 69\,210$$

M.14 The annual premium for the capital redemption policy is $P' = 100\,000/\ddot{s}_{\overline{50}|}$ at 6% = £324·93.

   The accumulation of premiums up to 31 December 2003 is $P'\ddot{s}_{\overline{19}|}$ at rate $i_1 (9\% \times 0 \cdot 625 = 5 \cdot 625\%)$.

The accumulation of premiums to 31 December 2034 is

$$A = P'\ddot{s}_{\overline{19}|i_1}(1 + i_2)^{31} + P'\ddot{s}_{\overline{31}|i_2}$$

where $i_2 = 9\% \times 0 \cdot 7 = 6 \cdot 3\%$. Thus

$$A = 324 \cdot 93 \left\{ \left[\frac{(1 \cdot 05625)^{20} - 1 \cdot 05625}{0 \cdot 05625} \right] (1 \cdot 063)^{31} \right.$$
$$\left. + \left[\frac{(1 \cdot 063)^{32} - 1 \cdot 063}{0 \cdot 063} \right] \right\}$$
$$= 324 \cdot 93(228 \cdot 186 + 95 \cdot 258) = \pounds 105\,096 \cdot 66$$

Expected profit = £5097.

M.15 *Account 1*

$$100\left[(1 \cdot 04)^4(1 \cdot 05)^4(1 \cdot 06)^4 + (1 \cdot 04)^2(1 \cdot 05)^4(1 \cdot 06)^4\right.$$
$$+ (1 \cdot 05)^4(1 \cdot 06)^4 \qquad + (1 \cdot 05)^2(1 \cdot 06)^4$$
$$+ (1 \cdot 06)^4 \qquad\qquad + (1 \cdot 06)^2]$$
$$= \pounds 876 \cdot 75$$

Account 2

Measure time in years from 1 January 1976. Then, in the notation of section 6.7, we have $i_0 = i_1 = 0 \cdot 09, i_2 = i_3 = 0 \cdot 11$, and $i_4 = i_5 = 0 \cdot 13$. The accumulation (to time 6) of six annual payments of £1 is (see equation 6.7.2):

$$A_0 = 1 \cdot 09 + 1 \cdot 09^2 + 1 \cdot 09^2(1 \cdot 11) + 1 \cdot 09^2(1 \cdot 11)^2$$
$$+ 1 \cdot 09^2(1 \cdot 11)^2(1 \cdot 13) + 1 \cdot 09^2(1 \cdot 11)^2(1 \cdot 13)^2$$
$$= 8 \cdot 5841$$

The accumulation is thus £858·41.

   Thus account 1 gives a higher accumulation by £18·34.

M.16 Let K be present value of capital:

$$K = 103[200a_{\overline{40}|} - 50(a_{\overline{30}|} + a_{\overline{20}|} + a_{\overline{10}|})] \qquad \text{at } 4\tfrac{1}{2}\% \qquad = 187\,444$$

Note that $g = 0{\cdot}05/1{\cdot}03$, $C = 515\,000$, $t_1 = 0{\cdot}25$.

By Makeham's formula, the price of the whole loan is

$$A = K + \frac{g(1 - t_1)}{i^{(4)}}(C - K) \qquad (i \text{ being } 4\tfrac{1}{2}\%)$$

$$= 187\,444 + \frac{0{\cdot}05(1 - 0{\cdot}25)}{1{\cdot}03 \times 0{\cdot}045^{(4)}}(515\,000 - 187\,444)$$

$$= 456\,890$$

Price per bond = £91·38.

Since the price per bond is less than £103, the yield decreases as the term increases. The yield on a particular bond, redeemed after n years, is found by solving the equation of value

$$91{\cdot}38 = 5(1 - 0{\cdot}25)a_{\overline{n}|}^{(4)} + 103v^n$$

If $i = 0{\cdot}045$, this gives

$$91{\cdot}38 = 5(1 - 0{\cdot}25)\frac{(1 - v^n)}{0{\cdot}045^{(4)}} + 103v^n$$

from which we obtain

$$v^n = 0{\cdot}3641$$

Since, at $4\tfrac{1}{2}\%$, $v^{22} > 0{\cdot}3641 > v^{23}$, the yield on a single bond will be less than $4\tfrac{1}{2}\%$ if and only if it is redeemed after 23 or more years. The required probability is thus

$$\frac{\text{no. of bonds redeemed at times 23 years and later}}{\text{total no. of bonds}}$$

$$= 1 - \frac{(50 \times 10) + (100 \times 10) + (150 \times 2)}{5000} = 0{\cdot}64$$

COMPOUND INTEREST TABLES   1 PER CENT

Constants	
Function	Value
i	0·010 000
$i^{(2)}$	0·009 975
$i^{(4)}$	0·009 963
$i^{(12)}$	0·009 954
δ	0·009 950
$(1 + i)$	1·010 000
$(1 + i)^{1/2}$	1·004 988
$(1 + i)^{1/4}$	1·002 491
$(1 + i)^{1/12}$	1·000 830
v	0·990 099
$v^{1/2}$	0·995 037
$v^{1/4}$	0·997 516
$v^{1/12}$	0·999 171
d	0·009 901
$d^{(2)}$	0·009 926
$d^{(4)}$	0·009 938
$d^{(12)}$	0·009 946
$i/i^{(2)}$	1·002 494
$i/i^{(4)}$	1·003 742
$i/i^{(12)}$	1·004 575
i/δ	1·004 992
$i/d^{(2)}$	1·007 494
$i/d^{(4)}$	1·006 242
$i/d^{(12)}$	1·005 408

n	$(1 + i)^n$	v^n	$s_{\overline{n}}$	$a_{\overline{n}}$	$(Ia)_{\overline{n}}$	n
1	1·010 00	0·990 10	1·0000	0·9901	0·9901	1
2	1·020 10	0·980 30	2·0100	1·9704	2·9507	2
3	1·030 30	0·970 59	3·0301	2·9410	5·8625	3
4	1·040 60	0·960 98	4·0604	3·9020	9·7064	4
5	1·051 01	0·951 47	5·1010	4·8534	14·4637	5
6	1·061 52	0·942 05	6·1520	5·7955	20·1160	6
7	1·072 14	0·932 72	7·2135	6·7282	26·6450	7
8	1·082 86	0·923 48	8·2857	7·6517	34·0329	8
9	1·093 69	0·914 34	9·3685	8·5660	42·2619	9
10	1·104 62	0·905 29	10·4622	9·4713	51·3148	10
11	1·115 67	0·896 32	11·5668	10·3676	61·1744	11
12	1·126 83	0·887 45	12·6825	11·2551	71·8238	12
13	1·138 09	0·878 66	13·8093	12·1337	83·2464	13
14	1·149 47	0·869 96	14·9474	13·0037	95·4258	14
15	1·160 97	0·861 35	16·0969	13·8651	108·3461	15
16	1·172 58	0·852 82	17·2579	14·7179	121·9912	16
17	1·184 30	0·844 38	18·4304	15·5623	136·3456	17
18	1·196 15	0·836 02	19·6147	16·3983	151·3940	18
19	1·208 11	0·827 74	20·8109	17·2260	167·1210	19
20	1·220 19	0·819 54	22·0190	18·0456	183·5119	20
21	1·232 39	0·811 43	23·2392	18·8570	200·5519	21
22	1·244 72	0·803 40	24·4716	19·6604	218·2267	22
23	1·257 16	0·795 44	25·7163	20·4558	236·5218	23
24	1·269 73	0·787 57	26·9735	21·2434	255·4234	24
25	1·282 43	0·779 77	28·2432	22·0232	274·9176	25
26	1·295 26	0·772 05	29·5256	22·7952	294·9909	26
27	1·308 21	0·764 40	30·8209	23·5596	315·6298	27
28	1·321 29	0·756 84	32·1291	24·3164	336·8212	28
29	1·334 50	0·749 34	33·4504	25·0658	358·5521	29
30	1·347 85	0·741 92	34·7849	25·8077	380·8098	30
31	1·361 33	0·734 58	36·1327	26·5423	403·5817	31
32	1·374 94	0·727 30	37·4941	27·2696	426·8554	32
33	1·388 69	0·720 10	38·8690	27·9897	450·6188	33
34	1·402 58	0·712 97	40·2577	28·7027	474·8599	34
35	1·416 60	0·705 91	41·6603	29·4086	499·5669	35
36	1·430 77	0·698 92	43·0769	30·1075	524·7282	36
37	1·445 08	0·692 00	44·5076	30·7995	550·3324	37
38	1·459 53	0·685 15	45·9527	31·4847	576·3682	38
39	1·474 12	0·678 37	47·4123	32·1630	602·8246	39
40	1·488 86	0·671 65	48·8864	32·8347	629·6907	40
41	1·503 75	0·665 00	50·3752	33·4997	656·9559	41
42	1·518 79	0·658 42	51·8790	34·1581	684·6095	42
43	1·533 98	0·651 90	53·3978	34·8100	712·6412	43
44	1·549 32	0·645 45	54·9318	35·4555	741·0408	44
45	1·564 81	0·639 05	56·4811	36·0945	769·7982	45
46	1·580 46	0·632 73	58·0459	36·7272	798·9037	46
47	1·596 26	0·626 46	59·6263	37·3537	828·3475	47
48	1·612 23	0·620 26	61·2226	37·9740	858·1200	48
49	1·628 35	0·614 12	62·8348	38·5881	888·2118	49
50	1·644 63	0·608 04	64·4632	39·1961	918·6137	50

COMPOUND INTEREST TABLES

Constants	
Function	Value
i	0·020 000
$i^{(2)}$	0·019 901
$i^{(4)}$	0·019 852
$i^{(12)}$	0·019 819
δ	0·019 803
$(1 + i)$	1·020 000
$(1 + i)^{1/2}$	1·009 950
$(1 + i)^{1/4}$	1·004 963
$(1 + i)^{1/12}$	1·001 652
v	0·980 392
$v^{1/2}$	0·990 148
$v^{1/4}$	0·995 062
$v^{1/12}$	0·998 351
d	0·019 608
$d^{(2)}$	0·019 705
$d^{(4)}$	0·019 754
$d^{(12)}$	0·019 786
$i/i^{(2)}$	1·004 975
$i/i^{(4)}$	1·007 469
$i/i^{(12)}$	1·009 134
i/δ	1·009 967
$i/d^{(2)}$	1·014 975
$i/d^{(4)}$	1·012 469
$i/d^{(12)}$	1·010 801

n	$(1 + i)^n$	v^n	$s_{\overline{n}\rvert}$	$a_{\overline{n}\rvert}$	$(Ia)_{\overline{n}\rvert}$	n
1	1·020 00	0·980 39	1·0000	0·9804	0·9804	1
2	1·040 40	0·961 17	2·0200	1·9416	2·9027	2
3	1·061 21	0·942 32	3·0604	2·8839	5·7297	3
4	1·082 43	0·923 85	4·1216	3·8077	9·4251	4
5	1·104 08	0·905 73	5·2040	4·7135	13·9537	5
6	1·126 16	0·887 97	6·3081	5·6014	19·2816	6
7	1·148 69	0·870 56	7·4343	6·4720	25·3755	7
8	1·171 66	0·853 49	8·5830	7·3255	32·2034	8
9	1·195 09	0·836 76	9·7546	8·1622	39·7342	9
10	1·218 99	0·820 35	10·9497	8·9826	47·9377	10
11	1·243 37	0·804 26	12·1687	9·7868	56·7846	11
12	1·268 24	0·788 49	13·4121	10·5753	66·2465	12
13	1·293 61	0·773 03	14·6803	11·3484	76·2959	13
14	1·319 48	0·757 88	15·9739	12·1062	86·9062	14
15	1·345 87	0·743 01	17·2934	12·8493	98·0514	15
16	1·372 79	0·728 45	18·6393	13·5777	109·7065	16
17	1·400 24	0·714 16	20·0121	14·2919	121·8473	17
18	1·428 25	0·700 16	21·4123	14·9920	134·4502	18
19	1·456 81	0·686 43	22·8406	15·6785	147·4923	19
20	1·485 95	0·672 97	24·2974	16·3514	160·9518	20
21	1·515 67	0·659 78	25·7833	17·0112	174·8071	21
22	1·545 98	0·646 84	27·2990	17·6580	189·0375	22
23	1·576 90	0·634 16	28·8450	18·2922	203·6231	23
24	1·608 44	0·621 72	30·4219	18·9139	218·5444	24
25	1·640 61	0·609 53	32·0303	19·5235	233·7827	25
26	1·673 42	0·597 58	33·6709	20·1210	249·3198	26
27	1·706 89	0·585 86	35·3443	20·7069	265·1380	27
28	1·741 02	0·574 37	37·0512	21·2813	281·2205	28
29	1·775 84	0·563 11	38·7922	21·8444	297·5508	29
30	1·811 36	0·552 07	40·5681	22·3965	314·1129	30
31	1·847 59	0·541 25	42·3794	22·9377	330·8915	31
32	1·884 54	0·530 63	44·2270	23·4683	347·8718	32
33	1·922 23	0·520 23	46·1116	23·9886	365·0393	33
34	1·960 68	0·510 03	48·0338	24·4986	382·3803	34
35	1·999 89	0·500 03	49·9945	24·9986	399·8813	35
36	2·039 89	0·490 22	51·9944	25·4888	417·5293	36
37	2·080 69	0·480 61	54·0343	25·9695	435·3119	37
38	2·122 30	0·471 19	56·1149	26·4406	453·2170	38
39	2·164 74	0·461 95	58·2372	26·9026	471·2330	39
40	2·208 04	0·452 89	60·4020	27·3555	489·3486	40
41	2·252 20	0·444 01	62·6100	27·7995	507·5530	41
42	2·297 24	0·435 30	64·8622	28·2348	525·8358	42
43	2·343 19	0·426 77	67·1595	28·6616	544·1869	43
44	2·390 05	0·418 40	69·5027	29·0800	562·5965	44
45	2·437 85	0·410 20	71·8927	29·4902	581·0553	45
46	2·486 61	0·402 15	74·3306	29·8923	599·5544	46
47	2·536 34	0·394 27	76·8172	30·2866	618·0850	47
48	2·587 07	0·386 54	79·3535	30·6731	636·6388	48
49	2·638 81	0·378 96	81·9406	31·0521	655·2078	49
50	2·691 59	0·371 53	84·5794	31·4236	673·7842	50

COMPOUND INTEREST TABLES

Constants			n	$(1+i)^n$	v^n	$s_{\overline{n}}$	$a_{\overline{n}}$	$(Ia)_{\overline{n}}$	n
Function	**Value**		1	1·030 00	0·970 87	1·0000	0·9709	0·9709	1
			2	1·060 90	0·942 60	2·0300	1·9135	2·8561	2
i	0·030 000		3	1·092 73	0·915 14	3·0909	2·8286	5·6015	3
$i^{(2)}$	0·029 778		4	1·125 51	0·888 49	4·1836	3·7171	9·1554	4
$i^{(4)}$	0·029 668		5	1·159 27	0·862 61	5·3091	4·5797	13·4685	5
$i^{(12)}$	0·029 595								
δ	0·029 559		6	1·194 05	0·837 48	6·4684	5·4172	18·4934	6
			7	1·229 87	0·813 09	7·6625	6·2303	24·1850	7
$(1+i)$	1·030 000		8	1·266 77	0·789 41	8·8923	7·0197	30·5003	8
$(1+i)^{1/2}$	1·014 889		9	1·304 77	0·766 42	10·1591	7·7861	37·3981	9
$(1+i)^{1/4}$	1·007 417		10	1·343 92	0·744 09	11·4639	8·5302	44·8390	10
$(1+i)^{1/12}$	1·002 466								
			11	1·384 23	0·722 42	12·8078	9·2526	52·7856	11
v	0·970 874		12	1·425 76	0·701 38	14·1920	9·9540	61·2022	12
$v^{1/2}$	0·985 329		13	1·468 53	0·680 95	15·6178	10·6350	70·0546	13
$v^{1/4}$	0·992 638		14	1·512 59	0·661 12	17·0863	11·2961	79·3102	14
$v^{1/12}$	0·997 540		15	1·557 97	0·641 86	18·5989	11·9379	88·9381	15
d	0·029 126		16	1·604 71	0·623 17	20·1569	12·5611	· 98·9088	16
$d^{(2)}$	0·029 341		17	1·652 85	0·605 02	21·7616	13·1661	109·1941	17
$d^{(4)}$	0·029 450		18	1·702 43	0·587 39	23·4144	13·7535	119·7672	18
$d^{(12)}$	0·029 522		19	1·753 51	0·570 29	25·1169	14·3238	130·6026	19
			20	1·806 11	0·553 68	26·8704	14·8775	141·6761	20
$i/i^{(2)}$	1·007 445								
$i/i^{(4)}$	1·011 181		21	1·860 29	0·537 55	28·6765	15·4150	152·9647	21
$i/i^{(12)}$	1·013 677		22	1·916 10	0·521 89	30·5368	15·9369	164·4463	22
			23	1·973 59	0·506 69	32·4529	16·4436	176·1002	23
i/δ	1·014 926		24	2·032 79	0·491 93	34·4265	16·9355	187·9066	24
			25	2·093 78	0·477 61	36·4593	17·4131	199·8468	25
$i/d^{(2)}$	1·022 445								
$i/d^{(4)}$	1·018 681		26	2·156 59	0·463 69	38·5530	17·8768	211·9028	26
$i/d^{(12)}$	1·016 177		27	2·221 29	0·450 19	40·7096	18·3270	224·0579	27
			28	2·287 93	0·437 08	42·9309	18·7641	236·2961	28
			29	2·356 57	0·424 35	45·2189	19·1885	248·6021	29
			30	2·427 26	0·411 99	47·5754	19·6004	260·9617	30
			31	2·500 08	0·399 99	50·0027	20·0004	273·3613	31
			32	2·575 08	0·388 34	52·5028	20·3888	285·7881	32
			33	2·652 34	0·377 03	55·0778	20·7658	298·2300	33
			34	2·731 91	0·366 04	57·7302	21·1318	310·6755	34
			35	2·813 86	0·355 38	60·4621	21·4872	323·1139	35
			36	2·898 28	0·345 03	63·2759	21·8323	335·5351	36
			37	2·985 23	0·334 98	66·1742	22·1672	347·9295	37
			38	3·074 78	0·325 23	69·1594	22·4925	360·2881	38
			39	3·167 03	0·315 75	72·2342	22·8082	372·6024	39
			40	3·262 04	0·306 56	75·4013	23·1148	384·8647	40
			41	3·359 90	0·297 63	78·6633	23·4124	397·0675	41
			42	3·460 70	0·288 96	82·0232	23·7014	409·2038	42
			43	3·564 52	0·280 54	85·4839	23·9819	421·2671	43
			44	3·671 45	0·272 37	89·0484	24·2543	433·2515	44
			45	3·781 60	0·264 44	92·7199	24·5187	445·1512	45
			46	3·895 04	0·256 74	96·5015	24·7754	456·9611	46
			47	4·011 90	0·249 26	100·3965	25·0247	468·6762	47
			48	4·132 25	0·242 00	104·4084	25·2667	480·2922	48
			49	4·256 22	0·234 95	108·5406	25·5017	491·8047	49
			50	4·383 91	0·228 11	112·7969	25·7298	503·2101	50

COMPOUND INTEREST TABLES

Constants	
Function	**Value**
i	0·040 000
$i^{(2)}$	0·039 608
$i^{(4)}$	0·039 414
$i^{(12)}$	0·039 285
δ	0·039 221
$(1 + i)$	1·040 000
$(1 + i)^{1/2}$	1·019 804
$(1 + i)^{1/4}$	1·009 853
$(1 + i)^{1/12}$	1·003 274
v	0·961 538
$v^{1/2}$	0·980 581
$v^{1/4}$	0·990 243
$v^{1/12}$	0·996 737
d	0·038 462
$d^{(2)}$	0·038 839
$d^{(4)}$	0·039 029
$d^{(12)}$	0·039 157
$i/i^{(2)}$	1·009 902
$i/i^{(4)}$	1·014 877
$i/i^{(12)}$	1·018 204
i/δ	1·019 869
$i/d^{(2)}$	1·029 902
$i/d^{(4)}$	1·024 877
$i/d^{(12)}$	1·021 537

n	$(1 + i)^n$	v^n	$s_{\overline{n}}$	$a_{\overline{n}}$	$(Ia)_{\overline{n}}$	n
1	1·040 00	0·961 54	1·0000	0·9615	0·9615	1
2	1·081 60	0·924 56	2·0400	1·8861	2·8107	2
3	1·124 86	0·889 00	3·1216	2·7751	5·4776	3
4	1·169 86	0·854 80	4·2465	3·6299	8·8969	4
5	1·216 65	0·821 93	5·4163	4·4518	13·0065	5
6	1·265 32	0·790 31	6·6330	5·2421	17·7484	6
7	1·315 93	0·759 92	7·8983	6·0021	23·0678	7
8	1·368 57	0·730 69	9·2142	6·7327	28·9133	8
9	1·423 31	0·702 59	10·5828	7·4353	35·2366	9
10	1·480 24	0·675 56	12·0061	8·1109	41·9922	10
11	1·539 45	0·649 58	13·4864	8·7605	49·1376	11
12	1·601 03	0·624 60	15·0258	9·3851	56·6328	12
13	1·665 07	0·600 57	16·6268	9·9856	64·4403	13
14	1·731 68	0·577 48	18·2919	10·5631	72·5249	14
15	1·800 94	0·555 26	20·0236	11·1184	80·8539	15
16	1·872 98	0·533 91	21·8245	11·6523	89·3964	16
17	1·947 90	0·513 37	23·6975	12·1657	98·1238	17
18	2·025 82	0·493 63	25·6454	12·6593	107·0091	18
19	2·106 85	0·474 64	27·6712	13·1339	116·0273	19
20	2·191 12	0·456 39	29·7781	13·5903	125·1550	20
21	2·278 77	0·438 83	31·9692	14·0292	134·3705	21
22	2·369 92	0·421 96	34·2480	14·4511	143·6535	22
23	2·464 72	0·405 73	36·6179	14·8568	152·9852	23
24	2·563 30	0·390 12	39·0826	15·2470	162·3482	24
25	2·665 84	0·375 12	41·6459	15·6221	171·7261	25
26	2·772 47	0·360 69	44·3117	15·9828	181·1040	26
27	2·883 37	0·346 82	47·0842	16·3296	190·4680	27
28	2·996 70	0·333 48	49·9676	16·6631	199·8054	28
29	3·118 65	0·320 65	52·9663	16·9837	209·1043	29
30	3·243 40	0·308 32	56·0849	17·2920	218·3539	30
31	3·373 13	0·296 46	59·3283	17·5885	227·5441	31
32	3·508 06	0·285 06	62·7015	17·8736	236·6660	32
33	3·648 38	0·274 09	66·2095	18·1476	245·7111	33
34	3·794 32	0·263 55	69·8579	18·4112	254·6719	34
35	3·946 09	0·253 42	73·6522	18·6646	263·5414	35
36	4·103 93	0·243 67	77·5983	18·9083	272·3135	36
37	4·268 09	0·234 30	81·7022	19·1426	280·9825	37
38	4·438 81	0·225 29	85·9703	19·3679	289·5433	38
39	4·616 37	0·216 62	90·4091	19·5845	297·9915	39
40	4·801 02	0·208 29	95·0255	19·7928	306·3231	40
41	4·993 06	0·200 28	99·8265	19·9931	314·5345	41
42	5·192 78	0·192 57	104·8196	20·1856	322·6226	42
43	5·400 50	0·185 17	110·0124	20·3708	330·5849	43
44	5·616 52	0·178 05	115·4129	20·5488	338·4189	44
45	5·841 18	0·171 20	121·0294	20·7200	346·1228	45
46	6·074 82	0·164 61	126·8706	20·8847	353·6951	46
47	6·317 82	0·158 28	132·9454	21·0429	361·1343	47
48	6·570 53	0·152 19	139·2632	21·1951	368·4397	48
49	6·833 35	0·146 34	145·8337	21·3415	375·6104	49
50	7·106 68	0·140 71	152·6671	21·4822	382·6460	50

COMPOUND INTEREST TABLES

Constants	
Function	Value
i	0·050 000
$i^{(2)}$	0·049 390
$i^{(4)}$	0·049 089
$i^{(12)}$	0·048 889
δ	0·048 790
$(1 + i)$	1·050 000
$(1 + i)^{1/2}$	1·024 695
$(1 + i)^{1/4}$	1·012 272
$(1 + i)^{1/12}$	1·004 074
v	0·952 381
$v^{1/2}$	0·975 900
$v^{1/4}$	0·987 877
$v^{1/12}$	0·995 942
d	0·047 619
$d^{(2)}$	0·048 200
$d^{(4)}$	0·048 494
$d^{(12)}$	0·048 691
$i/i^{(2)}$	1·012 348
$i/i^{(4)}$	1·018 559
$i/i^{(12)}$	1·022 715
i/δ	1·024 797
$i/d^{(2)}$	1·037 348
$i/d^{(4)}$	1·031 059
$i/d^{(12)}$	1·026 881

n	$(1 + i)^n$	v^n	$s_{\overline{n}|}$	$a_{\overline{n}|}$	$(Ia)_{\overline{n}|}$	n
1	1·050 00	0·952 38	1·0000	0·9524	0·9524	1
2	1·102 50	0·907 03	2·0500	1·8594	2·7664	2
3	1·157 62	0·863 84	3·1525	2·7232	5·3580	3
4	1·215 51	0·822 70	4·3101	3·5460	8·6488	4
5	1·276 28	0·783 53	5·5256	4·3295	12·5664	5
6	1·340 10	0·746 22	6·8019	5·0757	17·0437	6
7	1·407 10	0·710 68	8·1420	5·7864	22·0185	7
8	1·477 46	0·676 84	9·5491	6·4632	27·4332	8
9	1·551 33	0·644 61	11·0266	7·1078	33·2347	9
10	1·628 89	0·613 91	12·5779	7·7217	39·3738	10
11	1·710 34	0·584 68	14·2068	8·3064	45·8053	11
12	1·795 86	0·556 84	15·9171	8·8633	52·4873	12
13	1·885 65	0·530 32	17·7130	9·3936	59·3815	13
14	1·979 93	0·505 07	19·5986	9·8986	66·4524	14
15	2·078 93	0·481 02	21·5786	10·3797	73·6677	15
16	2·182 87	0·458 11	23·6575	10·8378	80·9975	16
17	2·292 02	0·436 30	25·8404	11·2741	88·4145	17
18	2·406 62	0·415 52	28·1324	11·6896	95·8939	18
19	2·526 95	0·395 73	30·5390	12·0853	103·4128	19
20	2·653 30	0·376 89	33·0660	12·4622	110·9506	20
21	2·785 96	0·358 94	35·7193	12·8212	118·4884	21
22	2·925 26	0·341 85	38·5052	13·1630	126·0091	22
23	3·071 52	0·325 57	41·4305	13·4886	133·4973	23
24	3·225 10	0·310 07	44·5020	13·7986	140·9389	24
25	3·386 35	0·295 30	47·7271	14·0939	148·3215	25
26	3·555 67	0·281 24	51·1135	14·3752	155·6337	26
27	3·733 46	0·267 85	54·6691	14·6430	162·8656	27
28	3·920 13	0·255 09	58·4026	14·8981	170·0082	28
29	4·116 14	0·242 95	62·3227	15·1411	177·0537	29
30	4·321 94	0·231 38	66·4388	15·3725	183·9950	30
31	4·538 04	0·220 36	70·7608	15·5928	190·8261	31
32	4·764 94	0·209 87	75·2988	15·8027	197·5419	32
33	5·003 19	0·199 87	80·0638	16·0025	204·1377	33
34	5·253 35	0·190 35	85·0670	16·1929	210·6097	34
35	5·516 02	0·181 29	90·3203	16·3742	216·9549	35
36	5·791 82	0·172 66	95·8363	16·5469	223·1705	36
37	6·081 41	0·164 44	101·6281	16·7113	229·2547	37
38	6·385 48	0·156 61	107·7095	16·8679	235·2057	38
39	6·704 75	0·149 15	114·0950	17·0170	241·0224	39
40	7·039 99	0·142 05	120·7998	17·1591	246·7043	40
41	7·391 99	0·135 28	127·8398	17·2944	252·2508	41
42	7·761 59	0·128 84	135·2318	17·4232	257·6621	42
43	8·149 67	0·122 70	142·9933	17·5459	262·9384	43
44	8·557 15	0·116 86	151·1430	17·6628	268·0803	44
45	8·985 01	0·111 30	159·7002	17·7741	273·0886	45
46	9·434 26	0·106 00	168·6852	17·8801	277·9645	46
47	9·905 97	0·100 95	178·1194	17·9810	282·7091	47
48	10·401 27	0·096 14	188·0254	18·0772	287·3239	48
49	10·921 33	0·091 56	198·4267	18·1687	291·8105	49
50	11·467 40	0·087 20	209·3480	18·2559	296·1707	50

COMPOUND INTEREST TABLES

	Constants	
Function	Value	
i	0·060 000	
$i^{(2)}$	0·059 126	
$i^{(4)}$	0·058 695	
$i^{(12)}$	0·058 411	
δ	0·058 269	
$(1 + i)$	1·060 000	
$(1 + i)^{1/2}$	1·029 563	
$(1 + i)^{1/4}$	1·014 674	
$(1 + i)^{1/12}$	1·004 868	
v	0·943 396	
$v^{1/2}$	0·971 286	
$v^{1/4}$	0·985 538	
$v^{1/12}$	0·995 156	
d	0·056 604	
$d^{(2)}$	0·057 428	
$d^{(4)}$	0·057 847	
$d^{(12)}$	0·058 128	
$i/i^{(2)}$	1·014 782	
$i/i^{(4)}$	1·022 227	
$i/i^{(12)}$	1·027 211	
i/δ	1·029 709	
$i/d^{(2)}$	1·044 782	
$i/d^{(4)}$	1·037 227	
$i/d^{(12)}$	1·032 211	

n	$(1 + i)^n$	v^n	$s_{\overline{n}}$	$a_{\overline{n}}$	$(Ia)_{\overline{n}}$	n
1	1·060 00	0·943 40	1·0000	0·9434	0·9434	1
2	1·123 60	0·890 00	2·0600	1·8334	2·7234	2
3	1·191 02	0·839 62	3·1836	2·6730	5·2422	3
4	1·262 4?	0·792 09	4·3746	3·4651	8·4106	4
5	1·338 23	0·747 26	5·6371	4·2124	12·1469	5
6	1·418 52	0·704 96	6·9753	4·9173	16·3767	6
7	1·503 63	0·665 06	8·3938	5·5824	21·0321	7
8	1·593 85	0·627 41	9·8975	6·2098	26·0514	8
9	1·689 48	0·591 90	11·4913	6·8017	31·3785	9
10	1·790 85	0·558 39	13·1808	7·3601	36·9624	10
11	1·898 30	0·526 79	14·9716	7·8869	42·7571	11
12	2·012 20	0·496 97	16·8699	8·3838	48·7207	12
13	2·132 93	0·468 84	18·8821	8·8527	54·8156	13
14	2·260 90	0·442 30	21·0151	9·2950	61·0078	14
15	2·396 56	0·417 27	23·2760	9·7122	67·2668	15
16	2·540 35	0·393 65	25·6725	10·1059	73·5651	16
17	2·692 77	0·371 36	28·2129	10·4773	79·8783	17
18	2·854 34	0·350 34	30·9057	10·8276	86·1845	18
19	3·025 60	0·330 51	33·7600	11·1581	92·4643	19
20	3·207 14	0·311 80	36·7856	11·4699	98·7004	20
21	3·399 56	0·294 16	39·9927	11·7641	104·8776	21
22	3·603 54	0·277 51	43·3923	12·0416	110·9827	22
23	3·819 75	0·261 80	46·9958	12·3034	117·0041	23
24	4·048 93	0·246 98	50·8156	12·5504	122·9316	24
25	4·291 87	0·233 00	54·8645	12·7834	128·7565	25
26	4·549 38	0·219 81	59·1564	13·0032	134·4716	26
27	4·822 35	0·207 37	63·7058	13·2105	140·0705	27
28	5·111 69	0·195 63	68·5281	13·4062	145·5482	28
29	5·418 39	0·184 56	73·6398	13·5907	150·9003	29
30	5·743 49	0·174 11	79·0582	13·7648	156·1236	30
31	6·088 10	0·164 25	84·8017	13·9291	161·2155	31
32	6·453 39	0·154 96	90·8898	14·0840	166·1742	32
33	6·840 59	0·146 19	97·3432	14·2302	170·9983	33
34	7·251 03	0·137 91	104·1838	14·3681	175·6873	34
35	7·686 09	0·130 11	111·4348	14·4982	180·2410	35
36	8·147 25	0·122 74	119·1209	14·6210	184·6596	36
37	8·636 09	0·115 79	127·2681	14·7368	188·9440	37
38	9·154 25	0·109 24	135·9042	14·8460	193·0951	38
39	9·703 51	0·103 06	145·0585	14·9491	197·1142	39
40	10·285 72	0·097 22	154·7620	15·0463	201·0031	40
41	10·902 86	0·091 72	165·0477	15·1380	204·7636	41
42	11·557 03	0·086 53	175·9505	15·2245	208·3978	42
43	12·250 45	0·081 63	187·5076	15·3062	211·9078	43
44	12·985 48	0·077 01	199·7580	15·3832	215·2962	44
45	13·764 61	0·072 65	212·7435	15·4558	218·5655	45
46	14·590 49	0·068 54	226·5081	15·5244	221·7182	46
47	15·465 92	0·064 66	241·0986	15·5890	224·7572	47
48	16·393 87	0·061 00	256·5645	15·6500	227·6851	48
49	17·377 50	0·057 55	272·9584	15·7076	230·5048	49
50	18·420 15	0·054 29	290·3359	15·7619	233·2192	50

COMPOUND INTEREST TABLES

	Constants			n	$(1 + i)^n$	v^n	$s_{\overline{n}}$	$a_{\overline{n}}$	$(Ia)_{\overline{n}}$	n

Function	Value
i	0·070 000
$i^{(2)}$	0·068 816
$i^{(4)}$	0·068 234
$i^{(12)}$	0·067 850
δ	0·067 659
$(1 + i)$	1·070 000
$(1 + i)^{1/2}$	1·034 408
$(1 + i)^{1/4}$	1·017 059
$(1 + i)^{1/12}$	1·005 654
v	0·934 579
$v^{1/2}$	0·966 736
$v^{1/4}$	0·983 228
$v^{1/12}$	0·994 378
d	0·065 421
$d^{(2)}$	0·066 527
$d^{(4)}$	0·067 090
$d^{(12)}$	0·067 468
$i/i^{(2)}$	1·017 204
$i/i^{(4)}$	1·025 880
$i/i^{(12)}$	1·031 691
i/δ	1·034 605
$i/d^{(2)}$	1·052 204
$i/d^{(4)}$	1·043 380
$i/d^{(12)}$	1·037 525

n	$(1 + i)^n$	v^n	$s_{\overline{n}}$	$a_{\overline{n}}$	$(Ia)_{\overline{n}}$	n
1	1·070 00	0·934 58	1·0000	0·9346	0·9346	1
2	1·144 90	0·873 44	2·0700	1·8080	2·6815	2
3	1·225 04	0·816 30	3·2149	2·6243	5·1304	3
4	1·310 80	0·762 90	4·4399	3·3872	8·1819	4
5	1·402 55	0·712 99	5·7507	4·1002	11·7469	5
6	1·500 73	0·666 34	7·1533	4·7665	15·7449	6
7	1·605 78	0·622 75	8·6540	5·3893	20·1042	7
8	1·718 19	0·582 01	10·2598	5·9713	24·7602	8
9	1·838 46	0·543 93	11·9780	6·5152	29·6556	9
10	1·967 15	0·508 35	13·8164	7·0236	34·7391	10
11	2·104 85	0·475 09	15·7836	7·4987	39·9652	11
12	2·252 19	0·444 01	17·8885	7·9427	45·2933	12
13	2·409 85	0·414 96	20·1406	8·3577	50·6878	13
14	2·578 53	0·387 82	22·5505	8·7455	56·1173	14
15	2·759 03	0·362 45	25·1290	9·1079	61·5540	15
16	2·952 16	0·338 73	27·8881	9·4466	66·9737	16
17	3·158 82	0·316 57	30·8402	9·7632	72·3555	17
18	3·379 93	0·295 86	33·9990	10·0591	77·6810	18
19	3·616 53	0·276 51	37·3790	10·3356	82·9347	19
20	3·869 68	0·258 42	40·9955	10·5940	88·1031	20
21	4·140 56	0·241 51	44·8652	10·8355	93·1748	21
22	4·430 40	0·225 71	49·0057	11·0612	98·1405	22
23	4·740 53	0·210 95	53·4361	11·2722	102·9923	23
24	5·072 37	0·197 15	58·1767	11·4693	107·7238	24
25	5·427 43	0·184 25	63·2490	11·6536	112·3301	25
26	5·807 35	0·172 20	68·6765	11·8258	116·8071	26
27	6·213 87	0·160 93	74·4838	11·9867	121·1523	27
28	6·648 84	0·150 40	80·6977	12·1371	125·3635	28
29	7·114 26	0·140 56	87·3465	12·2777	129·4399	29
30	7·612 26	0·131 37	94·4608	12·4090	133·3809	30
31	8·145 11	0·122 77	102·0730	12·5318	137·1868	31
32	8·715 27	0·114 74	110·2182	12·6466	140·8585	32
33	9·325 34	0·107 23	118·9334	12·7538	144·3973	33
34	9·978 11	0·100 22	128·2588	12·8540	147·8047	34
35	10·676 58	0·093 66	138·2369	12·9477	151·0829	35
36	11·423 94	0·087 54	148·9135	13·0352	154·2342	36
37	12·223 62	0·081 81	160·3374	13·1170	157·2612	37
38	13·079 27	0·076 46	172·5610	13·1935	160·1665	38
39	13·994 82	0·071 46	185·6403	13·2649	162·9533	39
40	14·974 46	0·066 78	199·6351	13·3317	165·6245	40
41	16·022 67	0·062 41	214·6096	13·3941	168·1833	41
42	17·144 26	0·058 33	230·6322	13·4524	170·6331	42
43	18·344 35	0·054 51	247·7765	13·5070	172·9772	43
44	19·628 46	0·050 95	266·1209	13·5579	175·2188	44
45	21·002 45	0·047 61	285·7493	13·6055	177·3614	45
46	22·472 62	0·044 50	306·7518	13·6500	179·4084	46
47	24·045 71	0·041 59	329·2244	13·6916	181·3630	47
48	25·728 91	0·038 87	353·2701	13·7305	183·2286	48
49	27·529 93	0·036 32	378·9990	13·7668	185·0085	49
50	29·457 03	0·033 95	406·5289	13·8007	186·7059	50

COMPOUND INTEREST TABLES

Constants	
Function	Value
i	0·080 000
$i^{(2)}$	0·078 461
$i^{(4)}$	0·077 706
$i^{(12)}$	0·077 208
δ	0·076 961
$(1 + i)$	1·080 000
$(1 + i)^{1/2}$	1·039 230
$(1 + i)^{1/4}$	1·019 427
$(1 + i)^{1/12}$	1·006 434
v	0·925 926
$v^{1/2}$	0·962 250
$v^{1/4}$	0·980 944
$v^{1/12}$	0·993 607
d	0·074 074
$d^{(2)}$	0·075 499
$d^{(4)}$	0·076 225
$d^{(12)}$	0·076 715
$i/i^{(2)}$	1·019 615
$i/i^{(4)}$	1·029 519
$i/i^{(12)}$	1·036 157
i/δ	1·039 487
$i/d^{(2)}$	1·059 615
$i/d^{(4)}$	1·049 519
$i/d^{(12)}$	1·042 824

n	$(1 + i)^n$	v^n	$s_{\overline{n}}$	$a_{\overline{n}}$	$(Ia)_{\overline{n}}$	n
1	1·080 00	0·925 93	1·0000	0·9259	0·9259	1
2	1·166 40	0·857 34	2·0800	1·7833	2·6406	2
3	1·259 71	0·793 83	3·2464	2·5771	5·0221	3
4	1·360 49	0·735 03	4·5061	3·3121	7·9622	4
5	1·469 33	0·680 58	5·8666	3·9927	11·3651	5
6	1·586 87	0·630 17	7·3359	4·6229	15·1462	6
7	1·713 82	0·583 49	8·9228	5·2064	19·2306	7
8	1·850 93	0·540 27	10·6366	5·7466	23·5527	8
9	1·999 00	0·500 25	12·4876	6·2469	28·0550	9
10	2·158 92	0·463 19	14·4866	6·7101	32·6869	10
11	2·331 64	0·428 88	16·6455	7·1390	37·4046	11
12	2·518 17	0·397 11	18·9771	7·5361	42·1700	12
13	2·719 62	0·367 70	21·4953	7·9038	46·9501	13
14	2·937 19	0·340 46	24·2149	8·2442	51·7165	14
15	3·172 17	0·315 24	27·1521	8·5595	56·4451	15
16	3·425 94	0·291 89	30·3243	8·8514	61·1154	16
17	3·700 02	0·270 27	33·7502	9·1216	65·7100	17
18	3·996 02	0·250 25	37·4502	9·3719	70·2144	18
19	4·315 70	0·231 71	41·4463	9·6036	74·6170	19
20	4·660 96	0·214 55	45·7620	9·8181	78·9079	20
21	5·033 83	0·198 66	50·4229	10·0168	83·0797	21
22	5·436 54	0·183 94	55·4568	10·2007	87·1264	22
23	5·871 46	0·170 32	60·8933	10·3711	91·0437	23
24	6·341 18	0·157 70	66·7648	10·5288	94·8284	24
25	6·848 48	0·146 02	73·1059	10·6748	98·4789	25
26	7·396 35	0·135 20	79·9544	10·8100	101·9941	26
27	7·988 06	0·125 19	87·3508	10·9352	105·3742	27
28	8·627 11	0·115 91	95·3388	11·0511	108·6198	28
29	9·317 27	0·107 33	103·9659	11·1584	111·7323	29
30	10·062 66	0·099 38	113·2832	11·2578	114·7136	30
31	10·867 67	0·092 02	123·3459	11·3498	117·5661	31
32	11·737 08	0·085 20	134·2135	11·4350	120·2925	32
33	12·676 05	0·078 89	145·9506	11·5139	122·8958	33
34	13·690 13	0·073 05	158·6267	11·5869	125·3793	34
35	14·785 34	0·067 63	172·3168	11·6546	127·7466	35
36	15·968 17	0·062 62	187·1021	11·7172	130·0010	36
37	17·245 63	0·057 99	203·0703	11·7752	132·1465	37
38	18·625 28	0·053 69	220·3159	11·8289	134·1868	38
39	20·115 30	0·049 71	238·9412	11·8786	136·1256	39
40	21·724 52	0·046 03	259·0565	11·9246	137·9668	40
41	23·462 48	0·042 62	280·7810	11·9672	139·7143	41
42	25·339 48	0·039 46	304·2435	12·0067	141·3718	42
43	27·366 64	0·036 54	329·5830	12·0432	142·9430	43
44	29·555 97	0·033 83	356·9496	12·0771	144·4317	44
45	31·920 45	0·031 33	386·5056	12·1084	145·8415	45
46	34·474 09	0·029 01	418·4261	12·1374	147·1758	46
47	37·232 01	0·026 86	452·9002	12·1643	148·4382	47
48	40·210 57	0·024 87	490·1322	12·1891	149·6319	48
49	43·427 42	0·023 03	530·3427	12·2122	150·7602	49
50	46·901 61	0·021 32	573·7702	12·2335	151·8263	50

COMPOUND INTEREST TABLES   9 PER CENT

	Constants		n	$(1 + i)^n$	v^n	$s_{\overline{n}\rceil}$	$a_{\overline{n}\rceil}$	$(Ia)_{\overline{n}\rceil}$	n
Function	**Value**		1	1·090 00	0·917 43	1·0000	0·9174	0·9174	1
			2	1·188 10	0·841 68	2·0900	1·7591	2·6008	2
i	0·090 000		3	1·295 03	0·772 18	3·2781	2·5313	4·9173	3
$i^{(2)}$	0·088 061		4	1·411 58	0·708 43	4·5731	3·2397	7·7510	4
$i^{(4)}$	0·087 113		5	1·538 62	0·649 93	5·9847	3·8897	11·0007	5
$i^{(12)}$	0·086 488								
δ	0·086 178		6	1·677 10	0·596 27	7·5233	4·4859	14·5783	6
			7	1·828 04	0·547 03	9·2004	5·0330	18·4075	7
$(1 + i)$	1·090 000		8	1·992 56	0·501 87	11·0285	5·5348	22·4225	8
$(1 + i)^{1/2}$	1·044 031		9	2·171 89	0·460 43	13·0210	5·9952	26·5663	9
$(1 + i)^{1/4}$	1·021 778		10	2·367 36	0·422 41	15·1929	6·4177	30·7904	10
$(1 + i)^{1/12}$	1·007 207								
			11	2·580 43	0·387 53	17·5603	6·8052	35·0533	11
v	0·917 431		12	2·812 66	0·355 53	20·1407	7·1607	39·3197	12
$v^{1/2}$	0·957 826		13	3·065 80	0·326 18	22·9534	7·4869	43·5600	13
$v^{1/4}$	0·978 686		14	3·341 73	0·299 25	26·0192	7·7862	47·7495	14
$v^{1/12}$	0·992 844		15	3·642 48	0·274 54	29·3609	8·0607	51·8676	15
d	0·082 569		16	3·970 31	0·251 87	33·0034	8·3126	55·8975	16
$d^{(2)}$	0·084 347		17	4·327 63	0·231 07	36·9737	8·5436	59·8257	17
$d^{(4)}$	0·085 256		18	4·717 12	0·211 99	41·3013	8·7556	63·6416	18
$d^{(12)}$	0·085 869		19	5·141 66	0·194 49	46·0185	8·9501	67·3369	19
			20	5·604 41	0·178 43	51·1601	9·1285	70·9055	20
$i/i^{(2)}$	1·022 015								
$i/i^{(4)}$	1·033 144		21	6·108 81	0·163 70	56·7645	9·2922	74·3432	21
$i/i^{(12)}$	1·040 608		22	6·658 60	0·150 18	62·8733	9·4424	77·6472	22
			23	7·257 87	0·137 78	69·5319	9·5802	80·8162	23
i/δ	1·044 354		24	7·911 08	0·126 40	76·7898	9·7066	83·8499	24
			25	8·623 08	0·115 97	84·7009	9·8226	86·7491	25
$i/d^{(2)}$	1·067 015								
$i/d^{(4)}$	1·055 644		26	9·399 16	0·106 39	93·3240	9·9290	89·5153	26
$i/d^{(12)}$	1·048 108		27	10·245 08	0·097 61	102·7231	10·0266	92·1507	27
			28	11·167 14	0·089 55	112·9682	10·1161	94·6580	28
			29	12·172 18	0·082 15	124·1354	10·1983	97·0405	29
			30	13·267 68	0·075 37	136·3075	10·2737	99·3017	30
			31	14·461 77	0·069 15	149·5752	10·3428	101·4452	31
			32	15·763 33	0·063 44	164·0370	10·4062	103·4753	32
			33	17·182 03	0·058 20	179·8003	10·4644	105·3959	33
			34	18·728 41	0·053 39	196·9823	10·5178	107·2113	34
			35	20·413 97	0·048 99	215·7108	10·5668	108·9258	35
			36	22·251 23	0·044 94	236·1247	10·6118	110·5437	36
			37	24·253 84	0·041 23	258·3759	10·6530	112·0692	37
			38	26·436 68	0·037 83	282·6298	10·6908	113·5066	38
			39	28·815 98	0·034 70	309·0665	10·7255	114·8600	39
			40	31·409 42	0·031 84	337·8824	10·7574	116·1335	40
			41	34·236 27	0·029 21	369·2919	10·7866	117·3311	41
			42	37·317 53	0·026 80	403·5281	10·8134	118·4566	42
			43	40·676 11	0·024 58	440·8457	10·8380	119·5137	43
			44	44·336 96	0·022 55	481·5218	10·8605	120·5061	44
			45	48·327 29	0·020 69	525·8587	10·8812	121·4373	45
			46	52·676 74	0·018 98	574·1860	10·9002	122·3105	46
			47	57·417 65	0·017 42	626·8628	10·9176	123·1291	47
			48	62·585 24	0·015 98	684·2804	10·9336	123·8960	48
			49	68·217 91	0·014 66	746·8656	10·9482	124·6143	49
			50	74·357 52	0·013 45	815·0836	10·9617	125·2867	50

Constants	
Function	**Value**
i	0·100 000
$i^{(2)}$	0·097 618
$i^{(4)}$	0·096 455
$i^{(12)}$	0·095 690
δ	0·095 310
$(1 + i)$	1·100 000
$(1 + i)^{1/2}$	1·048 809
$(1 + i)^{1/4}$	1·024 114
$(1 + i)^{1/12}$	1·007 974
v	0·909 091
$v^{1/2}$	0·953 463
$v^{1/4}$	0·976 454
$v^{1/12}$	0·992 089
d	0·090 909
$d^{(2)}$	0·093 075
$d^{(4)}$	0·094 184
$d^{(12)}$	0·094 933
$i/i^{(2)}$	1·024 404
$i/i^{(4)}$	1·036 756
$i/i^{(12)}$	1·045 045
i/δ	1·049 206
$i/d^{(2)}$	1·074 404
$i/d^{(4)}$	1·061 756
$i/d^{(12)}$	1·053 378

n	$(1 + i)^n$	v^n	$s_{\overline{n}}$	$a_{\overline{n}}$	$(Ia)_{\overline{n}}$	n
1	1·100 00	0·909 09	1·0000	0·9091	0·9091	1
2	1·210 00	0·826 45	2·1000	1·7355	2·5620	2
3	1·331 00	0·751 31	3·3100	2·4869	4·8159	3
4	1·464 10	0·683 01	4·6410	3·1699	7·5480	4
5	1·610 51	0·620 92	6·1051	3·7908	10·6526	5
6	1·771 56	0·564 47	7·7156	4·3553	14·0394	6
7	1·948 72	0·513 16	9·4872	4·8684	17·6315	7
8	2·143 59	0·466 51	11·4359	5·3349	21·3636	8
9	2·357 95	0·424 10	13·5795	5·7590	25·1805	9
10	2·593 74	0·385 54	15·9374	6·1446	29·0359	10
11	2·853 12	0·350 49	18·5312	6·4951	32·8913	11
12	3·138 43	0·318 63	21·3843	6·8137	36·7149	12
13	3·452 27	0·289 66	24·5227	7·1034	40·4805	13
14	3·797 50	0·263 33	27·9750	7·3667	44·1672	14
15	4·177 25	0·239 39	31·7725	7·6061	47·7581	15
16	4·594 97	0·217 63	35·9497	7·8237	51·2401	16
17	5·054 47	0·197 84	40·5447	8·0216	54·6035	17
18	5·559 92	0·179 86	45·5992	8·2014	57·8410	18
19	6·115 91	0·163 51	51·1591	8·3649	60·9476	19
20	6·727 50	0·148 64	57·2750	8·5136	63·9205	20
21	7·400 25	0·135 13	64·0025	8·6487	66·7582	21
22	8·140 27	0·122 85	71·4027	8·7715	69·4608	22
23	8·954 30	0·111 68	79·5430	8·8832	72·0294	23
24	9·849 73	0·101 53	88·4973	8·9847	74·4660	24
25	10·834 71	0·092 30	98·3471	9·0770	76·7734	25
26	11·918 18	0·083 91	109·1818	9·1609	78·9550	26
27	13·109 99	0·076 28	121·0999	9·2372	81·0145	27
28	14·420 99	0·069 34	134·2099	9·3066	82·9561	28
29	15·863 09	0·063 04	148·6309	9·3696	84·7842	29
30	17·449 40	0·057 31	164·4940	9·4269	86·5035	30
31	19·194 34	0·052 10	181·9434	9·4790	88·1186	31
32	21·113 78	0·047 36	201·1378	9·5264	89·6342	32
33	23·225 15	0·043 06	222·2515	9·5694	91·0550	33
34	25·547 67	0·039 14	245·4767	9·6086	92·3859	34
35	28·102 44	0·035 58	271·0244	9·6442	93·6313	35
36	30·912 68	0·032 35	299·1268	9·6765	94·7959	36
37	34·003 95	0·029 41	330·0395	9·7059	95·8840	37
38	37·404 34	0·026 73	364·0434	9·7327	96·8999	38
39	41·144 78	0·024 30	401·4478	9·7570	97·8478	39
40	45·259 26	0·022 09	442·5926	9·7791	98·7316	40
41	49·785 18	0·020 09	487·8518	9·7991	99·5551	41
42	54·763 70	0·018 26	537·6370	9·8174	100·3221	42
43	60·240 07	0·016 60	592·4007	9·8340	101·0359	43
44	66·264 08	0·015 09	652·6408	9·8491	101·6999	44
45	72·890 48	0·013 72	718·9048	9·8628	102·3172	45
46	80·179 53	0·012 47	791·7953	9·8753	102·8910	46
47	88·197 49	0·011 34	871·9749	9·8866	103·4238	47
48	97·017 23	0·010 31	960·1723	9·8969	103·9186	48
49	106·718 96	0·009 37	1057·1896	9·9063	104·3778	49
50	117·390 85	0·008 52	1163·9085	9·9148	104·8037	50

COMPOUND INTEREST TABLES

Constants			n	$(1 + i)^n$	v^n	$s_{\overline{n}}$	$a_{\overline{n}}$	$(Ia)_{\overline{n}}$	n
Function	**Value**		1	1·110 00	0·900 90	1·0000	0·9009	0·9009	1
			2	1·232 10	0·811 62	2·1100	1·7125	2·5241	2
i	0·110 000		3	1·367 63	0·731 19	3·3421	2·4437	4·7177	3
$i^{(2)}$	0·107 131		4	1·518 07	0·658 73	4·7097	3·1024	7·3526	4
$i^{(4)}$	0·105 733		5	1·685 06	0·593 45	6·2278	3·6959	10·3199	5
$i^{(12)}$	0·104 815								
δ	0·104 360		6	1·870 41	0·534 64	7·9129	4·2305	13·5277	6
			7	2·076 16	0·481 66	9·7833	4·7122	16·8994	7
$(1 + i)$	1·110 000		8	2·304 54	0·433 93	11·8594	5·1461	20·3708	8
$(1 + i)^{1/2}$	1·053 565		9	2·558 04	0·390 92	14·1640	5·5370	23·8891	9
$(1 + i)^{1/4}$	1·026 433		10	2·839 42	0·352 18	16·7220	5·8892	27·4109	10
$(1 + i)^{1/12}$	1·008 735								
			11	3·151 76	0·317 28	19·5614	6·2065	30·9011	11
v	0·900 901		12	3·498 45	0·285 84	22·7132	6·4924	34·3311	12
$v^{1/2}$	0·949 158		13	3·883 28	0·257 51	26·2116	6·7499	37·6788	13
$v^{1/4}$	0·974 247		14	4·310 44	0·231 99	30·0949	6·9819	40·9268	14
$v^{1/12}$	0·991 341		15	4·784 59	0·209 00	34·4054	7·1909	44·0618	15
d	0·099 099		16	5·310 89	0·188 29	39·1899	7·3792	47·0745	16
$d^{(2)}$	0·101 684		17	5·895 09	0·169 63	44·5008	7·5488	49·9582	17
$d^{(4)}$	0·103 010		18	6·543 55	0·152 82	50·3959	7·7016	52·7090	18
$d^{(12)}$	0·103 908		19	7·263 34	0·137 68	56·9395	7·8393	55·3249	19
			20	8·062 31	0·124 03	64·2028	7·9633	57·8056	20
$i/i^{(2)}$	1·026 783								
$i/i^{(4)}$	1·040 353		21	8·949 17	0·111 74	72·2651	8·0751	60·1522	21
$i/i^{(12)}$	1·049 467		22	9·933 57	0·100 67	81·2143	8·1757	62·3669	22
			23	11·026 27	0·090 69	91·1479	8·2664	64·4528	23
i/δ	1·054 044		24	12·239 16	0·081 70	102·1742	8·3481	66·4137	24
			25	13·585 46	0·073 61	114·4133	8·4217	68·2539	25
$i/d^{(2)}$	1·081 783								
$i/d^{(4)}$	1·067 853		26	15·079 86	0·066 31	127·9988	8·4881	69·9781	26
$i/d^{(12)}$	1·058 634		27	16·738 65	0·059 74	143·0786	8·5478	71·5911	27
			28	18·579 90	0·053 82	159·8173	8·6016	73·0981	28
			29	20·623 69	0·048 49	178·3972	8·6501	74·5043	29
			30	22·892 30	0·043 68	199·0209	8·6938	75·8148	30
			31	25·410 45	0·039 35	221·9132	8·7331	77·0347	31
			32	28·205 60	0·035 45	247·3236	8·7686	78·1693	32
			33	31·308 21	0·031 94	275·5292	8·8005	79·2233	33
			34	34·752 12	0·028 78	306·8374	8·8293	80·2017	34
			35	38·574 85	0·025 92	341·5896	8·8552	81·1090	35
			36	42·818 08	0·023 35	380·1644	8·8786	81·9498	36
			37	47·528 07	0·021 04	422·9825	8·8996	82·7282	37
			38	52·756 16	0·018 96	470·5106	8·9186	83·4485	38
			39	58·559 34	0·017 08	523·2667	8·9357	84·1145	39
			40	65·000 87	0·015 38	581·8261	8·9511	84·7299	40
			41	72·150 96	0·013 86	646·8269	8·9649	85·2982	41
			42	80·087 57	0·012 49	718·9779	8·9774	85·8226	42
			43	88·897 20	0·011 25	799·0655	8·9886	86·3063	43
			44	98·675 89	0·010 13	887·9627	8·9988	86·7522	44
			45	109·530 24	0·009 13	986·6386	9·0079	87·1630	45
			46	121·578 57	0·008 23	1096·1688	9·0161	87·5414	46
			47	134·952 21	0·007 41	1217·7474	9·0235	87·8897	47
			48	149·796 95	0·006 68	1352·6996	9·0302	88·2101	48
			49	166·274 62	0·006 01	1502·4965	9·0362	88·5048	49
			50	184·564 83	0·005 42	1668·7712	9·0417	88·7757	50

COMPOUND INTEREST TABLES

Constants	
Function	**Value**
i	0·120 000
$i^{(2)}$	0·116 601
$i^{(4)}$	0·114 949
$i^{(12)}$	0·113 866
δ	0·113 329
$(1 + i)$	1·120 000
$(1 + i)^{1/2}$	1·058 301
$(1 + i)^{1/4}$	1·028 737
$(1 + i)^{1/12}$	1·009 489
v	0·892 857
$v^{1/2}$	0·944 911
$v^{1/4}$	0·972 065
$v^{1/12}$	0·990 600
d	0·107 143
$d^{(2)}$	0·110 178
$d^{(4)}$	0·111 738
$d^{(12)}$	0·112 795
$i/i^{(2)}$	1·029 150
$i/i^{(4)}$	1·043 938
$i/i^{(12)}$	1·053 875
i/δ	1·058 867
$i/d^{(2)}$	1·089 150
$i/d^{(4)}$	1·073 938
$i/d^{(12)}$	1·063 875

n	$(1 + i)^n$	v^n	$s_{\overline{n}\rceil}$	$a_{\overline{n}\rceil}$	$(Ia)_{\overline{n}\rceil}$	n
1	1·120 00	0·892 86	1·0000	0·8929	0·8929	1
2	1·254 40	0·797 19	2·1200	1·6901	2·4872	2
3	1·404 93	0·711 78	3·3744	2·4018	4·6226	3
4	1·573 52	0·635 52	4·7793	3·0373	7·1647	4
5	1·762 34	0·567 43	6·3528	3·6048	10·0018	5
6	1·973 82	0·506 63	8·1152	4·1114	13·0416	6
7	2·210 68	0·452 35	10·0890	4·5638	16·2080	7
8	2·475 96	0·403 88	12·2997	4·9676	19·4391	8
9	2·773 08	0·360 61	14·7757	5·3282	22·6846	9
10	3·105 85	0·321 97	17·5487	5·6502	25·9043	10
11	3·478 55	0·287 48	20·6546	5·9377	29·0665	11
12	3·895 98	0·256 68	24·1331	6·1944	32·1467	12
13	4·363 49	0·229 17	28·0291	6·4235	35·1259	13
14	4·887 11	0·204 62	32·3926	6·6282	37·9906	14
15	5·473 57	0·182 70	37·2797	6·8109	40·7310	15
16	6·130 39	0·163 12	42·7533	6·9740	43·3410	16
17	6·866 04	0·145 64	48·8837	7·1196	45·8169	17
18	7·689 97	0·130 04	55·7497	7·2497	48·1576	18
19	8·612 76	0·116 11	63·4397	7·3658	50·3637	19
20	9·646 29	0·103 67	72·0524	7·4694	52·4370	20
21	10·803 85	0·092 56	81·6987	7·5620	54·3808	21
22	12·100 31	0·082 64	92·5026	7·6446	56·1989	22
23	13·552 35	0·073 79	104·6029	7·7184	57·8960	23
24	15·178 63	0·065 88	118·1552	7·7843	59·4772	24
25	17·000 06	0·058 82	133·3339	7·8431	60·9478	25
26	19·040 07	0·052 52	150·3339	7·8957	62·3133	26
27	21·324 88	0·046 89	169·3740	7·9426	63·5794	27
28	23·883 87	0·041 87	190·6989	7·9844	64·7518	28
29	26·749 93	0·037 38	214·5828	8·0218	65·8359	29
30	29·959 92	0·033 38	241·3327	8·0552	66·8372	30
31	33·555 11	0·029 80	271·2926	8·0850	67·7611	31
32	37·581 73	0·026 61	304·8477	8·1116	68·6126	32
33	42·091 53	0·023 76	342·4294	8·1354	69·3966	33
34	47·142 52	0·021 21	384·5210	8·1566	70·1178	34
35	52·799 62	0·018 94	431·6635	8·1755	70·7807	35
36	59·135 57	0·016 91	484·4631	8·1924	71·3894	36
37	66·231 84	0·015 10	543·5987	8·2075	71·9481	37
38	74·179 66	0·013 48	609·8305	8·2210	72·4604	38
39	83·081 22	0·012 04	684·0102	8·2330	72·9298	39
40	93·050 97	0·010 75	767·0914	8·2438	73·3596	40
41	104·217 09	0·009 60	860·1424	8·2534	73·7531	41
42	116·723 14	0·008 57	964·3595	8·2619	74·1129	42
43	130·729 91	0·007 65	1081·0826	8·2696	74·4418	43
44	146·417 50	0·006 83	1211·8125	8·2764	74·7423	44
45	163·987 60	0·006 10	1358·2300	8·2825	75·0167	45
46	183·666 12	0·005 44	1522·2176	8·2880	75·2672	46
47	205·706 05	0·004 86	1705·8838	8·2928	75·4957	47
48	230·390 78	0·004 34	1911·5898	8·2972	75·7040	48
49	258·037 67	0·003 88	2141·9806	8·3010	75·8939	49
50	289·002 19	0·003 46	2400·0182	8·3045	76·0669	50

COMPOUND INTEREST TABLES

Constants	
Function	**Value**
i	0·130000
$i^{(2)}$	0·126029
$i^{(4)}$	0·124104
$i^{(12)}$	0·122842
δ	0·122218
$(1+i)$	1·130000
$(1+i)^{1/2}$	1·063015
$(1+i)^{1/4}$	1·031026
$(1+i)^{1/12}$	1·010237
v	0·884956
$v^{1/2}$	0·940721
$v^{1/4}$	0·969908
$v^{1/12}$	0·989867
d	0·115044
$d^{(2)}$	0·118558
$d^{(4)}$	0·120369
$d^{(12)}$	0·121597
$i/i^{(2)}$	1·031507
$i/i^{(4)}$	1·047509
$i/i^{(12)}$	1·058269
i/δ	1·063676
$i/d^{(2)}$	1·096507
$i/d^{(4)}$	1·080009
$i/d^{(12)}$	1·069102

n	$(1+i)^n$	v^n	$s_{\overline{n}|}$	$a_{\overline{n}|}$	$(Ia)_{\overline{n}|}$	n
1	1·13000	0·88496	1·0000	0·8850	0·8850	1
2	1·27690	0·78315	2·1300	1·6681	2·4512	2
3	1·44290	0·69305	3·4069	2·3612	4·5304	3
4	1·63047	0·61332	4·8498	2·9745	6·9837	4
5	1·84244	0·54276	6·4803	3·5172	9·6975	5
6	2·08195	0·48032	8·3227	3·9975	12·579	6
7	2·35261	0·42506	10·4047	4·4226	15·5548	7
8	2·65844	0·37616	12·7573	4·7988	18·5641	8
9	3·00404	0·33288	15·4157	5·1317	21·5601	9
10	3·39457	0·29459	18·4197	5·4262	24·5059	10
11	3·83586	0·26070	21·8143	5·6869	27·3736	11
12	4·33452	0·23071	25·6502	5·9176	30·1421	12
13	4·89801	0·20416	29·9847	6·1218	32·7962	13
14	5·53475	0·18068	34·8827	6·3025	35·3257	14
15	6·25427	0·15989	40·4175	6·4624	37·7241	15
16	7·06733	0·14150	46·6717	6·6039	39·9880	16
17	7·98608	0·12522	53·7391	6·7291	42·1167	17
18	9·02427	0·11081	61·7251	6·8399	44·1113	18
19	10·19742	0·09806	70·7494	6·9380	45·9745	19
20	11·52309	0·08678	80·9468	7·0248	47·7102	20
21	13·02109	0·07680	92·4699	7·1016	49·3229	21
22	14·71383	0·06796	105·4910	7·1695	50·8181	22
23	16·62663	0·06014	120·2048	7·2297	52·2015	23
24	18·78809	0·05323	136·8315	7·2829	53·4789	24
25	21·23054	0·04710	155·6196	7·3300	54·6564	25
26	23·99051	0·04168	176·8501	7·3717	55·7402	26
27	27·10928	0·03689	200·8406	7·4086	56·7361	27
28	30·63349	0·03264	227·9499	7·4412	57·6502	28
29	34·61584	0·02889	258·5834	7·4701	58·4879	29
30	39·11590	0·02557	293·1992	7·4957	59·2549	30
31	44·20096	0·02262	332·3151	7·5183	59·9562	31
32	49·94709	0·02002	376·5161	7·5383	60·5969	32
33	56·44021	0·01772	426·4632	7·5560	61·1816	33
34	63·77744	0·01568	482·9034	7·5717	61·7147	34
35	72·06851	0·01388	546·6808	7·5856	62·2004	35
36	81·43741	0·01228	618·7493	7·5979	62·6424	36
37	92·02428	0·01087	700·1867	7·6087	63·0445	37
38	103·98743	0·00962	792·2110	7·6183	63·4099	38
39	117·50580	0·00851	896·1984	7·6268	63·7418	39
40	132·78155	0·00753	1013·7042	7·6344	64·0431	40
41	150·04315	0·00666	1146·4858	7·6410	64·3163	41
42	169·54876	0·00590	1296·5289	7·6469	64·5640	42
43	191·59010	0·00522	1466·0777	7·6522	64·7885	43
44	216·49682	0·00462	1657·6678	7·6568	64·9917	44
45	244·64140	0·00409	1874·1646	7·6609	65·1756	45
46	276·44478	0·00362	2118·8060	7·6645	65·3420	46
47	312·38261	0·00320	2395·2508	7·6677	65·4925	47
48	352·99234	0·00283	2707·6334	7·6705	65·6285	48
49	398·88135	0·00251	3060·6258	7·6730	65·7513	49
50	450·73593	0·00222	3459·5071	7·6752	65·8623	50

COMPOUND INTEREST TABLES

Constants			n	$(1+i)^n$	v^n	$s_{\overline{n}|}$	$a_{\overline{n}|}$	$(Ia)_{\overline{n}|}$	n
Function	**Value**		1	1·140 00	0·877 19	1·0000	0·8772	0·8772	1
			2	1·299 60	0·769 47	2·1400	1·6467	2·4161	2
i	0·140 000		3	1·481 54	0·674 97	3·4396	2·3216	4·4410	3
$i^{(2)}$	0·135 416		4	1·688 96	0·592 08	4·9211	2·9137	6·8094	4
$i^{(4)}$	0·133 198		5	1·925 41	0·519 37	6·6101	3·4331	9·4062	5
$i^{(12)}$	0·131 746								
δ	0·131 028		6	2·194 97	0·455 59	8·5355	3·8887	12·1397	6
			7	2·502 27	0·399 64	10·7305	4·2883	14·9372	7
$(1+i)$	1·140 000		8	2·852 59	0·350 56	13·2328	4·6389	17·7417	8
$(1+i)^{1/2}$	1·067 708		9	3·251 95	0·307 51	16·0853	4·9464	20·5092	9
$(1+i)^{1/4}$	1·033 299		10	3·707 22	0·269 74	19·3373	5·2161	23·2067	10
$(1+i)^{1/12}$	1·010 979		11	4·226 23	0·236 62	23·0445	5·4527	25·8095	11
			12	4·817 90	0·207 56	27·2707	5·6603	28·3002	12
v	0·877 193		13	5·492 41	0·182 07	32·0887	5·8424	30·6671	13
$v^{1/2}$	0·936 586		14	6·261 35	0·159 71	37·5811	6·0021	32·9030	14
$v^{1/4}$	0·967 774		15	7·137 94	0·140 10	43·8424	6·1422	35·0045	15
$v^{1/12}$	0·989 140		16	8·137 25	0·122 89	50·9804	6·2651	36·9707	16
			17	9·276 46	0·107 80	59·1176	6·3729	38·8033	17
d	0·122 807		18	10·575 17	0·094 56	68·3941	6·4674	40·5054	18
$d^{(2)}$	0·126 828		19	12·055 69	0·082 95	78·9692	6·5504	42·0814	19
$d^{(4)}$	0·128 905		20	13·743 49	0·072 76	91·0249	6·6231	43·5367	20
$d^{(12)}$	0·130 316		21	15·667 58	0·063 83	104·7684	6·6870	44·8770	21
			22	17·861 04	0·055 99	120·4360	6·7429	46·1088	22
$i/i^{(2)}$	1·033 854		23	20·361 58	0·049 11	138·2970	6·7921	47·2383	23
$i/i^{(4)}$	1·051 067		24	23·212 21	0·043 08	158·6586	6·8351	48·2723	24
$i/i^{(12)}$	1·062 649		25	26·461 92	0·037 79	181·8708	6·8729	49·2170	25
			26	30·166 58	0·033 15	208·3327	6·9061	50·0789	26
i/δ	1·068 472		27	34·389 91	0·029 08	238·4993	6·9352	50·8640	27
			28	39·204 49	0·025 51	272·8892	6·9607	51·5782	28
$i/d^{(2)}$	1·103 854		29	44·693 12	0·022 37	312·0937	6·9830	52·2271	29
$i/d^{(4)}$	1·086 067		30	50·950 16	0·019 63	356·7868	7·0027	52·8159	30
$i/d^{(12)}$	1·074 316		31	58·083 18	0·017 22	407·7370	7·0199	53·3496	31
			32	66·214 83	0·015 10	465·8202	7·0350	53·8329	32
			33	75·484 90	0·013 25	532·0350	7·0482	54·2701	33
			34	86·052 79	0·011 62	607·5199	7·0599	54·6652	34
			35	98·100 18	0·010 19	693·5727	7·0700	55·0220	35
			36	111·834 20	0·008 94	791·6729	7·0790	55·3439	36
			37	127·490 99	0·007 84	903·5071	7·0868	55·6341	37
			38	145·339 73	0·006 88	1030·9981	7·0937	55·8955	38
			39	165·687 29	0·006 04	1176·3378	7·0997	56·1309	39
			40	188·883 51	0·005 29	1342·0251	7·1050	56·3427	40
			41	215·327 21	0·004 64	1530·9086	7·1097	56·5331	41
			42	245·473 01	0·004 07	1746·2358	7·1138	56·7042	42
			43	279·839 24	0·003 57	1991·7088	7·1173	56·8579	43
			44	319·016 73	0·003 13	2271·5481	7·1205	56·9958	44
			45	363·679 07	0·002 75	2590·5648	7·1232	57·1195	45
			46	414·594 14	0·002 41	2954·2439	7·1256	57·2305	46
			47	472·637 32	0·002 12	3368·8380	7·1277	57·3299	47
			48	538·806 55	0·001 86	3841·4753	7·1296	57·4190	48
			49	614·239 46	0·001 63	4380·2819	7·1312	57·4988	49
			50	700·232 99	0·001 43	4994·5213	7·1327	57·5702	50

COMPOUND INTEREST TABLES

Constants			n	$(1+i)^n$	v^n	$s_{\overline{n}|}$	$a_{\overline{n}|}$	$(Ia)_{\overline{n}|}$	n
Function	**Value**		1	1·150 00	0·869 57	1·0000	0·8696	0·8696	1
			2	1·322 50	0·756 14	2·1500	1·6257	2·3819	2
i	0·150 000		3	1·520 87	0·657 52	3·4725	2·2832	4·3544	3
$i^{(2)}$	0·144 761		4	1·749 01	0·571 75	4·9934	2·8550	6·6414	4
$i^{(4)}$	0·142 232		5	2·011 36	0·497 18	6·7424	3·3522	9·1273	5
$i^{(12)}$	0·140 579								
δ	0·139 762		6	2·313 06	0·432 33	8·7537	3·7845	11·7213	6
			7	2·660 02	0·375 94	11·0668	4·1604	14·3528	7
$(1+i)$	1·150 000		8	3·059 02	0·326 90	13·7268	4·4873	16·9680	8
$(1+i)^{1/2}$	1·072 381		9	3·517 88	0·284 26	16·7858	4·7716	19·5264	9
$(1+i)^{1/4}$	1·035 558		10	4·045 56	0·247 18	20·3037	5·0188	21·9982	10
$(1+i)^{1/12}$	1·011 715								
			11	4·652 39	0·214 94	24·3493	5·2337	24·3626	11
v	0·869 565		12	5·350 25	0·186 91	29·0017	5·4206	26·6055	12
$v^{1/2}$	0·932 505		13	6·152 79	0·162 53	34·3519	5·5831	28·7184	13
$v^{1/4}$	0·965 663		14	7·075 71	0·141 33	40·5047	5·7245	30·6970	14
$v^{1/12}$	0·988 421		15	8·137 06	0·122 89	47·5804	5·8474	32·5404	15
			16	9·357 62	0·106 86	55·7175	5·9542	34·2502	16
d	0·130 435		17	10·761 26	0·092 93	65·0751	6·0472	35·8300	17
$d^{(2)}$	0·134 990		18	12·375 45	0·080 81	75·8364	6·1280	37·2845	18
$d^{(4)}$	0·137 348		19	14·231 77	0·070 27	88·2118	6·1982	38·6195	19
$d^{(12)}$	0·138 951		20	16·366 54	0·061 10	102·4436	6·2593	39·8415	20
$i/i^{(2)}$	1·036 190		21	18·821 52	0·053 13	118·8101	6·3125	40·9572	21
$i/i^{(4)}$	1·054 613		22	21·644 75	0·046 20	137·6316	6·3587	41·9737	22
$i/i^{(12)}$	1·067 016		23	24·891 46	0·040 17	159·2764	6·3988	42·8977	23
			24	28·625 18	0·034 93	184·1678	6·4338	43·7361	24
i/δ	1·073 254		25	32·918 95	0·030 38	212·7930	6·4641	44·4955	25
$i/d^{(2)}$	1·111 190		26	37·856 80	0·026 42	245·7120	6·4906	45·1823	26
$i/d^{(4)}$	1·092 113		27	43·535 31	0·022 97	283·5688	6·5135	45·8025	27
$i/d^{(12)}$	1·079 516		28	50·065 61	0·019 97	327·1041	6·5335	46·3618	28
			29	57·575 45	0·017 37	377·1697	6·5509	46·8655	29
			30	66·211 77	0·015 10	434·7451	6·5660	47·3186	30
			31	76·143 54	0·013 13	500·9569	6·5791	47·7257	31
			32	87·565 07	0·011 42	577·1005	6·5905	48·0911	32
			33	100·699 83	0·009 93	664·6655	6·6005	48·4188	33
			34	115·804 80	0·008 64	765·3654	6·6091	48·7124	34
			35	133·175 52	0·007 51	881·1702	6·6166	48·9752	35
			36	153·151 85	0·006 53	1014·3457	6·6231	49·2103	36
			37	176·124 63	0·005 68	1167·4975	6·6288	49·4204	37
			38	202·543 32	0·004 94	1343·6222	6·6338	49·6080	38
			39	232·924 82	0·004 29	1546·1655	6·6380	49·7754	39
			40	267·863 55	0·003 73	1779·0903	6·6418	49·9248	40
			41	308·043 08	0·003 25	2046·9539	6·6450	50·0579	41
			42	354·249 54	0·002 82	2354·9969	6·6478	50·1764	42
			43	407·386 97	0·002 45	2709·2465	6·6503	50·2820	43
			44	468·495 02	0·002 13	3116·6334	6·6524	50·3759	44
			45	538·769 27	0·001 86	3585·1285	6·6543	50·4594	45
			46	619·584 66	0·001 61	4123·8977	6·6559	50·5337	46
			47	712·522 36	0·001 40	4743·4824	6·6573	50·5996	47
			48	819·400 71	0·001 22	5456·0047	6·6585	50·6582	48
			49	942·310 82	0·001 06	6275·4055	6·6596	50·7102	49
			50	1083·657 44	0·000 92	7217·7163	6·6605	50·7563	50

COMPOUND INTEREST TABLES

Constants			n	$(1 + i)^n$	v^n	$s_{\overline{n}}$	$a_{\overline{n}}$	$(Ia)_{\overline{n}}$	n
Function	**Value**		1	1·200 00	0·833 33	1·0000	0·8333	0·8333	1
			2	1·440 00	0·694 44	2·2000	1·5278	2·2222	2
i	0·200 000		3	1·728 00	0·578 70	3·6400	2·1065	3·9583	3
$i^{(2)}$	0·190 890		4	2·073 60	0·482 25	5·3680	2·5887	5·8873	4
$i^{(4)}$	0·186 541		5	2·488 32	0·401 88	7·4416	2·9906	7·8967	5
$i^{(12)}$	0·183 714								
δ	0·182 322		6	2·985 98	0·334 90	9·9299	3·3255	9·9061	6
			7	3·583 18	0·279 08	12·9159	3·6046	11·8597	7
$(1 + i)$	1·200 000		8	4·299 82	0·232 57	16·4991	3·8372	13·7202	8
$(1 + i)^{1/2}$	1·095 445		9	5·159 78	0·193 81	20·7989	4·0310	15·4645	9
$(1 + i)^{1/4}$	1·046 635		10	6·191 74	0·161 51	25·9587	4·1925	17·0796	10
$(1 + i)^{1/12}$	1·015 309								
			11	7·430 08	0·134 59	32·1504	4·3271	18·5600	11
v	0·833 333		12	8·916 10	0·112 16	39·5805	4·4392	19·9059	12
$v^{1/2}$	0·912 871		13	10·699 32	0·093 46	48·4966	4·5327	21·1209	13
$v^{1/4}$	0·955 443		14	12·839 18	0·077 89	59·1959	4·6106	22·2113	14
$v^{1/12}$	0·984 921		15	15·407 02	0·064 91	72·0351	4·6755	23·1849	15
d	0·166 667		16	18·488 43	0·054 09	87·4421	4·7296	24·0503	16
$d^{(2)}$	0·174 258		17	22·186 11	0·045 07	105·9306	4·7746	24·8166	17
$d^{(4)}$	0·178 229		18	26·623 33	0·037 56	128·1167	4·8122	25·4927	18
$d^{(12)}$	0·180 943		19	31·948 00	0·031 30	154·7400	4·8435	26·0874	19
			20	38·337 60	0·026 08	186·6880	4·8696	26·6091	20
$i/i^{(2)}$	1·047 723								
$i/i^{(4)}$	1·072 153		21	46·005 12	0·021 74	225·0256	4·8913	27·0655	21
$i/i^{(12)}$	1·088 651		22	55·206 14	0·018 11	271·0307	4·9094	27·4641	22
			23	66·247 37	0·015 09	326·2369	4·9245	27·8112	23
i/δ	1·096 963		24	79·496 85	0·012 58	392·4842	4·9371	28·1131	24
			25	95·396 22	0·010 48	471·9811	4·9476	28·3752	25
$i/d^{(2)}$	1·147 723								
$i/d^{(4)}$	1·122 153								
$i/d^{(12)}$	1·105 317								

COMPOUND INTEREST TABLES

Constants			n	$(1 + i)^n$	v^n	$s_{\overline{n}}$	$a_{\overline{n}}$	$(Ia)_{\overline{n}}$	n
Function	**Value**		1	1·250 00	0·800 00	1·0000	0·8000	0·8000	1
			2	1·562 50	0·640 00	2·2500	1·4400	2·0800	2
i	0·250 000		3	1·953 13	0·512 00	3·8125	1·9520	3·6160	3
$i^{(2)}$	0·236 068		4	2·441 41	0·409 60	5·7656	2·3616	5·2544	4
$i^{(4)}$	0·229 485		5	3·051 76	0·327 68	8·2070	2·6893	6·8928	5
$i^{(12)}$	0·225 231								
δ	0·223 144		6	3·814 70	0·262 14	11·2588	2·9514	8·4657	6
			7	4·768 37	0·209 72	15·0735	3·1611	9·9337	7
$(1 + i)$	1·250 000		8	5·960 46	0·167 77	19·8419	3·3289	11·2758	8
$(1 + i)^{1/2}$	1·118 034		9	7·450 58	0·134 22	25·8023	3·4631	12·4838	9
$(1 + i)^{1/4}$	1·057 371		10	9·313 23	0·107 37	33·2529	3·5705	13·5575	10
$(1 + i)^{1/12}$	1·018 769								
			11	11·641 53	0·085 90	42·5661	3·6564	14·5024	11
v	0·800 000		12	14·551 92	0·068 72	54·2077	3·7251	15·3271	12
$v^{1/2}$	0·894 427		13	18·189 89	0·054 98	68·7596	3·7801	16·0418	13
$v^{1/4}$	0·945 742		14	22·737 37	0·043 98	86·9495	3·8241	16·6575	14
$v^{1/12}$	0·981 577		15	28·421 71	0·035 18	109·6868	3·8593	17·1853	15
d	0·200 000		16	35·527 14	0·028 15	138·1085	3·8874	17·6356	16
$d^{(2)}$	0·211 146		17	44·408 92	0·022 52	173·6357	3·9099	18·0184	17
$d^{(4)}$	0·217 034		18	55·511 15	0·018 01	218·0446	3·9279	18·3427	18
$d^{(12)}$	0·221 082		19	69·388 94	0·014 41	273·5558	3·9424	18·6165	19
			20	86·736 17	0·011 53	342·9447	3·9539	18·8471	20
$i/i^{(2)}$	1·059 017								
$i/i^{(4)}$	1·089 396		21	108·420 22	0·009 22	429·6809	3·9631	19·0408	21
$i/i^{(12)}$	1·109 971		22	135·525 27	0·007 38	538·1011	3·9705	19·2031	22
			23	169·406 59	0·005 90	673·6264	3·9764	19·3389	23
i/δ	1·120 355		24	211·758 24	0·004 72	843·0329	3·9811	19·4522	24
			25	264·697 80	0·003 78	1054·7912	3·9849	19·5467	25
$i/d^{(2)}$	1·184 017								
$i/d^{(4)}$	1·151 896								
$i/d^{(12)}$	1·130 804								

INDEX

Lightning Source UK Ltd.
Milton Keynes UK
07 November 2009

145930UK00001B/34/A